# Springer Series in Geomechanics and Geoengineering

**Editors**

Prof. Wei Wu
Institut für Geotechnik
Universität für Bodenkultur
Feistmantelstraße 4
1180 Vienna
Austria
E-mail: wei.wu@boku.ac.at

Prof. Ronaldo I. Borja
Department of Civil and Environmental Engineering
Stanford University
Stanford, CA 94305-4020
USA
E-mail: borja@stanford.edu

For further volumes:
http://www.springer.com/series/8069

Michael Zhengmeng Hou, Heping Xie,
and Patrick Were (Eds.)

# Clean Energy Systems in the Subsurface: Production, Storage and Conversion

Proceedings of the 3rd Sino-German Conference "Underground Storage of $CO_2$ and Energy", Goslar, Germany, 21–23 May 2013

Springer

*Editors*
Dr. Michael Zhengmeng Hou
Energie-Forschungszentrum Niedersachsen (EFZN)
Clausthal University of Technology (TUC)
Goslar
Germany

Patrick Were
Energie-Forschungszentrum Niedersachsen (EFZN)
Clausthal University of Technology (TUC)
Goslar
Germany

Dr. Heping Xie
President Office
Sichuan University
Chengdu
Sichuan
People's Republic China

ISSN 1866-8755 ISSN 1866-8763 (electronic)
ISBN 978-3-642-37848-5 ISBN 978-3-642-37849-2 (eBook)
DOI 10.1007/978-3-642-37849-2
Springer Heidelberg New York Dordrecht London

Library of Congress Control Number: 2013935695



Printed on acid-free paper

Springer is part of Springer Science+Business Media (www.springer.com)

# Preface

The worldwide consumption of energy is still largely dependent on conventional fossil resources - the main reason for global warming and climate change. The new energy policy paradigm in Germany (Energiewende: from fossil fuels to renewable energy) foresees the utilization of renewable resources to produce 50% of the German electrical power demand by 2030 and 80% by 2050. On one hand Germany has taken a lead in the extensive research for increased production and usage of renewable energies to promote its energy policy, the "Energiewende". However, the country still has a rather modest contribution of 3% to the global greenhouse gas emissions. On the other hand 68% of China's primary energy consumption is provided from coal alone, contributing about 20% of the global greenhouse gas emissions. Like Germany, China intends to reduce its emission intensity by 40–45% by 2020, based on the 2005 levels. In order to face this huge challenge, clean energy resources in the underground have to be considered in the near future for producing geothermal energy and unconventional gas. In this regard, geo-reservoirs are going to serve as large storage systems for the fluctuating renewable energy converted into e.g. compressed air, synthetic methane, hydrogen or heat. In addition, through the use of new technologies, geo-reservoirs will help to dampen the detrimental effects that result from the excessive use of fossil energy. These new technologies allow a switch from carbon intensive fossil fuels to low carbon energy carriers (e.g. natural gas) via $CO_2$ capture and utilization (CCU) or $CO_2$ capture, utilization and storage (CCUS) or as biochemical reactors to convert hydrogen and carbon dioxide into methane under favorable high pressure and high temperature conditions deep underground. Geo-reservoirs may thus be used for production, storage and conversion purposes in the context of the "Energiewende".

A significant understanding of the geo-reservoirs already exists and ample technical methods have already been developed and experience gained in recent decades, during the exploration and exploitation for e.g. ore minerals, coal, oil and gas. However, their use for producing tight and shale gas, storing the fluctuating renewable energy, converting hydrogen and carbon dioxide into methane gas is a new topic, and in many aspects not well understood. In this respect several institutes from Germany and China have formed a Sino-German Cooperation Group "Underground Storage of $CO_2$ and Energy", which is funded by the Sino-German Center for Research Promotion in Beijing (jointly financed by the National Natural Science Foundation of China (NSFC) and the Deutsche Forschungsgemeinschaft (DFG, German Research Foundation)).

Since its inception in 2007, the group has so far organized two international gatherings with the goal of bridging the gap between researchers, scientists and experts in both academia and industry as an important vehicle for information sharing and cooperation. In addition such meetings bring together experts from different countries, providing opportunities for closer understanding and collaboration. The first Sino-German Group meeting was a workshop "Relevant aspects on the underground storage of natural gas and $CO_2$", held on the 18–21 September 2007 in Goslar, Germany. The second Sino-German gathering was a conference "Underground storage of $CO_2$ and energy" held on the 6–7 July 2010 in Beijing and extended for another 2 days in Daqing, China. This year the group is meeting on the 21–23 May 2013 for the second time in Goslar, German, to convene its $3^{rd}$ Sino-German conference on the theme "Clean Energy Systems in the Underground: Production, Storage and Conversion (CLEANUP)".

This book is a collection of quality papers reflecting the magnitude of energy and time that has been invested in this project by the organizing committee, advisory board, peer-reviewers and the student editors to select and develop suitable scientific works from a diversity of authors. Papers in this volume are organized into the following 6 thematic sections covering a range of topics from current developments in $CO_2$ geologic sequestration research to reduce greenhouse emission including measures to monitor surface leakage, groundwater quality and integrity of caprock, while ensuring a sufficient supply of clean energy:

Section 1: Integrated Energy and Environmental Utilization of Geo-reservoirs: Law, Risk Management & Monitoring
Section 2: $CO_2$ for Enhanced Gas and Oil Recovery, Coal Bedded Methane and Geothermal Systems
Section 3: Trapping Mechanisms and Multi-Barrier Sealing Systems for Long-Term $CO_2$ Storage
Section 4: Coupled THMC-Processes and Numerical Modelling
Section 5: Rock Mechanical Behaviour in Consideration of Cyclic Loading, Dilatancy, Damage, Self-sealing and Healing
Section 6: Underground Storage and Supply of Energy

We are indebted to many people and institutions whose effort and support has been sought to make this conference a success. We are most thankful to the following: the Advisory Board, the International Scientific Board, the

Organizing Committee and the peer-reviewers for their great contribution to observe and maintain the high quality and standard of the manuscripts assembled herein for conference proceedings; Energie-Forschungszentrum Niedersachsen, Clausthal University of Technology, Sichuan University, Chinese Academy of Science, Helmholz Center, Northeast Petroleum University, Kiel University and the Sino-German Center for Research Promotion for supporting the conference in Goslar; Springer Verlag for its excellent efforts to publish the conference proceedings in due time; Mr. Tobias Kracke and the TUC students Can Zhang, Anna Elizabeth Kwokori, Richard Onyekachi Awo and James Baffoe for grammar corrections and formatting; the authors for their quality scientific contributions and the guests and participants for sacrificing their invaluable time and activities to make our conference a success.

Michael Zhengmeng Hou
Editor in Chief & Conference Chairman

Heping Xie
Conference Chairman

Patrick Were
Conference Coordinator

**May 2013, Clausthal-Zellerfeld, Germany & Chengdu, China**

# Organizing Institutions

Clausthal University of Technology

Energie-Forschungszentrum Niedersachsen

Sichuan University

Northeast Petroleum University

# Conference Committees

## Organizing Committee

### Conference Chair

| | |
|---|---|
| Michael Zhengmeng Hou | Clausthal University of Technology, Germany<br>Energie-Forschungszentrum Niedersachsen, Germany |

### Secretariat

| | |
|---|---|
| Karin Peya | Energie-Forschungszentrum Niedersachsen, Germany |

### Conference Coordinator

| | |
|---|---|
| Hongwei Zhou | China University of Mining and Technology, China |

### Conference Coordinator

| | |
|---|---|
| Patrick Were | Energie-Forschungszentrum Niedersachsen, Germany |

### Conference Coordinator

| | |
|---|---|
| Olaf Franz | LBEG Niedersachsen, Germany |

### Conference Coordinator

| | |
|---|---|
| Leonhard Ganzer | Energie-Forschungszentrum Niedersachsen, Germany |

### Conference Coordinator

| | |
|---|---|
| Chunhe Yang | Institute of Rock & Soil Mechanics CA, China |

## International Scientific Committee

## Advisory Committee

| | |
|---|---|
| Hans-Peter Beck | Chairman of EFZN in Goslar, Germany |
| Thomas Hanschke | President of Clausthal University of Technology, Germany |
| Yang Liu | President of Northeast Petroleum University, China |
| Heping Xie | President of Sichuan University, China |

# Contents

## Section I: Integrated Energy and Environmental Utilization of Geo-reservoirs: Law, Risk Management and Monitoring

## Section II: $CO_2$ for Enhanced Gas and Oil Recovery, Coal Bedded Methane and Geothermal Systems



## Section III: Trapping Mechanisms and Multi-barrier Sealing Systems for Long-Term $CO_2$ Storage

## Section IV: Coupled THMC-Processes and Numerical Modelling



## Section V: Rock Mechanical Behaviour in Consideration of Cyclic Loading, Dilatancy, Damage, Self-Sealing and Healing

# Regulation of Carbon Capture and Storage in China: Lessons from the EU CCS Directive

Tianbao Qin

Research Institute of Environmental Law (RIEL), Wuhan University, Luojia Shan, Wuhan, 430072, P.R. China

**Abstract.** Is CCS (Carbon Capture and Storage) a super hero who successfully deals with the issue of climate change or a dangerous monster who will bring a series of environmental risks and problems? In the context of global climate change, that is a hard question that has to be answered. As a big developing country with 75% of its electricity supply generated from thermal power China is at a crucial crossroad on how to respond to the issue of climate change. On one hand, China needs to develop its economy, this means that in the short term, China cannot escape its dependence on fossil fuels; and on the other hand, China is also facing strong international pressure on carbon emission reduction, this means that China will have to abate the use of fossil fuels. However, CCS gives China a new road and a brilliant hope especially in the context of China's national climate change law which is in the legislative process, although there is no special law to regulate CCS. The legislature should develop a reasonably clear legal framework for CCS to ensure environmental safety and sustainable development. In recent years, the European Union has been trying to play a leadership role in the global action response to climate change. As proof of its determination to deal with the issue of climate change, the EU has been seeking a substantial reduction of $CO_2$ emissions. For this purpose, the carbon capture and storage becomes a crucial technology. Therefore, the experience of the EU CCS legal framework will have more implication for China.

**Keywords:** Carbon Capture and Storage, carbon emission reduction, EU CCS legal framework, climate change.

## Introduction

As the biggest emitter of $CO_2$ in the world, China face great national and international pressures on combating climate change. CCS provides a possible solution for China to reduce its carbon emission. For the purpose of proposing a legal framework for China to regulate its CCS activities, this paper will first

M.Z. Hou et al. (Eds.): *Clean Energy Systems in the Subsurface*, SSGG, pp. 1–19.
DOI: 10.1007/978-3-642-37849-2_1 

introduce the development of CCS in the context of climate change, give an overview and make comments on the EU CCS directive; based on that, the paper will conclude with some implications of the EU CCS Directive for China.

## 1 Development of CCS in the Context of Climate Change

According to the provisions of the United Nations Framework Convention on Climate Change (UNFCCC), "Mitigation" and "adaptation" are the two main measures to deal with climate change. Due to the technical characteristics of the CCS and the practical needs of greenhouse gas emission reduction, not only can CCS be used as a mitigating measure, but can also be regarded as one of the ways of adapting to climate change. Moreover, under the current technical conditions, there are three technical measures for controlling greenhouse gas emissions: the first is energy conservation and an improvement in the efficiency of energy use. The second is the development and utilization of renewable energy and the third is the carbon capture and storage technology. However, considering the "Ceiling Effect" of energy conservation and renewable energy, therefore no choice but to embrace the last hope - CCS.

Furthermore, according to a research report from IEA, the contribution of CCS to the total emission reductions will be increased steadily year by year, 3% by 2020, and 10% from 2030 until 2050. The emission reduction contribution of this technology is expected to reach 19%, becoming the single largest share of emission reduction technology. Furthermore, the IPCC (International Panel on Climate Change) special report on carbon dioxide capture and storage said that CCS will be one of the very important methods for reducing the concentration of greenhouse gases in the atmosphere [1].

However, as any coin has two sides, CCS, just like a double-edged sword, will bring environmental and health risks, especially from the marine geological storage of carbon dioxide streams–the storage and disposal of carbon dioxide streams into sub-seabed geological formations. It is well known that any leakage of carbon dioxide in the deep-sea will directly harm marine life, and may also lead to acidification of seawater, which will change the entire marine environment. This is tantamount to a huge disaster for marine ecology. Moreover, the worst scenario is that a submarine explosion of carbon dioxide will even threaten the safety of human life.

Is CCS Moses Rod or Pandora's Box? Faced with this complex situation, it is not clear how effective the legislation will be, especially a CCS legal framework.

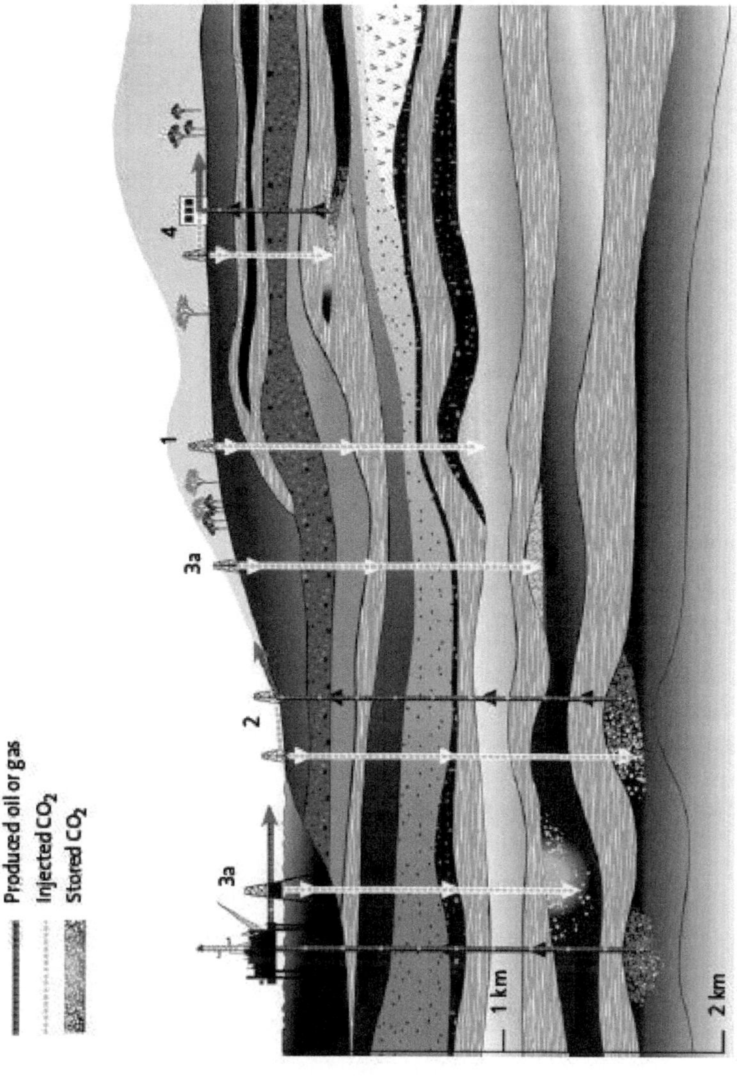


1 Depleted oil and gas reservoirs
2 Use of $CO_2$ in enhanced oil and gas recovery (EOR, EGR)
3 Deep saline formations (a: offshore, b: onshore)
4 Use of $CO_2$ to increase seam yield (enhanced coal bed methane recovery, ECBM)

**Fig. 1** Options for Geological $CO_2$ Storage [1]

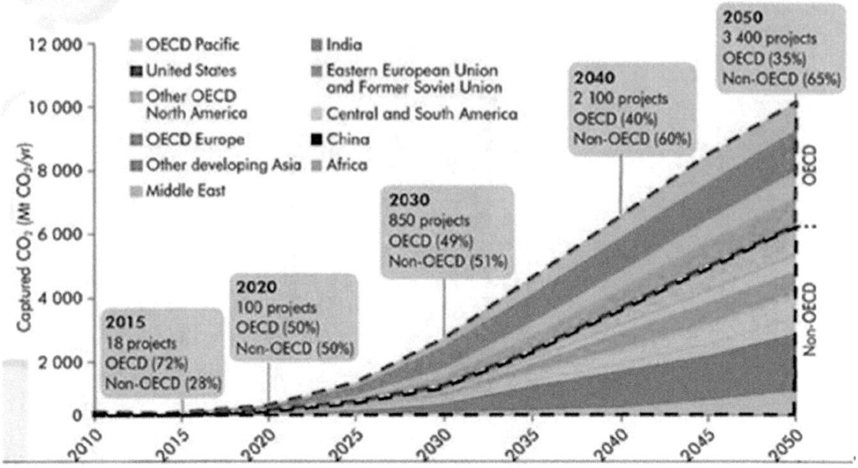


**Fig. 2** Development Prospects of CCS [2]

**Table 1** Typification of Risks in $CO_2$ Storage [3]

| Type of risk | Local risk for humans, animals and environment | Global risk for the climate |
|---|---|---|
| Spontaneous escape of $CO_2$ (»accident«) | Short-term, passing, massive impact; life-threatening in worst case | Release of the captured $CO_2$ amounts |
| Slow, gradual leakage from storage facility | Chronic and creeping threat to groundwater, flora and fauna in soil; possible danger for humans at point sources | Release of the captured $CO_2$ amounts |

## 2 Overview of EU CCS Directive

In recent years, the European Union has been trying to play a leadership role in the global response to climate change. As a proof of its determination to deal with the issue of climate change, the EU has been seeking a substantial reduction of $CO_2$ emissions. For this purpose, carbon capture and storage becomes a crucial technology.

## 2.1 *Background of EU CCS Directive*

EU is not only the pioneer of the research and development of CCS technology, but has also actively advocated promoting and improving the CCS legal framework. In October 2005, the European Union initiated *ECCP II*. In 2006, the European Union issued: *A European Strategy for Sustainable, Competitive and Secure Energy [COM (2006) 105 final] and Commission Communication: Sustainable power generation from fossil fuels: aiming for near-zero emission from coal after 2020[COM （2006） 843 final*]. In 2007, European Council put forward a resolution called *Integrating Energy and Climate into Planning* and on the 23rd of January 2008, the European Commission issued the policy called *the EU Climate and Energy Package* followed by *the CCS Directive draft*. Thereafter, *the EU Directive on the geological storage of carbon dioxide (Directive 2009/31/EC)* was definitively adopted by the Council of Ministers on the 6th of April 2009, and published in the Official Journal on the 5th of June 2009, coming into force twenty days later. At this point, the legal framework of the EU climate change policy had been established.

**Table 2** The Legal Framework of EU Climate Change

| | |
|---|---|
| Directive 2009/28/EC | The establishment of each Member State's National Renewable Energy goals, which are legally binding. |
| Decision No.406/2009/EC | The establishment of agreement for joint efforts- Each Member State agreed to limit carbon emissions targets in 2020, but the range of carbon emissions has not been regulated in the ETS Directive. |
| Directive 2009/29/EC | Modify and strengthen the ETS Directive, the establishment of a reserve fund of 300 million euros to support CCS and renewable energy demonstration projects. |
| Directive 2009/31/EC | The establishment of a legal framework for the safe application of CCS and its long-term development. |

After the policy of climate change emerged the following trends were observed: the first was to enlarge the control range of greenhouse gas and expand the EUETS. The second was to set up a regime of responsible distribution, the third was to stipulate the binding goals of renewable energy which put more emphasis s on biomass fuel and the final one was to establish new rules of storage and capture or environmental subsidy.

## 2.2 *Main Content of EU CCS Directive*

CCS Directive provides the legal framework to deal with climate change, and to ensure that the geological storage of carbon dioxide is not harmful to the environment or human health. The CCS directive includes a preamble (51 paragraphs), 41 Articles, and 2 Annex. The Directive forms part of the EU's Climate Change Package, developed in the context of the recognized need for developed nations to achieve greenhouse gas emission reductions of 30% by 2020 and 60-80% by 2050. Preliminary estimates indicate that up to 160 million tons of $CO_2$ could be stored by 2030, accounting for some 15% of the European Union's required reductions. The Preamble to the Directive describes CCS as a "bridging technology", which should not serve as an incentive to increase the share of fossil fuel power plants. The major detailed provisions of CCS Directive are as follows:

1. Site selection and exploration.
2. Storage permits: applications, contents and conditions.
3. Operation.
4. Closure, post-closure and transfer of liability
5. Compliance, penalties and liability (what happens if something goes wrong?)
6. Removal of legal barriers
7. Rights to use of third party
8. General provisions.

The underpinning aim of the Directive is the "environmentally safe storage of $CO_2$, meaning the permanent containment of $CO_2$ in such as way as to prevent or, where this is not possible, to eliminate as far as possible any negative effects and any risk to the environment and human health.

The Directive focuses primarily on the storage aspect of CCS, but it does briefly address the capture and transport elements. Member States retain the right to disallow storage in their territories in whole or in part, although those that choose to permit storage must carry out an assessment of their region's potential $CO_2$ storage capacity.

Importantly, the Directive provides certainty as to the legality of CCS activities by removing CCS from the scope of EU Waste law and EU Water law. The Directive is often described as "enabling" legislation, opting not to make CCS mandatory but to provide the necessary regulatory framework upon which CCS deployment could be fostered. Through amendments to the EU's Emission Trading Scheme (ETS) however, efforts have been made to give incentives for investment in CCS.

## 2.3 *Implementation of EU CCS Directive*

The CCS Directive applies to geological storage of $CO_2$ within the territory of the Member States, their exclusive economic zones and on their continental shelves, thus envisaging storage on both onshore and offshore. Member States had until 25 June 2011 to transpose it into their respective national laws.

**Table 3** Attitude of the EU Member States on CCS [4]

| | |
|---|---|
| Sweden | Sweden according to the sequestration of Carbon dioxide gas capacity is to provide the appropriate carbon tax credits promotions. |
| Belgium | Belgium accepts the specified use of funds. |
| Denmark | Denmark advocates that it should not be forced to develop CCS. |
| Netherlands | The Netherlands wants further research on the CCS legal framework and any problems that may be encountered in the implementation process. |
| Poland | Poland believes CCS will reduce energy efficiency and increase the cost of energy production and tests have not been carried out for industrial applications. So it is not ready to develop CCS. |

Among EU member states, the United Kingdom have being making great effort to develop CCS. The 2008 Energy Act determined the legal status of CCS, whilst the Energy Act 2010 officially launched the commercial operation of the CCS. In 2008, the British Government officially introduced the idea of the geological storage of carbon dioxide in the "Energy Act" The Energy Act (2010) provides a legal framework for CCS including mandatory social price support mechanisms and the establishment of new CCS legal system to achieve market-based incentives for the development of CCS . According to the Energy Act 2010, four CCS commercial scale demonstration projects will be constructed in the British coal-fired power plants to achieve full CCS power generation capacity by 2025 and will make the UK become the global CCS R&D center.

Generally speaking, the most crucial feature of the Energy Act 2010 is the market-based incentives mechanism for CCS.

**Table 4** Three Basic Systems for Market-Based Incentive Mechanisms in the UK [5]

| | |
|---|---|
| 1. | CCS Tax System |
| 2. | CCS demonstration projects funded in the form of a contract |
| 3. | CCS demonstration projects selected through market competition |

Meanwhile, the situation of CCS development is relatively subtle in Germany. Several natural-gas fields are reaching the end of their production phase and would thus become available, in principle, in the next few years for storing $CO_2$. The overall storage capacity in aquifers and depleted natural-gas repositories together amounts to between 40 and 130 times the annual $CO_2$ emissions from German power plants (approximately 350 Mt/yr).

**Table 5** Estimates of $CO_2$-Storage Capacities [6]

| Storage option | Capacity (in bn t $CO_2$) | | |
|---|---|---|---|
| | Global | Europe | Germany |
| Depleted gas reservoirs | 675–900 | 31–163 | 3 |
| Depleted oil reservoirs/EOR | | 4–65 | 0.1 |
| Aquifers | 1,000–10,000 | 1–47 | 12–28 |
| Unminable coal seams/ECBM | 3–200 | 0–10 | 0.4–1.7 |
| Source | IPCC 2005 | Hendriks *et al.* 2004 | Christensen *et al.* 2004 |

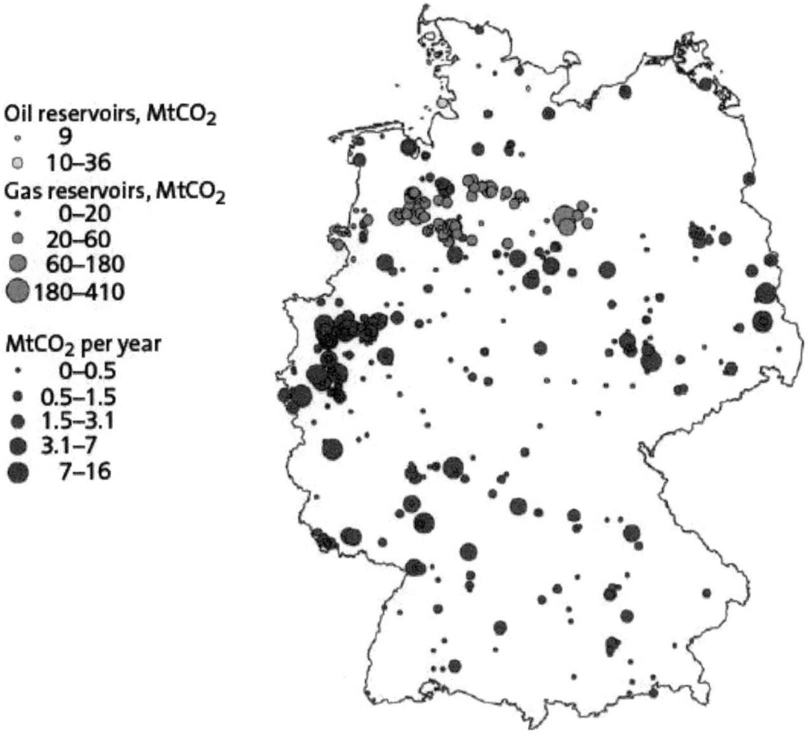


**Fig. 3** Locations of Gas or Oil Repositories and $CO_2$ Sources in Germany [7]

On the policy and legal level, in September 2007, Germany released a report on "the Development and Prospects of German Carbon Capture and Storage". In April 2009, according to the EU CCS Directive, the German Federal Government adopted the "Draft Act on Carbon Capture and Storage (CCS) in the Federal Republic of Germany". In June 2009, the Federal Parliament temporarily postponed the legislation taking into account factors such as the maturity of the technology, the source of funds and especially public opposition. In October 2009, the new German Federal Government indicated that the "National Carbon Capture and Storage Regulation (CCS-Regulation)" would be enacted before the end of 2010.

**Table 6** Regulatory Areas Relevant for the CCS Technology Chain [8]

| Process | Regulatory areas affected | Relevant laws/ ordinances | Assessment |
|---|---|---|---|
| *Capture* | | | |
| | Pollution control | BImSchG/BImSchV | Construction of capture systems subject to approval |
| | Handling of waste | KrW-/AbfG | Must be applied if $CO_2$ is classified as waste |
| *Transport* | | | |
| | Transport of waste | KrW-/AbfG | Must be applied if $CO_2$ is classified as waste |
| Marine transport | Hazardous goods on sea | GGVSee | Must be applied |
| Pipeline | Environmental compatibility | UVPG | Lays down general protection standard |
| | Safety of pipelines | RohrFLtgV | Must be applied |
| *Injection and storage* | | | |
| | Mining and similar underground activities | BBergG/UVP-V | Not applicable in current form |
| | Handling of waste | KrW-/AbfG | Must be applied if $CO_2$ is classified as waste |
| | Pollution control | BImSchG/BImSchV | Provisions applicable to systems not subject to approval |
| | Water/groundwater protection | WHG/GrWV | Discharge of $CO_2$ subject to approval |
| | Soil protection | BBodSchG | Might apply |

The question of whether the theoretical capacity potential for $CO_2$ storage and can be economically tapped or indeed be used is dependent on a number of geological details, on economic, legal, and political conditions, and on social acceptance. The best illustration is that Germany had attempted to implement a small-scale (30MW) Oxyfuel Power Plant CCS integration project, but eventually gave up the idea after the local public strongly resisted it. Therefore, generally speaking, public acceptance is the biggest obstacle that the Germany government will have to deal with in developing a roadmap for a CCS legal framework. On the other hand, the good news is that Germany has now enacted a national CCS law (Kohlendioxid-Speicherungsgesetz v. 17.8.2012, Bundesgesetzblatt 2012 Teil I S.

1726) to implement the CCS directive. This means that a comprehensive regulatory framework has been developed and coordinated at EU level. The national CCS law has opened a new window of opportunity for developing CCS in that the legislature has drawn up a clear legal framework for the geological storage of carbon dioxide streams to ensure environmental safety and sustainable development. However, it is doubtful if the public will really accept it. This will depend on the implementation of the CCS law in future.

# 3 Comments on EU CCS Directive

Currently, the EU member states are increasingly aware of the complexity of EU environmental law and policy structure. There are many overlapping environmental legislation and policies including the so-called regulatory "fragmentation" phenomenon. The CCS directive is a good example of the integration of scattered regulations into a unified CCS legal framework. It has also become the legal basis for the EU member states to promote and regulate the development of CCS technology.

However, from the perspective of the development of CCS in EU countries, to achieve commercial operations for the carbon capture and storage technology they still need to overcome various kinds of obstacles. These include the following aspects: first is technological security, second is the economic cost and third is the position of the individual governments and the acceptance by the general public. The existing EU CCS still has the following disadvantages: (1) there are no clear definitions on the related environmental and climate damage liability. (2) Though the impact assessment of CCS has been included in relevant laws and regulations of the European Union, there is still a lack of operability. On the other hand, the EU CCS legal framework has some important features and is worthy of our attention.

## *3.1 Complete Legal Framework*

There is no doubt that EU CCS directive is the core of CCS legal framework. However, this legal framework does not only include the EU CCS Directive but also contains the EU EIA Directive and EU's Emission Trading Scheme (ETS). The EU EIA Directive is to ensure that the CCS project is safe for environment and human health, through amendments to the EU's Emission Trading Scheme (ETS). Efforts have been made to incentivize investment in CCS. Besides, some other directives of the European Union, such as Directive 85/337/EC on the CCS equipment requirements, Amendment Ordinance of Directive 2000/60/EC, set out the provisions for the $CO_2$ storage site. The relevant provisions of Directive 2004/35/EC (to extend provisions for the $CO_2$ storage site) also involve $CO_2$ capture and storage.

## 3.2 EIA of the Whole Process of CCS

The Directive focuses primarily on the storage aspect of CCS, but it does briefly address the capture and transport elements. The underpinning aim of the Directive is the “environmentally safe” storage of $CO_2$,-meaning the permanent containment of $CO_2$ in such as way as to prevent or, where this is not possible, eliminate as far as possible negative effects and any risk to the environment and human health. To Achieve these aims, the whole process of CCS–capture, transport, storage should be subjected to an Environmental Impact Assessment.

Although from another perspective the impact assessment of CCS has been included in relevant laws and regulations of the European Union, the lack of operability still exists.

## 3.3 Market-Based Incentives Mechanism for CCS

The funding problems and economic cost are important obstacles to the development of CCS. The EU CCS legal framework attaches great importance to the market-based incentives mechanism, through amendments to the EU's Emission Trading Scheme (ETS). However, efforts have been made to incentivize investment in CCS. The most crucial feature of UK Energy Act 2010 is the market-based incentives mechanism for CCS. There are three basic system of market-based incentives mechanism in UK Energy Act 2010: CCS tax system, CCS demonstration projects funded in the form of a contract and CCS demonstration projects selected through market competition.

Furthermore, a New Entrants Reserve (NER) will be established in the ETS, from which 300 million allowances (corresponding to 300 million tonnes of $CO_2$ emissions or their cash equivalent) have been earmarked for the benefit of early projects in CCS and similarly innovative and currently non-commercial low-carbon technologies. "NER300" is a financing instrument managed jointly by the European Commission, European Investment Bank and Member States, so-called because Article 10(a) 8 of the revised Emissions Trading Directive 2009/29/EC contains the provision to set aside 300 million allowances (rights to emit one tonne of carbon dioxide) in the New Entrants’ Reserve of the European Emissions Trading Scheme for subsidising installations of innovative renewable energy technology and carbon capture and storage (CCS). The allowances will be sold on the carbon market and the money rose - which could be 2.4 bn EUR if each allowance is sold for 8 EUR - will be made available to the projects as they operate [9].

However, there is a crucial problem with the NER300–the fund associated with the price of carbon. Due to the current European debt crisis and economic depression, carbon prices have fallen sharply and this decline in carbon price has led to the fund shrinking by 50%. The fluctuation of carbon price and decrease of NER300 fund will exert a negative influence on the EU CCS development. Thus, NER300 fund should adopt a completely independent mode–the fund should decouple from the carbon price. In other words, if the EU wants to develop CCS, they should abandon uncertainties for future competition.

## *3.4 EU CCS Project Network to Advance CCS Demonstration [10]*

A number of exciting initiatives look set to advance CCS demonstration. Amongst them, the CCS Project Network which was proposed by the Commission in late 2008 to coordinate the demonstration of CCS and provide participating projects with a European identity. Communication actions, facilitation of knowledge sharing, public engagement to raise awareness of the potential of CCS and international cooperation, are the main tasks of the CCS Project Network. The European CCS Demonstration Project Network was established in 2009 by the European Commission to accelerate the deployment of safe, large-scale and commercially viable CCS projects. The Network that has been formed is a community of leading demonstration projects which is committed to sharing knowledge and experiences, and is united in working towards the goal of achieving safe and commercially viable CCS. The knowledge gained will be disseminated to other projects, stakeholders and the public to help gain acceptance of the technology and to support CCS to achieve its full potential as a vital technique in the fight against climate change.

- **Lack of public acceptance and public participation guideline**

The EU Commission has conducted an Internet-based consultation »Capturing and storing $CO_2$ underground – should we be concerned? « Among the 800 or so participants – nearly all climate/energy specialists 80 % of whom are CCS insiders – the question of whether CCS could be regarded as ranking equal to other options for reducing greenhouse gases met with a divided response: 52 % said »yes«, 46 % »no« (2 % don't know). By contrast, the proposition» Nuclear energy is the better solution for low-$CO_2$ power than CCS« was rejected by 62 %, with 30 % agreeing and 8 % abstentions. A high level of support (70 %) was found for the following statements: »Before 2020, all new fossil-fuel power plants built should be ›capture ready‹, »All ›capture-ready‹ plants should be retrofitted soon after 2020« and »From 2020 onwards all new coalfired power plants should be built with CCS«. There was even more agreement (over 75 %) on the question of whether the EU should support 12 full-scale demonstration projects by 2015 [10, 11].

Public acceptance is one of the crucial factors which will have great influence on the development of CCS. The EU CCS legal framework lacks public participation guidelines which would direct member states to establish CCS public participation procedure and mechanism. There are two interpretations of public participation that are relevant: The authority should establish sound and fair-minded public participation procedure to ensure the citizen's right to information and the right to participate and, the government should actively advocate the implementation of CCS projects to ensure that the public get the right information.

# 4 Implications of EU CCS Directive for China

CCS in China is an important component in achieving future climate-protection goals. A broadly based launch of $CO_2$ capture and storage is rather unlikely in the short to medium term, however, at least with the current underlying conditions [11]. In the face of the immense dynamics of power-plant extensions, CCS must be launched as quickly as possible, since otherwise the window of opportunity will close and remain so for many decades. In China, there appear to be a number of promising candidate sites for possible $CO_2$ storage, including some in regions with a high number of emission sources (power plants). But, whether these sediments are really suitable for $CO_2$ storage still requires in-depth investigation [8, 13].

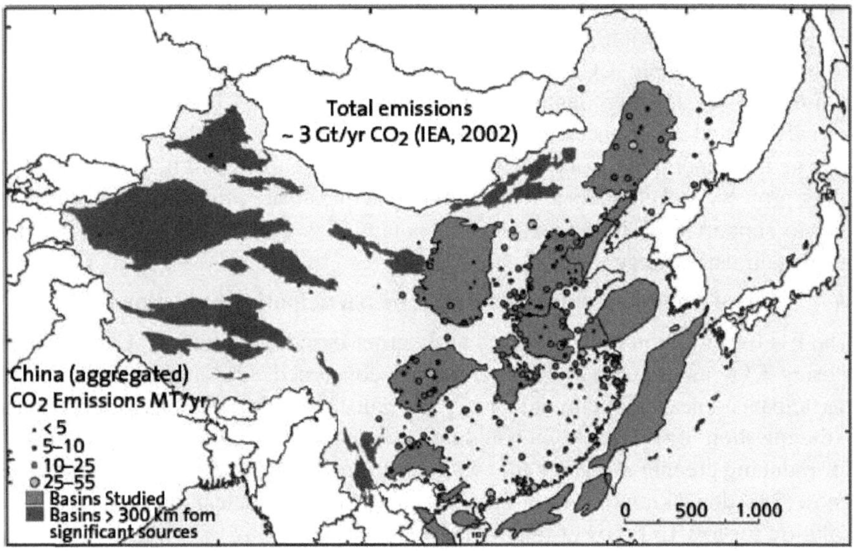


**Fig. 4** Geographic Location of Stationary Sources of $CO_2$ Emissions and Sediment Basins in China [8]

## *4.1 Basic Situation*

### 4.1.1 Policy Aspects

Chinese government has issued National medium- and long-term program for science and technology development (2006-2020), China's Policies and Actions for Addressing Climate Change 2011 and the roadmap of carbon capture, utilization and storage (CCUS) technology development in China [14].

### 4.1.2 Practical Aspect

The Chinese Huaneng Group, Shenhua Group and other enterprises carrying out the CCS demonstration projects of the whole process, have built the world's largest carbon dioxide capture project for coal-fired power plants.

**Table 7** CCS Demonstration Projects in China [15]

| PROJECT | CAPTURE | UTILIZATION / STORAGE | SCALE | STATUS QUO |
|---|---|---|---|---|
| Capture pilot project of Beijing thermal power plant by Huaneng Group | Post-combus -tion capture | Utilization in food industry | 3,000T/Y | Operating |
| Capture pilot project of Shanghai Shidongkou power plant by the Huaneng Group | Post-combus -tion capture | Utilization in food industry | 120,000T/Y | Operating |
| $CO_2$-EOR demonstration project at Jilin oilfield by PetroChina | Gas-$CO_2$ separation | EOR | 1,000,000T/y | Construction |
| $CO_2$-made biodegradable plastic project from CNOOC | Gas separation | $CO_2$-made biodegradable plastic | 2,100T/Y | Operating |
| Tianjin IGCC power plant demonstration project by the Huaneng Group | IGCC and Pre-combus-tion capture | EOR | \ | Construction |

**Table 7** (*Continued*)

| | | | | |
|---|---|---|---|---|
| Clean energy technology demonstration in Lianyungang | IGCC and Pre-combustion capture | Storage in Saline aquifers | 1,000,000T/Y | Preparation |
| Microalgae -carbon -sequestration bioenergy demonstration project by ENN | Coal-chemical -flue gas capture | Biological storage | 320,000T/Y | Construction |
| CCS demonstration project by the Shenhua Group | Capture in coal liquefaction plant | Storage in Saline aquifers | 100,000T/Y | Operating |
| CCS and EOR demonstration project at Shengli oilfield by Sinopec | Post-combustion capture | EOR | 30,000T/Y | Operating |

#### 4.1.3 Legal Aspect

For improving energy efficiency, development and utilization of energy-saving technologies and renewable energy sources, China has revised the *Renewable Energy Law* in 2007 and the *Energy Conservation Law* in 2009. However, there are no laws to date covering the supervision of CCS projects. There are several corporations in China that are already planning concrete CCS projects, some of which are at an advanced stage. However, without early amendments to the current law, these demonstration projects will not be effectively controlled. Thus there is an urgent need for legislation, especially a comprehensive CCS regulatory framework.

#### 4.1.4 International Cooperation

The Chinese government and the United Kingdom have signed a Memorandum of Understanding on Cooperation on near Zero Emissions Coal Power Generation Technology through CCS.

**Table 8** Commenced and Planned CCS Projects (Selection) [8]

| When | What | Where | Who |
|---|---|---|---|
| 2008 | 30-MWth oxyfuel power plant (commissioning) | Schwarze Pumpe/D | Vattenfall |
| 2010 | 475-MW gas-fired power plant with $CO_2$ capture for EOR | Peterhead/GB | BP |
| 2011 | 860-MW gas-fired power plant with $CO_2$ capture for EOR | Halten/Norway | Shell, Statoil |
| 2011 | IGCC with $CO_2$ capture and storage, southern North Sea | GB | E.on UK |
| 2014 | 450-MW IGCC with $CO_2$ capture and storage in saline aquifer | D | RWE |
| 2014 | 275-MW IGCC with $CO_2$ capture and storage in saline aquifer | US | »Future-Gen« project |

## *4.2 Key Points*

### 4.2.1 Establish a Comprehensive CCS Legal Framework

EU CCS legal framework does not only include the EU CCS Directive but also contains the EU EIA Directives, EU's Emission Trading Scheme (ETS) and other directive of the European Union. In the context of the progress of China's national climate change law through the legislative process since there is no special law to regulate CCS, the legislature should make clear legal institutional arrangements for CCS. It is also important to connect the CCS legislation to existing laws such as EIA law.

Due to the cross-sectoral and cross-regional characteristic of CCS, it is hard for Chinese authorities to establish a sound supervision and regulatory framework. There are three choices for the government. The first mode is to expand the scope of the application of the existing regulatory framework; the second mode is to rely on market mechanisms and minimize government intervention; the third mode is to establish a special regulatory framework for CCS. Although the last way entails the higher cost of legislation, it will be more suitable for CCS.

### 4.2.2 Improve the Whole Process EIA of CCS

The CCS projects, plans, programs and even the policies should be subjected to an EIA. As it is known that, to some extent, CCS will bring environmental and human health risk, the whole process of CCS–capture, transport and storage should also be subjected to an EIA to ensure environmental safety.

### 4.2.3 Establish Market-Based Incentives Mechanism for CCS

The funding problem and economic cost are important obstacles to the development of CCS. The EU CCS legal framework attaches great importance to the market-based incentives mechanism. The Chinese CCS legal framework should do likewise and make all stakeholders–governments, enterprises, Energy Companies and beneficiaries actively participate in the development of CCS to broaden the sources of funding. Using the BOT (build-operate-transfer) mode to develop CCS project is a possible option. CCS can be regarded as an environmental protection infrastructure in the field of $CO_2$ emission reduction. Government grants the private sector (including foreign companies) franchises for a certain period through contracts and permit enterprises to finance and operate CCS projects. They can also recover their investment and make a profit from the CCS projects. At the expiry of the concession period, the CCS projects should be transferred to the authorities so that the government will be in charge of the sequestration sites in the long run.

### 4.2.4 Perfect the Public Participation Mechanism

Public perception can have considerable and unexpected effects on CCS projects. They will bring long-term risks to safety, health and the environment. These risks are particularly prone to triggering public unrest and possibly resistance. When public participation is discussed, there are two interpretations aspects thereto: The authority should establish sound and fair-minded public participation procedure to ensure the citizen's right to information and the right to participate and, the government should actively advocate the implementation of CCS projects to ensure that the public get the right information. One important prerequisite for acceptance is the creation of transparency by providing comprehensive information both about the aims of CCS in general and about concrete intentions and projects. To avoid crises of acceptance and trust, an open-ended process of dialogue should be initiated between industry, stakeholders, science and the public at an early stage [8, 16].

## 5 Conclusion

CCS is a crucial technology for dealing with the climate change issue. The important implication of EU CCS Directive to China include the establishment of a comprehensive CCS legal framework, improvements to the CCS EIA process, the establishment of a market-based incentives mechanism for CCS and perfecting the public participation mechanism.

As any coin has two sides, CCS, just like a double-edged sword, will bring environmental and health risks. So, CCS is Moses Rod or Pandora's Box? The answer to this question depends on how the issue of CCS legal framework is dealt with.

## References


1. Metz, B., Davidson, O., de Coninck, H., Loos, M., Meyer, L.: IPCC special report on carbon dioxide capture and storage, Geneva (2005)
2. IEA, IEA CCS Technology Roadmap (2009)
3. Blohm, M., Ginzky, H., Erdmenger, C., Beckers, R., Briem, S., Clausen, U., Lohse, C., Marty, M., Rechenberg, J., Schäfer, L., Sternkopf, R.: Technische Abscheidung und Speicherung von $CO_2$ – nur eine Übergangslösung. Positionspapier des Umweltbundesamtes zu möglichen Auswirkungen, Potenzialen und Anforderungen (Technical $CO_2$ capture and storage – only a transition solution. Position paper of the Federal Environment Agency on possible implications, potentials and requirements), p. 58 (2006), Dessau http://www.umweltdaten.de/publikationen/fpdf-l/3074.pdf (September 11, 2007 )
4. Grünwald, R.: TAB: Greenhouse Gas–Bury it into Oblivion: Options and Risks of $CO_2$ Capture and Storage. Technology Assessment Studies Series No. 2, p. 14 (2009)
5. Author's elaboration
6. Jung, M., Kleßmann, C.: $CO_2$-Abscheidung und -Lagerung bei Kraftwerken ($CO_2$ capture and storage at power plants), p. 12. Ecofys Germany GmbH (2007)
7. Cremer, C., Schmidt, S.: FhG-ISI (Fraunhofer-Institut für System- und Innovationsforschung), Modellierung von Szenarien der Marktdiffusion von CCS-Technologien (Modelling of scenarios for the market diffusion of CCS technologies) (2007)
8. Grünwald, R.: TAB: Greenhouse Gas–Bury it into Oblivion: Options and Risks of $CO_2$ Capture and Storage (2009)
9. Finance for installations of innovative renewable energy technology and CCS in the EU (2012), http://www.ner300.com (accessed December 15, 2012)
10. The European CCS demonstration project network background, http://www.ccsnetwork.eu (accessed December 15, 2012)
11. Communication from the Commission to the European Council and the European Parliament. Sustainable power generation from fossil fuels: aiming for near-zero emissions from coal after 2020. COM 843 final (2006, 2007), http://ec.europa.eu/energy/energy_policy/doc/16_communication_fossil_fuels_en.pdf (Augest 21, 2007)
12. OECD/IEA 2004, Prospects for $CO_2$ capture and storage. Energy Technology Analysis, 62 p. (2004)
13. Grünwald, R.: Greenhouse Gas–Bury it into Oblivion: Options and Risks of $CO_2$ Capture and Storage. Technology Assessment Studies Series No. 2, p. 73 (2009)
14. IOSC. China's Policies and Actions for Addressing Climate Change (2011), http://www.gov.cn/english/official/2011-11/22/content_2000272.htm
15. Author's elaboration
16. Grünwald, R.: Greenhouse Gas–Bury it into Oblivion: Options and Risks of $CO_2$ Capture and Storage. Technology Assessment Studies Series No. 2, p. 18 (2009)

# Legal Framework for the Coordination of Competing Uses of the Underground in Germany

Hartmut Weyer

Institute of German & International Mining and Energy Law,
Clausthal University of Technology, Clausthal-Zellerfeld, Germany

**Abstract.** Suitable geological sites for underground storage are limited. Therefore it is important to coordinate the combined use of underground for different storage purposes (in particular gas, oil, compressed air and $CO_2$) as well as for other uses (in particular geothermal energy, mining). This paper proposes to examine the legal framework in Germany as well as in the EU CCS directive. Land use in Germany is subject to regional planning. Yet the existing plans only rudimentarily cover underground sites. So far a decision on competing uses of these sites therefore has to be taken mainly within the framework of the authorization procedures for underground uses. In general two coordination mechanisms can be distinguished: the conferral of an exclusive right on the company for exploration / production / $CO_2$ storage purposes, which excludes its use by other companies for the same purpose, and the authorization, in the absence of such exclusive rights, r within the operations plan procedure according to German mining law. This paper gives an overview of the authorization regimes which exist for different uses of underground which have been little explored in comparison. The recent German $CO_2$ Storage Act is taken into account. The result is that for most cases there are no general statutory criteria giving priority to specific uses of underground. Therefore an authority will have to base its decision on a case-by-case analysis. Due to the lack of guidance on the weight that has to be attached to the different purposes it is difficult to predict the outcome of the selection decision. The resulting legal uncertainty may hamper investments, yet granting priority to a specific use needs detailed analysis of the competing projects which in general cannot be accomplished by legal means.

**Keywords:** underground usage, EU CCS directive, coordination mechanisms, authorization regimes, German $CO_2$ Storage Act.

## 1 Introduction: Need and Mechanisms for the Coordination of the Use of the Underground

Underground sites are used for an increasing number of different purposes, including in particular the storage of gas, oil, compressed air and $CO_2$ as well as

M.Z. Hou et al. (Eds.): *Clean Energy Systems in the Subsurface*, SSGG, pp. 21–28.
DOI: 10.1007/978-3-642-37849-2_2 

geothermal and mining activities. Since suitable geological sites are limited, it is important to coordinate their use for these different purposes. This paper proposes to examine the legal framework in Germany as well as in the European CCS directive with respect to the purposes mentioned [1].

German law does not contain explicit statutory provisions on the priority of certain uses of underground. Only to a limited extent are there provisions giving preference to certain underground uses for specific regions in Germany. They are n a particular section of the German law on regional planning and will be discussed in section 2.

In the majority of cases, the selection decision has to be taken on a case-by-case basis within the framework of the authorization procedures for given projects. Three different authorization regimes will be addressed in the present context:

- Mining and geothermal activities are regulated in Germany to a large extent by the Federal Mining Act (Bundesberggesetz (BBergG)).
- $CO_2$ storage is regulated by a separate statute, the Federal $CO_2$ Storage Act (Kohlendioxid-Speicherungsgesetz (KSpG)), which shows some similarities to Mining Law but is based on the European CCS directive.
- Other underground storage facilities are (only) subject to certain provisions of the Federal Mining Act (BBergG).

Before the mining of so-called free minerals (bergfreie Bodenschätze) and geothermal energy can start, the competent *authority* is supposed to grant an initial license for exploration or production purposes respectively. At the second stage, the technical realization of the exploration / production project will requires a separate permit from the mining authority, the so-called operations plan (Betriebsplan). For underground storage activities other than $CO_2$ storage only such an operations plan, but not a prior license is needed. For $CO_2$ storage, the technical realization of an exploration / storage project requires a similar authorization.

The following description first treats the exclusive rights concerning certain natural resources, geothermal energy or storage sites for $CO_2$. This approach applies to competing underground uses for the same purpose (e.g. two projects on geothermal energy). In this case the selection among competing projects is decided on the basis of this exclusive right (section 3. below). If no prior licencing is provided for or if competing projects concern the use of underground sites for different purposes (e.g. projects for geothermal energy and for underground storage), the selection decision will be taken within the authorization procedure, which is often the operations plan procedure (section 4. below).

## 2 Coordination by Means of Regional Planning

Land use in Germany is subject to regional planning. Yet the existing plans basically restrict themselves to the use of the surface and only rudimentarily cover underground projects. In general, there is no decision as to its use for the purpose of underground storage or geothermal energy, and only to a limited extent does regional planning cover mining projects.

For the future, a more detailed regional underground planning regime is conceivable. In particular, the concept of "space", as used in the Federal Regional Planning Act (Raumordnungsgesetz (ROG)), is not limited to the surface [2-4]. The principles for regional planning with regard to natural resources (§ 2 no. 4 clause 4 ROG) and to the storage of climate-damaging substances (§ 2 no. 6 clause 8 ROG) prove that provisions on the use of underground sites are not excluded, which is also consistent with provisions for mining in existing plans. Yet the above-mentioned principles as such – without further elaboration – are not sufficient to support a specific selection decision [5]. It is possible , according to the Federal Regional Planning Act, to define regions where priority is given to specific uses of the land or where such uses are not admitted outside these regions (Vorrang-, Vorbehalts- oder Eignungsgebiete, § 8 (7) ROG).

Such selection decisions within the context of regional planning are being discussed in particular for $CO_2$ storage. The State of Lower-Saxony has expressed its intention to examine the need for further regional planning at State level in order to take account of the growing importance of subjacent geological structures for purposes of energy or climate policy [6]. According to the Energy Concept of the German government, an atlas for geothermal energy is being drawn up, in particular, to depict possible conflicts with $CO_2$ storage [7, 8]. The new Federal $CO_2$ Storage Act provides that the German States can designate regions, in which $CO_2$ storage is admitted or prohibited, taking into account in particular other possible uses of the potential storage site, geological specialties of the region and other public interests (§ 2 (5) KSpG). This possibility served as a compromise to gain the consent of German States opposing the storage of $CO_2$ on their territory. In fact, pursuant to Act 4 para 1 of the EU CCS directive Germany might have excluded the storage of $CO_2$ on its territory in general [9, 10].

# 3 Coordination by Means of Exclusive Rights

## *3.1 Mining and Geothermal Energy*

A company aiming at underground exploration for a certain natural resource, i.e. a free mineral or geothermal energy, has to apply for an exploration license (Erlaubnis) to the competent mining authority. For production purposes the company has to apply for a production license (Bewilligung or Bergwerkseigentum). These licenses, within their geographical and temporal limits, confer the exclusive right for the activity and resource concerned, e.g. an exploration license for natural gas excludes any other exploration activities for natural gas within the field covered by the license and for the duration of the license, save with the consent of the license holder.

Given the limited number of suitable sites, this may lead to a race for exploration / production licenses. In the case of competing applications, the mining authority has to give priority to the applicant that, with regard to his work program and economic power, seems best fitted to provide a useful and systematic exploration or production (§ 14 (2) BBergG). If a company already holds an exploration license (Erlaubnis zur Aufsuchung zu gewerblichen Zwecken) and later requests a production license, it is always given priority, due to the investments already made (§ 14 (1) BBergG). As a result, in the fields of mining and geothermal energy, the selection decision among competing projects aiming to use underground sites for the same purpose is often taken when issuing an exploration license, but at the latest when issuing a production license. An additional selection decision during the operations plan procedure therefore is not needed.

In order to prevent the permanent blockage of reserves, licenses may only be granted for a limited time period, but can be prolonged. The initial maximum time limit for exploration licenses is five years. Production licenses in general have an initial maximum duration of 50 years. One important element for the decision about the prolongation of a license is the extent to which the holder has complied with his work program [11]. As to the geographical extent of the license it is noteworthy that German mining law only allows for a horizontal delimitation but not for a vertical delimitation of the license area. Therefore it is not possible to issue separate licenses for the same natural resource, including geothermal energy, at different depths. Though this question is being discussed under the heading of "floor ownership" (Stockwerkseigentum), a change is not foreseeable. Unless a natural resource, in particular geothermal energy, is produced in relation to building purposes on certain premises, it is not subject to an authorization procedure and therefore is not excluded by a licence granted for the same natural resource produced at greater depth (§ 4 (2) no. 1 BBergG).

The coordination mechanism of exclusive rights means that the selection decision concerning a certain activity and natural resource is taken in advance for a given time and space. This gives a chance to gain priority at an early stage.

## *3.2 $CO_2$ Storage*

In the field of $CO_2$ storage, the Federal $CO_2$ Storage Act does not follow the two-stage approach of the Federal Mining Act, distinguishing between a prior exploration / production license on the one hand and the subsequent operations plan on the other. Yet similar to mining law the $CO_2$ Storage Act, in accordance with the EU CCS Directive, attaches an exclusive right to the exploration / storage permit. Thus no parallel explorations or storage activities for CO2 are admitted during the period of validity of the permit (§ 7 (5) KSpG, § 12 (4) KSpG).

In the case of competing applications for an exploration permit, the competent authority has to decide first which application has an exploration program that best fulfills the legal requirements of § 7 (1) KSpG, e.g. concerning the financial potential of the applicant or the protection of the environment. Should applications for an exploration permit be equivalent, priority is given to the application that first meets the requirements (§ 8 (1) KSpG). As to storage permits, the holder of an exploration permit has priority over any other applicant (§ 12 (4) KSpG). Thus the selection decision among competing projects for $CO_2$ Storage is generally taken when issuing an exploration permit, similar to the situation for mining licences.

Exploration permits are limited to the time necessary for an orderly exploration and may only be prolonged once. The maximum time limit in any case is the end of the year 2015 (§ 9 (1) KSpG). For storage permits the $CO_2$ Storage Act does not require a time limit, but the permit has to define inter alia the total quantity of $CO_2$ to be stored. Different from the normal situation in mining law, a storage site s cannot be used after closure.

Also the geographical extent of the exploration or storage permit for $CO_2$ is defined differently from mining law. The exploration permit is granted for activities within a specified exploration field (similar to mining law) but is limited in depth (§ 3 no. 16 KSpG). Moreover the permit is restricted to designated layers of rock within the exploration field (§ 8 (5) KSpG). The exclusive right established by the storage permit is similarly limited to the storage site within the designated layers of rock (cf. § 12 (4) KSpG, art. 6 para 1 CCS Directive).

### *3.3 Underground Storage Other Than $CO_2$ Storage*

The situation is different for underground storage activities other than $CO_2$ storage. This concerns in particular the storage of natural gas, oil, hydrogen or compressed air. German law neither provides for an exploration or storage license as in the case of natural resources, nor for an exploration or storage permit establishing an exclusive right as in the case of $CO_2$ storage. Therefore competing projects for the same purpose are not excluded by means of an exclusive right.

In consequence, competing underground uses of the underground – also for the same purpose – have to be coordinated within the operations plan procedure (section 4.3. below).

## 4 Coordination in the Absence of Exclusive Rights

### *4.1 Mining and Geothermal Energy*

The exclusive right granted by an exploration / production license for certain natural resources (free minerals or geothermal energy) only regulates the conflict with other uses of the underground for the same purpose, i.e. concerning the same activity and natural resource. Yet conflicts can also arise with regard to projects for different purposes, e.g. for the production of natural gas on the one hand and geothermal energy on the other. In principle such conflicts will already be considered by the mining authority when deciding about an exploration or production license. Should the conflict not have been dealt with in the licensing procedure, this will at least have to be done in the context of the operations plan procedure.

Within the licensing procedure the mining authority pursuant to § 11 BBergG (as appropriate in conjunction with § 12 BBergG) has to examine in particular whether granting the license could endanger a sensible and systematic exploration and production of natural resources (no. 8), whether natural resources of public interest would be adversely affected (no. 9) or whether overriding public interest excludes exploration throughout the whole field (No. 10). Thus no. 8 and no. 9 are limited to conflicts with the exploration or production of other natural resources whereas no. 10 takes account of other conflicts as well, in particular underground storage.

At the stage of the operations plan procedure the mining authority similarly has to examine pursuant to § 55 (1) no. 4 BBergG whether this would entail an impairment of natural resources that are of public interest, thus taking account conflicts with the exploration or production of natural resources. Conflicts with underground storage or other underground uses can be considered pursuant to § 48 (2) BBergG which takes account of any overriding public interest. The operations plan procedure applies not only to activities with regard to free minerals and geothermal energy but also with regard to other natural resources belonging to the landowner.

The approach means that the mining authority normally has to decide on a case-by-case basis which competing project should be given priority. It will therefore be important what weight the authority attaches to different purposes from the perspective of public interest, e.g. concerning the reduction of $CO_2$ emissions, the increase of power generation from renewable energy sources, the security of gas supply or the supply of other natural resources. Due to the lack of statutory rules it is difficult to find general guidance on the weight that has to be attached to such different purposes. The outcome of the selection decision therefore is hard to foresee.

Though the resulting legal uncertainty may hamper investments, it is difficult to see how priority could be granted to specific uses of underground facilities without a detailed analysis of the specific situation, e.g. concerning the benefits of the projects for climate protection, security of energy supply etc., the environmental impact of the projects, the availability of the natural resources concerned and so forth. Such an analysis in general cannot be accomplished at a statutory level. In particular, the environmental impact assessment foreseen by EU and German law can only take place in depth within the authorization procedure. Yet it might be possible and worth considering having statutory provisions giving guidance on particular aspects of the selection decision like the weight attached to different purposes (e.g. $CO_2$ storage, geothermal energy etc.).

## *4.2 $CO_2$ Storage*

The $CO_2$ Storage Act takes a similar approach in dealing with competing uses for underground storage. Pursuant to § 7 (1) no. 3 KSpG (as appropriate in conjunction with § 13 (1) KSpG) an exploration / storage permit for $CO_2$ may only be issued if an impairment of natural resources and of other possible uses for underground storage that are of public interest can be excluded. In this respect the German legislator has clearly stated that not all natural resources and possible uses of the underground have priority per se, but only those which are of greater importance for the public good than the permanent storage of $CO_2$. Special weight is attached to such natural resources and underground uses that – like $CO_2$ storage – serve the aim of climate protection, e.g. compressed air storage and geothermal energy [12]. In addition, the mining authority may only grant an exploration permit if it is not contrary to other statutes of public law and if there is no other prevailing public interest (§ 8 (1) no. 8 KSpG). For storage permits the corresponding provision is limited to opposing statutes of public law (§ 13 (1) no. 8 KSpG) which for the purpose of underground uses does not seem to apply.

In consequence, the selection decision between conflicting uses for underground sights follows rules similar to those of the Federal Mining Act. It is up to the competent authority to determine and weigh the public interest in the competing projects. There is no general guidance as to the weight attached to $CO_2$ storage in comparison to other underground uses, particularly if they are also relevant for climate protection.

## *4.3 Underground Storage Other Than $CO_2$ Storage*

Underground storage of gas, oil, hydrogen or compressed air is regulated by the Federal Mining Act, but does not follow the two stage approach applicable to free minerals and geothermal energy. Yet such projects in general will still need an authorization by way of the operations plan procedure. Thus competing uses of the underground will be coordinated by the mechanism described above with regard to the operations plan procedure for natural resources. The mining authority therefore normally has to decide on a case-by-case basis which competing project should be given priority. This applies not only to conflicts with other uses of underground storage for different purposes but, since no exclusive rights are established, also to uses for the same purpose like e.g. competing gas storage projects.

# 5 Résumé

Conflicting underground use is of increasing importance.

So far German Regional Planning Law covers the use of underground facilities only to a very limited extent.

In the case of free minerals, geothermal energy and $CO_2$ storage, conflicts with other underground uses for the same purpose are regulated by the mechanism of exclusive rights. This gives a chance of gaining priority at an early stage for a given time and space.

Conflicts with other uses of the underground for different purposes are decided on a case-by-case basis by the competent authority. There is little general guidance on the weight that has to be attached to such different purposes and therefore it is difficult to predict the outcome of the selection decision. The same holds true for competing projects of underground storage (other than $CO_2$) even if they serve the same purpose.

Though the resulting legal uncertainty may hamper investments, granting priority to a specific use of the underground needs a detailed analysis of the competing projects which in general cannot be accomplished at the statutory level. Yet it might be possible and worth considering to have statutory provisions giving guidance on partial aspects of the selection decision like the weight attached to different purposes (e.g. $CO_2$ storage, geothermal energy etc.).

## References

1. Weyer, H., Oppelt, C.: Geothermie: Notwendigkeit einer spezifischen Förderpolitik. In: Müller, T. (ed.) 20 Jahre Recht der Erneuerbaren Energien, pp. 660–687 (2012)
2. Deutscher Bundestag. Bundestags-Drucksache 16/10292: safeguarding sites for $CO_2$ storage by regional planning (2008)
3. Franke, P.: Die Einlagerung von $CO_2$ in unterirdischen geologischen Formationen unter besonderer Berücksichtigung des Bergrechts. In: Kühne, G., Ehricke, U. (eds.) Bergrecht Zwischen Tradition und Moderne, pp. 99–137 (2010)
4. Dietrich, L.: Nutzungskonflikte unter Tage. In: Kühne, G., Ehricke, U. (eds.) Bergrecht Zwischen Tradition und Moderne, pp. 139–167 (2010)
5. Dietrich, L., Schäperklaus, S.: Der Raum wird knapp: über die Steuerbarkeit von Nutzungskonflikten unter Tage. Zeitschrift für Erdöl, Erdgas und Kohle 125(1), 20–26 (2009)
6. Notice of 29 April 2009 – 303.1-20 302/25-2-1, `http://www.ml.niedersachsen.de/live/live.php?navigation_id=26909&article_id=90404&_ps-mand=7` (accessed December 21, 2010)
7. Deutscher Bundestag. Bundestags-Drucksache 16/13083, p. 8 (2009)
8. Deutscher Bundestag. Bundestags-Drucksache 17/3049, p. 10 (2010)
9. Weyer, H.: Legal Framework for CCS in the EU and Germany. In: Hou, Z., Xie, H., Yoon, J.S. (eds.) Underground Storage of $CO_2$ and Energy, pp. 3–8. CRC Press/Balkemap, London (2010)
10. Sellner, D.: Das CCS-Gesetz – Gesetzgebung im schwierigen Terrain. In: Appel, I., Hermes, G., Schönberger, C. (eds.) Öffentliches Recht im offenen Staat: Festschrift für Rainer Wahl zum 70, p. 595. Geburtstag (2011)
11. Verwaltungsgerichtsh of Mannheim. Zeitschrift für Neues Energierecht (ZNER), p. 499 (2010)
12. Deutscher Bundestag. Bundestags-Drucksache 17/5750, p. 40 (2011)

# Chances and Risks of Geologic $CO_2$ Storage

Jochen Großmann[1] and Andreas Dahmke[2]

[1] Grossmann Ingenieur Consult GmbH,
Tiergarten Strasse 48,
01219, Dresden. Germany
[2] Institut für Geowissenschaften,
CAU, Kiel, Angewandte Geologie,
Ludewig-Meyn-Str. 10, 24118 Kiel, Germany

**Abstract.** Hardly any other technology in Germany has been as controversial right from the start as $CO_2$ storage in underground geological formations, often referred to as CCS (carbon capture and storage or sequestration). Opponents and proponents of the technology appear seemingly irreconcilable; many discussions become emotional while some scientific principles have not even been sufficiently investigated. This applies both to the risks as well as the opportunities of underground $CO_2$ storage.

The use of $CO_2$ has been common in the international E&P industry and the chemical industry for many decades. However, up to this point neither a secondary process to increase yields, nor the process itself has ever been the focus of public interest.

As a consequence of the $CO_2$ storage process, $CO_2$ will become the main process substance and much larger amounts will be handled. In this framework, risk assessments are needed for handling the $CO_2$ move into the foreground and, ultimately, into the focus of public interest.

This paper focuses particularly on the systematic risk analysis of contact with $CO_2$ during $CO_2$ storage. The potential risks are systematized and examples of impact calculations are illustrated. The focus lies in the consideration of impact scenarios for the release of $CO_2$ into groundwater and into the atmosphere, including the resulting hazards. There are no hazard scenarios thus far and practically no reliable predictive models or risk assessments. At times, this results in potential dangers being dealt with irrationally. Existing risks of CCS and enhanced gas recovery (EGR) will be presented and strategies to systematically reduce risk and deal with disruptions in normal operations will be discussed. Current research findings on the release and dispersal of $CO_2$ in aquifers and in the atmosphere as well as risk management are incorporated into the depictions.

**Keywords:** CS, risk assessment, risk concept, groundwater protection, atmospheric protection.

M.Z. Hou et al. (Eds.): *Clean Energy Systems in the Subsurface*, SSGG, pp. 29–38.
DOI: 10.1007/978-3-642-37849-2_3 

# 1 Introduction

## 1.1 Chances for Geologic $CO_2$ Storage

The safety and environmental impact of $CO_2$ storage is first considered primarily in relation to the rising atmospheric $CO_2$ partial pressure, the induced climate changes and the resulting global and regional environmental risks. Ultimately, the primary goal of CCS is to reduce the release of $CO_2$ in the atmosphere. A global risk assessment must therefore include the progress towards realising this in the evaluation, namely its contribution to climate change reduction.

A resilient comparison of global risks is difficult because the regional effects of accelerated climate change with diverse economic and social interactions cannot be currently described in a serious manner within the exceedingly complex global ecosystem. The comparison is further complicated by the fact that the benefit to regional climate risks by any protective measure, not just CCS, is not expressed in terms of immediate or clearly-quantifiable minimization of the risks of global climate change. Regional acceptance, which in a first approximation is expressed as the ratio of the expected benefits to the region in question compared to the expected risk to that region, must therefore a priori not be too large for many pure climate protection measures.

There is widespread agreement in the scientific community, however, that climate change has already started and the issue now is the limitation of the consequences. Expressed in terms of a risk assessment, this means that the likelihood of climate change is already almost 100% and only the amount of impact can be reduced by slowing climate change, giving a greater period of time to implement the necessary adaptation processes.

A key target is the prevention of a global temperature increase of more than 2 °C, accompanied by an average atmospheric $CO_2$ concentration of approximately 350 ppm (Kyoto Protocol; United Nations [2]). Against the backdrop of a growing world population with a growing energy requirement, various policy scenarios for the restructuring of the global energy system are being discussed to achieve the announced value of 350 ppm (current global atmospheric value of 385 ppm) during the next 30 years and then long term stabilization. All policy scenarios have in common the assumption of a massive expansion of renewable energy systems and energy efficiency, while energy supplies based on coal, oil and natural gas will be gradually reduced. The role of nuclear energy and CCS for fossil and renewable carbon compounds is variably assessed. Based on the results of nearly all global policy scenarios, CCS is essential for the reduction of $CO_2$ immissions from coal and subordinately natural gas use. Nationally, the need for CCS can be assessed differently (excerpt from unpublished 2010 Environmental Expert Report) whereas technological development, the development of safety standards and the assessment of ecological and economic sustainability of CCS remain a global challenge.

# 2 Risk Analysis

## 2.1 *Assessment Methodology*

Safety and environmental impact of $CO_2$ storage must be considered on two different scales:

- globally as part of the indirect contribution of CCS and EGR to implement climate protection goals
- locally and therefore storage site-based within the framework of possible direct influences of $CO_2$ storage on local protected resources

The following remarks relate solely to the immediate, storage site-related environmental impacts of CCS and EGR. The risk assessments for $CO_2$ are valid for EGR, mutatis mutandis, but there may be additional EGR hazards through handling natural and associated gases. These risks are, however, the same as arise in conventional natural gas production and storage and are considered in this context.

The basis for a systematic risk analysis is a clear definition of the objective to be investigated. In Figure 1, the authors' proposals for a risk-based system for division of underground storage for $CO_2$ are visualized using an idealized profile of the North German underground systems.

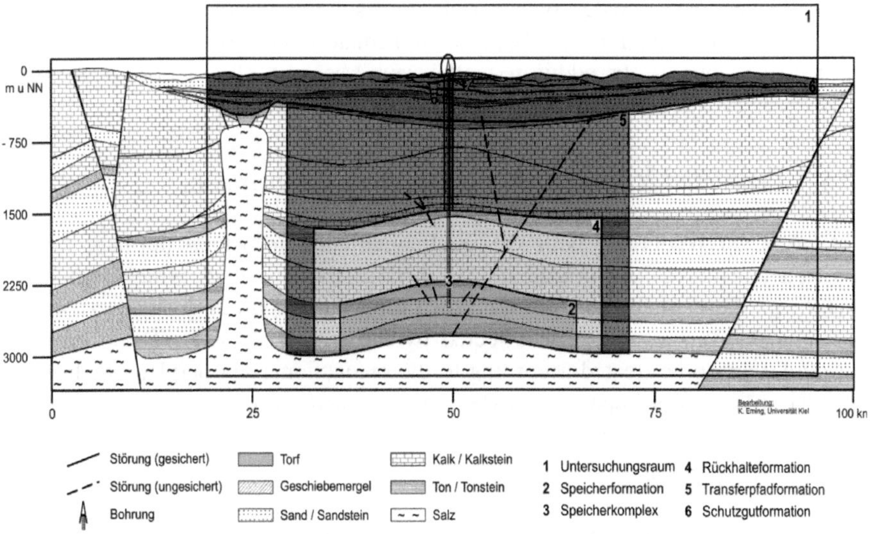


**Fig. 1** Visualization of the spatial definitions for underground $CO_2$ storage (here in a saline pore formation) in an idealized geological profile of Northern Germany. The spatial scales are used for rough indication of the size of the respective monitoring spaces.

## *2.2 Storage Site or Formation*

The storage site or reservoir can be defined as the geological space foreseen for $CO_2$ storage in a gas, liquid or fluid phase (referred to as $CO_2$ phase) or in a solution provided by the operator and verified by means of numerical modeling. The storage formation consists of reservoir rock and caprock.

## *2.3 Containment Formations*

One or more containment formations overlying the storage site can be defined by the operator, which in turn are formed from reservoir rock and semi-impermeable caprock and as such form a multi-barrier system for the actual storage site. The massive transfer of $CO_2$ from the storage formation into the containment formation is not part of normal operation, but is allowed to assist in the control of disruptions to normal operation and triggers additional and intensified monitoring activities and, if necessary, intervention.

## *2.4 Transfer Path Formations*

The transfer path formations lie between the overlying near-surface resources (groundwater, soil, air and humans) and the underlying containment formations. $CO_2$ leakage into transfer path formations is not permitted and constitutes a failure, but this situation results in no immediate danger or even impact on the protected resources. In addition to the reinforcement of monitoring activities, further intervention measures should be considered.

## *2.5 Protected Resource Formations and Resources*

As protected resources, both the geological units in which defined, legally-protected resources are considered as well as the resources which themselves make up the formation. Furthermore, protected resources located above the geological layers can also be affected by the storage of $CO_2$. These protected resources are to be treated according to the precautionary principle, which means that hazards and / or impacts are to be prevented and avoided.

## *2.6 Study Area*

In addition to the EU Directive and in line with the approach in the German Environmental Impact Assessment Act (UVPG) [3], the study area must first be defined. The study area represents the area (geological space plus the potentially-affected aboveground resources of air, flora, and fauna and human, cultural and material resources), which is significantly influenced, both directly and indirectly, in its physical-chemical properties during normal operations. It will be demonstrated that for this study area no significant adverse environmental impacts are encountered during storage operations. The methodology of this demonstration is taken from the German Air Pollution Control Act and the UVPG [3, 4].

This means that, for example, the hydrostatic and rock mechanical pressure propagation resulting from $CO_2$ injection must be recorded, which seems sensible in view of existing and future potentially competing uses and also the potential rise of highly mineralized formation waters. The terms “storage site integrity” and “storage site safety”, used in the CCS law, also apply to the study area and are to be applied.

## 3 Risk Assessment

During unintended operation of the storage formation, hazards to various environmental resources can arise. Improper operating conditions cannot be excluded. This situation can be caused by a variety of causes such as material supply, unforeseen processes and above all human error. Current studies on causes of accidents in chemical plants have shown that human error is the most common cause of accidents.

As part of the German Accident Law [6], extensive experience in the systematic assessment of risks and dealing with related hazards has been gained. It is proposed that risk management for the storage of $CO_2$ be implemented in accordance with the basic methodology of the 4$^{th}$ Federal Immission Control Ordinance and the Major Accident Ordinance [5].

The process is carried out in the following steps:

- systematic study of the hazard sources (occupational and environmental safety hazards), as well as selection of potential hazard scenarios and their probability of occurrence
- quantification / calculation of scenarios
- protected resource assessment, evaluation of the potential hazard
- risk assessment based on probability of occurrence and the expected extent of damage
- derivation of any necessary intervention measures (monitoring, protective measures, preventative measures)

For the evaluation of scenarios on the effects of these hazard sources, accident law distinguishes between the following cases:

➔ unreasonably excluded hazards (emergency response planning, measures for prevention of hazard scenarios)

➔ reasonably excluded hazards (disaster prevention planning, measures to limit impacts)

➔ "exceptional" hazards that defy any experience and predictability (disaster prevention planning - usually no responsive measures, falls within the range of acceptable residual risk).

# 4 Pertinent Environmental Risks during Underground $CO_2$ Storage

Following this classification from Major Accident Law, the environmental impacts of underground $CO_2$ storage can be grouped as follows (example representation):

## *4.1 Unreasonably Excluded Hazards*

- land uplifting or subsidence above that forecasted as a result of intended operations
- influence of the groundwater aquifer, especially the salt/fresh water boundary through the rise of formation water beyond that forecasted as a result of intended operations
- significant leakage of other gases such as methane in protected resources
- increased seismic activity during the injection and post-injection phases
- $CO_2$ leakage into groundwater and/or the atmosphere from active and old borings (including eruptive release of $CO_2$)
- release of $CO_2$ from above-ground facilities during the injection of $CO_2$

## *4.2 Reasonably Excluded Hazards*

- significant $CO_2$ leakage from the reservoir formation into the protected resource areas including atmosphere, for example through faults and fractures
- unintended future development of the $CO_2$ storage formation as a result of other economic usage aspects
- significant release of $CO_2$ from above-ground facilities during the injection of $CO_2$.

## *4.3 Exceptional Hazards that Defy Predictability*

- eruptive release of $CO_2$ from the storage formation through rock faults and fractures, for example, as a result of strong quakes, meteorite impacts or volcanic events
- strong earthquakes or localized rockbursts
- complete destruction of surface facilities

The classification is a measure of the probability of occurrence of such events. The more likely an event is and the greater the possibly associated negative effects are, the more steps must be taken to detect and prevent these events. In all cases, even for the reasonably-excluded hazards, impact analyses must be conducted in order to be prepared for such events when disaster strikes.

In order not to deviate too far from the intended framework of this assessment, only the following incidents and resulting environmental risks will be considered:

- release and dispersal of $CO_2$ in the atmosphere
- effects of underground $CO_2$ storage on the "groundwater" resource

## 4.4 $CO_2$ Release and Dispersal in the Atmosphere

In the case of incidents, $CO_2$ can follow various pathways to reach near-surface compartments. Depending on the scenario, different compartments are affected, which could on the one hand lead to an adverse change, but on the other hand provide a means of detection.

In the $CO_2$ storage literature, $CO_2$ release is often reduced only to the immediate release from the storage area. As part of the overall assessment of CCS and EGR, the surface facilities should also be considered. These are also part of the storage complex, as defined in Directive 2009/31/EG of the European Parliament from 23.04.2009 on the geological storage of carbon dioxide [1], but until now have been of secondary importance in the risk assessment literature. Surface facilities can play a crucial role, especially in the context of short-term releases into the atmosphere.

The relevant release scenarios for a $CO_2$ release to the atmosphere can be described as follows:

- Release from the storage area
  - release from active borings
  - release from old borings
  - release by density driven advective transport from the reservoir through existing or newly-emerged (for various reasons) pathways (fractures, for example)
- Release from aboveground facilities
  - leaks at storage and transport facilities
  - release during filling of equipment for surface facilities, e.g. filling of storage tanks for temporary storage of $CO_2$ via tanker or pipeline
  - release during injection
  - release from surface facilities via safety devices, e.g. blow-out valves

Depending on the conditions of the release, the $CO_2$ released is either completely vaporized and disperses as a cloud of gas or a portion of the $CO_2$ rains down as an icy pool of liquid. $CO_2$ then evaporates from the pool and flows away as a heavy gas cloud.

The release mechanisms to be considered are, in essence:

- free-jet dispersal (with and without the influence of crosswind)
- direct, impulse-free gas dispersal
- gas dispersal with sublimation source terms
- gas dispersal following heavy gas discharge
- impulse-affected release of $CO_2$ from the liquid phase

In the presentation, several examples of calculations for different release scenarios from surface facilities and from the storage area are given.

Overall, the calculations show that no blanket statements as to the hazardous areas resulting from eruptions from a bore hole are possible. The possible hazardous situations must be determined in a much more event- and location-specific manner. Based on this approach, however, site-specific prognoses and thus the introduction of specific monitoring and intervention measures are possible.

$CO_2$ is an invisible, odorless gas. Upon release, it is very difficult to detect. GICON has developed a process to make released $CO_2$ visible whereby the color is also a measure of the concentration, and thus of the hazard potential. Initial tests have been successfully carried out. With such a methodology, a very important disadvantage in $CO_2$ risk management can be removed – the hazard can now be made easily and immediately perceptible.

$CO_2$ immediately after release | $CO_2$ swirling in the wind

**Fig. 2** Trials of $CO_2$ release with visualization of $CO_2$

## 5 Impacts of Underground $CO_2$ Storage on "Groundwater" Resources

Basically, two different hazards to the "groundwater" resource resulting from underground $CO_2$ storage should be considered:

- displacement of highly-mineralized formation water in aquifers that are used as a drinking water supply, and salinization of these aquifers
- direct ascent of the stored, gas phase $CO_2$ from depths of about 800 m into aquifers

The temporal development and extent of displacement of formation waters are also dependent on the injection rate, on the mixing and solution kinetics of the $CO_{2fluid}$ in the formation water of the storage formation, on the horizontal and vertical pressure propagation and on the related vertical formation water flow.

Very simplistic worst-case estimates of displaced formation water volumes caused by long-term $CO_2$ injection show that the envisaged initial injection rates of 1-10 million tons of $CO_2$ per year can in principle represent a groundwater hazard, especially since mixing ratios (formation water / groundwater) of 1:1000 can already lead to an excess of drinking water threshold limit values. However, still not considered in this highly simplistic assessment is that during an immission of formation water into the groundwater, stratification and not a complete mixing would occur due to the differences in density. The volumes associated with formation water immission volume or pressure build up within the shallow aquifer and should be easily recognized, so that any necessary intervention measures can be taken. Against this backdrop, the simple mixing calculations appear unrealistic or represent a worst-case approach. On the other hand, this is proof of the need for a sensitive monitoring system in the near-surface aquifers and development of effective and efficient intervention measures.

Depending on the leak rate, the $CO_{2gas}$ first rises in the aquifer from the leak source with a vertical component along narrow paths to the next aquiclude or in unconfined aquifers to the unsaturated zone. The development of limited, discrete routes to the surface results from the fact that the $CO_{2gas}$ represents the non-wetting phase relative to the groundwater and therefore the groundwater must be displaced. The size and distribution of the $CO_{2gas}$ vertical pathways in the aquifer is therefore also determined by distribution of capillary intrusion pressure in addition to permeability. Due to the internal sediment stratification or discrete sediment bodies with lower permeability and/or a higher capillary intrusion pressure, the vertical rise can experience a relevant horizontal deflection or form a separate gas phase body. At the interface to the overlying aquiclude, the $CO_{2gas}$-phase is enriched, spreads at appropriately high leakage rates, even against the groundwater flow, and accumulates at anticline structures. Only in unconfined aquifers is $CO_{2gas}$ enriched in the soil gas of the unsaturated zone. Even assuming virtually punctiform immission, a larger dispersal but higher dilution is assumed due to the faster transport processes.

There is dissolution of the $CO_{2gas}$-phase in groundwater as $CO_{2gel}$ always occurs at the contact area between the gas and aqueous phase and is diluted by dispersion processes during transport. Only at very low leakage rates in the resource formation or by low migration rates relative to the thickness of each aquifer does a complete dissolution of the $CO_{2gas}$ occur in groundwater. The basic processes between $CO_{2gas}$, the weakly mineralized groundwater and the rock matrix are known and have been relatively well examined for some time. The dissolution of $CO_{2gas}$ in groundwater initially leads to a lowering of the pH value, an increase of the $HCO_3^-$ and $CO_3^{2-}$ concentrations and of activities that can effect ion-pair-forming or -complexing or act on other dissolved species. Under certain circumstances, mineral precipitation or mineral transformation processes can be triggered in the aquifer via the pH decrease, the increase in ionic strength, the increase in activity and by complexation processes(in the broad sense) ion exchange and mineral dissolution processes.

In the presentation, the processes and their monitoring and forecasting tools are displayed by means of a recent field study of the reactive transport behavior of $CO_2$ in an aquifer.

## References

1. Directive 2009/31/EC of the European Parliament and of the Council of 23 April 2009 on the Geologic Storage of Carbon Dioxide. Official Journal of the European UnionL140, 114–135 (2009)
2. United Nations, Kyoto Protocol to the United Nations Framework Convention on Climate Change, December 10, 1997, U.N. Doc FCCC/CP/1997/7/Add.1, 37 I.L.M. 22 (1998)
3. Bundesjustizministerium. German Environmental Impact Assessment Act - Official German citation: Gesetz über die Umweltverträglichkeitsprüfung in der Fassung der Bekanntmachung vom 24 February 2010 (BGBl. I S. 94), das zuletzt durch Artikel 2 des Gesetzes vom 17 August 2012 (BGBl. I S. 1726) geändert worden ist (2012)
4. Bundesjustizministerium. German Air Pollution Control Act - Official German citation: Bundes-Immissionsschutzgesetz in der Fassung der Bekanntmachung vom 26 September 2002 (BGBl. I S.3830), das zuletzt durch Artikel 2 des Gesetzes vom 27 Juni 2012 (BGBl. I S. 1421) geändert worden ist (2002)
5. Bundesjustizministerium. 4th Federal Immission Control Ordinance - Official German citation: Vierte Verordnung zur Durchführung des Bundes-Immissionsschutzgesetzes (Verordnung über genehmigungsbedürftige Anlagen) in der Fassung der Bekanntmachung vom 14 März 1997 (BGBl. I S. 504), die zuletzt durch Artikel 7 des Gesetzes vom 17 August 2012 (BGBl. I S. 1726) geändert worden ist (2012)
6. Bundesjustizministerium. Major Accident Ordinance - Official German citation: Störfall-Verordnung in der Fassung der Bekanntmachung vom 8 Juni 2005 (BGBl. I S. 1598), die zuletzt durch Artikel 5 Absatz 4 der Verordnung vom 26 November 2010 (BGBl. I S. 1643) geändert worden ist (2010)

# Supercritical $CO_2$-Corrosion in Heat Treated Steel Pipes during Carbon Capture and Storage CCS

Anja Pfennig[1], Phillip Zastrow[1], and Axel Kranzmann[2]

[1] Department of Mechanical Engineering,
HTW University of Applied Sciences Berlin,
Wilhelminenhofstraße 75A, Gebäude C, 12459 Berlin, Germany
[2] Federal Institute of Materials Research and Testing, BAM, Berlin, Germany

**Abstract.** Heat treatment of steels used for engineering a saline aquifer Carbon Capture and Storage (CCS) site may become an issue if handled trivially. Thus its influence on local corrosion (pitting) needs to be considered to guarantee reliability and safety during the injection of compressed emission gases (mainly containing $CO_2$) into deep geological rock formations. 13% Chromium steel injection pipes heat treated differently (X46Cr13, 1.4034 with 0.46% Carbon and X20Cr13, 1.4021 with 0.20% Carbon) were tested in the laboratory under supercritical $CO_2$ at 100 bar and 60 °C. Independent of the exposure time, the fewest pits were found on hardened steels with martensitic microstucture. For steels with similar Cr-content the higher C-content in 1.4034 resulted in fewer pits and lower maximum intrusion depth compared to 1.4021.

**Keywords:** steel, heat treatment, pit corrosion, CCS, $CO_2$-storage.

**Abbreviations:** CCS, carbon capture and storage; $CO_2$, carbon dioxide; X20Cr13, high alloyed steel with 13% Cr and 0.20% C – steel-no.: 1.4021; X46Cr13, high alloyed steel with 13% Cr and 0.46% C – steel-no.: 1.4034; $p_{Ksp,}$ log molar solubility; µm, micron; L, litre; mg, milligram; kg, kilogram; g, gram; HT, heat treatment.

## 1 Introduction

### *1.1 Engineering a Carbon Capture and Storage Site*

In engineering a geological on-shore saline aquifer CCS-site (CCS Carbon Capture and Storage [1-3]), corrosion of the injection pipe steels may become an issue especially when emission gases, from the combustion processes of power plants, are compressed into deep geological layers [4-8]. Typically, 42CrMo4 (1.7225, AISI 4140) is used for casing, and the X46Cr13 (1.4034, AISI 420 C) or

M.Z. Hou et al. (Eds.): *Clean Energy Systems in the Subsurface*, SSGG, pp. 39–51.
DOI: 10.1007/978-3-642-37849-2_4 

X20Cr13 (1.4021, AISI 420 J) steel is used for injection pipes. . Figure 1 shows a schematic drawing of a saline aquifer CCS-site in the Northern German Basin and possible steels used for the injection.

Saline formations are the most favored storage sites in Germany [9], because of their large potential storage volume and their common occurrence. Unfortunately, relatively little is known about them. The escape of $CO_2$ back into the atmosphere may be a problem in saline aquifer storage, especially around the water level within the borehole. Carbon dioxide corrosion may cause failure of pipelines and wells as the oil and gas industry are aware [4, 6, 10, 11].

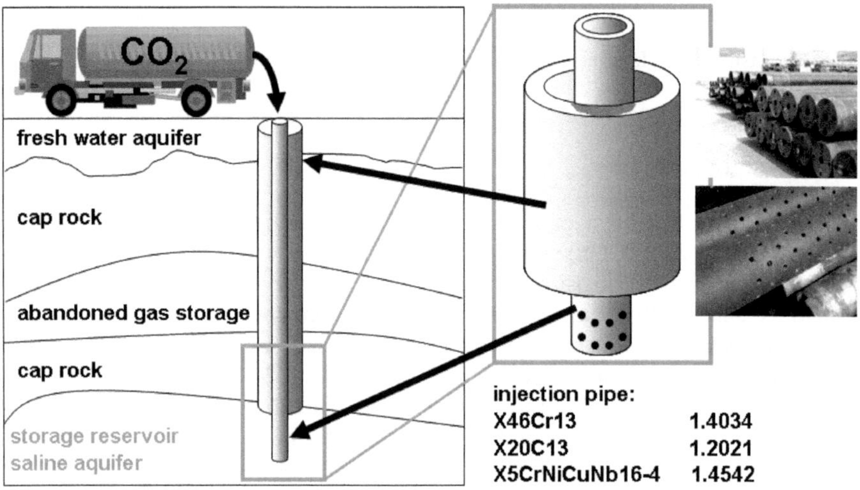


**Fig. 1** Schematic drawing, casing and tubing of the CCS-site Northern German Basin

Corrosion phenomena are most likely to occur at the interface between the caprock and the reservoir where the brine may be in contact with the injection pipe**,** If $CO_2$ is injected into the aquifer; it will not contact the aquifer water until near the bottom of the injection well. So internal corrosion will depend largely on the source of the injected gas, its composition and the presence of water and dissolved salts. Here, corrosion of the injection pipe in $CO_2$-rich aquifer water (2) may be a possibility when at injection intervals, the aquifer water flows back into the injection pipe and then forms phase boundaries (3) [12].

## *1.2 Corrosion Phenomena in $CO_2$-Saturated Aquifer Water*

It is known from thermal energy production, that the $CO_2$-corrosion is sensitively dependent on alloy composition, contamination of alloy and media, environmental conditions like temperature, $CO_2$ partial pressure, flow conditions and protective corrosion scales [6-8; 13-20]. Considering different environments, aquifer waters and pressures, the temperature regime (40 °C to 60 °C) under which the analysis

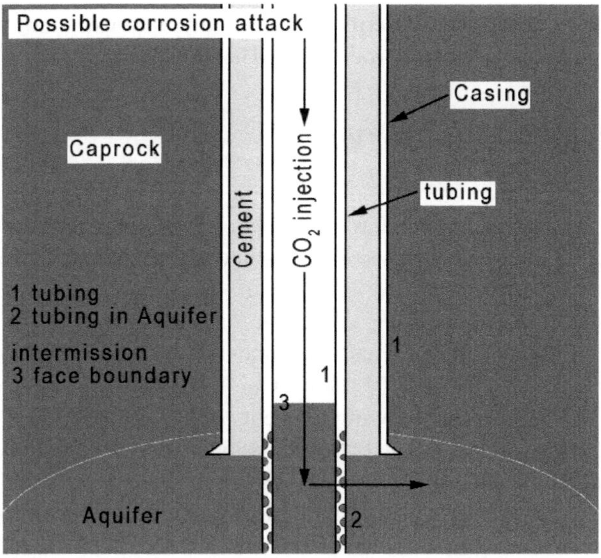


**Fig. 2** Schematic drawing of possible corrosive attack of the bottom of a well at a CCS-site (adjusted from [12])

was carried out is the critical temperature region well known for corrosion processes as shown by Pfennig et al. [21-23] and various other authors [10, 24-28]. Here a maximum corrosion rate of about 4.7 mm/year was found for a pit intrusion depth of 13% Cr steel X46Cr13. This may have been predicted by the rather conservative Norsok-Model used in the oil and gas industry to calculate surface corrosion rates of carbon steels [29].

Generally, steels that are exposed to a $CO_2$-environment precipitate slow growing surface layers mainly comprised of $FeCO_3$ (siderite) [4, 8, 30]. After the $CO_2$ is dissolved, a corrosive environment is created and the solubility of FeCO3 in water is reduced. (pKsp = 10.54 at 25 °C [28, 31). As a result of the anodic iron dissolution a siderite corrosion layer grows on the alloy surface which is also found in pits of locally corroded samples [21, 23]:

$$CO_2\,(g) + H_2O\,(l) \rightarrow H^+ + HCO_3^-\,(aq) \quad (1)$$

cathodic: $$2\,HCO_3^- + 2\,e^- \rightarrow 2\,CO_3^{2-} + H_2 \quad (2)$$

anodic: $$Fe \rightarrow Fe^{2+} + 2e^- \quad (3.1)$$

$$Fe^{2+} + CO_3^{2-} \rightarrow FeCO_3 \quad (3.2)$$

$$Fe^{2+} + 2\,HCO_3^- \rightarrow Fe(HCO_3)_2 \quad (4.1)$$

$$Fe(HCO_3)_2 \rightarrow FeCO_3 + CO_2 + H_2O \quad (4.2)$$

These reactions have been described in detail by various authors [7, 21] and a precipitation model has been introduced by Han et. al [30].

### *1.3 Influence of Heat Treatment*

The influence of heat treatment, that is: temperature and time of austenitisizing, cooling rate as well as temperature and time of annealing, has been illustrated by various authors. The presence and amount of retained austenite as a microstructural component resulting from applying heat treatments has a beneficial effect on the pitting corrosion resistance of 13%-chromium steels (13CrNiMo) [32]. A higher Ni and Cr content in the heat treated steels improve the corrosion resistance [32, 33]. Hou et al. introduce a method of empirically calculating the influence of alloying elements in heat treated steels [34]. Cvijović and G. Radenković showed that the corrosion resistance of duplex steels with chromium contents even as high as 22-27% varied with the solidification mode and annealing condition [35]. In general, raising the annealing temperature lowers the pitting potential of lean duplex stainless steels [35-37]. The lowest potentials corresponding to the transition from metastable to stable pitting, are observed for annealing at 900 °C, while a maximum improvement of corrosion stability can be achieved by annealing at 1200 °C [35]. The better corrosion resistance of martensitic stainless steels with 13% Cr at a higher austenitizing temperature (980-1050 °C) is related to the dissolution of carbides [37-39]. The precipitation of Cr-rich M23C6 and M7C3 carbides reduced the resistance of passive film and pitting corrosion [37] and had a high impact on mechanical properties due to secondary hardening [38]. The influence of heat treatment on the microstructure and mechanical properties is well known [37, 40]. However for C-Mn (carbon) steels in a $H_2S$-containing NaCl solution, the martensitic microstructure has the highest corrosion rate up to two orders of magnitude higher than ferritic or ferritic-bainitic microstructures due to the fact that martensitic grain boundaries are more reactive [40].

This work was carried out to assess the influence of heat treatment and microstructure of steels on the local corrosion behaviour. This knowledge may be used to estimate the corrosion phenomena during $CO_2$-injection into aquifer water reservoirs and predict the reliability of steels used in on-shore CCS sites.

## 2 Materials and Methods

Usually the $CO_2$ is meant to be injected either as a liquid or as a supercritical phase. In these laboratory experiments, the samples were kept in the liquid phase ($CO_2$-saturated aquifer water) to simulate the injection intervals, where aquifer water may flow back into the injection pipe. This may happen whenever there is pressure. The steel quality in the injection tube was not maintained.

## 2.1 Materials

For laboratory experiments - X20Cr13 (1.4021, AISI 420 J) and X46Cr13 (1.4034, AISI 420 C) were used as injection pipes. The alloy composition was analyzed by spark emission spectrometry SPEKTROLAB M (Spektro). Exposure tests in CO2-saturated aquifer brine were carried out using samples made of thermally treated steel specimens with 8 mm thickness, 20 mm width, 50 mm length (Table 1). Austenitizing prior to exposure was done at 950 °C, 1000 °C and 1050 °C for 30 min, 60 min and 90 min respectively. A hole of 3.9 mm diameter was used for sample positioning for the X46Cr13 (0.46 mass% C) and X20Cr13 (0.2 mass% C) steel with the same chromium content, but different carbon content. It is well known, that the corrosion rate increases with increasing carbon content. Therefore both specimens were analyzed to evaluate the influence of the carbon content of the steel.

**Table 1** Heat treatment of samples used in exposure experiments

| material | abbr. | heat treatment | temperature [°C] | hold time [min] | cooling medium |
|---|---|---|---|---|---|
| | HT1 | normalizing | 785 | 30 | air |
| **X20Cr13** | HT2 | hardening | 1000 | 30 | oil |
| **1.4021** | HT3 | hardening + tempering 1 | 1000 / 600 | 30 | oil |
| | HT4 | hardening + tempering 2 | 1000 / 670 | 30 | oil |
| | HT5 | hardening + tempering 3 | 1000 / 755 | 30 | oil |
| | HT1 | normalizing | 785 | 30 | air |
| **X46Cr13** | HT2 | hardening | 1000 | 30 | oil |
| **1.4034** | HT3 | hardening + tempering 1 | 1000 / 600 | 30 | oil |
| | HT4 | hardening + tempering 2 | 1000 / 670 | 30 | oil |
| | HT5 | hardening + tempering 3 | 1000 / 755 | 30 | oil |

## 2.2 Exposure Conditions

Both the $CO_2$ injected into the borehole of the CCS-test site at the Northern German Basin and the $CO_2$ used throughout these experiments is of very high purity( 99.995 vol.-%). The brine (known to be similar to the Stuttgart Aquifer [42]) was synthesized in a strictly orderly way to avoid precipitation of salts and carbonates. This led to an unusually high initial pH of about 9.2. The pH of the synthesized aquifer water after the experiments was between 5.2 and 5.6 revealing that the brine had been totally saturated with $CO_2$. The exposure of the samples between 700 h to 8000 h was recreated in autoclaves under identical conditions to those at the geological site at 60 °C at 100 bar for comparison and as well as for ambient pressure – each material was placed in a separated reaction vessel (Figure 3). Samples of each base metal were positioned within both the vapour phase and the liquid phase. The brine (known to be similar to the Stuttgart Aquifer [41]: $Ca^{2+}$: 1760 mg/L, $K^{2+}$: 430 mg/L, $Mg^{2+}$: 1270 mg/L, $Na^{2+}$: 90,100 mg/L, $Cl^{-}$: 143,300 mg/L, $SO_4^{2-}$: 3600 mg/L, $HCO^{3-}$: 40 mg/L) was synthesized in a strictly orderly way to avoid precipitation of salts and carbonates. Flow control (3 NL/h) of the technical $CO_2$ (purity 99,995 vol.-%) into the brine at ambient pressure was achieved by a capillary meter GDX600_manufactured by QCAL Messtechnik GmbH, Munic.

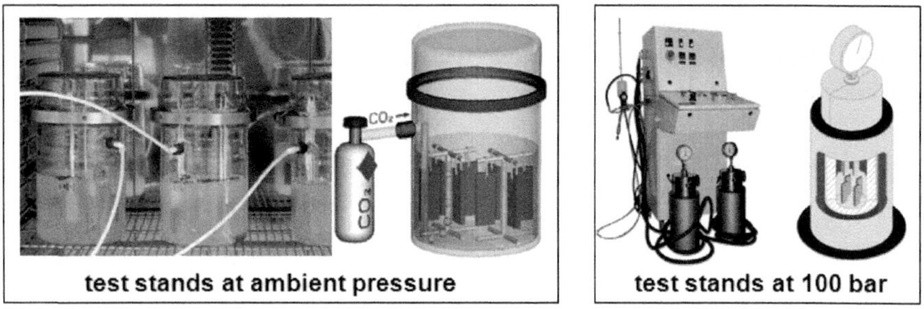


**Fig. 3** Experimental Setup for long exposure of steel samples to $CO_2$-saturated saline aquifer brine

## 2.3 *Sample Preparation*

Before the corrosion tests were conducted, the surfaces of the steels were activated by grinding them under water with SiC-Paper down to 120 µm. After the corrosion tests, the samples were cut partly for scale analysis of the corrosion layer and prepared partly for kinetic analysis after the scale was etched. Descaling of the samples was performed by exposure to 37% HCl. Then parts of the samples were embedded in a cold resin (Epoxicure, Buehler), cut and polished from 180 µm to 1200 µm firstly using SiC-Paper under water and then finished with diamond paste to 6 µm and 1 µm.

## 2.4 *Analysis*

Different optical and electron microscopy techniques were performed on the specimens to investigate the layer structures and morphology of the samples. X-ray diffraction was carried out in a URD-6 (Seifert-FPM) with an automatic slit adjustment using CoKα-radiation, step 0.03°nd count 5 sec. Phase analysis was performed by matching peak positions automatically with PDF-2 (2005) powder patterns. Structures that were likely to precipitate from the steels were chosen mainly from the ICSD and refined to fit the raw-data-files using AUTOQUAN ®

by Seifert FPM. To characterise the pitting corrosion, 3-D-images were realized using the double optical system Microproof TTV by FRT. Kinetics of the corrosion were determined by comparing the corrosion rates calculated from the change in the mass of the samples before and after corrosion testing according to DIN 50 905 parts 1-4.

$$corrosion\ rate\left[\frac{mm}{year}\right]=\frac{8.76\cdot 10^{4}\cdot weight\ loss[g]}{area[cm^2]\cdot density\left[\frac{g}{cm^3}\right]\cdot time[hour]} \tag{5}$$

The measurement of the layer thicknesses and residual pipe wall thicknesses was performed by the line scan method using the semi-automatic analyzing program Analysis Docu ax-4 by Aquinto. The scale thickness for each exposure time was measured by taking a set of 10 to 20 micrographs. Then a set of 100 line scans was placed over the scale giving the average scale thickness of the sample.

## 3 Results and Discussion

Specimens exposed to the brine form a carbonate layer as stated in equations 1 to 4, which is demonstrated by the reflection changes of the surfaces (from a polished metal to dull reddish) (Figure 4 and 5). These non-uniform corrosion layers are different in thickness but formed all over the surface and indicate their dependence on the environmental condition, which is either the water-saturated supercritical $CO_2$ or the liquid $CO_2$-saturated aquifer phase.

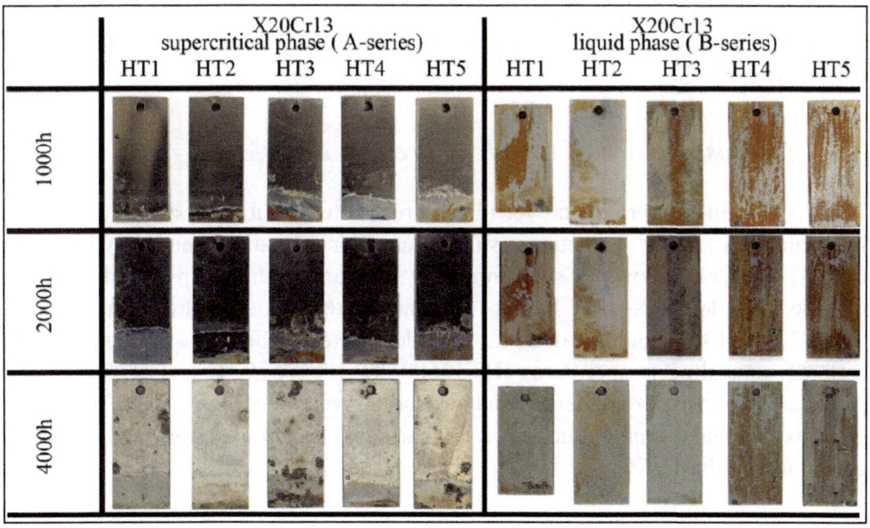


**Fig. 4** Sample surfaces of steel qualities (X20Cr13 with pits) after 4000 hours of exposure at 60 °C and 100 bar to water-saturated supercritical $CO_2$ or liquid $CO_2$-saturated aquifer phase

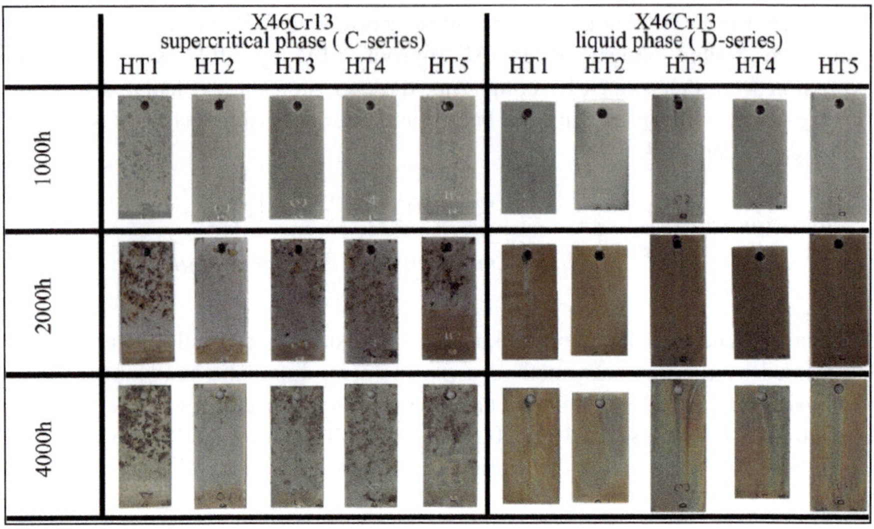


**Fig. 5** Sample surfaces of steel qualities (X46Cr13 with pits) after 4000 hours of exposure at s60 °C and 100 bar to water-saturated supercritical $CO_2$ or liquid $CO_2$-saturated aquifer phase

The multi-layered carbonate/oxide structure is described in detail by Pfennig and Bäßler [21]. It reveals siderite $FeCO_3$, goethite α-FeOOH, mackinawite FeS and akaganeite $Fe_8O_8(OH)_8Cl_{1.34}$ as well as the spinal-phases of various compositions. Carbides, such as $Fe_3C$, were identified within the corrosion layer, similar to the high-temperature corrosion phenomena [42]. The pits were covered with the same precipitates of the corrosion products as those formed on the surface elsewhere [23].

## *3.1 Microstructure of the Heat Treated Base Metal*

$X_{20}Cr_{13}$ exhibits a normalized ferritic microstructure, but also seems to contain remaining martensitic formations (slightly acicular formations marked with green circles). $X_{46}$Cr13 shows a soft annealed structure with encapsulated globular carbides. Hardened $X_{20}Cr_{13}$ has an acicular martensitic structure with carbides, while X46Cr13 is dominated by Cr-carbides. After hardening and tempering at different annealing temperatures (HT3=600 °C, HT4=675 °C and HT5=755 °C) $X_{20}Cr_{13}$ will consist of tempered acicular martensite with carbides whereas $X_{46}Cr_{13}$ exhibits hardened microstructure with carbides in combination with residual austenite or delta-ferrite.

## *3.2 Kinetics*

The differently heat treated samples were examined using optical methods to predict the number and depth of the pits. Kinetics were obtained from the weight loss, according to DIN 50 905 (figure 6-7), following exposure to the $CO_2$-saturated aquifer water. Since corrosion rates are usually given for surface corrosion rates, only these rates apply to the overall surface but results that only considered the pit corrosion layers only will serve as a starting point for a comparison of the steel qualities and the heat treatment methods.

According to Figure 6, the hardened steel exhibits the lowest corrosion rate of less than 0.002 mm/a after 4000 hrs of exposure. The other heat treatments behaved very similarly with rates around 0.005 mm/a after 4000 hrs. HT5 was the exception showing not only the highest corrosion rate of 0.017 mm/a after 4000 hrs, but was also the only corrosion rate to increase over time. This may be due to the small grain size rather than the high amount of grain boundaries acting as ionic sinks and initiating corrosion. It could also be due to the amount of carbides precipitated within the martensitic matrix which increases with the temperature of the heat treatment. Carbides are generally more susceptible to corrosion and enhance local corrosion and base metal degradation.

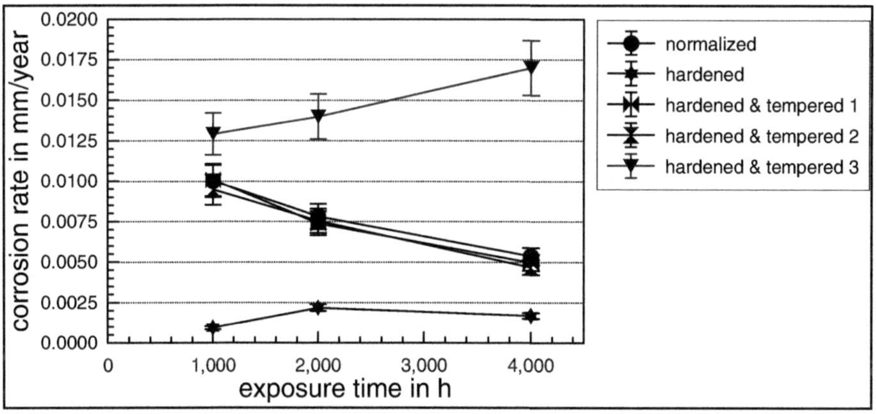


**Fig. 6** Influence of heat treatment: Corrosion rate summed up for all differently heat treated samples of X20Cr13 and X46Cr13 after 4000 hours of exposure to aquifer brine water at 60 °C and 100 bar

Figure 7 stresses that the specimens in the supercritical phase have lower corrosion rates compared to specimens in the liquid phase. The highest corrosion rates in the liquid phase are accompanied by the formation of large siderite crystal sizes and the lowest corrosion rates lead to small crystal sizes in the supercritical phase. This can be explained as followed: firstly the diffusion is much slower in the liquid phase due to the higher density than in the supercritical phase. Here fast nucleation takes place leading to a greater amount of nuclei that are limited in growth once a continuous scale has been formed on the metal surface. Secondly, the easier access of $CO_2$ through the porous corrosion layer and the higher $CO_2$-partial pressure in the supercritical phase also leads to enhanced nucleation and therefore smaller siderite crystals. These smaller crystal sizes develop into a continuous pacifying scale protecting the base metal from further fast degradation.

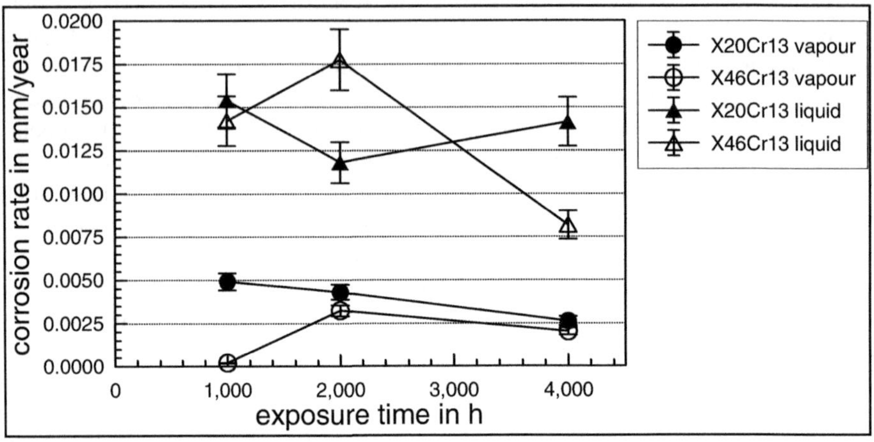


**Fig. 7** Influence of atmosphere: Corrosion rate summed up for all samples of X20Cr13 and X46Cr13 after 4000 hours of exposure to either water-saturated $CO_2$-phase (vapour) or $CO_2$-saturated aquifer water at 60 °C and 100 bar

Maximum intrusion depths after normalizing, hardening and tempering 2 and 3 are below 50 μm. Maximum intrusion depths for X20Cr13 after 1400 hours of exposure were about 35 μm. For these 3 types of heat treatments, the exposure time does not influence the pit penetration depths. The 3 samples, hardened and hardened-tempering exhibited an increase in pit penetration depth with exposure time. X20Cr13 shows a maximum penetration depth (~250 μm) after hardening. These specific high penetration depths could be due to local non-homogeneties of the microstructure. Pit depths measured on X46Cr13 are not susceptible to exposure time or heat treatment, but are rather a phenomenon of the steel's particular composition. This is mainly due to the higher Cr- and Ni-content that build up reservoirs of free Cr to precipitate a protecting layer of Chromium oxide on the steel surface. In combination with slow diffusion kinetics in the supercritical phase (compared with diffusion rates of ionic species in the liquid phase) this corrosion protection is even enhanced. The heat treatment has little influence on the maximum penetration depth, that is: there is no heat treatment available for lowering corrosion rate. For the 13Cr steels (X20Cr13 and X46Cr13) normalizing and hardening+tempering1 show less intrusion (in general 18-25 μm) than the other heat treatments.

# 4 Conclusions

During the normal storage procedure, $CO_2$ is supposedly injected while in liquid or supercritical phase. When there are intermissions in the injections, the water level may rise into the injection pipe leading to the precipitation of corrosion products and formation of pits as stated. Pit growth cannot be calculated as easily as surface corrosion rates, because of its unpredictability. Therefore it is not possible to give corrosion rates and lifetime predictions for pit corrosion in CCS technology. Summarizing the kinetic results, the preferred heat treatment for obtaining the least corrosive attack is hardening X20Cr13 and X46Cr13. Annealing at too high temperatures (650 °C) will lead to the precipitation of high amounts of ferrous carbides resulting in high corrosion rates irrespective of the exposure time. Although pit growth rates will be above 0.1 mm/year and below 1 mm/year it may still be possible that the analysed steel qualities are suitable for injection pipes in a CCS environment if monitored closely. In the future a detailed microstructure analysis will be necessary.

**Acknowledgements.** This work was supported by the FNK (Fachkonferenz für wissenschaftlichliche Nachwuchskräfte) of the Applied University of Berlin, HTW and by IMPACT (EU-Project EFRE 20072013 2/21).

# References

1. Thomas, D.C.: Capture and Separation of Carbon Dioxide form Combustion Sources, $CO_2$ Capture Project, vol. 1. Elsevier Ltd., UK (2005) ISBN 0080445748
2. van den Broek, M., et al.: Effects of technological learning on future cost and performance of power plants with $CO_2$ capture. Progress in Energy and Combustion Science 35, 457–480 (2009)
3. GeoForschungszentrum Potsdam, $CO_2$-SINK – drilling project, description of the project PART 1: 1-39 (2006)
4. Nešić, S.: Key issues related to modelling of int. corrosion of oil and gas pipelines. Corrosion Science 49, 4308–4338 (2007)
5. Hurter, S., Labregere, D., Berge, J., Desitter, A.: Impact of Mutual Solubility of $H_2O$ and $CO_2$ on Injection Operations for Geological Storage of $CO_2$. In: International Conference of the Properties of Water and Steem ICPWS, Berlin, September 8-11 (2008)
6. Zhang, L., Yang, J., Sun, J.S., Lu, M.: Effect of pressure on wet $H_2S/CO_2$ corrosion of pipeline steel, No. 09565. In: NACE Corrosion 2008 Conference and Expo, New Orleans, Louisiana, USA, March 16-20 (2008)
7. Mu, L.J., Zhao, W.Z.: Investigation on Carbon Dioxide Corrosion Behaviors of 13Cr Stainless Steel in Simulated Strum Water. Corrosion Science 52, 82–89 (2010)
8. Bonis, M.: Weight loss corrosion with $H_2S$: From facts to leading parameters and mechanisms, Paper No. 09564. In: NACE Corrosion 2008 Conference and Expo, New Orleans, Louisiana, USA, March 16-20 (2008)

9. IPCC, Special Report Carbon Dioxide Capture and Storage Technical Summary (August 09, 2012) (2012), http://www.ipcc.ch/pdf/special-reports/srccs/srccs_technicalsummary.pdf
10. Seiersten, M.: Material selection for separation, transportation and disposal of $CO_2$. NACE Corrosion Paper number: 01042 (2001)
11. Cui, Z.D., Wu, S.L., Zhu, S.L., Yang, X.J.: Study on corrosion properties of pipelines in simulated produced water saturated with supercritical $CO_2$. Applied Surface Science 252, 2368–2374 (2006)
12. Pfennig, A., Kranzmann, A.: Reliability of pipe steels with different amounts of C and Cr during onshore carbon dioxide injection. International Journal of Greenhouse Gas Control 5, 757–769 (2011)
13. Enerhaug, J.: A study of localized corrosion in super martensitic stainless steel weldments. Thesis submitted to the Norwegen University of Science and Technology (NTNU), Trondheim (2002)
14. Neubert, V.: Beanspruchung der Förderrohrtour durch korrosive Gase, VDI-Berichte Nr. 2026 (2008)
15. Kirchheiner, R., Wölpert, P.: Qualifizierung metallischer Hochleistungswerkstoffe für die Energieumwandlung in geothermischen Prozessen, VDI-Berichte Nr. 2026 (2008)
16. Zhang, H., Zhao, Y.L., Jiang, Z.D.: Effects of temperature on the corrosion behaviour of 13Cr martensitic stainless steel during exposure to $CO_2$ and $Cl^-$ environment. Material Letters 59, 3370–3374 (2005)
17. Alhajji, J.N., Reda, M.R.: The effect of alloying elements on the electrochemical corrosion of low residual carbon steels in stagnant $CO_2$-saturated brine. Corrosion Science 34(11), 1899–1911 (1993)
18. Choi, Y.S., Nešić, S.: Corrosion behaviour of carbon steel in supercritical $CO_2$-water environments, No. 09256. In: NACE Corrosion 2008 Conference and Expo, New Orleans, Louisiana, USA, March 16-20 (2008)
19. Jiang, X., Nešić, S., Huet, F.: The Effect of Electrode Size on Electrochemical Noise Measurements and the Role of Chloride on Localized $CO_2$ Corrosion of Mild Steel, 09575. In: NACE Corrosion 2008, USA, March 16-20 (2008)
20. Ahmad, Z., Allam, I.M., Abdul Aleem, B.J.: Effect of environmental factors on the atmospheric corrosion of mild steel in aggressive sea coastal environment. Anti Corrosion Methods and Materials 47, 215–225 (2000)
21. Pfennig, A., Bäßler, R.: Effect of $CO_2$ on the stability of steels with 1% and 13% Cr in saline water. Corrosion Science 51(4), 931–940 (2009)
22. Yevtushenko, O., Bäßler, R., Pfennig, A.: Corrosion behaviour of Cr13steel in CO2 saturated brine with high chloride concentration. Materials and Corrosion 62, 517–521 (2012)
23. Pfennig, A., Kranzmann, A.: The role of pit corrosion in engineering the carbon storage site Ketzin, Germany. WIT Transactions on Ecology and the Environment 126, 109–118 (2010)
24. Nyborg, R.: Controlling Internal Corrosion in Oil and Gas Pipelines. Business Briefing: Exploration & Production: The Oil & Gas Review (2), 70–74 (2005)
25. Carvalho, D.S., et al.: Corrosion rate of iron and iron-chromium alloys in $CO_2$-medium. CorSci. 47, 2974–2986 (2005)
26. Linter, B.R., Burstein, G.T.: Reactions of pipeline steels in $CO_2$ solutions. Corrosion Science 41, 117–139 (1999)

27. Wu, S.L., Cui, Z.D., Zhao, G.X., Yan, M.L., Zhu, S.L., Yang, X.J.: EIS study of the surface film on the surface of carbon steel form supercritical carbon dioxide corrosion. Applied Surface Science 228, 17–25 (2004)
28. Brown, B., et al.: $CO_2$ corrosion in the presence of trace amounts of $H_2S$. Corrosion 04736, 1–28 (2004)
29. $CO_2$ corrosion calculation model (August 9, 2012), http://www.standard.no/PageFiles/1178/M-506d1r2.pdf
30. Han, J., Yang, Y., Nešić, S., Brown, B.N.: Roles of passivation and galvanic effects in localized CO2 corrosion of mild steel. 08332. In: NACE Corrosion 2008, New Orleans, Louisiana, USA, March 16-20 (2008)
31. Banaś, J., Lelek-Borkowska, U., Mazurkiewicz, B., Solarski, W.: Effect of $CO_2$ and $H_2S$ on the composition and stability of passive film on iron alloy in geothermal water. Electrochimica Acta 52, 5704–5714 (2007)
32. Bilmes, P.D., Llorente, C.L., Méndez, C.M., Gervasi, C.A.: Microstructure, heat treatment and pitting corrosion of 13CrNiMo plate and weld metals. Corrosion Science 51(4), 876–882 (2009)
33. Bülbül, S., Sun, Y.: Corrosion behaviours of high Cr-Ni cast steels in HCl solution. Journal of Alloys and Compounds 498, 143–147 (2010)
34. Hou, B., Li, Y., Li, Y., Zhang, J.: Effect of alloy elements on the anti-corrosion properties of low alloy steel. Bull. Material Sci. 23(3), 189–192 (2000)
35. Cvijović, Z., Radenković, G.: Microstructure and pitting corrosion resistance of annealed duplex stainless stell. Corrosion Science 48, 3887–3906 (2006)
36. Zhang, L., Zhang, W., Jiang, Y., Deng, B., Sun, D., Li, J.: Influence of annealing treatment on the corrosion resistance of lean duplex stainless steel 2101 (2009)
37. Choi, Y.S., Kim, J.G., Park, Y.S., Park, J.Y.: Austenitzing treatment influence on the electrochemical corrosion behaviour of 0.3C-14Cr-3Mo martnesitic stainless steel. Materials Letters 61, 244–247 (2007)
38. Isfahany, A.N., Saghafian, H., Borhani, G.: The effect of heat treatment on mechanical properties and corrosion behaviour of AISI420 martensitic stainless steel. Journal of Alloys and Compounds 509, 3931–3936 (2011)
39. Park, J.Y., Park, Y.S.: The effects of heat-treatment parameters on corrosion resistance and phase transformation of 14Cr-3Mo martensitic stainless steel. Materials Science and Engineering A 449-451, 1131–134 (2007)
40. Lucio-Garciaa, M.A., et al.: Effect of heat treatment on $H_2S$ corrosion of a micro-alloyed C–Mn steel. Corrosion Science 51, 2380–2386 (2009)
41. Förster, et al.: Baseline characterization of the $CO_2$SINK geological storage site at Ketzin, Germany. Environmental Geosciences 13(3), 145–161 (2006)
42. Kranzmann, D., Huenert, H., Rooch, I., Urban, W., Schulz, W., Österle.: Reactions at the interface between steel and oxide scale in wet $CO_2$ containing atmospheres. In: NACE Corrosion Conference & Expo, Atlanta (2009)

# Operational Reservoir Monitoring at the $CO_2$ Pilot Storage Site Ketzin, Germany

Sebastian Köhler[1,*], Jochen Zemke[1], Winfried Becker[2], Jürgen Wiebach[2], Axel Liebscher[3], Fabian Möller[3], and Andreas Bannach[4]

[1] Untergrundspeicher-und Geotechnologie-Systeme GmbH (UGS), Berliner Chaussee 2, D-15749 Mittenwalde, Germany
[2] VNG Gasspeicher GmbH (VGS), Maximilianallee 2, D-04129 Leipzig, Germany
[3] Helmholtz Centre Potsdam, GFZ German Research Centre for Geosciences, Telegrafenberg, D-14473 Potsdam, Germany
[4] ESK GmbH, Halsbrücker Str. 34, D-09599 Freiberg, Germany

**Abstract.** In June 2008, a continuous injection of $CO_2$ into the subsurface of Ketzin (approx. 25 km West of Berlin, Germany) was commenced for research purposes. By May 20112 A total of 61,396 t $CO_2$ had been injected into the deep underground in a sandstone aquifer without any safety-related disturbances and abiding by the Federal Mining Law regulations. The initial project layout using 1 injection well and 2 observation wells was further developed in 2011 by drilling a shallow well for monitoring in the upper groundwater horizon and in 2012 by drilling another observation well for additional monitoring and the extraction of $CO_2$ flooded samples from the reservoir. The monitoring concept developed consisted of continuous measurements of surface and underground pressure and temperature (P/T), wellbore logging campaigns including reservoir saturation and wellbore integrity measurements, gas-, hydro-, and geochemical sampling, as well as differently scaled geoelectric and seismic measurements. Besides its scientific benefits, the monitoring concept provided all the data needed for a safe and reliable operation. Furthermore, the data provided important insights into the reservoir behaviour, which can be transferred to other locations. Compared to the operational engineering of underground storage of natural gas, which has been an established secure technology for decades, differences were found in the detection of the gas distribution and the installation design of the P/T tools. The long-term experiences in natural gas storage of the operator VGS and the subcontractors UGS and ESK turned out to be essential for the effective and secure integration of best practise operational reservoir monitoring at the $CO_2$ storage pilot site Ketzin. Based on this the next step should be to transfer the experiences gained within the scope of this project to an industrial scale, on-shore $CO_2$ storage project as soon as possible.

**Keywords:** $CO_2$ storage, reservoir monitoring, reservoir simulation, continuous measurements.

---

* Corresponding author.

M.Z. Hou et al. (Eds.): *Clean Energy Systems in the Subsurface*, SSGG, pp. 53–63.
DOI: 10.1007/978-3-642-37849-2_5 

## 1 Introduction

Successful geological storage of $CO_2$ in deep saline formations is crucial for the industrial scale implementation of this climate friendly technology (Bachu 2000, Zemke et al. 2003, Bradshaw et al. 2007). Pilot sites for the geological storage of $CO_2$ in saline aquifers thus form a central part of most technology roadmaps. Natural gas storage in saline aquifers has been a well proven technology for many decades and a lot of the operational and technological experiences gained there may be transferred to the storage of $CO_2$. However, different physicochemical properties of $CO_2$ and natural gas together with the potential of chemical reactivity of injected $CO_2$ within the storage complexes prevent direct transfer from one to the other. There is, therefore, every indication that a general need exists for first-hand operational and technological experiences which can be gained at active $CO_2$ injection sites. The paper in hand presents a brief overview of the operational and reservoir engineering experiences from the Ketzin pilot site in Germany, which is the first European on-shore pilot site for $CO_2$ storage in saline aquifers and is still the only active German storage project (Würdemann et al. 2010, Martens et al. 2011, 2012). It is focussing on the injection related monitoring applied at Ketzin and its applicability for future $CO_2$ storage operations.

## 2 Project and Site Description

Under the entrepreneurship of the German Research Centre for Geosciences (GFZ) the pilot site Ketzin is operated by the VNG Gasspeicher GmbH (VGS), the owner of the mineral rights at the Ketzin site. The subcontractors Untergrundspeicher- und Geotechnologie-Systeme GmbH (UGS) and ESK GmbH (ESK) are responsible for the operational and reservoir engineering. Additionally, UGS is responsible for the regulatory reporting to the Federal Mining Authority (Landesamt für Bergbau, Geologie und Rohstoffe Brandenburg, LBGR). With their long-term experience in natural gas storage the operator VGS and the subcontractors UGS and ESK guarantee a successful integration of the operational reservoir monitoring at the $CO_2$ storage pilot site.

In June 2008, a continuous injection of $CO_2$ was commenced, for research purposes, into the subsurface at the pilot site for geological $CO_2$ storage near the town Ketzin in Brandenburg, Germany (approx. 25 km West of Berlin). The pilot site is located at the south-east flank of a roughly west-south-west to east-north-east elongated double anticline ("Roskow-Ketzin anticline"), which was formed above a salt pillow at 1,500 to 2,000 m depth (for details of the site geology see Förster et al. 2010, Norden et al. 2010). The reservoir was formed by sandstone layers in the upper parts of the Upper Triassic Stuttgart formation at a depth of about 630 to 650 m. The reservoir is overlain by more than 165 m of clay- and mudstones of the Upper Triassic Weser and Arnstadt formations, which form the first seal of the multibarrier system at the Ketzin site. The Arnstadt formation is overlain by a succession of permeable and impermeable lithologies of the Upper

Triassic Exter formation and Lower Jurassic formations up to the base of the Tertiary at depth of about 150 m. The base of the Tertiary is formed by Rupelian Clay, which separates the shallow freshwater horizons from the deep saline formations and therefore forms the ultimate seal at the Ketzin site.

## 3 Technical Facilities

Prior to the injection, three wells (Ktzi 200, Ktzi 201 and Ktzi 202) -were drilled in 2007 to depths of approx. 800 m and were completed with permanently installed monitoring devices ("smart casing" concept, Prevedel et al. 2008, 2009, Schmidt-Hattenberger et al. 2011). The well Ktzi 201 serves as a combined injection and observation well whereas the wells Ktzi 200 and Ktzi 202, which are located at distances of 50 and 112 m from the well Ktzi 201 serve as observation wells (Figure 1). All wells have a 5 ½" production casing and are connected to the reservoir via slotted liners with filter screens. In the injection and observation well, Ktzi 201, a 3 ½" injection tubing is installed at a depth of 560 m. In 2012, a third observation well Ktzi 203, which is located at a distance of ~ 25 m from the well Ktzi 201, was drilled to allow additional monitoring and to recover core samples from the reservoir and cap rocks exposed to $CO_2$. Above zone monitoring is done with a shallow observation well P300, which reaches the first aquifer above the main seal within the Exter formation at a depth of 446 m.

The $CO_2$ at the Ketzin pilot site is delivered by trucks at ~ 18 bar/-18 °C, stored in surface tanks, and then pre-conditioned in the injection facility to the desired injection conditions using plunger pumps and ambient air and electrical heaters. The injection facility allows maximum injection rates of 3,250 kg/h (set point value). Typical injection rates during continuous operation are either ~ 1,600 kg/h or ~ 3,200 kg/h.

## 4 Injection History

During the commissioning phase of the pilot site prior to the onset of continuous injection, the injection facility was intensely tested and the first operational experience was gained by the injection of small amounts of $CO_2$. These operational tests included several start, shut-in and re-start runs as well as the test of $N_2$ admission during the shut-in time. The commissioning phase was completed on June 30$^{th}$, 2008, when the pilot site entered the test run phase with continuous injection of $CO_2$ (Figure 1).

The test run phase ended on September 24th, 2008 and since then the facility works under normal operation on a 24 hours/7 days basis. Due to the drilling of the well Ktzi 203, the injection was stopped in May 2012 and will be re-started at the beginning of 2013. Since the start of injection a total of 61,396 tons of $CO_2$ have been injected. Until March 2010, the mean injection rate was ~ 1,500 t $CO_2$/month (= total of ~ 32,000 t $CO_2$) and from March 2010 to May 2012, the overall injection rate was lowered to ~ 1,100 t $CO_2$/month (= total of ~ 30,000 t $CO_2$).

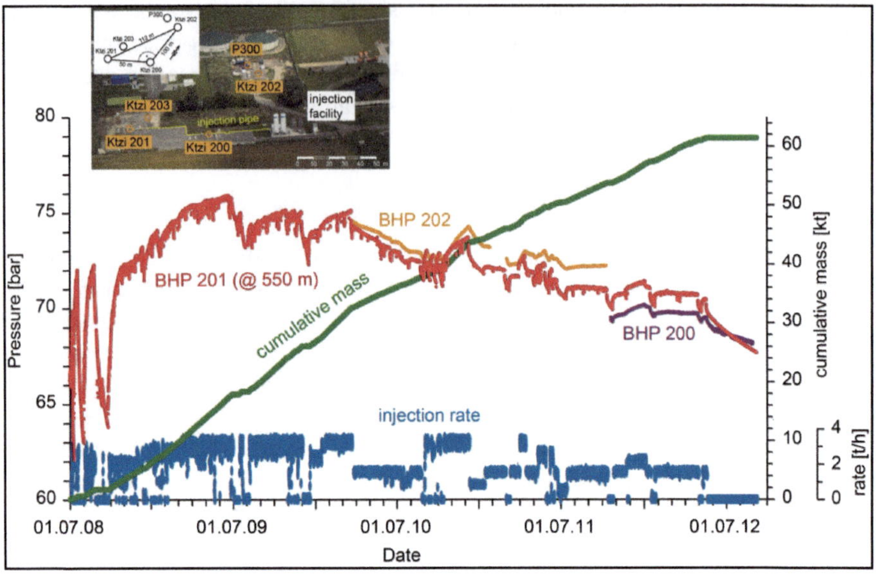


**Fig. 1** Recorded development of downhole pressures in the wells Ktzi 200, Ktzi 201 and Ktzi 202, injection rates and cumulative mass of injected $CO_2$ since the start of continuous injection in June 2008. The small inlet shows site infrastructure and outlay of the four deep wells and one shallow well.

# 5 Operational $CO_2$ Storage Monitoring

## *5.1 General Approach*

The monitoring methods applied to underground storage of natural gas and those applied to $CO_2$ storage show overall similarities but also significant differences in certain aspects. The experiences gained from the operational reservoir monitoring on natural gas storage as well as on $CO_2$ storage clearly confirm that the choice of monitoring activities has to depend on the local geological and technical conditions. The standard monitoring methods applied to underground storage of natural gas establish a basis for the development of suitable monitoring systems for $CO_2$ storage at the Ketzin pilot site. Comparable to the underground storage of natural gas, pressure and temperature monitoring of the injection facility, the wells and the reservoir together with the control of the injection rate form an integral part of the monitoring for $CO_2$ projects.

However, based on the non-cyclic operation and the different chemical and physical properties of $CO_2$ compared to natural gas, the monitoring systems have to be adjusted. The detection of the distribution of the $CO_2$ in the reservoir, especially, requires the application of other procedures and methods. Test development and verification of these methods are fundamental objectives of the R&D work done at the Ketzin pilot site (Liebscher et al. 2012, Martens et al. 2012).

## 5.2 Well Monitoring

For the purposes of the operational and reservoir engineering the following operational data are monitored and recorded (Möller et al. 2012): pressure and temperature conditions, and flow rate at the injection facility, wellhead pressure (WHP) at all three wells, bottom hole pressure (BHP) and bottom hole temperature (BHT) in the well Ktzi 201 via a permanently installed P-T sensor at the end of the injection tubing at a depth of 550 m, and BHP in the wells Ktzi 202 (March 2010 to October 2011) and Ktzi 200 (since October 2011) via a wireline pressure sensor.

In order to optimize the data base within the scope of the reservoir monitoring on the Ketzin site, it has been necessary to install additional pressure sensors in the observation wells Ktzi 200 and 202 at top-reservoir-depth. Pressure-temperature logging and distributed temperature sensing show complicated 2-phase fluid conditions within the observation wells that preclude extrapolation of recorded wellhead pressures to reservoir pressure (Henninges et al. 2011). The in-well pressure sensors then allow for direct recording of the reservoir pressure and thereby also for investigation of the pressure interferences between the different wells. The 2-phase fluid conditions within the observation wells are not only observed by P-T measurements and theoretically deduced, but are also visualized by spectacular borehole video inspections in the observation wells Ktzi 200 and Ktzi 202. These video inspections show the transition from the 2-phase gas-dominated $CO_2$ conditions in the upper part of the well to the 2-phase liquid-dominated $CO_2$ conditions in the lower part, to occur at depths of ~ 290 m (Ktzi 200) and ~ 270 m (Ktzi 202), consistent with the P-T measurements.

## 5.3 Reservoir Simulation

As a responsible operational manager for the $CO_2$ injection, UGS is obliged by the Federal Mining Authority to report half-yearly on the analysis of the overall storage performance and after every 20 kt of injected $CO_2$ to provide a storage forecast for the next 20 kt. Performing this task of data sampling and integration from the different monitoring methods and operational experiences, GFZ provides UGS with a static and dynamic reservoir model that is continuously adjusted and updated by GFZ according to the information available for the Ketzin pilot site (Martens et al. 2012). This model forms the basis for the dynamic 3D simulations with ECLIPSE 100 (Mark of Schlumberger) for storage analyses and storage forecasts, evaluation of potential gas leaks and adhering to limiting parameters for the injection operation. Prior to the start of the injection, a first reservoir model was used to determine the feasibility of injecting $CO_2$ at the pre-selected site and of recording the arrival times of the $CO_2$ plume at the two observation wells (Kempka et al. 2010). In addition, hydraulic test data from pre injection tests (Wiese et al. 2010) as well as results from a modified isochronal test during $CO_2$ injection are incorporated into the models.

A revised simulation model for the $CO_2$ migration within the Stuttgart formation was set up by the modelling group at GFZ in 2012. For the geological model which is in use for reservoir monitoring, the following points have been substantially revised:

- facies within the Stuttgart formation
- integration of the faults
- geometry of the grid (vertical / horizontal)

History matching was carried out on the basis of the updated geological model with a dynamic flow simulation of the past injection period. Therefore the data base up to the end of June 2012 (61,369 tonnes of $CO_2$ injected) was taken into account in terms of the following outcomes:

- reservoir pressure at Ktzi 201
- $CO_2$ arrival times at the observation wells Ktzi 200 and Ktzi 202.

Regardless of some details in the comparison of the modelled and the measured pressure data, the simulation results show a very representative picture of the pressure development within the injection horizon.

Another important match parameter for the verification of the dynamic model is the arrival time of $CO_2$ at the observation wells Ktzi 200 and Ktzi 202 (Zimmer et al. 2011). For this purpose the following values are available for history matching:

- observation well Ktzi 200 - 21 days after the start of injection
- observation well Ktzi 202 - 271 days after the start of injection.

The reservoir simulation performed, based on the current geologic model, resulted in the following arrival times at the observation wells:

Ktzi 200: 21 days simulated vs. 21 days detected
Ktzi 202: 254 days simulated vs. 271 days detected.

Thus, in Ktzi 200, the $CO_2$ arrival time can be simulated accurately. At Ktzi 202, an early arrival by 17 days (after 254 days) occurs during the simulation. The final results of the history matching and the simulated horizontal and vertical distribution of the $CO_2$ plume in the Stuttgart formation for the current amount of 61,396 t $CO_2$ injected until the end of June 2012 is shown graphically in Figures 2 and 3.

During the geological modelling adjustment, zones with higher porosities and permeabilities were located east of the injection point and at the top of the Stuttgart formation which lead to increased concentrations of gas in these area. The migration of carbon dioxide within the Stuttgart formation in a north-western direction (top of the structure Ketzin) is reflected in the vertical profiles (Figures 3a and 3b).

Advanced geologic modelling and flow simulation techniques are used to develop the model for the target interval of $CO_2$ injection. Intensive simulation studies are carried out to gather information about the areal pressure and $CO_2$ plume distribution in the structure for the period in time when the forecasted volume will be injected.

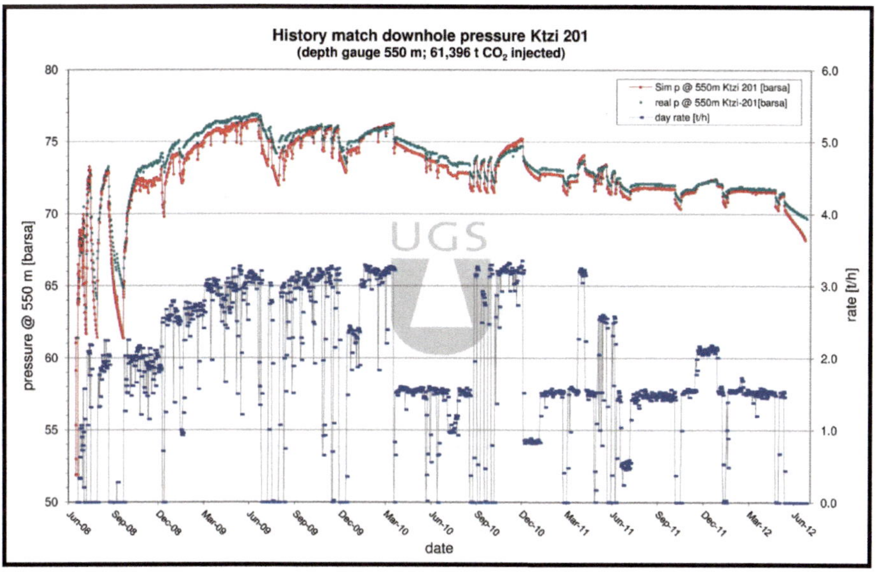


**Fig. 2** Injection rate (blue curve) and measured downhole pressure in the injection well, Ktzi 201, at 550 m depth (green curve). The red curve is the simulated downhole pressure in the injection well, Ktzi 201, at 550 m depth. History matching was performed on June 30th, 2012 after the injection of 61.396 t $CO_2$.

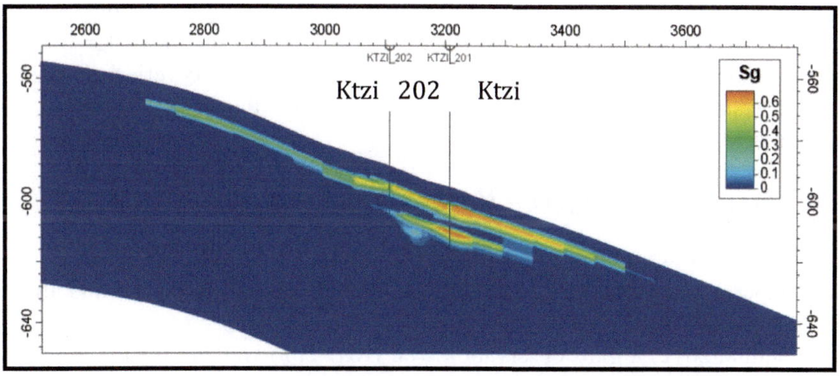


**Fig. 3a** $CO_2$ saturation in the Stuttgart Formation after 61,396 t $CO_2$ injected @ 30.06.2012 - North / South cross section 5x exaggerated (Units: x-axis [m]; z-axis [MASL]; Sg [%])

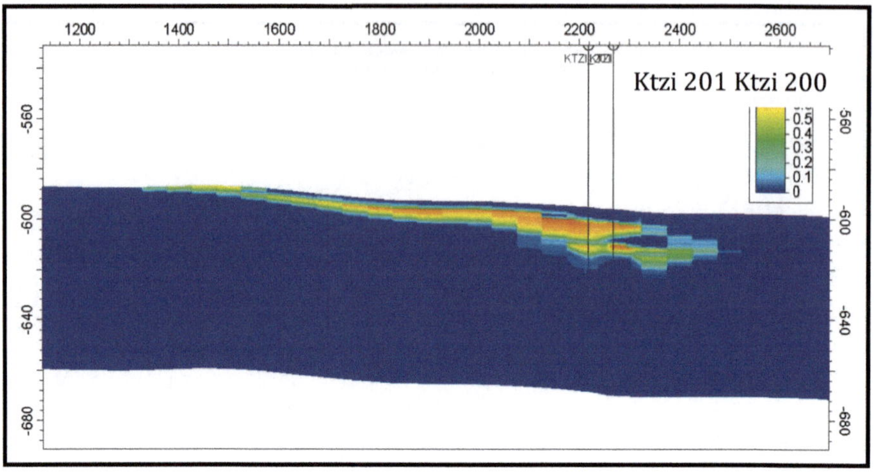


**Fig. 3b** $CO_2$ saturation in the Stuttgart Formation after 61,396 t $CO_2$ injected @ 30.06.2012 - West / East cross section 5x exaggerated (Units: x-axis [m]; z-axis [MASL]; Sg [%])

The modelling, after more than four years of operation, shows a very good history matching between the forecasted pressure development and the recorded data and indicates the amount of $CO_2$ that could be injected in the target interval without violating regulatory constraints. Modelling itself proved to be a valuable and accepted tool for controlling the injection and, furthermore, for all permitting procedures and regulatory issues. In addition it can be stated that intensive operational reservoir monitoring and high level data integration together with advanced reservoir simulation methods are the key preconditions for a secure operation on every future CCS site, even if the actual results reflect the current project status and accordingly an optimization of the reservoir model still continues.

## *5.4 State of the Project – Ready for the Industrial Scale*

Using 1 injection and 2 observation wells, a total of 61,396 t $CO_2$ was injected deep underground until May 2012 without safety-related disturbances and abiding by the Federal Mining Law regulations.

In autumn2012, a third observation well (Ktzi 203) was installed at a distance of ~ 25 m from the injection well Ktzi 201. The injection pipeline had to be partially removed partially for the drilling work. At this time, the $CO_2$ injection was interrupted. After Ktzi 203 is finished, it is intended to carry on the injection with a rate of ~ 1,000 tons per month.

After more than 4 years of $CO_2$ injection experiences in operational reservoir monitoring at the pilot site Ketzin, the consortium of engineers and scientists engaged in the different projects (e.g., $CO_2$MAN and the precursor project $CO_2$SINK) are prepared for the next step. The combination of different monitoring procedures, such as:

- continuous underground P/T measuring,
- sequential gas-, hydro, microbiological and geochemical sampling,
- soil gas sampling
- different geophysical methods of measurement, i.e.
  - geoelectric,
  - seismic and
  - gas saturation

takes place at the storage site. Comparing these measures with the standards applied for natural gas storage (Table 1) it becomes clear that at the Ketzin site many more methodologies are used.

**Table 1** Monitoring methods at the Ketzin site compared to the standard methods of natural gas storage in aquifers

| Methodology | Ketzin | Natural gas storage |
|---|---|---|
| Bottom hole pressure | X | rarely |
| Wellhead pressure | X | x |
| Flow rate | X | x |
| Temperature | X | only wellhead |
| Reservoir saturation | X | x |
| Repeated 3D seismic | X | - |
| Reservoir modelling | X | x |
| Well integrity | X | x |
| Gas sampling | X | x |
| Geochemical sampling | X | - |
| Soil gas sampling | X | - |
| Microbiological sampling | X | only exceptional |
| Geoelectric monitoring | X | - |
| Elevation monitoring (terrestic or InSar) | X | x |

As there will be an evaluation of the effectiveness of each single method regarding cost/benefit relation and, even more important, the impact on the local community it is expected that the complete programme of Ketzin will not be transferred to industrial projects. It seems likely that some methods will be used only once for site characterisation or documenting the base line and others will become standard, independent of the size or location of the project.

The pilot site for $CO_2$ storage at Ketzin is the only on-shore laboratory in Europe that addresses the main issues of $CO_2$ injection and storage at a scientific scale, forming the basis for CCS implementation on an industrial scale.

Despite the current discussions in central Europe about the feasibility of large scale CCS projects in terms of economics and political framework, it has to be stated that the project at Ketzin delivered valuable experiences. All three main parties involved in future projects (authorities, general public and companies/institutions) will benefit strongly from these experiences and therefore any future project should be based on the reliable results gained at the Ketzin pilot site. The next step should be transferring the experiences gained to an industrial scale on-shore $CO_2$ storage project as soon as possible.

## References

1. Bachu, S.: Sequestration of $CO_2$ in geological media, criteria and approach for site selection in response to climate change. Energy Conversion Management 41, 953–970 (2000)
2. Bradshaw, J., Bachu, S., Bonijoly, D., Burruss, R., Holloway, S., Christensen, N.P., Mathiassen, O.M.: $CO_2$ storage capacity estimation, issues and development of standards. International Journal of Greenhouse Gas Control 1, 62–68 (2007)
3. Förster, A., Schöner, R., Förster, H.J., Norden, B., Blaschke, A.W., Luckert, J., Beutler, G., Gaupp, R., Rhede, D.: Reservoir characterization of a $CO_2$ storage aquifer. The Upper Triassic Stuttgart Formation in the Northeast German Basin. Marine Petrol Geol. 27, 2156–2172 (2010)
4. Kempka, T., Kühn, M., Class, H., Frykman, P., Kopp, A., Nielsen, C.M., Probs, P.: Modelling of $CO_2$ arrival time at Ketzin – Part I. International Journal of Greenhouse Gas Control 4(6), 1007–1015 (2010)
5. Henninges, J., Liebscher, A., Bannach, A., Brandt, W., Hurter, S., Köhler, S., Möller, F.: $CO_2$SINK Group, P-T-$\rho$ and two-phase fluid conditions with inverted density profile in observation wells at the $CO_2$ storage site at Ketzin (Germany). Energy Procedia 4, 6085–6090 (2011)
6. Martens, S., Liebscher, A., Möller, F., Würdemann, H., Schilling, F., Kühn, M., Ketzin Group: Progress report on the first European on-shore $CO_2$ storage site at Ketzin (Germany) – Second year of injection. Energy Procedia 4, 3246–3253 (2011)
7. Möller, F., Liebscher, A., Martens, S., Schmidt-Hattenberger, C., Kühn, M.: Yearly operational datasets of the $CO_2$ storage pilot site Ketzin, Germany. Scientific Technical Report (2012), doi:10.2312/GFZ.b103-12066
8. Norden, B., Förster, A., Vu-Hang, D., Marcelis, F., Springer, N., Le Nir, I.: Lithological and Petrophysical Core-Log Interpretation in $CO_2$SINK, the European $CO_2$ On-shore Research Storage and Verification Project. SPE Reserv. Eval. & Engin. 13, 179–192 (2010)
9. Prevedel, B., Wohlgemuth, L., Henninges, J., Krüge, K., Norden, B., Förster, A., $CO_2$SINK Drilling Group: The $CO_2$SINK boreholes for geological storage testing. Scientific Drilling 6, 32–37 (2008)
10. Prevedel, B., Wohlgemuth, L., Legarth, B., Henninges, J., Schütt, H., Schmidt-Hattenberger, C., Norden, B., Förster, A., Hurter, S.: The $CO_2$ SINK boreholes for geological $CO_2$-storage testing. Energy Procedia 1, 2087–2094 (2009)
11. Schmidt-Hattenberger, C., Bergmann, P., Kießling, D., Krüger, K., Rücker, C., Schütt, H., Ketzin Group: Application of a Vertical Electrical Resistivity Array (VERA) for Monitoring $CO_2$ Migration at the Ketzin Site: First Performance Evaluation. Energy Procedia 4, 3363–3370 (2011)

12. Stoewer, M., Zemke, J., Feldrappe, H.: Site selection criteria for $CO_2$ storage sites and implication for the exploration. In: Proceeding of the Sino-German Conference on Underground Storage of $CO_2$ an Energy, China, Balkema, July 6-13 (2010)
13. Wiese, B., Böhner, J., Enachescu, C., Würdemann, H., Zimmermann, G.: Hydraulic characterisation of the Stuttgart formation at the pilot test site for $CO_2$ storage, Ketzin, Germany. International Journal of Greenhouse Gas Control 4(6), 960–971 (2010)
14. Würdemann, H., Möller, F., Kühn, M., Heidug, W., Christensen, N.P., Borm, G., Schilling, F., $CO_2$ SINK Group: $CO_2$SINK – From site characterisation and risk assessment to monitoring and verification, one year of operational experience with the field laboratory for $CO_2$ storage at Ketzin, Germany. International Journal of Greenhouse Gas Control 4, 938–951 (2010)
15. Zemke, J., Stöwer, M., Arnold, C., Becker, W., May, F., Gerling, P., Krull, P.: $CO_2$ Sequestration in Germany – General Conditions and first field studies. In: Proceedings of the 22nd World Gas Conference, Tokyo, June 1-5, p. 12 (2003)
16. Zimmer, M., Erzinger, J., Kujawa, C., $CO_2$SINK Group: The gas membrane sensor (GMS). A new method for gas measurements in deep boreholes applied at the $CO_2$SINK site. International Journal of Greenhouse Gas Control 5, 995–1001 (2011)

# Modelling of Kinetic Interface Sensitive Tracers for Two-Phase Systems

Alexandru Tatomir, Friedrich Maier, Mario Schaffer,
Tobias Licha, and Martin Sauter

Geoscience Centre, Dept. Applied Geology,
University of Göttingen, Goldschmidtstr. 3, 37077, Germany

**Abstract.** This article presents a mathematical model for interface sensitive tracer transport used for the evaluation of the interface between two fluid-phases (i.e. $CO_2$ and brine) with general applicability in a series of engineering applications: oil recovery, vapour-dominated geothermal reservoirs, contaminant spreading, $CO_2$ storage, etc. Increasing the $CO_2$ storage efficiency in brine deep geological formations requires better injection strategies to be developed which could be accomplished with better tools for quantification of the fluid-fluid interfaces. The $CO_2$ residual and solubility trapping are highly influenced by the interfaces separating the phases. An increase in the interface area is expected to produce an increase in the solubility trapping. However, standard multi-phase models do not account for the specific fluid-fluid interface area. A new class of reactive tracers is used for the characterization of interfacial areas between supercritical $CO_2$ and brine. The tracer is injected in the $CO_2$ and migrates to the interface where it undergoes a hydrolysis reaction in contact with water. A mathematical model is constructed based on volume-averaged properties (saturation, porosity, permeability, etc.) at the macroscale. The fluid phases are described with an extended form of the Darcy equation based on thermodynamic principles and complemented with relations for relative permeability and saturation and a specific equation for interfacial area. The kinetic mass transfer effects between the two phases are highly dependent on the interface area, and are captured with an approach introduced by [1]. The mathematical model is tested with a simple numerical example.

**Keywords:** two phase flow in porous media, kinetic interface sensitive tracers, specific interfacial area.

## 1 Motivation

Flow, transport and phase transfer processes in porous media occur in a variety of environmental, biological or industrial systems. The correct understanding of these processes is important for applications that range from kilometre scale

M.Z. Hou et al. (Eds.): *Clean Energy Systems in the Subsurface*, SSGG, pp. 65–74.
DOI: 10.1007/978-3-642-37849-2_6 

problems (e.g. ground water management, contaminant spreading, petroleum industry, geothermal energy or $CO_2$ storage in geo-reservoirs, etc.) to millimetre or micrometre scale problems (e.g. pharmacokinetics, fuel cells, etc.). Many of these applications involve two, or three fluid phases, where mass transfer processes through the interfaces take place and involve changes in phase compositions.

A very recent problem dealing with two phase flow is the storage of $CO_2$ in deep saline aquifers. The $CO_2$ storage (CCS) in geological formations is one technology that can be used to reduce the greenhouse gas emissions [2–4]. After injecting the $CO_2$, four main trapping mechanisms in the storage formation have been identified [2]: 1) structural and stratigraphic trapping; 2) capillary or residual trapping; 3) solubility trapping; and 4) geochemical trapping. In order to increase the $CO_2$ storage efficiency in deep geological formations, injection strategies have to be optimized and a correct assessment of the $CO_2$ plume development is necessary. It is assumed that the size of the $CO_2$-brine interface determines the rate of dissolution of $CO_2$ into the brine (solubility trapping). Thus, an increase in the interfacial area would increase the dissolution of $CO_2$. In general, the interfacial area is expected to increase during the injection of $CO_2$ due to mixing, spreading and dispersion. However, the real plume development has to be determined experimentally in the laboratory and in the field, and with the help of numerical models. Standard multiphase models consider relations between capillary pressure and saturation and ignore the interfacial area between fluids. Recent advances in experimental setups and modelling approaches try to overcome this limitation. Currently there is very limited experimental data which show the pressure-saturation-interfacial area relationship for both drainage and imbibition. No experimental data is conclusive for the verification of theoretical model predictions which suggest that there are different interfacial areas for drainage and imbibition [5]. Most of the available relations have been derived from pore network models.

For $CO_2$ storage, the main questions to be answered are: what is happening to the $CO_2$ in the subsurface and how can it be monitored during and after injection? Monitoring techniques (e.g. tracers, well logs, seismic profiling, time-lapse 3D seismic imaging, soil and water chemistry, electrical and electromagnetic techniques, etc.) have been applied in the oil industry and for enhanced oil/gas recovery projects, where the timing of $CO_2$ injection relative to production is vital for the success of the operation. Nevertheless, direct techniques for time-dependent monitoring are limited in availability at present. Therefore, new tools, capable of describing and characterizing the spatial and temporal evolution of the plume and its interface in the reservoir are required.

Conventional tracer tests with water soluble tracer substances are useful for characterizing the reservoir, finding the flow and transport parameters, identifying possible leakage passages and predicting the integrity of the well. A new class of reactive tracers, termed as kinetic interface sensitive (KIS) tracers has been proposed [6]. These tracers have the potential to characterize the interfacial areas between supercritical $CO_2$ and brine and may thus provide additional information on: the effect of pressure stimulation on mixing; plume spreading, characterization of interfacial area and area change; identification of fingering effects; estimation of residual $scCO_2$ saturation; assessment of storage capacity and effectiveness; optimization of operational strategies, management of injection wells. The results of the first laboratory experiments are promising. To date, however, no model exists, which couple the prediction of the interfacial areas with the expected behaviour of KIS tracers.

In contrast to partitioning tracers, which are used to quantify the amount of saturation of immiscible hydrophobic phases (e.g. NAPLs or $CO_2$) in porous multiphase systems, KIS tracers exploit the reaction kinetics of a hydrolysis reaction at the interface for describing the interface development with time. Further details on tracer development and target molecular design can be found in Schaffer et al. (2012) [6]. The basic concept of these tracers can be summarized as follows: 1) the tracer is injected together with the $CO_2$ in the brine saturated aquifer; 2) the tracer adsorbs and saturates the interface; 3) the tracer reacts at the interface with water to highly water soluble reaction products; and 4) the reaction products can be measured in the water phase. The larger the interface (reaction surface) between both phases, the higher is the mass of the reaction products in the water phase. Thus, it is possible to establish a connection between time, concentration in water phase, and interfacial area size.

The focus of this work is to create the first mathematical model that is able to describe the behaviour of KIS tracers in two-fluid-phase porous media systems ($CO_2$-brine) and to account for the fluid-fluid interfacial areas. The aim is to use numerical modelling to design better tracers, experimental setups and better injection strategies capable of maximizing the interfacial area and capable of increasing the long-term trapping effectiveness.

## 2 Mathematical Model

### 2.1 Extension of the Two-Phase Flow Standard Model

The standard approach to model multi-phase flow in porous media is based on the extended Darcy's law:

$$v_\alpha = -K \frac{k_{r\alpha}}{\mu_\alpha} (\nabla p_\alpha - \rho_\alpha g), \tag{1}$$

where $v_\alpha$ is the apparent velocity of the phase $\alpha$, $K$ is the intrinsic permeability, $k_{r\alpha}$ is the relative permeability, $\mu_\alpha$ is the dynamic viscosity, $p_\alpha$ is the pressure, $\rho_\alpha$ is the density and $g$ is the gravitational acceleration. The conservation of mass in a multiphase flow system with respect to the volume is then written:

$$\frac{\partial (S_\alpha \phi \rho_\alpha)}{\partial t} - \nabla \cdot \left( \rho_\alpha K \frac{k_{r\alpha}}{\mu_\alpha} (\nabla p_\alpha - \rho_\alpha g) \right) - \rho_\alpha q_\alpha = 0, \tag{2}$$

where $S_\alpha$ is the saturation of phase $\alpha$, $\phi$ is the porosity.

The classical coupling relations in a two phase flow porous medium system are:

$$S_w + S_n = 1, \tag{3}$$

$$p_n - p_w = p_c, \tag{4}$$

where $p_c$ represents the capillary pressure, ($w$) wetting, ($n$) non-wetting phase.

A more fundamental approach of the two-fluid-phase flow in porous media can be derived from the thermodynamic principles. The movement of phases or interfaces is controlled by the gradient in the Gibbs free energy [7]. This interfacial area-based model is, indeed, able to describe the hysteretic behaviour of the classical model as well as kinetic interphase mass and energy transfer in a physically based way [1], [8], [9].

$$\tau_\alpha = \phi S_\alpha \rho_\alpha (\nabla G_\alpha - g), \tag{5}$$

$$\tau_{\alpha\beta} = \phi a_{\alpha\beta} \Gamma_{\alpha\beta} (\nabla G_{\alpha\beta} - g). \tag{6}$$

Here the $\tau_\alpha$ and $\tau_{\alpha\beta}$ represent friction forces applied on phase $\alpha$ or interface $\alpha\beta$, $G_\alpha$ is the Gibbs free energy density, $a_{\alpha\beta}$ is the specific interfacial area, $\Gamma_{\alpha\beta}$ is the mass density and $G_{\alpha\beta}$ is the Gibbs free energy density of the $\alpha\beta$- interface. Considering that the medium is non-deformable, fluids are incompressible, interfacial mass density $\Gamma_{\alpha\beta}$ is constant, the influence of the solid on the interface is negligible and following the second law of thermodynamics as shown by [7] results:

$$\frac{\partial (a_{\alpha\beta})}{\partial t} + \nabla \cdot (a_{\alpha\beta} v_{\alpha\beta}) = E_{\alpha\beta} \quad \text{, with} \qquad v_{\alpha\beta} = -K_{\alpha\beta} \cdot \nabla a_{\alpha\beta} \tag{7}$$

where $E_{\alpha\beta}$ is the rate of production of specific interfacial area.

## 2.2 Specific Interfacial Area

The measurement of interfacial area, $a_{\alpha\beta}(S_w, p_c)$ is a hard task especially for non-isothermal models where the solid phase has to be also taken into account. Most of the literature describes the calculation of the interfacial area by means of pore-network models. The pore network models are a very useful tool to get an insight in the fundamental understanding of the physics of flow and transport in porous media. The recent improvements in the imaging techniques such as X-ray computed tomography (CT) and magnetic resonance imaging (MRI), lead to more detailed information about the structure of the porous media. The basic idea of the pore network theory is to create a simplified network of pore geometries while preserving the essential features of the void spaces. Joekar Niasar et al. (2010) [15] suggested a bi-quadratic relationship between interfacial area, capillary pressure and fluid saturation:

$$a_{\alpha\beta}(S_w, p_c) = a_{00} + a_{10}S_w + a_{01}p_c + a_{11}S_w p_c + a_{20}S_w^2 + a_{02}p_c^2 \tag{8}$$

A different formulation of the $p_c - S_w - a_{wn}$ quadratic surface which prevents negative values in the capillary pressure and saturation can be done in the form:

$$a_{\alpha\beta}(S_w, p_c) = a_0 \cdot (S_w)^{a1} \cdot (1 - S_w)^{a2} \cdot (p_c^{\max} - p_c)^{a3} \tag{9}$$

This formulation has the advantage that it keeps the capillary pressures and saturations within the physical range, therefore, being more realistic. Figure 1 shows the comparison of the two approaches to represent $p_c - S_w - a_{wn}$ surfaces.

Table 1 reviews the coefficients used in the literature for macroscale modelling of capillary pressure-saturation-interfacial area by: 1) Niessner and Hassanizadeh (2008) [1] (Figure 1a), 2) Joekar Niasar et al. (2008) [10], 3) Ahrenholtz et al. (2011) [8], 4) new approach of Joekar-Niasar.

**Table 1** Coefficients for the capillary pressure-saturation – interfacial area in literature

| | $a_{00}$ ($m^{-1}$) | $a_{10}$ ($m^{-1}$) | $a_{01}$ ($m^{-1}$ $Pa^{-1}$) | $a_{11}$($m^{-1}$ $Pa^{-1}$) | $a_{20}$($m^{-1}$) | $a_{02}$($m^{-1}$ $Pa^{-2}$) |
|---|---|---|---|---|---|---|
| 1 | -313.6 | 5535 | 0.085 | -3937 | -0.307 | -5.0E-06 |
| 2 | 849 | 3858 | -0.224 | -3992 | 0.006 | 1.283E-05 |
| 3 | -71.64 | 4.103E+03 | 0.3626 | -0.4505 | -3.733E+03 | -3.514E-05 |
| 4 | 5.630 | 0.4797 | -7.692E-04 | -4.376 | 1.606E-04 | 3.169E-08 |

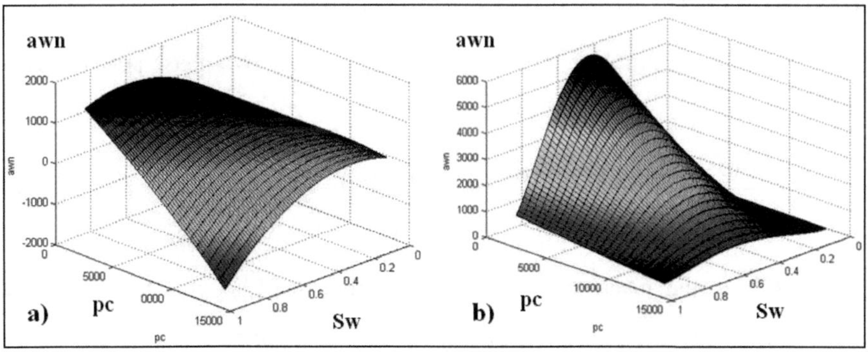


**Fig. 1** Capillary pressure - saturation - specific interfacial area ($p_C - S_w - a_{wn}$) surfaces according to a) Niessner and Hassanizadeh (2008) [1] and to b) equation (9), which maintains the capillary pressure and saturations in physical ranges

## 2.3 Capillary Pressure

Generally, the capillary pressure–saturation relationship in standard two-phase models is given by one of the following formulations as a function of saturation: Brooks-Corey, Van Genuchten or a linear law e.g. Helmig (1997) [11]. In the following the Brooks-Corey approach is going to be used:

$$p_c(S_w) = p_d S_e^{-\frac{1}{\lambda}} \quad \text{with } S_e = \frac{S_w - S_{wr}}{1 - S_{wr}} \tag{10}$$

However, the capillary pressure is a function not only of saturation, but also a function of interfacial area: $p_c = p_c(S_w, a_{wn})$ and it has been shown [10] that it is not unique.

## 2.4 Relative Permeability

According to Niessner and Hassanizadeh (2008) [1] the dependence of relative permeability on the interfacial area is small and therefore standard formulations should be sufficient. Based on the *Burdine-Brooks-Corey* approach we can write the relative permeability relations:

$$k_{rw} = S_e^{\frac{2+3\lambda}{\lambda}}, \text{and } k_{rn} = (1 - S_e)^2 \left(1 - S_e^{\frac{2+\lambda}{\lambda}}\right) \tag{11}$$

## 2.5 KIS Tracer Modelling at the Interface

The governing equation for KIS tracer migration in the two-phase porous media system is:

$$\frac{\partial (S_n \phi C_n^T)}{\partial t} - \nabla \cdot (C_n^T v_n + D_n^T \nabla C_n^T) - Q_{n \to w}^T = 0, \tag{12}$$

where $D_\alpha^T$ is the macroscale dispersion coefficient, $Q_{n \to w}^T$ is the interphase mass transfer rate of tracer $T$ from phase $n$ ($CO_2$) to $w$ (brine). $D_\alpha^T$ is the sum of the mechanical or kinematic dispersion and the effective diffusion coefficient. For the numerical modelling of the interfacial sensitive tracers several approaches could be considered: effective rate coefficient [12], local equilibrium assumption, dual domain approach, or non-equilibrium approach [9]. The main targets in the molecular design of the KIS tracer are substances which decay in the non-wetting phase following the linear zero order reaction kinetics of the hydrolysis reaction. This makes the interphase mass transfer dependent on the interfacial area:

$$Q^T_{\alpha\rightarrow\beta} = k^T_{\alpha\rightarrow\beta} a_{\alpha\beta}\,, \tag{13}$$

where $k^T_{\alpha\rightarrow\beta}$ is the mass transfer rate coefficient.

## 3 Numerical Simulations

We perform a simple numerical experiment to simulate the evolution of interfacial area in a column. The results could be used later for comparisons with laboratory experiments. Compressed air is injected into a fully water saturated porous medium. The pressure at the inlet is kept constant at 200 kPa. The outlet pressure is controlled and stays at 100 kPa. (Air represents the non-wetting phase and water is the wetting phase.) The domain is 30 cm long and has a 3 cm diameter. The interfacial area is calculated with the formulation from equation (9) having a0=1, a1 = 2, a2 = 2, a3 = 1.2 (see Figure 1b). These parameters should correspond to medium-sized sands. Intrinsic permeability is $K$=1.0e-12 m$^2$, porosity 0.2. Brooks Corey parameters from equation (10) are $p_d$ =1000, $\lambda$=2. Initial conditions are $S_w$=1.0, $p_w$=100 kPa.

The system of equations is solved with the free open-source numerical toolbox DuMux (www.dumux.org)[13] . For the spatial discretization the box method is used (see [11]), which combines the advantages of finite element and finite volume methods, being locally mass conservative and applicable to unstructured grids. The time discretization uses the fully implicit Euler scheme known for its stability.

The results (Figure 2) depict the spatial and temporal distributions of the non-wetting phase saturation and the specific interfacial area. Immediately after the air begins to enter the domain, the interfacial-area has a sharp increase which continues until the air front reaches the outlet (Figure 2(a)(b)). The Dirichlet boundary at the outlet has an accumulation-like effect, producing an increase of non-wetting phase saturation in the whole domain. However, the interfacial area is beginning to decrease as the water is completely pushed out from the porous medium (Figure 2(c)).

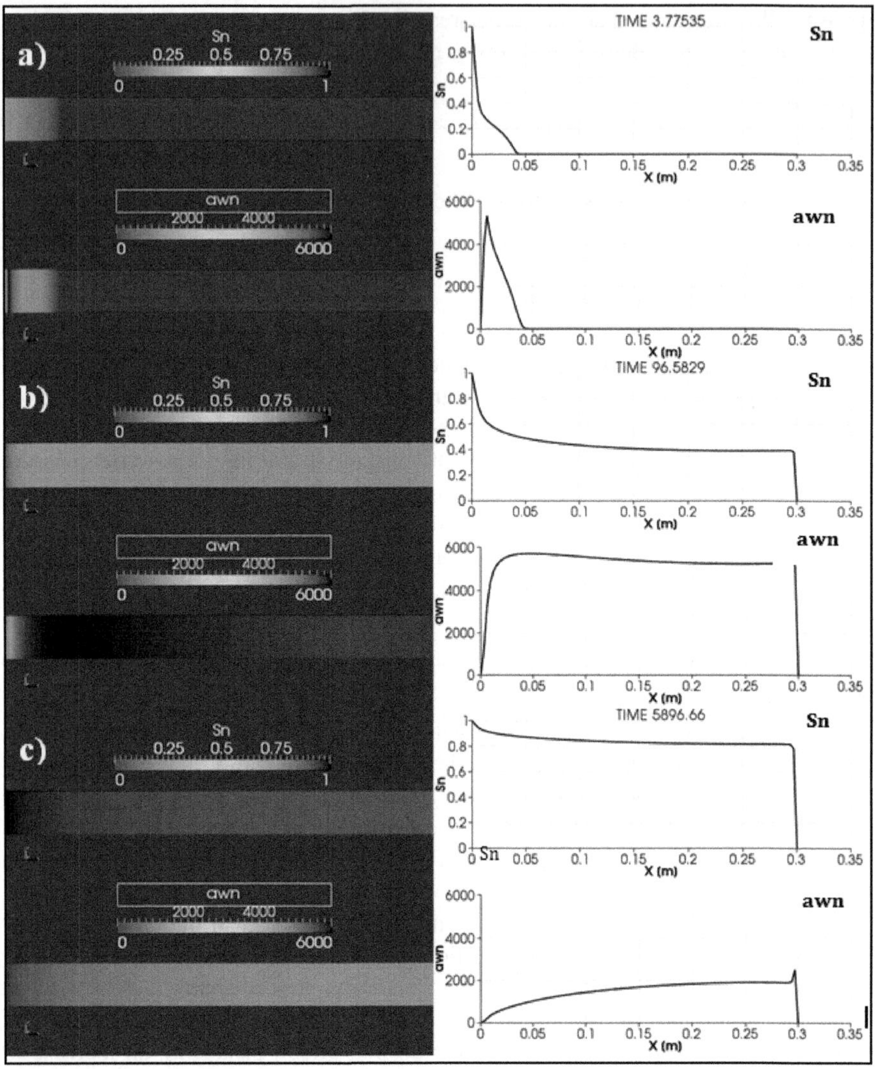


**Fig. 2** Column experiment: spatial distribution of non-wetting saturation (upper-left), interfacial area (lower left) and plot over line passing through points A1(0.0, 0.015)-A2(0.30, 0.015). a) Air begins to infiltrate into the column; b) Air reaches the outlet and starts to accumulate in the domain; c) interfacial area decreases as water is pushed out of the column.

## 4 Conclusions

We have shown a first mathematical model that accounts for interface sensitive tracer transport in two-fluid-phases in porous media has been demonstrated.

The main features of the models are that it includes additional balance equations for the interfaces.

The relations for capillary pressure – saturation – specific interfacial area per unit volume of porous medium $p_c - S_w - a_{wn}$ were calculated according to pore network modelling results. The interfacial area is represented with a quadratic surface and accounts for non-unique capillary pressure values.

The model is aimed to provide an extension to the laboratory work of developing kinetic interface sensitive tracers [6], [14]. The amount of interfacial area will be later compared with the laboratory results and could be used as a validation for pore network models. Future work includes constructing and running dynamic laboratory experiments and development of the system of equations towards a multi-compositional approach, and of the numerical simulator. The multi-compositional model will be able to handle the rate of dissolution of $CO_2$ into water and of water in the $CO_2$ phase together with the hydrolysis reaction of tracer at the interface.

## References


1. Niessner, J., Hassanizadeh, S.M.: A model for two-phase flow in porous media including fluid-fluid interfacial area. Water Resour. Res. 44, 10 (2008)
2. IPCC. IPCC special report on carbon dioxide capture and storage. In: Metz, B., Davidson, O., de Coninck, H.C., Loos, M., Meyer, L.A. (eds) Prepared by Working Group III of the Intergovernmental Panel on Climate Change. Cambridge University Press (2005)
3. Bachu, S.: Screening and ranking of sedimentary basins for sequestration of $CO_2$ in geological media in response to climate change. Environmental Geology 44(3), 277–289 (2003)
4. Klara, S.M., Srivastava, R.D., McIlvried, H.G.: Integrated collaborative technology development program for $CO_2$ sequestration in geologic formations–United States Department of Energy R&D. Energy Conversion and Management 44(17), 2699–2712 (2003)
5. Schaefer, C.E., DiCarlo, D.A., Blunt, M.J.: Experimental measurement of air-water interfacial area during gravity drainage and secondary imbibition in porous media. Water Resources Research 36(4), 885–890 (2000)
6. Schaffer, M., Maier, F., Licha, T., Sauter, M.: Development of Kinetic Interface Sensitive Tracers (KIS-Tracer) for Supercritical Carbon Dioxide Injections into Deep Saline Aquifers. In: EGU General Assembly Conference Abstracts, vol. 14, p. 2048 (2012)
7. Hassanizadeh, S.M., Gray, W.G.: Mechanics and thermodynamics of multiphase flow in porous media including interphase boundaries. Advances in Water Resources 13(4), 169–186 (1990)
8. Ahrenholz, B., Niessner, J., Helmig, R., Krafczyk, M.: Pore-scale determination of parameters for macroscale modeling of evaporation processes in porous media. Water Resources Research 47(7) (2011)

9. Niessner, J., Hassanizadeh, S.: Modeling Kinetic Interphase Mass Transfer for Two-Phase Flow in Porous Media Including Fluid–Fluid Interfacial Area. Transport in Porous Media 80(2), 329–344 (2009)
10. Joekar-Niasar, V., Hassanizadeh, S., Leijnse, A.: Insights into the Relationships Among Capillary Pressure, Saturation, Interfacial Area and Relative Permeability Using Pore-Network Modeling. Transport in Porous Media 74(2), 201–219 (2008)
11. Helmig, R.: Multiphase Flow and Transport Processes in the Subsurface: A Contribution to the Modeling of Hydrosystems, 1st edn. Springer (1997)
12. Zhang, H., Schwartz, F.W.: Simulating the in situ oxidative treatment of chlorinated ethylenes by potassium permanganate. Water Resources Research 36(10), 3031–3042 (2000)
13. Flemisch, B., Darcis, M., Erbertseder, K., Faigle, B., Lauser, A., Mosthaf, K., Müthing, S., Nuske, P., Tatomir, A., Wolff, M., Helmig, R.: DuMux: DUNE for multi-{phase, component, scale, physics, ...} flow and transport in porous media. Advances in Water Resources (2011) (in press) (corrected proof)
14. Schaffer, M., Maier, F., Licha, T., Sauter, M.: A new generation of tracers for the characterization of interfacial areas during supercritical carbon dioxide injections into deep saline aquifers: kinetic interface sensitive tracers (KIS tracer). International Journal of Greenhouse Gas Control (2012) (submitted)
15. Joekar-Niasar, V., Prodanović, M., Wildenschild, D., Hassanizadeh, S.M.: Network model investigation of interfacial area, capillary pressure and saturation relationships in granular porous media. Water Resources Research 46(W06526), 18 (2010), doi:10.1029/2009WR008585

# A Brief Overview of Isotope Measurements Carried Out at Various CCS Pilot Sites Worldwide

Martin Nowak[1], Anssi Myrttinen[1], Robert van Geldern[1], Veith Becker[1,2], Bernhard Mayer[2], and Johannes A.C. Barth[1]

[1] GeoZentrum Nordbayern, Friedrich-Alexander-Universität Erlangen Nürnberg, Angewandte Geologie, Schlossgarten 5, 91054 Erlangen, Germany
[2] Applied Geochemistry Group, Department of Geoscience, University of Calgary, 2500 University Drive N.W., Calgary, Alberta, Canada T2N 1N4

**Abstract.** About 1800 geochemical measurements including more than 1000 isotope analyses have been published as a result of geochemical monitoring programs established at several $CO_2$ storage and enhanced oil and gas recovery projects worldwide. These projects are briefly discussed here in order to compare sampling techniques to obtain fluid and gas samples for chemical and isotopic analyses. In all the projects, changes of stable isotope ratios of $CO_2$ and dissolved inorganic carbon in samples obtained pre- and post-injection were used to quantify solubility and ionic trapping of $CO_2$ via stable isotope mass balances. Further applications include, monitoring of underground $CO_2$ migration and early detection of potential $CO_2$ leaks into overlying formations. Other benefits of these stable isotope tracers include a better understanding of water-rock-gas interactions with $CO_2$ under supercritical conditions and often highly saline reservoir fluids that are present in the storage formations. While the results of these projects need further laboratory and experimental confirmation, further increase in field applications of stable isotope tracer techniques are anticipated with the introduction of new portable laser stable isotope mass spectrometers.

**Keywords:** Carbon capture and storage (CCS), enhanced oil and gas recovery (EOR and EGR), stable isotope monitoring, down-hole sampling.

## 1 Introduction

Well-based monitoring has been applied for several decades in the oil and gas industry to provide information for economic and safe extraction of hydrocarbons from the subsurface [1]. However, for verifying the storage of $CO_2$ in carbon capture and storage (CCS) and enhanced oil or gas recovery (EOR and EGR) projects these monitoring methods need to be further refined because $CO_2$ is mostly injected in a supercritical form resulting in potential geochemical rock-water-brine interactions and pressure changes [2,3]. This may lead to a

M.Z. Hou et al. (Eds.): *Clean Energy Systems in the Subsurface*, SSGG, pp. 75–87.
DOI: 10.1007/978-3-642-37849-2_7 

mobilization of organic matter, trace elements and mineral dissolution that may affect subsurface geochemical compositions. Most downhole and surface sampling techniques applied so far focus on the geochemical assessment of liquid and gas samples and more recently also on stable isotope measurements [4-8].

Stable isotope ratios of carbon, oxygen and hydrogen are applied routinely to quantify interactions between the atmosphere, biosphere and geosphere [9]. The principle of stable isotope applications is that the naturally occurring variations of isotope abundances are useful tracers for phase changes, mobility and mixing of materials. For instance, if stable isotope ratios of fresh-water or dissolved carbon, remain unaltered, they can indicate various sources and enable mixing calculations. On the other hand, stable isotope ratios may change due to environmental processes including evaporation, diffusion as well as chemical or biological turnover of organic matter. If these isotope ratio changes (i.e. fractionations) are systematic, they can help to differentiate and quantify processes. By convention, natural stable isotope ratios are expressed in a per mille (‰) deviation from an internationally accepted standard, e.g. for the $^{13}C/^{12}C$ ratio according to the following equation:

$$\delta^{13}C \quad [‰] = (R_{sample}/R_{reference} - 1) \times 1000 \tag{1}$$

In this case, the international reference material is known as Vienna Pee Dee Belemnite (V-PDB).

At several CCS, EOR or EGR sites, isotope measurements have been applied successfully to monitor the distribution and mixing of injected $CO_2$ in the subsurface (Fig. 1). These isotope measurements include the determination of stable carbon ($^{13}C/^{12}C$) as well as oxygen isotope ratios ($^{18}O/^{16}O$) in order to quantify solubility trapping as dissolved inorganic carbon (DIC) and/or to trace

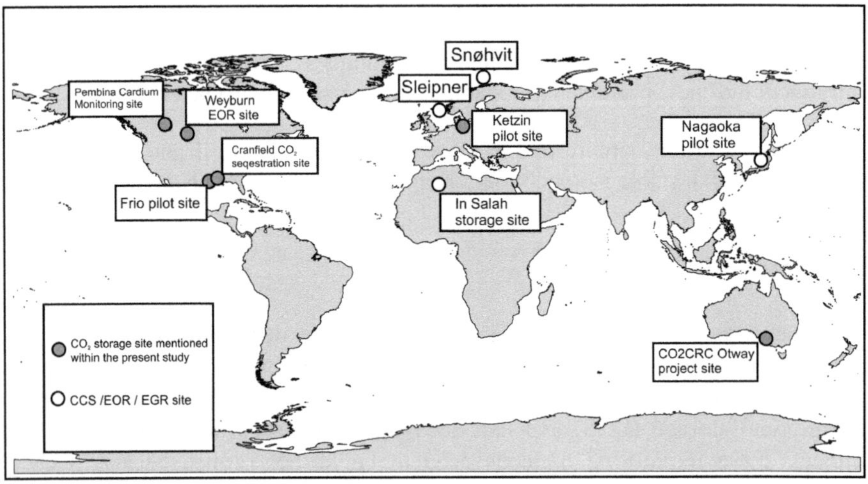


**Fig. 1** Overview over selected CCS / EOR / EGR projects world-wide [11]

$CO_2$ breakthrough at observation wells [10,11,4]. Stable isotope values also allow the tracing of $CO_2$ migration [12] and can indicate saturation conditions of porous media with free-phase-$CO_2$ [13, 14]. The objective of this manuscript is to review and discuss selected published stable isotope applications in $CO_2$ storage projects worldwide. The goal is to review differences in sampling and monitoring strategies in order to reveal advantages and limitations of this innovative tracer approach and to outline the usefulness of these methods for future applications.

# 2 Overview of Isotope Applications in CCS Projects

## *2.1 The Weyburn Study*

In Weyburn (Saskatchewan, Canada) isotope measurements were performed within the framework of a large scale monitoring program for a $CO_2$ enhanced oil recovery project to obtain baseline values from a carbonate reservoir before injection and to monitor the movement and the fate of $CO_2$ in the reservoir over a 10-year injection period [15]. After the start of $CO_2$ injection, continuous stable isotope monitoring of casing gas $CO_2$ revealed the movement of $CO_2$ in the reservoir. Furthermore, $\delta^{13}C$ measurements on dissolved inorganic carbon (DIC) enabled quantification of $CO_2$ dissolution (i.e. solubility trapping) [2,10]. Initial sampling took place between August 2000 and March 2001 and yielded 87 samples from production observation wells. Within this time span, more than 350 million $m^3$ of $CO_2$ was injected as miscible fluid, in order to enhance oil production [5]. Within the first monitoring campaign, the $\delta^{13}C$ value of DIC in fluid samples showed hardly any variance, although the injected $CO_2$ was depleted in $^{13}C$ compared to the baseline values (Tab. 1) [2]. This was attributed to calcite dissolution within the carbonate reservoir [16]. Further isotope studies that were conducted over a 40-month period after the commencement of $CO_2$ injection revealed increases in DIC concentrations and decreases in $\delta^{13}C$ values in the reservoir DIC [10]. Isotope mass balances allowed the assessment of the extent, to which increasing DIC was caused by solubility trapping of injected $CO_2$ and by mineral dissolution (Tab. 2) [10]. In this project that was conducted in a fully operational oil field, gas and fluid samples were collected at wellheads. This implies that the gas samples represented casing gas that consisted of a mixture of reservoir gases ascending to the wellhead and gases exhaled from the wellbore fluids. A total of 29 samples were analyzed for DIC (i.e. $HCO^{3-}$) concentrations and corresponding $\delta^{13}C$ values (Tab. 1), from the four wells reported in [10] but many more analyses were conducted throughout the 10-year duration of this project [15].

## *2.2 The Frio Experiment*

During the Frio experiment in Texas, an integrated monitoring program was performed to investigate water-rock-interactions caused by 1600 tonnes of $CO_2$

that was injected into a saline sandstone aquifer [17]. The experiment aimed to explore the potential of injecting larger amounts of $CO_2$ and to evaluate associated geochemical reactions. For this purpose, an injection well was constructed and screened at about 1500 m depth below the ground surface. Advancements in sampling of fluids and gas were achieved by installing a U-tube sampling system in an observation well at a lateral distance of 30 m from the injection well. The U-tube sampler consists of a loop of tubing that forms a U with its terminations at the surface. Beneath the apex of the U-tube, a check valve allows an inlet of formation fluid through a filtered screen. By applying gas pressure at one of the U-legs (driving leg), the check valve can be closed, if the pressure exceeds the hydrostatic pressure of the formation. The loop can be filled with the reservoir fluid by releasing the pressure and opening the check valve, allowing the fluid to rise within the tubing until it reaches the hydrostatic level of the formation. Subsequently, a supply of inert gas under high pressure to the driving leg will close the check valve again and push the fluid out of the sample leg. The amount of sampled fluid depends on the depth and diameter of the tubing [18]. The system is able to keep fluid samples at original reservoir pressures and minimizes gas losses during sampling. It also excludes oxygen contamination from the atmosphere and degassing of dissolved $CO_2$ from the aqueous fluid, which may lead to rapid changes in sample chemistry and isotope ratios, especially $\delta^{13}C$ DIC of the formation fluid [18, 6]. Together with other down-hole and surface sampling techniques, baseline, injection and post-injection monitoring amounted to 100 gas and fluid samples during the experiment [6]. The $CO_2$ injection was accomplished in 10 days. Within this period, 41 liquid and more than 15 gas samples were obtained and analyzed for their chemical and isotopic compositions [18]. In addition, a quadrupole mass spectrometer was applied to detect tracer gases including perfluorcarbon (PFT`s), noble gases (Kr, Xe) and $SF_6$ that were co-injected with the $CO_2$ stream [17]. Isotope baseline values were reported for $\delta^{13}C$ of DIC, as well as $\delta^{18}O$ of $H_2O$ and injected $CO_2$ (Tab. 1) [14]. Carbon isotope ratios of DIC and oxygen isotopes of the reservoir brine measured during and after $CO_2$ injection revealed a clear shift of fluid and gas samples towards $\delta^{13}C$ and $\delta^{18}O$ values of the injected $CO_2$ (Tab. 1). Therefore, the isotope data was used to trace the dissolution of $CO_2$ and its conversion into DIC. Oxygen isotope measurements on $CO_2$ and reservoir fluids also helped to quantify saturation of the sandstone formation with the free $CO_2$-phase (Tab. 2). Furthermore, isotope analyses revealed the migration of $CO_2$ into higher stratigraphic sections of the reservoir [14, 17].

## 2.3 *The Pembina Cardium Pilot Study*

The Pembina Cardium $CO_2$ monitoring pilot site in Alberta, Canada offered new possibilities to carry out stable isotope monitoring during an EOR pilot project [19]. In this project, approximately 75,000 tonnes of $CO_2$ were injected between 2005 and 2008 into a siliciclastic reservoir at a depth of 1650 m below the ground surface. Injection was accomplished by two injection wells that were surrounded by six observation wells [12]. The process started with the injection of supercritical $CO_2$ as the only phase and changed to an alternating water-gas-regime after two years of injection [19]. After establishing the baseline values of the reservoir and the injected $CO_2$, a three-year monitoring program was established, with monthly monitoring campaigns [12]. Fluid and gas-samples were collected at the wellheads of the observation wells. Analyses comprised measurements of the isotope composition of $CO_2$, DIC and water (Tab. 1). Oxygen isotope measurements on $H_2O$ were conducted to estimate the pore-space saturation with free phase $CO_2$ (Tab. 2) [13]. This approach relies on the expected oxygen isotope equilibration between $H_2O$ and $CO_2$ [20, 13]. The presence of large amounts of supercritical $CO_2$ may change the oxygen isotope ratios of the reservoir water if the $\delta^{18}O$ value of the injected $CO_2$ is distinct from that of the water. The successful application of this method was demonstrated in this project, although previous water flooding of the reservoir complicated the distribution of $\delta^{18}O$ values at the study site [13].

Further studies at the Pembina Cardium site focussed on carbon isotope measurements of $CO_2$ in casing gas [12] combining $\delta^{13}C$ analyses with $CO_2$ flux measurements at several observation wells. This combination was used to reveal the proportions of injected $CO_2$ from gas fluxes at producing wells, as baseline $\delta^{13}C$ values of injected $CO_2$ had a $\delta^{13}C$ value of about -4.6 ‰. This was more than 10 ‰ higher than the casing gas values before injection (Tab. 1) [12]. Mass balance calculations were used to establish mixing models of residual casing gas with injected $CO_2$ (Tab. 2). The obtained values matched well with the measured elevated $CO_2$ concentrations within the wellbore with elevated $\delta^{13}C$ values and indicated the production of injected $CO_2$ at the sampled wells. This revealed preferential migration of the injected $CO_2$ along intra-reservoir fractures following a local main stress regime. Wells that were located along these pathways showed the highest $CO_2$ fluxes with the most elevated $\delta^{13}C$ values [12].

## *2.4 The Cranefield Experiment*

Isotope investigations have also been conducted at the EOR site in Cranfield, Mississippi, USA [11]. The reservoir investigated in this study consists of a depleted oil and gas field with fluvial sandstone- and conglomerate units. Production was abandoned in the 1960´s and the reservoir was not affected by extensive water flooding after this time. Therefore, the site was considered suitable for geochemical monitoring because the reservoir properties were anticipated to have re-equilibrated within this timespan [11]. In 2008, EOR operations began with extensive $CO_2$ flooding. Until 2011, 2.5 million tonnes of supercritical $CO_2$ were injected into the reservoir via 24 injection wells at depths of 3050 m. The $CO_2$ was transported via a pipeline from a nearby natural reservoir with an average $\delta^{13}C$ value of -2.6 ‰ (Tab. 1) [11]. Baseline acquisition was accomplished by the collection of brines from boreholes before the injection started. Sampling campaigns covered a time span of 21 months, during which gas and fluid samples were collected at the wellheads of producing wells [11]. In 2009, the acquisition of the samples was expanded due to the addition of two observation wells that were equipped with U-tube samplers. With this setup 11 brine and 830 gas samples were collected within one month.

Baseline samples revealed that $CO_2$ was dissolved in the reservoir brine displaying $\delta^{13}C$ values of –8 to –10 ‰ with concentrations ranging from 1 to 5 mole % of dissolved gas in the brine. After commencement of the $CO_2$ injection, gas produced at monitored wells revealed higher $CO_2$ concentrations accompanied with enriched $\delta^{13}C$ values compared to baseline values. Isotope mass-balances established plausible mixing models of the injected end-members that matched well with field observations (Tab. 2) [11].

## 2.5 The Otway Project

At the $CO_2$CRC Otway project in Australia more than 65.000 tonnes of a $CO_2$-$CH_4$ mixture were injected into a depleted natural gas-field. The injection site is located 25 km northwest of Port Campbell, Victoria. After $CO_2$ injection commenced in March 2008, 124 tonnes of $CO_2$ were injected per day over a time period of 17 months [21]. The injected gas was taken from $CO_2$-rich suspended exploratory well and piped 2.25 km to a newly drilled injection well. $CO_2$ was injected into a saline sandstone aquifer at a depth of about 2000 m [4]. Natural and noble gas tracers were used to verify $CO_2$ injection in order to detect the breakthrough of the $CO_2$ plume [4]. Sampling was conducted with a modified U-tube sampling system that comprised check valves at three different levels [21]. Gas samples were collected from the uppermost U-tube, situated inside the gas cap of the reservoir, whereas fluid samples of the formation brine were collected from the two lower levels [4]. Gas samples from the gas cap were lifted up in the U-tube by formation pressure whereas pressurized nitrogen was applied to lift water samples from the two lower reservoir levels. Carbon isotope ratios of produced gas samples as well as degassed $CO_2$ gained from reservoir brines were analyzed and compared [4]. The injected $CO_2$ was depleted in $^{13}C$ by 4.5 ‰ when compared to residual reservoir gas (Tab. 1). This difference was applied as a natural tracer for $CO_2$ detection that coincided with increases of molar amounts of $CO_2$ in produced gas samples at the observation well 150 days after the start of injection [4]. Isotope mass balance equations were set up to calculate the amount of isotopically distinct $CO_2$ that would be required to cause the observed isotope shifts at the observation wells (Tab. 2). Mass balances also helped to determine the proportion of injected and residual $CO_2$ within the wellbore (Tab. 2) [4]. In this project close to 100 samples were analyzed for their carbon stable isotope ratios.

**Table 1** Overview of isotope measurements conducted at selected pilot sites for CCS

| Site & Location | t of $CO_2$ injected [kt] | Isotope system monitored | Number of samples presented | Sampling method | Average isotope value $CO_2$ injected [$\delta^{13}C$ in ‰ V-PDB $\delta^{18}O$ in ‰ V-SMOW] | Averge isotope value before injection (DIC, $CO_2$, $H_2O$) [$\delta^{13}C$ in ‰ V-PDB $\delta^{18}O$ in ‰ V-SMOW] | Average isotope value after injection (DIC, $CO_2$, $H_2O$) [$\delta^{13}C$ in ‰ V-PDB $\delta^{18}O$ in ‰ V-SMOW] | References |
|---|---|---|---|---|---|---|---|---|
| Weyburn, Saskatchewan, Canada | 150600 | $\delta^{13}C_{CO2}$; $\delta^{13}C_{HCO3-}$ | 116 water and gas samples | Well head sampling | $\delta^{13}C$ $CO_2$ :-20.4 | $\delta^{13}C$ $CO_2$: -16 to -22<br>$\delta^{13}C$ DIC: -0.4 to -6.6 | $\delta^{13}C$ $CO_2$: -9 to -17<br>$\delta^{13}C$ DIC: -11.2 to -15.5 | [2]<br>[10] |
| Frio pilot site, Texas, USA | 1.6 | $\delta^{13}C_{CO2}$; $\delta^{13}C_{HCO3-}$ $\delta^{18}_{H2O}$ $\delta D_{H2O}$ | 40 water samples with at least 15 gas samples | Gas samples at Well head; Fluid samples at U-tube f | $\delta^{13}C$ $CO_2$ : 34 to -44 | $\delta^{13}C$ DIC - 5<br>$\delta^{18}O$ $H_2O$: 0.8<br>$\delta^{13}O$ $CO_2$: 9 | $\delta^{13}C$ DIC - 33<br>$\delta^{18}O$ $H_2O$: -11.1<br>$\delta^{13}O$ $CO_2$: -43 | [14]<br>[6] |
| Pembina Cardium monitoring site Alberta, Canada | 75 | $\delta^{13}C_{CO2}$ $\delta^{18}O_{CO2}$ $\delta^{13}C_{HCO3-}$ $\delta^{18}_{H2O}$ $\delta D_{H2O}$ | More than 135 gas samples<br>More than 151 fluid samples | Gas and water samples at well head | $\delta^{13}C$ $CO_2$ : -4.6<br>$\delta^{13}C$ $CO_2$ : -28.6 | $\delta^{13}C$ $CO_2$: -16.4 to -23.5<br>$\delta^{18}O$ $H_2O$: -14.5 to -17.1 | $\delta^{13}C$ $CO_2$: -5 to -12.3<br>$\delta^{18}O$ $H_2O$: -13.7 to -14.7 | [13]<br>[12] |
| Cranfield $CO_2$ sequestration site, Mississippi, USA | 2500 | $\delta^{13}C_{CO2}$ | More than 850 gas samples<br>More than 35 fluid samples fluid | Gas and water samples at well head, as well as U-tube and downhole sampler | $\delta^{13}C$ $CO_2$ : -2.6 | $\delta^{13}C$ $CO_2$: -10.5 | $\delta^{13}C$ $CO_2$: -3 | [11] |
| CO2CRC Otway project, Victoria, Australia | 65 | $\delta^{13}C$ $CO_2$ | About 100 gas samples | U-tube | $\delta^{13}C$ $CO_2$ : -6.5 | $\delta^{13}C$ $CO_2$: -11.5 | $\delta^{13}C$ $CO_2$: -6.5 | [4] |
| Ketzin pilot site, Germany | 65 | $\delta^{13}C_{CO2}$ $\delta^{18}O_{CO2}$ $\delta^{13}C_{HCO3-}$ | More than 70 fluid samples<br>More than 300 gas samples | Rising tube and GMS for gas samples<br>U-tube and downhole sampler for fluid samples | $\delta^{13}C$ $CO_2$ : - 3 to -31 | $\delta^{13}C$ DIC: - 4.2 to -8.4<br>$\delta^{18}O$ $H_2O$: -5.5 to -5.2 | $\delta^{13}C$ DIC: - 20<br>$\delta^{18}O$ $H_2O$: -5.5 to -12.0 | [22]<br>[24] |

**Table 2** Overview of isotope mass balances used at various pilot sites for corresponding isotope system; square brackets indicate fractions or concentrations

| Site & Location | Purpose / isotope system used | Mass balance equation | Ref. |
|---|---|---|---|
| Frio pilot site, Texas, USA | $\delta^{18}O$ $CO_2$ and $H_2O$ / saturation with free phase $CO_2$ | $\frac{X_{brine}}{X_{CO2}} = \frac{\delta^{18}O_{inj.\ CO2\ after\ injection} - \delta^{18}O_{inj.\ CO2\ baseline}}{\delta^{18}O_{H2O\ before\ injection} - \delta^{18}O_{H2O\ after\ injection}}$ | [14] |
| Pembina Cardium monitoring site Alberta, Canada | $\delta^{13}C$ $CO_2$ /proportion of in. $CO_2$ at produced gas | $\delta^{13}C_{CO2\ measured} = \frac{\delta^{13}C_{CO_2\ injected} \times [CO_{2\ injected}] + \delta^{13}C_{CO_2\ baseline} \times [CO_{2\ baseline\ in\ wellbore}]}{[CO_{2\ measured}]}$ * | [12] |
| | $\delta^{18}O$ $CO_2$ and $H_2O$ / fraction of O derived from inj. $CO_2$ | $X^{O}_{derived\ from\ injected\ CO2} = \frac{\delta^{18}O_{H2O\ before\ injection} - \delta^{18}O_{H2O\ after\ injection}}{\delta^{18}O_{H2O\ before\ injection} + \varepsilon - \delta^{18}O_{CO2}}$ ** | [13] |
| Cranfield $CO_2$ sequestration site, Mississippi, USA | $\delta^{13}C$ $CO_2$ /mixing between formation $CO_2$ and $CO_2$ from injectate | $\delta^{13}C_{sample} =$ $\frac{\delta^{13}C_{formation\ gas} \times [CO_{2\ injectate}] \times [CO_{2\ formation}] - \delta^{13}C_{injectate} \times [CO_{2\ injectate}] \times [CO_{2\ formation}]}{[CO_{2\ injectate}] - [CO_{2\ formation}]} \times \frac{100}{[CO_{2\ sample}]} + \frac{\delta^{13}C_{injectate} \times [CO_{2\ injectate}] - \delta^{13}C_{formation} \times}{[CO_{2\ injectate}] - [CO_{2\ formation}]}$ *** | [11] |
| CO2CRC Otway project, Victoria, Australia | $\delta^{13}C$ $CO_2$ / proportion of inj. $CO_2$ to initial dissolved gas | $\frac{X_{CO2\ injected}}{X_{CO2\ initial\ dissolved}} = \frac{\delta^{13}C_{CO2\ measured} - \delta^{13}C_{CO2\ baseline}}{\delta^{13}C_{CO2\ injected} - \delta^{13}C_{CO2\ baseline}}$ | [4] |
| Ketzin pilot site, Germany | $\delta^{13}C$ DIC / quantifying ionic trapping | $\%DIC\ from\ CO2\ dissolution = \frac{\delta^{13}C_{DIC\ after\ inj.} - \delta^{13}C_{DIC\ baseline}}{\delta^{13}C_{CO2(g)inj.} - \delta^{13}C_{DIC\ baseline}} \times 100$ | [24] |
| | $\delta^{13}C$ $CO_2$ / proportion of injected $CO_2$ | $\delta^{13}C_{CO2\ measured} = \delta^{13}C_{CO2\ injected} \times \frac{[CO_{2\ injected}]}{[CO_{2\ residual}]} + \delta^{13}C_{baseline} \times \left(1 - \frac{[CO_{2\ injected}]}{[CO_{2\ residual}]}\right)$ **** | |

* square brackets indicate fractions of the component where $[CO_{2\ measured}] = 1$

** ε is isotope enrichment factor for $CO_2$ -water at reservoir temperature

*** square brackets indicate $CO_2$ content in the gas in %

**** square brackets indicate fractions of the gas component

## 2.6 *The Ketzin Pilot Site*

The Ketzin pilot site is the longest operating onshore test injection site for CCS in Europe. Injection of $CO_2$ began in 2008 into a saline aquifer at approximately 640 m depth below the ground surface. The purpose of this ongoing project was to establish sound scientific knowledge about $CO_2$ behaviour in the subsurface as well as to test and develop appropriate monitoring strategies [22]. The site has a pre-injection facility with two 50-ton storage tanks, one injection well and two observation wells at 50 and 100 m distance from the injector. A third observation well that reaches the reservoir will be completed by the end of 2012. At first, sampling of reservoir brine was achieved by downhole samplers. Also a newly developed gas membrane sensor (GMS) combined with a portable quadrupole mass-spectrometer allowed in situ and real-time analyses of dissolved gases within the reservoir brine [8]. After the supercritical $CO_2$ displaced most of the reservoir brine around the observation wells, a rising tube was installed that allowed continuous gas sampling from the reservoir. The rising tube produces about 8 liters of $CO_2$ per hour, supported by the reservoir pressure [22].

Stable isotope measurements had been_conducted since the beginning of the $CO_2$ injection in 2008. Investigations started with a detailed geochemical baseline characterization of the reservoir and overlying formations, comprising $\delta^{13}C$ and $\delta^{18}O$ data of brine DIC and $H_2O$ (Tab. 1) [23]. After this baseline characterization, stable carbon isotopes of the dissolved inorganic carbon have been determined to confirm the $CO_2$ arrival at two observation wells as well as to quantify solubility trapping within the reservoir brine using isotope mass balances (Tab. 2) [24]. In addition, in 2011, a further amount of more than 1500 tons of isotopically different $CO_2$ from an oxyfuel pilot plant was injected over a time period of approximately 2 months. This $CO_2$ had a $\delta^{13}C$ value that was 2 ‰ more positive than that of the previously injected $CO_2$. This caused an isotope shift at one observation well that coincided with the breakthrough of simultaneously injected noble gas tracers [22].

Further $CO_2$ gas monitoring in the reservoir, including the road tankers that supply the site with $CO_2$, revealed also that during normal injection processing two isotopically very different kinds of $CO_2$ were injected. This difference is attributed to the different sources of the $CO_2$. The first source provides the vast majority of all injected $CO_2$ and is an oil refinery that produces technical grade $CO_2$ with an average $\delta^{13}C$ value of about –30 ‰. The other source is a natural $CO_2$ accumulation that provides $CO_2$ with a $\delta^{13}C$ value of about -3‰. The isotopically distinct $CO_2$ from the natural source is injected in smaller amounts and sporadically at time intervals of several days to weeks. Recent studies at Ketzin focus on exploiting these differences in $\delta^{13}C$ values between technical and natural source $CO_2$, which amount to a maximum of 28 ‰ (Tab. 1). Different isotope compositions generated from the sporadic charging with isotopically distinct natural source $CO_2$ into the pre-injection tanks can be characterized before injection and should subsequently be identified at an observation well. If so, this would provide an excellent additional tracer because the isotopically distinct $CO_2$

from the natural source that has been injected alternately since beginning of injection provides an opportunity to validate migration models by using isotope mass balances (Tab. 2).

In Addition to the gas tracer experiments, isotope investigations are applied in order to support leakage detection of the above reservoir zone. In 2011, an observation well was completed within the closest stratigraphic overlying aquifer [22]. The well contains several down-hole geophysical devices as well as a U-tube sampler. It is part of a new monitoring concept, which aims at detecting potential leakage of the first barrier as soon as possible [25]. Within this concept, frequent stable isotopes monitoring of the formation brine DIC is a substantial part. Leaking $CO_2$ from the reservoir is assumed to be detectable with the aid of stable carbon isotopes, because it would change the $\delta^{13}C$ value of the DIC in a systematic manner [25].

Analytical and sampling methods were continuously improved at Ketzin. Modifying a traditional isotope ratio mass spectrometer and its autosampler with additional He-dilution allows determination of samples with up to 100 % $CO_2$ content without further pre-dilution in the laboratory. In addition, sampling in the field was simplified by flushing standard laboratory sampling vials with 100 % $CO_2$ gas samples that are subsequently analyzed in the laboratory. This enabled isotope measurements of more than 260 samples within a time period of three months. Overall, 74 fluid samples and 310 gas samples were analyzed for their carbon and oxygen isotopic signature within the project period at Ketzin.

## 3 Conclusions

While a number of ongoing and recently started projects were not considered in this paper, this review has shown that, worldwide, a total of more than 1000 isotope measurements were carried out during $CO_2$ storage projects for various purposes. The main use of these measurements is to improve the understanding of gas - water - rock interaction under elevated pressure and temperature conditions, $CO_2$ migration in the subsurface as well as leakage detection.

This review demonstrates that stable isotope monitoring represents an important tool in $CO_2$ storage and enhanced hydrocarbon recovery projects. This technique requires detailed planning of sampling strategies and needs to account for degassing of fluids before sampling. Therefore, down-hole sampling techniques such as the U-tube system or rising tubes for gas samples should be preferred. However, as has been shown, data obtained from wellhead sampling can also provide sufficiently accurate information, when corrections for geochemical alteration are applied. This seems especially important for isotope applications in large scale CCS or commercial EOR/EGR projects, because they usually comprise of numerous injection, production and/or observation wells, which can not be equipped with complex down hole sampling tools as it is usually done at smaller scale pilot sites. However, there is a lack of detailed studies that compare isotope measurements from wellhead sampling with down-hole techniques.

Like many other monitoring methods, isotope studies also require detailed baseline monitoring before $CO_2$ injection starts in order to facilitate comparisons of geochemical changes. Furthermore, sufficient baseline data of reservoir isotope ($\delta^{13}C$, $\delta^{18}O$) and concentration values, injection rates and isotope measurements during and after injection allow the application of highly useful isotope mass balances. The latter are the most commonly applied techniques in stable isotope utilizations. They allow quantifications of several key-processes accompanying $CO_2$ injection into the subsurface including ionic trapping, plume development and $CO_2$ migration.

The above stable isotope techniques are limited by low spatial resolution, caused by a sparse number of observation wells. This is also often combined by insufficient temporal resolution with too low sampling frequencies. However, new developments in mass spectrometry with new generation mobile laser-based instruments are promising. They simplify measurements of $\delta^{13}C$ and $\delta^{18}O$ in both, gases and fluids. These new sensors are applicable in the field and allow on-line measurements. With this, stable isotope measurements can become an increasingly applied method and a substantial part of complementary monitoring systems for CCS.

**Acknowledgements.** The German Federal Ministry for Education and Research (BMBF) provided funding for this study within the GEOTECHNOLOGIEN Program (GEOTECH-2047) in the project CO2MAN (03G0760B). Furthermore, background information for this manuscript was also assembled within the BMBF Project CO2ISO-Label (03G0801A).

## References

1. Freifeld, B.M., Daley, T.M., Hovorka, S.D., Henninges, J., Underschultz, J., Sharma, S.: Recent advances in well-based monitoring of $CO_2$ sequestration. Energy Procedia 1(1), 2277–2284 (2009)
2. Emberley, S., Hutcheon, I., Shevalier, M., Durocher, K., Mayer, B., Gunter, W.D., Perkins, E.H.: Monitoring of fluid-rock interaction and $CO_2$ storage through produced fluid sampling at the Weyburn $CO_2$-injection enhanced oil recovery site, Saskatchewan, Canada. Appl. Geochem. 20(6), 1131–1157 (2005), doi:10.1016/J.Apgeochem.2005.02.007
3. Kharaka, Y.K., Cole, D.R., Thordsen, J.J., Kakouros, E., Nance, H.S.: Gas-water-rock interactions in sedimentary basins: $CO_2$ sequestration in the Frio Formation, Texas, USA. J. Geochem. Explor. 89(1-3), 183–186 (2006), doi:10.1016/J.Gexplo.2005.11.077
4. Boreham, C., Underschultz, J., Stalker, L., Kirste, D., Freifeld, B., Jenkins, C., Ennis-King, J.: Monitoring of $CO_2$ storage in a depleted natural gas reservoir: Gas geochemistry from the CO2CRC Otway Project, Australia. Int. J. Greenh. Gas Con. 5(4), 1039–1054 (2011), doi:10.1016/J.Ijggc.2011.03.011
5. Emberley, S., Hutcheon, I., Shevalier, M., Durocher, K., Gunter, W.D., Perkins, E.H.: Geochemical monitoring of fluid-rock interaction and $CO_2$ storage at the Weyburn CO2-injection enhanced oil recovery site, Saskatchewan, Canada. Energy 29(9-10), 1393–1401 (2004), doi:10.1016/J.Energy.2004.03.073

6. Kharaka, Y.K., Thordsen, J.J., Hovorka, S.D., Nance, H.S., Cole, D.R., Phelps, T.J., Knauss, K.G.: Potential environmental issues of CO2 storage in deep saline aquifers: Geochemical results from the Frio-I Brine Pilot test, Texas, USA. Appl. Geochem. 24(6), 1106–1112 (2009), doi:10.1016/J.Apgeochem.2009.02.010
7. Shi, J.Q., Sinayuc, C., Durucan, S., Korre, A.: Assessment of carbon dioxide plume behaviour within the storage reservoir and the lower caprock around the KB-502 injection well at In Salah. Int. J. Greenh. Gas Con. 7, 115–126 (2012), doi:10.1016/J.Ijggc.2012.01.002
8. Zimmer, M., Erzinger, J., Kujawa, C.: The gas membrane sensor (GMS): A new method for gas measurements in deep boreholes applied at the CO2SINK site. Int. J. Greenh. Gas Con. 5(4), 995–1001 (2011)
9. Ghosh, P., Brand, W.A.: Stable isotope ratio mass spectrometry in global climate change research. International Journal of Mass Spectrometry 228(1), 1–33 (2003)
10. Raistrick, M., Mayer, B., Shevalier, M., Perez, R.J., Hutcheon, I., Perkins, E., Gunter, B.: Using chemical and isotopic data to quantify ionic trapping of injected carbon dioxide in oil field brines. Environ. Sci. Technol. 40(21), 6744–6749 (2006), doi:10.1021/Es060551a
11. Lu, J.M., Kharaka, Y.K., Thordsen, J.J., Horita, J., Karamalidis, A., Griffith, C., Hakala, J.A., Ambats, G., Cole, D.R., Phelps, T.J., Manning, M.A., Cook, P.J., Hovorka, S.D.: $CO_2$-rock-brine interactions in Lower Tuscaloosa Formation at Cranfield $CO_2$ sequestration site, Mississippi, USA. Chem. Geol. 291, 269–277 (2012), doi:10.1016/J.Chemgeo.2011.10.020
12. Johnson, G., Mayer, B., Shevalier, M., Nightingale, M., Hutcheon, I.: Tracing the movement of CO2 injected into a mature oilfield using carbon isotope abundance ratios: The example of the Pembina Cardium $CO_2$ Monitoring project. Int. J. Greenh. Gas Con. 5(4), 933–941 (2011), doi:10.1016/J.Ijggc.2011.02.003
13. Johnson, G., Mayer, B., Nightingale, M., Shevalier, M., Hutcheon, I.: Using oxygen isotope ratios to quantitatively assess trapping mechanisms during $CO_2$ injection into geological reservoirs: The Pembina case study. Chem. Geol. 283(3-4), 185–193 (2011), doi:10.1016/J.Chemgeo.2011.01.016
14. Kharaka, Y.K., Cole, D.R., Hovorka, S.D., Gunter, W.D., Knauss, K.G., Freifeld, B.M.: Gas-water-rock interactions in Frio Formation following $CO_2$ injection: Implications for the storage of greenhouse gases in sedimentary basins. Geology 34(7), 577–580 (2006), doi:10.1130/G22357.1
15. Mayer, B., Shevalier, M., Nightingale, M., Kwon, J.S., Johnson, G., Raistrick, M., Hutcheon, I., Perkins, E.: Tracing the movement and the fate of injected $CO_2$ at the IEA GHG Weyburn-Midale $CO_2$ monitoring and storage project (Saskatchewan, Canada) using carbon isotope ratios. Int. J. Greenh. Gas Con. (submitted)
16. Shevalier, M., Durocher, K., Perez, R.J., Hutcheon, I., Mayer, B., Perkins, E., Gunter, B.: Geochemical monitoring of gas-water-rock interaction at the iea Weyburn $CO_2$ Monitoring and Storage Project, Saskatchewan, Canada. In: Rubin, E.S., Keith, D.W., Gilboy, C.F., et al. (eds.) Greenhouse Gas Control Technologies 7, pp. 2135–2139. Elsevier Science Ltd., Oxford (2005)
17. Hovorka, S.D., Benson, S.M., Doughty, C., Freifeld, B.M., Sakurai, S., Daley, T.M., Kharaka, Y.K., Holtz, M.H., Trautz, R.C., Nance, H.S., Myer, L.R., Knauss, K.G.: Measuring permanence of $CO_2$ storage in saline formations: The Frio experiment. Environmental Geosciences 13(2), 105–121 (2006)

18. Freifeld, B.M., Trautz, R.C., Kharaka, Y.K., Phelps, T.J., Myer, L.R., Hovorka, S.D., Collins, D.J.: The U-tube: A novel system for acquiring borehole fluid samples from a deep geologic $CO_2$ sequestration experiment. J. Geophys. Res.-Sol. Ea. 110(B10) (2005), doi:Artn B10203
19. Shevalier, M., Nightingale, M., Johnson, G., Mayer, B., Perkins, E., Hutcheon, I.: Monitoring the reservoir geochemistry of the Pembina Cardium $CO_2$ monitoring project, Drayton Valley, Alberta. Energy Procedia 1(1), 2095–2102
20. Vogel, J.C., Grootes, P.M., Mook, W.G.: Isotopic fractionation between gaseous and dissolved carbon dioxide. Zeitschrift für Physik 230(3), 225–238 (1970)
21. Stalker, L., Boreham, C., Underschultz, J., Freifeld, B., Perkins, E., Schacht, U., Sharma, S.: Geochemical monitoring at the CO2CRC Otway Project: tracer injection and reservoir fluid acquisition. Energy Procedia 1, 2119–2125 (2009)
22. Martens, S., Kempka, T., Liebscher, A., Lüth, S., Möller, F., Myrttinen, A., Norden, B., Schmidt-Hattenberger, C., Zimmer, M., Kühn, M.: Europe's longest-operating on-shore $CO_2$ storage site at Ketzin, Germany: a progress report after three years of injection. Environmental Earth Sciences 67(2), 323–334 (2012)
23. Förster, A., Norden, B., Zinck-Jörgensen, K., Frykman, P., Kulenkampff, J., Spangenberg, E., Erzinger, J., Zimmer, M., Kopp, J., Borm, G., Juhlin, C., Cosma, C.G., Hurter, S.: Baseline characterization of the CO2SINK geological storage site at Ketzin, Germany. Environmental Geosciences 13(3), 145–161 (2006)
24. Myrttinen, A., Becker, V., van Geldern, R., Wurdemann, H., Morozova, D., Zimmer, M., Taubald, H., Blum, P., Barth, J.A.C.: Carbon and oxygen isotope indications for CO(2) behaviour after injection: First results from the Ketzin site (Germany). Int. J. Greenh. Gas Con. 4(6), 1000–1006 (2010), doi:10.1016/J.Ijggc.2010.02.005
25. Wiese, B., Zimmer, M., Nowak, M., Pilz, P.: Above-zone well-based hydraulic and geochemical monitoring of the $CO_2$ reservoir in Ketzin, Germany. Environmental Earth Sciences (Special Issue Geoenergy) (2012) (submitted)

# Use of a $CO_2$ Geological Storage System to Develop Geothermal Resources: A Case Study of a Sandstone Reservoir in the Songliao Basin of Northeast China

Yan Shi[1,2], Fugang Wang[1,*], Yanlin Yang[1], Hongwu Lei[1], Jin Na[1], and Tianfu Xu[1]

[1] Key Lab of Groundwater Resources and Environment, Ministry of Education, Jilin University, Changchun, China
[2] Jilin Architecture and Civil Engineering Institute, Changchun, China

**Abstract.** The concept of a carbon dioxide ($CO_2$) plume geothermal (CPG) system, which uses the $CO_2$ geological storage system to develop geothermal resources has been proposed recently. On the basis of the geological structures and geothermal conditions in the Songliao Basin in North East China, a three-dimensional model of a sandstone layer with a temperature of 120°C was developed using TOUGHREACT. Numerical simulations for operating both the geothermal system using $CO_2$ only and that using water only were carried out, and the results compared. A number of comprehensive processes including flow (under gravity, buoyancy and injection pressure), and heat transfer have been considered. A cold fluid (20°C) injection well and a hot fluid production well were specified in the geothermal reservoir. The heat extraction rates for $CO_2$ and water as heat transfer medium were analysed and compared. Modeling results indicate that compared to the CPG system, the variation range of temperature of the water system reduces by about 50% during the same simulation period. This phenomenon is caused by the favorable properties of $CO_2$ such as low density and viscosity. The heat extraction rate for $CO_2$ significantly increased by nearly 200% more than water in this sandstone reservoir under the present simulation conditions. Thus, the fundamental understanding and the scientific base for the future CPG development could be provided by this research.

**Keywords:** $CO_2$ plume geothermal, low-temperature pore media, heat production, Songliao Basin, Numerical simulation.

---

* Corresponding author.

M.Z. Hou et al. (Eds.): *Clean Energy Systems in the Subsurface*, SSGG, pp. 89–103.
DOI: 10.1007/978-3-642-37849-2_8 

# 1 Introduction

The excess emission of carbon dioxide is one of the main factors causing global warming (Holloway 2001, West et al. 2005, Wigley et al 1996). $CO_2$ geological sequestration is one effective technology recognized in the international community to significantly reduce greenhouse gas emissions and cope with climate change (Gentzis 2000, Gough 2008, Holloway 2005). The sandstone saline aquifers at 1000-3000 m depth in sedimentary basins are considered the foremost space for carbon dioxide geological sequestration (Gunter et al. 2000, Hitchon et al 1999, Izgec et al. 2008, Xu et al. 2008), and abundant geothermal resources are contained in this part of the medium. Coupling geothermal energy exploitation with $CO_2$ sequestration would further improve the economic viability of carbon capture and storage (CCS) (Randolph and Saar 2011).

In recent years, a number of scholars have conducted studies on the fluid flow, heat transfer and water-rock interaction. The geothermal energy production can be increased by using $CO_2$ as a heat transmission fluid instead of water (Wolf et al. 2008, Brown 2000, Pruess 2006 and 2008, Xu et al. 2008). The concept of an enhanced geothermal system using $CO_2$ as heat transmission fluid ($CO_2$-EGS) was initially proposed by Brown (2000). Pruess (2006, 2008) investigated the fluid dynamics and heat transfer issues of the $CO_2$-EGS, and evaluated the mass flow and heat extraction rates. Since then, chemically reactive transport modeling to investigate water-rock interaction in a $CO_2$-EGS system has been performed by Xu et al. (2004, 2005, 2008, and 2010).

However, EGS are typically generated by hydro-fracturing, so-called hot dry rock, and its commercial progress is facing technical and economic challenges. The artificial fracture is difficult and costly to generate. Actual operation of EGS may induce seismicity because the critical fracture stresses of geological formations are intentionally exceeded. The EGS has encountered considerable socio-political resistance, exemplified by the termination of several EGS projects such as the Swiss Basel EGS project during the year 2009 (Randolph et al. 2011, Glanz 2009). Therefore, research on how to use natural and stored $CO_2$ reservoirs to extract the deep geothermal resource is imperative. Randolph et al. (2011) presented a concept of a $CO_2$ plume geothermal (CPG) system which involves pumping $CO_2$ into deep, naturally porous and permeable geological formations where $CO_2$ displaces native formation fluid. In their study, the economic benefits of CPG with respect to CCS in particular and the potential for CPG to offset some of the costs associated with CCS were a primary concern. However, major issues such as how to estimate $CO_2$-water multiphase fluid displacement and heat transfer processes, and the influence of heat extraction rate have not been addressed for this coupled system.

The objective of this paper is to explore the heat transfer process when the injected $CO_2$ is heated by the sandstone reservoir; evaluate the heat extraction rate of $CO_2$ or water as heat medium under complex geological conditions. Some basic data and the technical method can be provided to study geothermal resources in a typical sedimentary basin coupled $CO_2$ geological storage.

The paper is organized as follows. First, the geological structure and thermal storage conditions of four members of the Quantou Formation in the Songliao Basin are discussed , from which the initial parameters of the numerical model are obtained. Then, the solution approach to non-isothermal flow under the action of gravity and buoyancy is explained. A three-dimensional model of Carbon dioxide-Plume Geothermal Systems in a sedimentary basin natural geological reservoir is developed. After that, two cases are developed, the water-only (base) case and $CO_2$-only case. With the base case, water at a surface temperature of 20°C is injected into the reservoir, where the processes of cold - hot water single-phase displacement and heat transfer occur. For the $CO_2$-only case, supercritical $CO_2$ at a temperature of 20°C is injected, resulting in the processes of cold-hot gas single-phase displacement and heat transfer. Results obtained from the two cases are discussed in detail. Finally, some findings and conclusions are summarized.

## 2 Reservoir Characterization

### *2.1 Reservoir Properties*

Target reservoir should not only be suitable for $CO_2$ geological storage but also meet the temperature requirements for the geothermal development. The location of the target reservoir is shown in Figure 1. The central depression of the Songliao Basin contains many sets of storage reservoir and cap rock, which are widely distributed. The geothermal gradient and flow are higher in this region (Hou et al. 2009, Compile group of Petroleum geology of Jilin oilfield records 1993). The stratigraphic column in Figure 2 shows various reservoir units and the corresponding depths and thicknesses (Zhang et al. 2011). So a target reservoir in the central depression at 2000-3000 m deep is preliminarily determined. A total of five sets of reservoir and cap-rock combinations can be divided, namely: the shallow combination, the upper combination, the central combination, the lower combination and the deep combination (Hou et al. 2009, Gao et al. 1997, Lou et al. 1998, Chi et al. 2002). The lower combination means Quantou formation is regarded as a reservoir and the Qingshankou formation is cap rock. In this paper, the 100 m thick sandstone of the Quan3 member of the lower combination is chosen as the target geothermal reservoir. The porosity and permeability of the reservoir is shown in Table 1.

### *2.2 Thermal Properties*

In the Songliao basin, the geothermal gradient is high and the geothermal flow is abundant. Large deep faults are developed at the basin bottom. These become excellent channels for the deep heat supply source, providing an important control

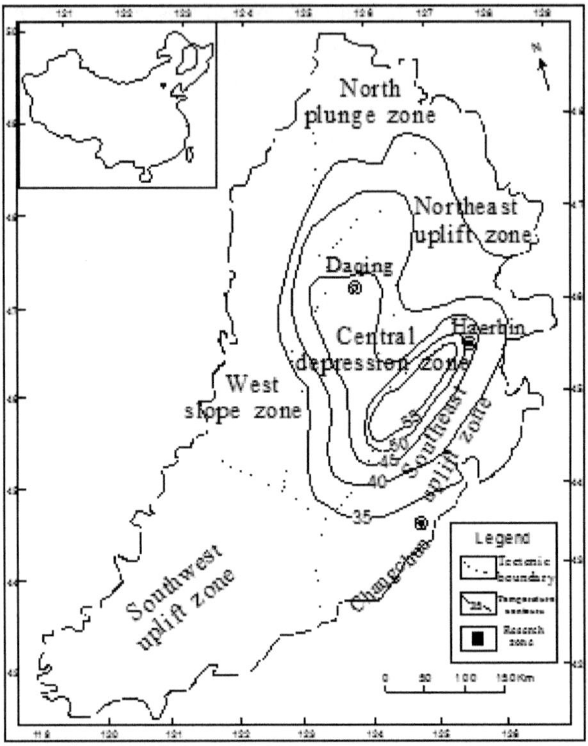


**Fig. 1** Tectonic unit zoning of Songliao Basin

action to the shallow ground temperature field (Zhai 2011). According to the 17 geothermal drillings in the central depression in the Songliao Basin, the geothermal gradient ranges from approximately 2.62 to 6.25 °C/100 m and the temperature at the depth of 800 m ranges from 29.8 to 42.8 °C (Wu et al. 1991 and 1990; Ren et al. 2001, Tan et al. 2001, Li et al. 1995, Zhou et al. 2007). Based on the measured value of the geothermal gradient and temperature at 800 m depth, the temperature at 2500 m depth could range from 79.4 to 148.25 °C. So the initial temperature in this simulation is 120 °C.

## 3 Model Setup

All simulations were done with the non-isothermal multiphase multicomponent reactive solute transport simulator TOUGHREACT (Xu et al. 2006). The program can deal with one-, two-, or three-dimensional porous and fractured media with physical and chemical heterogeneity and any number of chemical species present in liquid, gas and solid phases. The program has been applied to a large number of

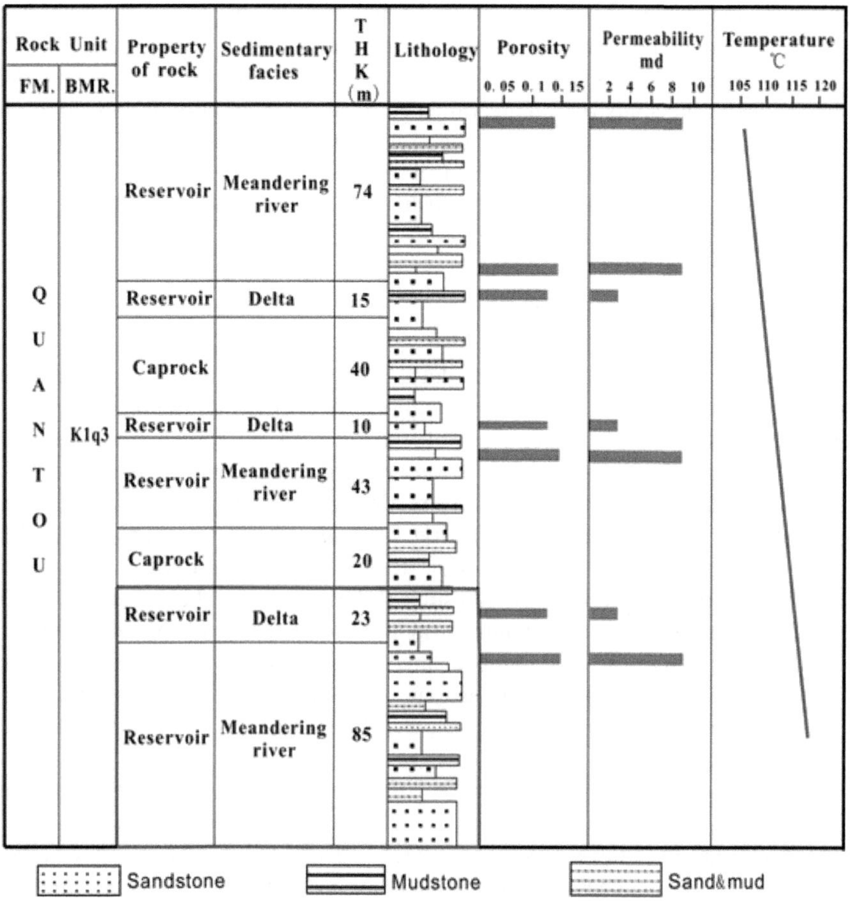


**Fig. 2** Stratigraphic chart for the Quantou Formation of Songliao Basin (Zhang et al. 2011)

EGS and CCS projects, such as the United States Nevada state Desert Peak EGS project, the French Solutz EGS site, Japan's Ogachi EGS project, the United States Frio $CO_2$ geological sequestration, and the Australian Otway $CO_2$ geological sequestration.

## 3.1 *The Mesh Making Method*

The irregular mesh named Tyson polygons is adopted in this paper. Its principle is that a polygon is built around the each sampling point . Compared to an arbitrary point outside the Tyson polygon, the distance from the point of the Tyson polygon to the sample point is less. The properties of this mesh are able to generate the

TOUGH2 mesh. The flexibilities of modifying mesh points and encrypting mesh are the prominent advantages of the Delaunay triangulation mesh (Gelder 1995).

The process of mesh making is as follows: Firstly, the 3-D geometrical dimensions of the study region are validated; then the encryption points, lines and areas, are ensured, after which the Delaunay triangulation is constructed. Subsequently, the height of each layer of the Tyson polygons is designed, translated into a mesh unit and finally written out according to the TOUGH MESH file format (Yang et al. 2012).

## 3.2 The Test Model

The object of constructing a test model is to determine properly the simulation region and give accurate boundary conditions. A geothermal doublet consists of two wells: a hot fluid production well, and a cold fluid reinjection well. To properly determine the simulation region of the geometric model is the basis for defining boundary conditions and constructing a geological model. It is a strong guarantee that the subsequent prediction simulation is accurate and reasonable. Therefore, this paper used a large-scale test model to determine the geometric model. According to the simulation results of the test model, the simulation region can be divided into the temperature response area and the temperature constant region to determine the maximum change range in reservoir temperature. The interface of the temperature response area and the temperature constant region is the boundary to the base case model, and the first boundary condition is given here.

In accordance with the characteristics of the target reservoir, a 3-D large scale test model with an average thickness of 100 m and a two dimensional size of 10000 m×5000 m is developed. The model is divided into 5 layers in a vertical direction and the thickness of each layer is 20 m; each layer is divided into 418 elements, amounting to 2090 elements and 7717 connections in total. The injection interval is located at the (4500, 2500, -90) coordinate, whereas the production interval is located at coordinate (5500, 2500, -10). Initially the test model is given the most adverse conditions for $CO_2$ migration. For example, a higher injection pressure (2 times the initial pressure), higher porosity and permeability (porosity=0.3 and permeability=$1.0\times10^{-13}m^2$) were adopted in the test model. Other parameters are set in reference to the actual hydrogeological condition of the research region. Supercritical $CO_2$ at a temperature of 20°C is injected into the geothermal reservoir and a hot fluid is produced from the production well. The influence boundary of the temperature is monitored by simulation. The test model mesh and the simulation results are shown in Figure 3. The temperature field in the x-y direction is different at different depth. Simulation results after 40 years show that the maximum thermal influence radii in each layer are all less than 2000 m.

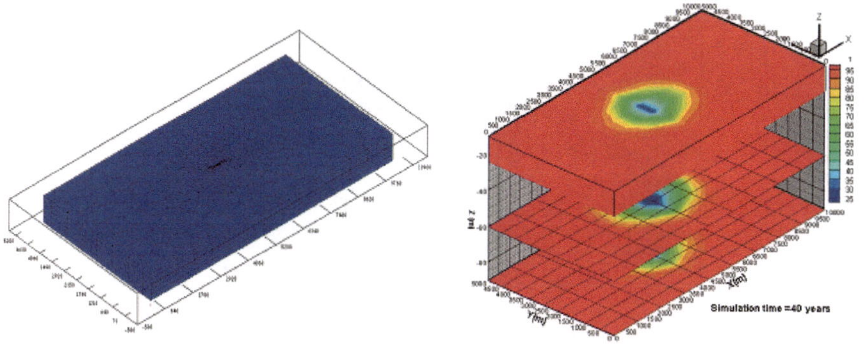


(a) (b)

**Fig. 3** (a) The test model meshes (b) The temperature field distribution at t=40 yr for the test model

## *3.3 Geometric Model and Boundary Conditions*

Based on the above geological structure and geothermal reservoir properties, a 3-D injection-production $CO_2$ plume geothermal model is established in a sandstone reservoir. The vertical spatial heterogeneity of porosity and permeability is considered. The various reservoir units and the corresponding depths and thicknesses are shown in Figure 2. An injector-producer doublet configuration with a basic area of 2000 m×2000 m×100 m is finally chosen. Vertically the model domain is divided into 20 layers, each layer is 5 m thick. The upper 20 m interval has a permeability of $3.41\times 10^{-3}$ $\mu m^2$ and a porosity of 12.5%, while the lower 80 m has a permeability of $8.72\times10^{-3}$ and a porosity of 14.3%. For the test model calculation result, the computational grid needs to cover only 1/4 of the domain as seen in Figure 4. The horizontal distance from the production to the injection interval is 650 m, the injection point is located at (0, 0,-97.5), and the production point is located at (650, 0,-27.5). Delaunay triangulation is used here. The region between injection and production point is considered as an encrypting grid, in order to obtain a detailed description of the flow displacement and heat transfer processes. Subdivision of the peripheral region out injection-production locations is sparse. So there are 392 grids of Tyson polygon in each layer, giving a total of 7840 grids. The production well is set as a "virtual well" with fixed pressure to extract heat. It is connected to the grids of the upper 4 layers, the heat extraction depth is 35 m. Formation parameters, initial conditions, and production-injection specifications are given in Table 1.

Considering the thermal compensation effect from the basal terrestrial heat flow, the semi-infinite thermal boundary was used (Vinsome and Westerveld, 1980). The simulation results of the test model the maximum thermal influence radii are all less than 2000 m. So the interfaces of the temperature constant region

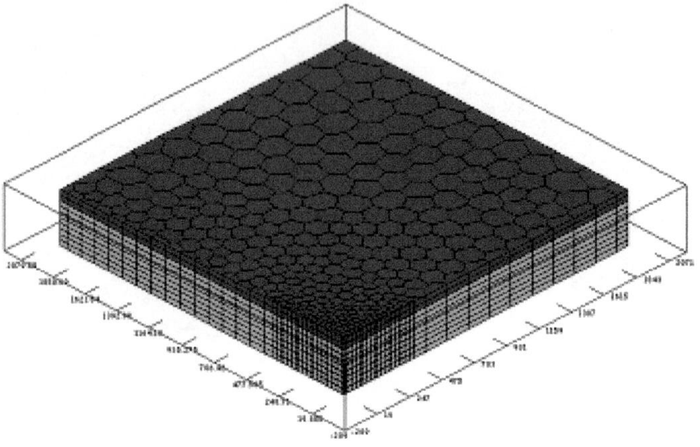

**Fig. 4** The numerical model meshes

**Table 1** The basic simulation parameters

| Simulation parameters | value |
|---|---|
| **Hydrogeological and thermal parameters** | |
| Thickness, $H$ | 100 m |
| Rock grain density, ρ | 2600 kg/m$^3$ |
| Rock specific heat, $c$ | 1000 J/kg·℃ |
| Rock thermal conductivity, $\lambda$ | 2.51 W/m·⍰ |
| Porosity of meandering river sedimentary facies , $\phi_1$ | 0.143 |
| Permeability of meandering river sedimentary facies , $k_1$ | $8.72\times10^{-15}$ m$^2$ |
| Porosity of delta sedimentary facies, $\phi_2$ | 0.125 |
| Permeability of delta sedimentary facies , $k_2$ | $3.41\times10^{-15}$ m$^2$ |
| Ratio of vertical to horizontal Permeability | 0.1 |
| **Initial condition** | |
| Water saturation in base case, $S_l$ | 1.0 |
| Gas saturation in CO2-only case, $S_g$ | 1.0 |
| Temperature, $T_{in}$ | 120 ⍰ |
| Pressure, $P_{in}$ | $25.0\times10^6$ Pa |
| **Production/injection** | |
| Injector–producer distance, $L$ | 650m |
| Injection temperature, $T_{inj}$ | 20⍰ |
| Injection pressure, $P_{inj}$ | $P_{in}$ + 100 bar |
| Production pressure, $P_{Pro}$ | $P_{Pro}$ - 100 bar |

and response region are located at x = 2000 m and y=2000 m, the first boundary conditions t = 120 °C are given to the two interfaces. The top of the model is regarded as an impermeable boundary for fluid flow.

# 4 Results and Discussion

## *4.1 Distribution of Temperature Field*

In the base case (water only), the cold water is injected into the sandstone reservoir. The change of temperature field for this case is shown in Figure 5. In the $CO_2$-only case, the reservoir is completely filled with hot $CO_2$. When cold $CO_2$ is injected into the hot reservoir, the cold $CO_2$ will be heated and then hot $CO_2$ will be extracted from the production well. Figure 6 shows the reservoir temperature field. The initial conditions are similar to the base case, but the plume characteristics of its temperature distribution are different. The temperature distribution in the base case presents obvious thermal hysteresis as compared to the $CO_2$-only case. This is due to the fact that the $CO_2$ density and viscosity is lower than that of water and the buoyancy effect is more obvious. Compared to the $CO_2$-only case system, the change span of the temperature field reduces to approximately 50% during the same simulation period as the water system. This phenomenon is caused by the favorable properties of $CO_2$ such as lower density and viscosity. So the supercritical $CO_2$ has more advantages as a heat transfer fluid than water.

## *4.2 Production Behaviour*

The heat extraction (HE) rate is the most important inspection parameter to measure the heat producing capability of the production well. It is calculated as:

HE=F (h-hinj) (Pruess 2008). Here, F is the mass flow of production well, (kg/s); h is the specific enthalpy of heat medium extracted from production well, (J/kg); hinj is the specific enthalpy of heat medium injected at T = 20 °C, (J/kg).

Under the same conditions, the heat extraction rate of $CO_2$-only case is higher. The heat extraction rate decreases from the initial 95.6 MW to 41.7 MW in the first 5 years, and then slowly reduces to 15.2 MW after 40 years as shown in Figure 7. For comparison, the heat extraction performance of the base case system was also simulated under the same temperature and pressure conditions and reservoir parameters. The heat extraction rate in the base case decreased from the initial 31.0 MW to 23.7 MW at 5 years and then slowly reduces to 8.2 MW after 40 years. The heat extraction rate for $CO_2$ is significantly increased and is nearly 200% larger than water in a sandstone reservoir. The mass flow of the base case is also smaller than $CO_2$-only case system as illustrated in Figure 5 so the production behavior of the $CO_2$-only case system is better than the base case system.

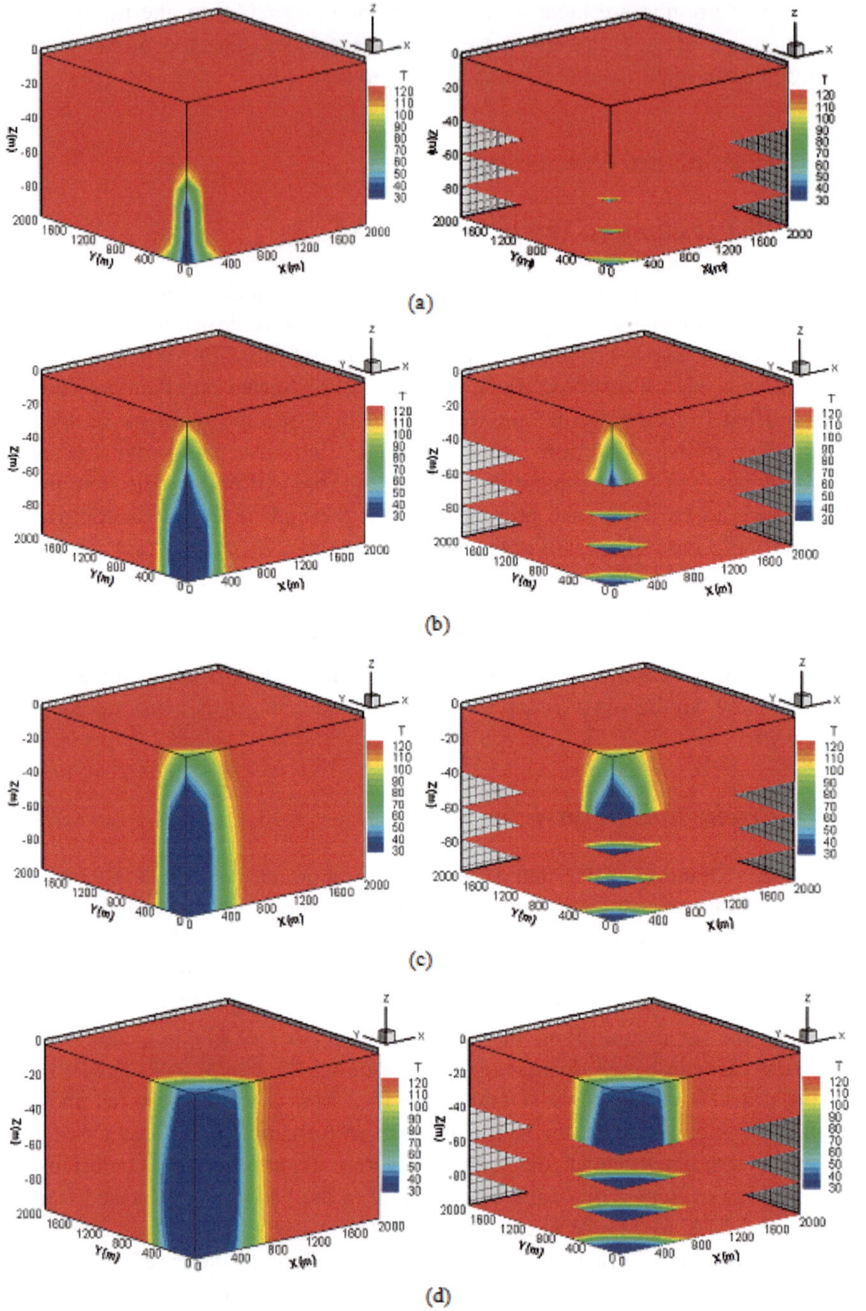


**Fig. 5** temperature distribution in reservoir for Water-only case (a) t=1 year, (b) t=5 years, (c) t=10 years, (d) t=40 years

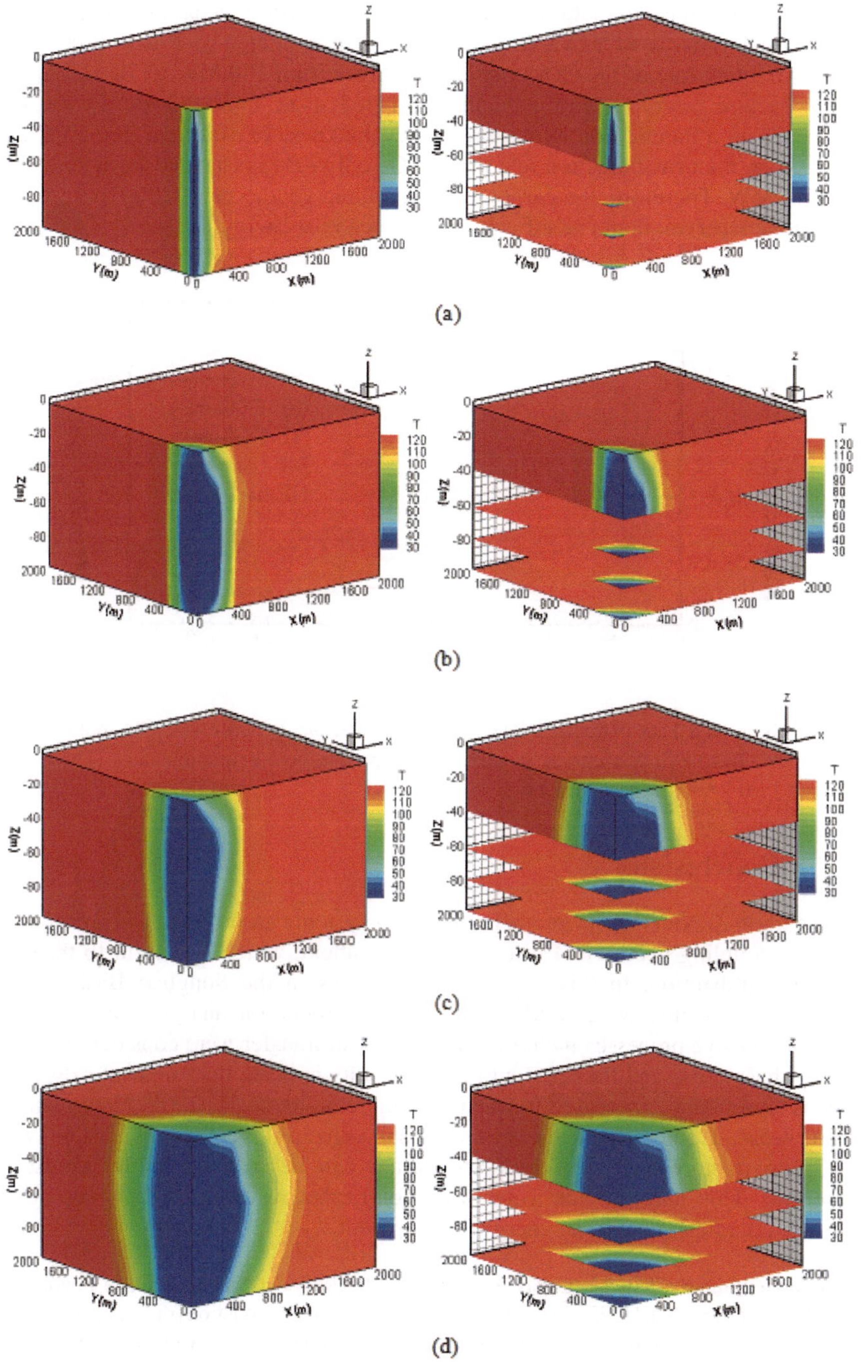


**Fig. 6** Temperature distribution in reservoir for $CO_2$-only case (a) t=1 year, (b) t=5 years, (c) t=10 years, (d) t=40 years

Comparing the simulation results to the production behavior of enhanced geothermal systems with $CO_2$ as working fluid investigated by Pruess (2008), the heat extraction rate of the $CO_2$-only case changes from 100 MW to 70 MW in the EGS fracture reservoir with a thickness of 305 m at T=200°C and P=200 bar. However, in a natural porous reservoir with thickness of 100 m at T=120°C and P=200 bar, the heat extraction rate of the $CO_2$-only case is changed from 95.6 MW to 15.2 MW. Therefore, the carbon dioxide plume geothermal (CPG) system in a sandstone reservoir could be efficient to produce heat energy.

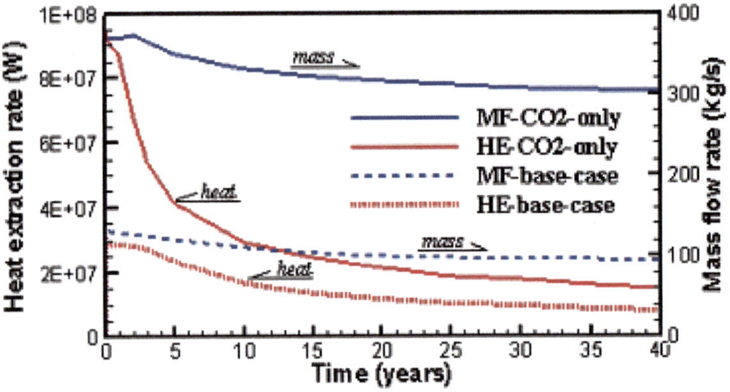


**Fig. 7** Simulated heat extraction rate for base (water –only) case and $CO_2$-only case at P=250bar, T=120°C. The curves labeled "HE" are the heat extraction rates. The curves labeled "MF" are the mass flow rate.

## 5 Conclusions

The carbon dioxide plume geothermal (CPG) system can promote $CO_2$ geological storage coupling geothermal resources development, which is helpful to the $CO_2$ resource utilization. In this paper, the conditions in the Songliao Basin in the northeast of China were used for numerical simulation analyses of the CPG. Comprehensive processes including flow and heat transfer were considered for the water (base) case and the $CO_2$-only case. Results indicate that the heat extraction rate is significantly increased for the $CO_2$-only case by nearly 200% more than the water case. Compared to the conventional enhanced geothermal system with artificial fractures, efficient heat production can be realized in the sandstone reservoir. This may provide a new area for further research on $CO_2$ geological storage coupling geothermal energy development.

**Acknowledgments.** This work was supported by the National Natural Science Foundation of China (Grant No.41272254), The Ph. D. Programs Foundation of the Ministry of Education of China (Grant No.20110061110057), Graduate Innovation Fund of Jilin University (No.20121069), and the Doctoral Interdisciplinary Scientific Research Project of Jilin University (No. 2012DC014).

## References


1. Holloway, S.: Storage of fossil fuel-derived carbon dioxide beneath the surface of the Earth. Annual Review Energy Environment 26, 145–166 (2001), doi:10.1146/annurev.energy.26.1.145
2. West, J.M., Pearce, J., Bentham, M., Maul, P.: Issue profile: environmental issues and the geological storage of $CO_2$. European Environment 15, 250–259 (2005), doi:10.1002/eet.388
3. Wigley, T.M.L., Richels, R., Edmonds, J.A.: Economic and environmental choices in the stabilization of atmospheric $CO_2$ concentrations. Nature 379, 240–243 (1996)
4. Gentzis, T.: Subsurface sequestration of carbon dioxide-an overview from an Alberta (Canada) perspective. International Journal of Coal Geology 43(1-4), 287–305 (2000), doi:10.1016/S0166-5162(99)00064-6
5. Gough, C.: State of the art in carbon dioxide capture and storage in the UK: An experts' review. International Journal of Greenhouse Gas Control 2, 155–168 (2008), doi:10.1016/S1750-5836(07)00073-4
6. Holloway, S.: Underground sequestration of carbon dioxide - a viable greenhouse gas mitigation option. Energy 30, 2318–2333 (2005), doi:10.1016/j.energy.2003.10.023
7. Gunter, W.D., Perkins, E.H., Hutcheon, I.: Aquifer disposal of acid gases: modeling of water-rock reactions for trapping of acid wastes. Applied Geochemistry 15, 1085–1095 (2000), doi:10.1016/S0883-2927(99)00111-0
8. Hitchon, B., Gunter, W.D., Gentzis, T., Bailey, R.T.: Sedimentary basins and greenhouse gases: a serendipitous association. Energy Conversion and Management 40, 825–843 (1999), doi:10.1016/S0196-8904(98)00146-0
9. Izgec, O., Demiral, B., Bertin, H., Akin, S.: $CO_2$ injection into saline carbonate aquifer formation. I. laboratory investigation. Transport in Porous Media 72, 1–24 (2008), doi:10.1007/s11242-007-9132-5
10. Xu, Z., Chen, D., Zeng, R.: The leakage risk assessment and remediation options of $CO_2$ geological storage. Geolo. Gical. Review 54(3), 373–386 (2008)
11. Randolpha, J., Saar, M.: Coupling carbon dioxide sequestration with geothermal energy capture in naturally permeable, porous geologic formations: Implications for CO2 sequestration. Energy Procedia 4, 2206–2213 (2011)
12. Wolf, K.H.A.A., Willemsen, A., Bakker, T.W., Wever, A.K. T., Gilding, D.T.: The development of a multi-purpose geothermal site in urban area. In: Proceedings of the SPE-70th EAGE Conference and Exhibition, Rome, Italy, vol. 2, pp. 1018–1102 (2008)
13. Brown, D.: A hot dry rock geothermal energy concept utilizing supercritical $CO_2$ instead of water. In: Proceedings of the 25th Workshop on Geothermal Reservoir Engineering, pp. 233–238. Stanford University (2000)
14. Pruess, K.: Enhanced geothermal systems (EGS) using $CO_2$ as working fluid – a novel approach for generating renewable energy with simultaneous sequestration of carbon. Geothermics 35, 351–367 (2006)
15. Pruess, K.: On production behavior of enhanced geothermal systems with $CO_2$ as working fluid. Energy Conversion and Management 49(6), 1446–1454 (2008)
16. Xu, T., Apps, A.J., Pruess, K.: Numerical studies of fluid-rock interactions in Enhanced Geothermal Systems (EGS) with $CO_2$ as working fluid. In: Proceedings of 33rd Workshop on Geothermal Reservoir Engineering, January 28-30, Stanford University, California (2008)

17. Xu, T., Apps, J.A., Pruess, K.: Numerical simulation to study mineral trapping for $CO_2$ disposal in deep aquifers. Appl. Geochem. 19, 917–936 (2004)
18. Xu, T., Apps, J.A., Pruess, K.: Mineral sequestration of carbon dioxide in a sandstone-shale system. Chemical Geology 217(3-4), 295–318 (2005), doi:10.1016/j.chemgeo.2004.12.015
19. Xu, T., Pruess, K.: Reactive transport modeling to study fluid-rock interactions in enhanced geothermal systems (EGS) with $CO_2$ as working fluid. Conference Paper Presented at the World Geothermal Congress, Bali, Indonesia, April 25-29 (2010)
20. Glanz, J.: Deep in bedrock, clean energy and quake fears, New York Times (June 2009)
21. Randolph, J.B., Martin, O.: Coupling carbon dioxide sequestration with geothermal energy capture in naturally permeable, porous geologic formations: Implications for $CO_2$ sequestration. Energy Procedia 4, 2206–2213 (2011)
22. Hou, Q., Feng, Z., Feng, Z.: Continental petroleum geology of Songliao Basin, vol. 3, pp. 277–283. Petroleum industry press, Beijing (2009)
23. Compile group of Petroleum geology of Jilin oilfield records, Chinese petroleum geological records (Volume 2)-Daqing, Jilin oilfield, vol. 3. Petroleum industry press, Beijing (1993)
24. Zhang, S., Cui, K., Zhang, C.: Controlling factors and distribution patterns of lithologic pools in the fluvial facies of the 3rd and 4th members of the Quantou Formation in the Songliao Basin. Oil & Gas Geology 32(3), 411–419 (2011)
25. Gao, R., Chai, X.: The forming factors and distribution rules of giant oil and gas fields in Songliao basin, vol. 1, pp. 246–255. Petroleum industry press, Beijing (1997)
26. Lou, Z., Cai, X., Gao, R.: The fluid history and analysis on generation of oil & gas reservoir in Songliao Basin, pp. 69–90. Guizhou Science & Technology Publishing House, Guiyang (1998)
27. Chi, Y., Yun, J., Meng, Q.: The deep structure and dynamics and oil accumulation in Songliao Basin, vol. 6, pp. 236–239. Petroleum industry press, Beijing (2002)
28. Zhai, Z., Shi, S., Zhu, H.: Discussion on utilization of oil field production water type geothermal resource take Daqing Oil field as an example. Journal of Natural Resources 26(3), 382–387 (2011)
29. Wu, Q.: The relationship between the geothermal field and the formation migration and accumulation of oil in Songliao Basin. Acta Petrolei. Sinica 11(1), 9–15 (1990)
30. Wu, Q.: The geothermal field in Songliao Basin. Journal of Seismological Research 14(1), 31–39 (1991)
31. Ren, Z., Xiao, D., Chi, Y.: Restoration of the palaeogeotherm in Songliao Basin. P G O D D 20(1), 13–14 (2001)
32. Tan, S., Shi, Y., Zhao, Y.: The formation and prospective evaluation of geothermal resources in the Songliao Basin. World Geology 20(2), 155–160 (2001)
33. Li, Z.: Evolutionary features of mantle heat flux in Songliao Basin. Geotectonica et Metallogenia 19(2), 104–112 (1995)
34. Zhou, Q., Feng, Z., Men, G.: The study on nowadays geothermal characteristics and the relationship of the gas generation in Xujiaweizi fault depression in northern Songliao Basin, vol. 37, pp. 177–188. Science in China Press (2007)
35. Xu, T., Nicolas, E.S., Pruess, K.: TOUGHREACT-a simulation program for non-isothermal multiphase reactive geochemical transport in variably saturated geologic media: Applications to geothermal injectivity and $CO_2$ geological sequestration. Computers & Geosciences 32, 145–165 (2006)

36. Gelder, A.V.: Efficient computation of polygon area and polyhedron. Graphics Gem., 4th edn. Academic Press (1995)
37. Yang, Y., Xu, T., Wang, F.: Toughvisual: a user-friendly pre-processing and post-processing graphical interface for TOUGHREACT. In: Proceedings TOUGH Symposium 2012, Lawrence Berkeley National Laboratory, Berkeley, California, September 17-19 (2012)
38. Vinsome, K.W., Westerveld, J.: A simple method for predicting cap and base rock heat losses in thermal reservoir simulators. Journal of Canadian Petroleum Technology 19(3), 87–90 (1980)

# Identification of Growth Barriers for Exploitation of Geothermal Energy in China

Mingxing Bai[1,2], Parimal A. Patil[2], and Catalin Teodoriu[2]

[1] Northeast Petroleum University, 163318, Daqing, China
[2] Institute of Petroleum Engineering, Clausthal University of Technology, 38678, Clausthal-Zellerfeld, Germany

**Abstract.** Renewable resources play an important role in the world nowadays and geothermal energy is one of them. It is known to be one of the clean and reliable sources of energy. Although the importance of geothermal energy was realized a long time ago, its exploitation is still far behind due to certain reasons which are emphasized in this paper. This paper gives an overview of the development of geothermal energy worldwide, identifies the barriers which hinder its development, proposes ways to overcome the barriers and finally gives some recommendations for faster exploitation of geothermal energy in China.

**Keywords:** geothermal energy, technical barrier, financial barrier, environmental impact, geothermal energy in China.

## 1 Introduction

As a result of rising demand for fossil energy worldwide and the high emission of $CO_2$ and other greenhouse gases, it is becoming increasingly important to utilize alternative forms of energy to meet the dramatic energy demand around the world. Among these alternative forms of energy, geothermal energy has a special position. Compared to other renewable energy resources, geothermal energy is extensively and permanently available, clean, and reliable (Erdlac 2006, Li et al. 2007, Zhang et al. 2009). Therefore exploitation of geothermal energy has been given more and more attention in the past decade. The global capacity of installed geothermal power plants is currently around 11,224 MW (Jennejohn et al. 2012). This accounts for approximately 50% of the installed electrical power output in the "new" renewable energy sector. It has been estimated by the International Energy Agency, that geothermal electricity generation could reach 1,400 TWh per year by 2050 which will reduce $CO_2$ emission by almost 800 megatons (Mt) per year (International Energy Agency 2010). However, there are still plenty of obstacles to overcome before this goal is reached, for instance, technical and financial barriers, government policies, public acceptance, etc. This paper aims to discuss the experiences of geothermal energy exploitation in the world in terms of these barriers, and give recommendations for the geothermal development in

M.Z. Hou et al. (Eds.): *Clean Energy Systems in the Subsurface*, SSGG, pp. 105–116.
DOI: 10.1007/978-3-642-37849-2_9 

China. For this, a prediction of total geothermal energy production worldwide until 2050 has been carried out considering three scenarios, viz., minimum growth without subvention, optimistic growth, and maximum growth. A detailed description of geothermal energy development in China is given, the challenges are identified and the instruments to overcome the challenges are proposed.

## 2 Exploitation of Geothermal Energy Worldwide

Geothermal activities generally fall into two categories: deep geothermal energy and shallow geothermal energy. Heat stored at a depth of 0 – 400 m is categorized as shallow geothermal energy. Shallow geothermal energy is perfect for cooling and heating buildings. Geothermal heat extracted from a depth in excess of 400 m is categorized as deep geothermal energy (DGE). It can either be used directly for heating purposes or to supply the energy for electricity generation. The heat can be extracted by Hot–Dry–Rock (HDR) process or by drilling into hydro-geothermal formation.

In 1904, the world's first commercial geothermal power plant was built in Italy. Several countries were soon to follow their example. In 1919 the first geothermal wells were drilled at Beppu in Japan, followed by wells drilled at The Geysers, California, USA in 1921. In 1958 a small geothermal power plant began operating in New Zealand, in 1959 another began in Mexico, in 1960 in the USA. During the past few years several geothermal projects have been implemented in Germany and many others are still in the process of development, for example, Neustadt–Glewe (250 MW), Landau (2.5–3 MW), Unterhaching (3 MW), Bruchsal (500 kW), Simbach–Braunau (150 kW), GeneSys Project in Hannover etc (Kreuter 2011).

Using 2009 data, Figure 1 shows the electricity generation in China using different fuels. It can be clearly seen from the graph that up to the present day, coal is the dominant resource for the generation of electricity in China. Hydro thermal energy comes next in play for producing power. Although oil and gas are the dominant players in the energy market, only small amounts are used for producing power. The renewable energy contribution to power generation is around 17% with the contribution from geothermal energy being very minor. The primary resources e.g., coal, oil and gas, are declining. Thus, the exploitation of renewable energies is becoming necessary.

The total growth of geothermal energy till the end of 2010 is shown in Figure 2. After 1970s, a reasonable amount of geothermal energy was produced. It can be seen from Figure 2 that until 2000 there was a steady growth but after 2000 the growth reduced. There may be a number of reasons for this, but it is clear that due to a subvention around 2002, the growth suddenly stimulated. If similar behavior can be expected after 2010, China will have a capacity of more than 16 GW up to 2020 which is the maximum growth that can be expected. The optimistic growth will give an installed capacity in 2050 of around 16 GW. The trend in the development of geothermal energy in past decades suggests that its share of the energy market will increase steeply in future.

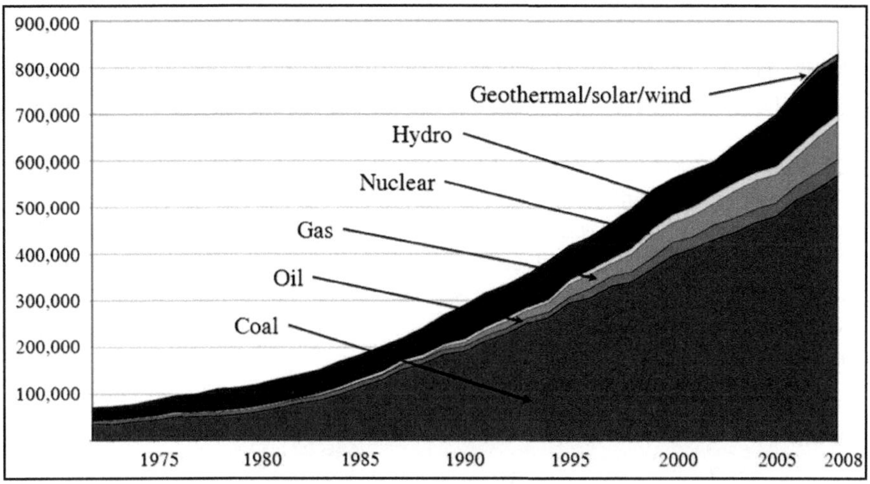


**Fig. 1** Electricity generation (GWh) by fuel using 2008 data (IEA,2010)

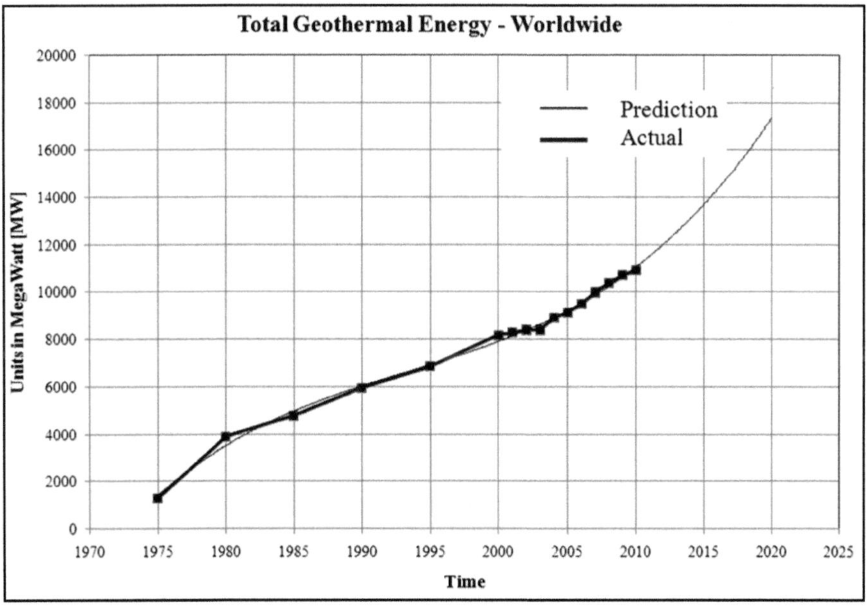


**Fig. 2** Total geothermal energy production worldwide and its prognosis till 2020 (IEA 2010)

## 3 Geothermal Energy in Europe and Development Barriers Identification

The geothermal energy distribution in Europe is shown in Figure 3. The areas with high enthalpy include Iceland, Italy, France, Greece, and Germany. The areas with low enthalpy include Ireland, Norway, Sweden and UK. Many countries in the European Union (EU) have realized the importance and challenges of deep geothermal energy exploitation. Countries with favorable geothermal conditions, e.g., the circum-pacific countries, Iceland, and Italy, have been exploiting geothermal energy for many decades. The EU plays a central role in promoting energy development from renewable sources, which has contributed to sustained growth in the geothermal sector until today, both for direct district heating, and for electricity production (Holm et al. 2011).

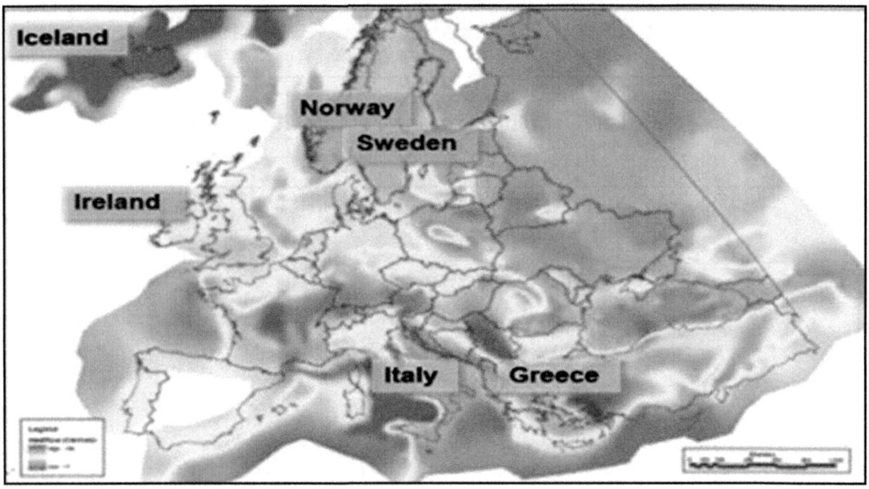


**Fig. 3** Geothermal resources in Europe (Jones and Ligtenberg 2008)

The European Investment Bank (EIB) has been instrumental in pursuing European geothermal objectives, especially in funding EGS research and development. The EU – funded GEOFAR (Geothermal Finance and Awareness in European Regions) project, which was launched by 8 partners from five European countries Germany, Greece, France, Spain and Portugal, to service the development and promote the financing of geothermal projects as part of the Intelligent Energy Europe (IEE) program. The European Geothermal Energy Council (EGEC) has set a target to install a total of 5,000 MW capacity of electricity power generation from geothermal sources over Europe by 2020, increasing it to 15,000 MW by 2030. The goal for the geothermal sector is to contribute 5% of the total energy production in Europe by 2030 (Holm et al. 2011, Berrill 2009, Wendel and Hiegl 2010). Table 1 shows the market potential up to 2015.

**Table 1** Market potential till 2015 (e.terras AG 2010)

| Country | Location | Capacity | Investment |
|---|---|---|---|
| Germany | Rhine valley | > 500 | 2000 Mio. € |
| | Bavaria | > 150 MWe | 675 Mio. € |
| Europe | Greece | > 500 MWe | 1500 Mio. € |
| | Turkey | > 1000 MWe | 3000 Mio. € |
| | Hungary | > 500 MWe | 1000 Mio. € |
| Worldwide | | > 8000 MWe | 16-32 Mrd. $ |
| | 25% low temp. | > 2000 MWe | 4-8 Mrd. $ |
| | 75% steam | > 6000 MWe | 12-24 Mrd. $ |

Geothermal energy has so many advantages over other renewable and non–renewable energy sources.; However, there are several barriers which hinder its wider use. These are summarized as technical barriers, financial barriers, environmental impact and public acceptance.

## 4 Technical Barriers

The typical technical barriers to recover geothermal energy mainly include:

- Lengthy process for applying for permits
- High chances of failure during exploitation
- High cost of geothermal wells
- Limited availability of drilling rigs
- Lack of skilled professionals
- Inadequate Research and Development support compared to other renewable, fossil fuel & nuclear energy sources
- Lack of industrial standards to define resources and associated risks

Geothermal projects are usually divided into five different phases, i.e., resource study, exploration, pre-feasibility, feasibility, and design & construction. Before actual drilling starts, legal procedure such as acquiring permits needs to be followed. During this phase, the lack of clear process to gain access to the land and rights to explore and tap the underground heat not only make the process complex but also lengthy lasting for several years (Ministry of Economic Development 2010, Falcone & Teodoriu 2008).

High chances of failure are associated with drilling the first well due to inadequate data on potential geothermal resources. Reduction in drilling costs can have a very positive impact on geothermal development as drilling of geothermal wells is a major factor in project costs, and can consume as much as 40% – 60% of total project cost. High drilling costs are due to drilling hard rock at high temperatures. Figure 4 shows the actual and predicted well costs as a function of

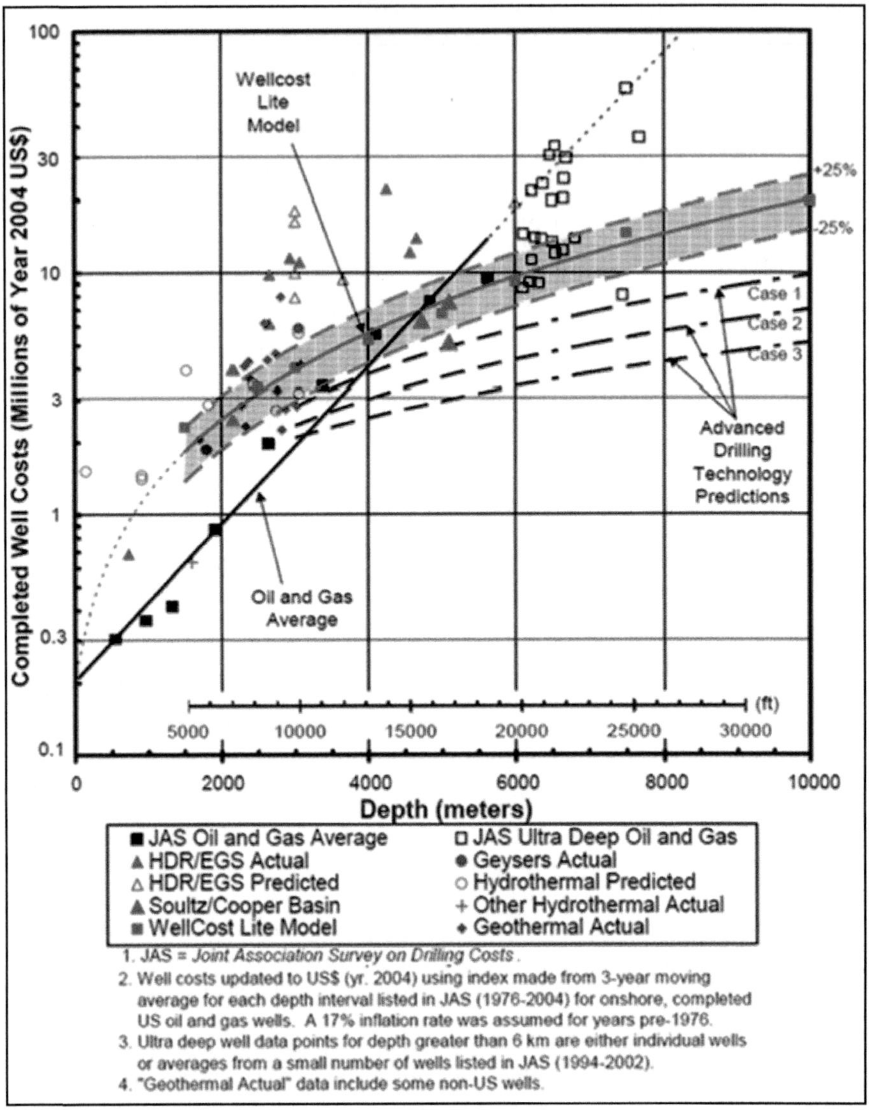


**Fig. 4** Predicted and actual drilling costs versus depth (after IEA, 2010)

well depth. It can be clearly seen that new drilling technologies need to be incorporated to lower the current well cost. Therefore, it is very difficult to transfer knowledge and expertise from oil and gas projects to operate in a technically challenging field.

One possible action to address the above problems is to increase co–operation between government, academia and industry. Government must play a key role by providing continued financial support with clear sets of legal procedures.

There should be a common platform to discuss critical issues throughout the development of any geothermal project. This will give an advantage to personnel to develop their expertise and accordingly modify available rigs.

## 5 Financial Barriers

The financial barriers which exist during the development of geothermal project can be described as follows:

- High CAPEX and long payback time
- Competition with natural gas
- Relatively long lead time from concept to production
- Uncertainty /volatile tax incentive

One of the main financial hurdles is the high Capital Expenditure (CAPEX) and long payback period. Most of the Capital Expenditure is spent on the wellbore drilling and power plant construction. These expenses occur during the initial stage of the project; hence, they are also called up-front costs. The cash flow prediction is based mainly on uncertain parameters which make a feasible project looks like a non–feasible one. The most important factors impacting on the high uncertainty and risk include (a) resource grade, (b) temperature gradient, (c) geological parameters, (d) total well cost, (e) total investment cost, etc. The less important impacting factors yet still important include (a) power plant design and total cost, (b) stimulation cost, (c) well configurations (singlet, duplet, multilateral wells), (d) operating cost, (e) flow rate, etc. To overcome the problem of high CAPEX, many European countries, e.g., France, Spain, Germany, have tried to use public subsidies and venture loans. On the other hand, the long payback time, compared to other types of energy, e.g., natural gas makes it risky for a high investment in geothermal projects. Many investors are, therefore, unwilling to invest. As such, bank facilities will be used to cover the high investment costs. The weakness of bank facilities is that it is difficult to provide loans in the early stage of exploration without an insurance mechanism.

Another typical financial barrier is the low outcome of the exploitation. For this, a feed-in tariff policy and tax reduction is commonly used. Tax reduction can on one hand promote increased capital investment, but on the other hand, will affect mainly the operational phase when revenues are generated. Likewise, the feed-in tariff is good because it secures income over a long-term period, but it also has disadvantages because it comes into action at a later stage of the project. People need to be aware of the fact that, the cost of producing electricity will be lower over time because the price and availability of the fuel is stable and predictable, which is also advantageous. A more detailed description of the financial instruments can be found in the reference (Wendel & Hiegl 2010).

## 6 Environmental Impacts and Public Acceptance

Besides the technical and financial barriers, the environmental impacts of geothermal recovery also play a vital role, due to the fact that this is closely related to public acceptance of the project. The potential environmental impacts include:

Gaseous and solids emission

Water and noise pollution

Induced seismicity

The typical adverse gases discharged during the production of geothermal energy mainly include $CO_2$, $H_2S$, and methane etc. In many countries $CO_2$ and $H_2S$ are strictly prohibited. During the process of drilling, stimulation and production, the liquid always has high TDS (Total Dissolved Solid), especially for high temperature reservoir. Some of the solids are poisonous, e.g., arsenic and boron. They pose a potential damage to the ground waters and vegetation. Besides, these operations can also generate high noise which affects the lives of local residents (MIT 2006).

## 7 Geothermal Outlook in China and Development Challenges

Central eastern China sedimentary basin holds about 49.17 billion cubic meters of underground hot water resources with a potential energy equivalent to 1.854 billion tons of common coal which is a safe, sustainable, clean, low carbon, renewable resource. Most of the geothermal resources are mainly concentrated in Tibet, Sichuan Province, and Yunnan Province, which are along the Himalayan geothermal belt. Figure 5 shows the distribution of high and low geothermal temperatures in China. Figure 6 shows China's share of total primary energy supply in 2009. It clearly shows that coal, oil and natural gas dominate the energy market and geothermal only accounts for a small portion of China's energy system (IEA Energy Statistics 2009).

Table 2 shows the information on China's energy consumption from the Energy Information Administration in USA. The table shows that renewable energy has a small share of the total energy. However, it has a rapid growth with an annual average increase of 6.4 percent in the BAU (Business-as-usual) case, 3.3 percent in the low-growth case and 4.1 percent in the high-growth case. The Chinese Academy of Sciences has estimated that installed geothermal electric power generation capacity has the potential to reach 70 GW by 2050, and if new technology is used, installed capacity could even double (Hexun News 2011). If coal based power generation was replaced with geothermal power generation then carbon dioxide emissions could be reduced by 1 billion tons per year by 2050, but if replaced by natural gas power generation emissions could be reduced by 500 million tons per year.

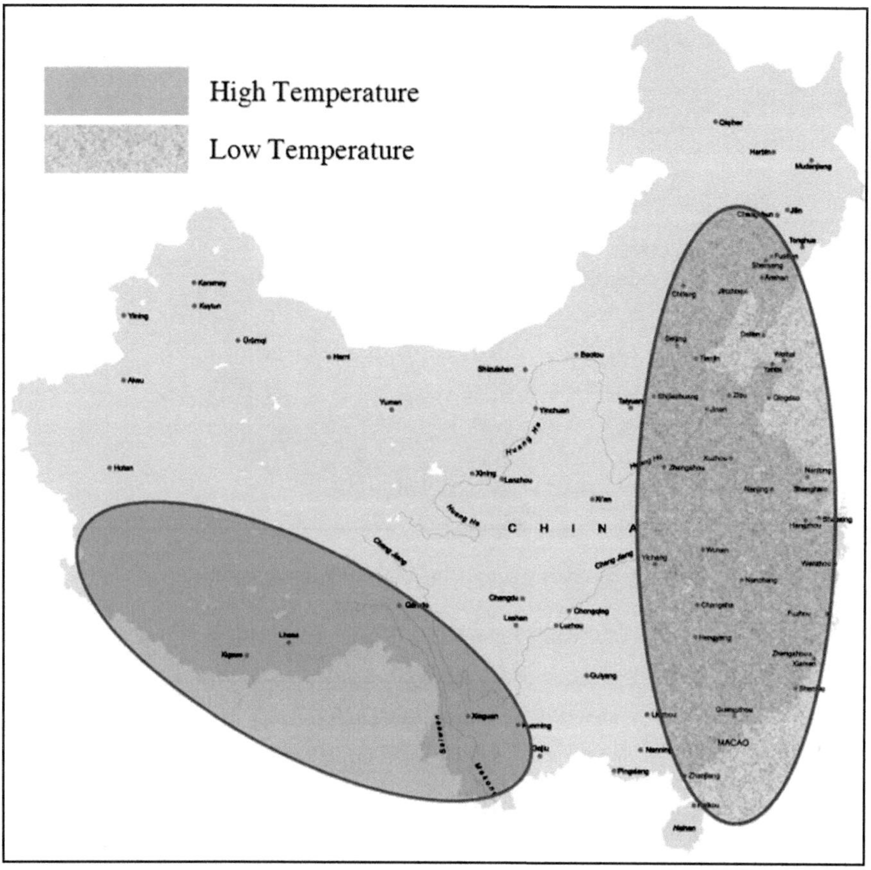


**Fig. 5** Geothermal energy distribution in China (After Hexun News, 2011)

According to an article from Hexun News (2011), although China began developing and exploiting geothermal resources in the 1970s, and full scale development of geothermal resources began early, it still faces many problems today. First of them is the bottleneck in geothermal technology application. Most of China's geothermal resources fall under the moderate to low temperature categories (ca. 50–150 °C). Most high temperature (> ca. 150 °C) geothermal resources are concentrated in areas like Yangbajing in Tibet and Tengchong in the west of Yunan province. China needs to develop its technology for low and moderate temperatures in order to reduce its dependency on foreign technology. Secondly, China has not done much technological research on the exploitation of geothermal resource, this call for analyses and better understanding of the geothermal resources and their capacities. Thirdly, there is a lack of manufacturing and application standards for geothermal heat pumps which is resulting in failed attempts to reach energy conservation and emission reduction

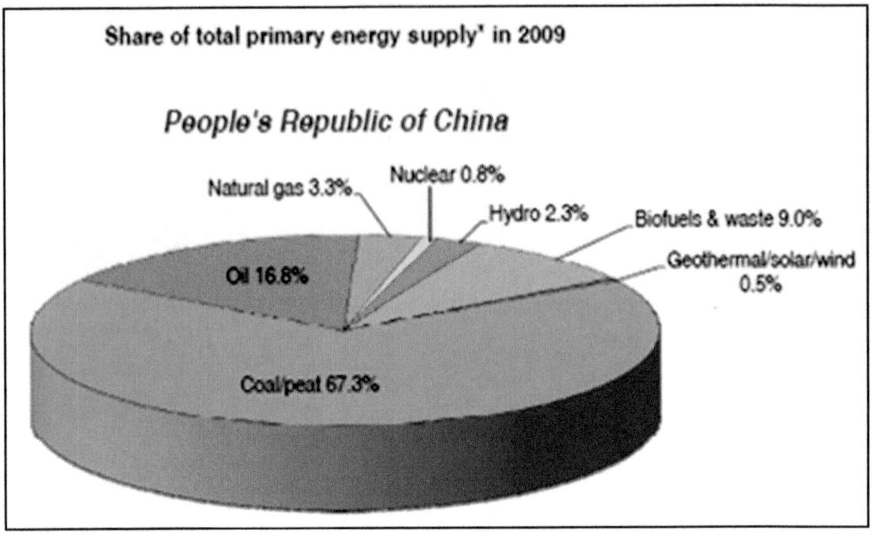


**Fig. 6** Share of total primary energy supply in 2009 (IEA Energy Statistics 2009)

**Table 2** EIA Projections of China´s Energy Use (Wang et al. 2004)

| | Energy consumption 2025 | | | Annual Growth 2000 - 2025 | | |
|---|---|---|---|---|---|---|
| Fuel | Low | BAU | High | Low | BAU | High |
| Oil, Mbpd | 7.4 | 10.9 | 12.9 | 1.7 % | 3.3% | 4.0% |
| Electricity, TWh | 2,418 | 3,596 | 4,289 | 2.6 % | 4.3% | 5.1% |
| Gas, Tcf | 4.2 | 6.1 | 7.5 | 6.2 % | 7.9% | 8.8% |
| Coal, 1,799 Mt | 1,799 | 2,917 | 3,476 | 1.1 % | 3.2% | 3.9% |
| Nuclear, TWh | 139 | 154 | 178 | 9.2 % | 9.7% | 10.4% |
| Renewable, Quad | 5.9 | 6.4 | 7.2 | 3.3 % | 6.4% | 4.1% |
| Total Energy, Mtoe | 1,520 | 2,288 | 2,708 | 1.8 % | 3.5% | 4.2% |

requirements. This is unfavorable for the promotion of large scale development of the geothermal market. On the other hand, government subsidies and management systems are not adequate. There is a need to follow strict supervision rules during construction of geothermal projects and an improper, incompetent management system is leading to inefficient use of the subsidies available for energy conservation. Finally, China lacks the required talent in the field of geothermal development. The entire scientific research community in China is not large enough to carry out extensive research.

## 8 Conclusions and Recommendations

Geothermal energy is playing a more and more important role in the energy market. However, there are still some barriers which hinder the faster development of geothermal energy. This paper has discussed the typical barriers including the technical barriers, financial barriers and environmental impacts. The experience in Europe has been introduced in many other parts of the world, proposing many instruments to overcome these barriers.

Since China has a vast distribution of geothermal reserves, a lot needs to be done to speed up the exploitation of geothermal energy and to secure China's growing power demand. A large amount of investment is required throughout the development process, thus financial support is needed to carry out geothermal research and exploration considering the challenges and potential benefits. The Chinese government through its policies is supporting geothermal heating and the heat pump manufacturing industry but more efforts are required for proper structuring of resource development. For instance, a subsidy towards the overall investment cost, a fixed feed–in tariff, reduction of tax, etc. As to the technical barriers, an advanced technology transfer from the traditional petroleum sector to the geothermal sector might encourage a faster development of geothermal technologies (Falcone and Teodoriu 2008, Reinicke et al. 2010). These technologies include the areas of geology and geophysics, drilling and completion, reservoir modeling, formation stimulation and so on. A trend toward an advanced geothermal production technology includes: new drilling technologies making the use of smaller and less complex equipment possible, new drilling processes to increase the rate of penetration and to improve productive drilling time, new technologies to allow for a smaller initial borehole diameter. These areas will merit further research in China.

## References

1. Berrill, P.: Geothermal energy has 'plentiful potential' for Europe. Recharge (2009), http://www.rechargenews.com/energy/geothermal/article174059.ece
2. e.terras AG, Presentation on 'Geothermal basic' (2010) (unpublished)
3. Falcone, G., Teodoriu, C.: Oil and gas expertise for geothermal exploitation: The need for technology transfer. In: SPE 113852, presented at the SPE Europec/EAGE Annual Conference and Exhibition, Rome, Italy (2008)
4. Hexun News, China's Geothermal: overview and current problems (2011), http://deblockconsulting.com/blog/china-news/chinas-geothermal-overview-and-current-problems/ (retrieved)
5. Holm, A., Blodgett, L., Jennejohn, D., Gawell, K.: Geothermal Energy: International Market Update. Geothermal Energy Association (2011)
6. International Energy Agency, Share of total primary energy supply in 2009 (2009), http://www.iea.org/stats/pdf_graphs/CNTPESPI.pdf (retrieved)

7. International Energy Agency (2010) http://www.iea.org/stats/pdf_graphs/CNELEC.pdf (retrieved)
8. Jennejohn, D., Hines, B., Gawell, K., Blodgett, L.: Geothermal: International market overview report. Geothermal Energy Association (2012), http://www.geo-energy.org/pdf/reports/2012-GEA_International_Overview.pdf (retrieved)
9. Jones, G.L., Ligtenberg, H.: Geothermal energy in Europe. In: 3rd International Professional Geology Conference, Flagstaff, Arizona, USA, September 21-24 (2008)
10. Erdlac Jr., R.J., Trentham, R.C., Armour, L., Lee, R., Miller, B.: Adding to the geothermal energy portfolio from deep sedimentary basins. Geothermal Resources Council Transactions 30, 875–883 (2006)
11. Kreuter, H.: Deep Geothermal Project in Germany Status and Future Development. Federal Ministry of Economic and Technology, presented in Paris (May 2011)
12. Li, K., Zhang, L., Ma, Q., Liu, M., Ma, J., Dong, F.: Low temperature geothermal resources at Huabei Oilfield, China. GRC Trans. 31 (2007)
13. Ministry of Economic Development, NewZealand. Geothermal Energy: Summary of emerging technologies and barries to development (2010)
14. Reinicke, K.M., Oppelt, J., Ostermeyer, G.P., Overmeyer, U., Teodoriu, C., Thomas, R.: Enhanced Technology transfer for geothermal exploitation through a new research concept: the geothermal energy and high-performance drilling research program: gebo. In: SPE 134436 presented at the SPE Annual Technical Conference and Exhibition, Florence, Italy (September 2010)
15. US Department of Energy, The future of geothermal energy – The future of impact of enhanced geothermal systems (EGS) on the United States in the 21st century. An assessment by an MIT-led interdisciplinary panel (2006), http://web.mit.edu/mitei/research/studies/documents/geothermal-energy/geothermal-energy-full.pdf
16. Wang, Y., Skeer, J., Mohamad, A.B.H., Doi, N., Kim, J.H., Bonou, M., Fuerte, D.A., Huang, C.H., Lin, C.Y., Tantiwisarn, V.: Energy in China: transportation electric power and fuel markets. Asia Pacific Energy Research Center (2004), http://www.ieej.or.jp/aperc/pdf/CHINA_COMBINED_DRAFT.pdf
17. Wendel, M., Hiegl, M.: GEOFAR-Financing geothermal energy in European regions. In: Proceedings World Geothermal Congress 2010 (2010)
18. Zhang, L., Liu, M., Li, K.: Estimation of geothermal reserves in oil and gas reservoirs. In: SPE 120031 presented at the SPE Western Regional Meeting, San Jose, California, USA (March 2009)

# Numerical Analysis of Parameters Affecting Hydraulic Fracture Re-orientation in Tight Gas Reservoirs

Jonas Wegner, Birger Hagemann, and Leonhard Ganzer

Institute of Petroleum Engineering, Clausthal University of Technology,
Agricolastraße 10, 38678 Clausthal-Zellerfeld, Germany

**Abstract.** In tight gas formations where the low matrix permeability prevents successful and economic production rates, hydraulic fracturing is required to produce a well at economic rates. The initial fracture opens in the direction of minimum stress and propagates in the direction of maximum stress. As production from the well and its initial fractures declines, re-fracturing treatments are required to accelerate recovery. The orientation of the following hydraulic fracture depends on the actual stress-state of the formation in the vicinity of the wellbore. Previous investigations by Elbel and Mack [1] demonstrated that the stress alters during depletion and a stress reversal region appears. This behavior causes a different fracture orientation of the re-fracturing operation.

For the investigation of re-fracture orientation a two-dimensional reservoir model has been designed using the software package COMSOL Multiphysics. The model represents a fractured tight gas reservoir of infinite thickness. A coupled simulation of fluid flow and geomechanics is realized by the use of Biot's theory of poroelasticity.

The simulation shows that the poroelastic behavior develops an elliptical shaped stress reversal region around the fracture, if the difference between minimum and maximum horizontal stresses is small. The time dependent analysis indicates that the dimension of the region initially extends quickly until it reaches its maximum. Subsequently, the stress reversal region shrinks slowly until it finally disappears. The reservoir characteristics influence the dimension and time development of the stress reversal region in this process. Different case studies shown in this work illustrate the sensitivity of various parameters to this behavior and the orientation of the resulting fractures.

The success of re-orientated fractures depends on the actual dimension of the stress reversal region. Therefore the work aims to help predict the optimum timing of re-fracturing in order to maximize production. Further, the paper will show numerical investigations for vertical, horizontal and multifractured wells. The findings will support effective field development in tight gas reservoirs.

**Keywords:** tight gas, re-fracturing, stress-state, stress altering, fracture orientation.

M.Z. Hou et al. (Eds.): *Clean Energy Systems in the Subsurface*, SSGG, pp. 117–130.
DOI: 10.1007/978-3-642-37849-2_10 

# 1 Introduction

Tight gas reservoirs are characterized by very low matrix permeability. The German Society for Petroleum and Coal Science and Technology (DGMK) defined tight gas reservoirs as having an average effective gas permeability below 0.6 mD. Porosities are commonly below 10 %. The drainage area of tight gas wells would be very small in the absence of fractures. The contribution of hydraulic fractures is to expand the drainage area to an elliptical shape around the fracture and make economic production possible [2]. As production from the well and its initial fracture declines, re-fracturing treatments are often required to accelerate recovery. In some cases, the second fracture treatment does not re-open the initial fracture, but produces an additional fracture which has a different orientation compared to the first one [2]. This effect is called re-orientation of hydraulic fractures and was confirmed by surface tiltmeter measurements [3].

The re-orientation can be explained by an altered stress distribution in the formation. Elbel and Mack [1] demonstrated that stress altering can be induced by production from the reservoir. The theory relies on small initial differences between the maximum and minimum horizontal stresses, which often appear in tight gas reservoirs [4]. The behavior in the near wellbore region, which depends highly on the borehole stresses, well perforations and fluid injection, is not considered in this paper. Therefore, it will be assumed that the fracture and re-fractures start to propagate dependent on the global stress state.

The creation of a re-oriented fracture is very useful, because it will connect the well to a less depleted region of the reservoir. The occurrence and the optimum timing of re-fracture treatments are not yet well understood.

# 2 Fracture Re-orientation Concept

The initial fracture treatment creates a fracture which propagates into the formation, perpendicular to the minimum effective stress [5]. This behavior is illustrated in Fig. 1, where the initial minimum effective stress is aligned with the y-axis and the initial fracture develops parallel to the x-axis. During depletion of the reservoir the stress distribution will be altered due to the change in pore pressure. This behavior is based on the Biot's concept wherefore the effective stress has to be taken into account [2]. The effective stress is equal to the total stress minus Biot's coefficient multiplied by pore pressure:

$$\sigma_{eff} = \sigma - \alpha * p \tag{1}$$

This equation denotes that the stress alters equally in all directions and would only be valid if the rock is prevented from moving. Consideration of the strain effects indicates that the stress reduction is higher in the direction parallel to the initial fracture [6]. Therefore it is possible that the initial small horizontal stress differences can be overcome. A region of stress reversal arises around the fracture which is illustrated by the grey ellipse in Fig. 1 [1]. As production continues, the area influenced by the induced stress difference spreads out [1]. The magnitude of induced stress differences at the fracture first increases and later decreases because of the propagation of tension [1]. Therefore the right point in time for the re-fracture treatment has to be identified. A re-fracturing treatment produces a fracture which starts to propagate into the formation perpendicular to the initial fracture [1], if a region of stress reversal is present. At the isotropic stress point (Fig. 1), where the stresses in x- and y-direction are equal, the fracture initiates a turn in direction until it propagates parallel to the initial fracture [1].

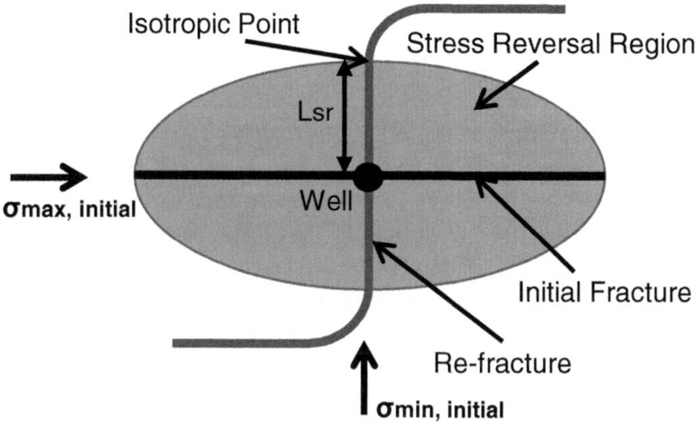


**Fig. 1** Schematic showing idealized orthogonal fracture propagation (Redrawn after [6])

An extending region of stress reversal results in an increasing distance of propagation in the perpendicular direction [6]. Therefore investigations on the sensitivity of model parameters were performed to understand and predict this behavior.

## 3 Coupled Fluid Flow and Geomechanics Modeling

The software package COMSOL Multiphysics® was used for the generation of the reservoir model. COMSOL allows the simulation of physical processes, which can be described by differential equations. Different physics interfaces are available for the simulation of common processes, such as transport phenomena, electromagnetic field theory and solid mechanics, but the definition of customized differential equations is also possible. The solution of equations is based on the finite elements method. Therefore the strength of COMSOL is the ability to couple different physical processes.

This ability was used in this work to couple the simulation of fluid flow with rock mechanics. Therefore an available physics interface was used, which is based on Biot's linear theory of poroelasticity. The mathematical formulation describes the porous medium as a solid rock framework, which behaves in a linear elastic way and a freely moving fluid in the pore space, which is described by Darcy's law [7]. These two systems interact using two mechanisms: An increase of pore pressure results in a distension of the rock and a compression of the rock results in an increasing pore pressure if the fluid cannot exit the rock [7]. Mathematically the coupling is described by two constitutive equations.

The first constitutive equation relates to the solid matrix. It describes the behavior between stress and strain [8]:

$$\sigma = C\varepsilon - \alpha pI \tag{2}$$

where σ is the total stress, C is the elasticity matrix, ε is the strain, α is the Biot's coefficient, p is the pore pressure and I is the unit matrix. The elasticity matrix C is a function of poroelastic material constants. In this work Young's modulus E and Poisson's ratio v are used to define the elasticity. These two properties suffice to fully describe the elastic behavior in all directions.

The second constitutive equation is related to the pore fluid and describes the relation of incremental fluid content to volumetric strain and pore pressure [8]:

$$\partial\xi = \alpha\varepsilon_{vol}\frac{\partial p}{M} \tag{3}$$

where ζ is the fluid content, $\varepsilon_{vol}$ is the volumetric strain and M is the Biot modulus. The total formulation results in a linear relationship between stress, strain, fluid content increment and fluid pressure [9].

## 4 Single Fractured Vertical Well Model

A reservoir model is implemented in two dimensions which represent a reservoir of infinite thickness. It has a rectangular shape with a dimension of 1600 m in x-direction and 1200 m in y-direction. The size is selected after the findings of Roussel and Sharma [10]. They determined that the boundaries have no impact on the size of the stress reversal region, if the distance from the fracture to the boundaries is at least three times the fracture half-length [10]. The permeability is 0.01 mD and the porosity is 10 % over the total area. The temperature is assumed to be constant at 110 °C. The initial fracture has a half-length of 200 m and is shaped as a straight line in x-direction. The reservoir geometry is illustrated in Fig. 2, where the well has only a symbolic meaning and is not discretized in the model.

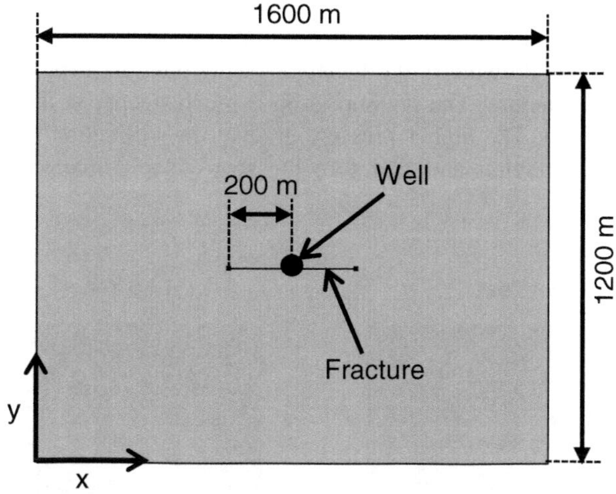


**Fig. 2** Reservoir geometry

## *4.1 Initial and Boundary Conditions*

The initial pressure for Darcy's law is defined as 300 bar. The four outer boundaries of the rectangular model area are defined without a cross flow. The inner boundary for the fluid flow is set to the fracture with a constant pressure of 50 bar. This condition relates to a constant bottom hole flowing pressure and a connected infinite conductivity fracture.

For the solid mechanical part of the model the in-situ stress distribution is defined as the outer boundaries of the reservoir. The maximum horizontal stress acts in x-direction relative to the initial fracture distribution. Therefore a stress of 39.5 MPa is applied to the left and right edge of the reservoir. The minimum horizontal stress is set at the top and bottom edge with a value of 38.5 MPa. All stresses act inwards to generate a compressed reservoir condition. To complete the boundary conditions a so called "roller" is defined to the fracture. This boundary condition originates from a roller bearing, where a movement in one direction is prevented while a movement in the perpendicular direction is possible. In this case the fracture cannot move in the y-direction but a movement in the x-direction is enabled.

## *4.2 Simulation Results*

In Fig. 3 the fluid pressure is shown after one and five years of production with a constant bottom hole flowing pressure. The Brown color represents the initial pressure of 300 bar and dark blue indicates the well pressure. It can be seen that the drainage area spreads to form an elliptical shape around the fracture. Later the reservoir boundaries are also influenced by the pressure drop. The essential finding from this drawing is the behavior of the pressure drop in the area above and below the fracture. The pressure drop is much steeper in the y-direction than in the x-direction. The higher pressure gradient in y-direction results in a higher strain in this direction and therefore the stress also remains higher in the y-direction.

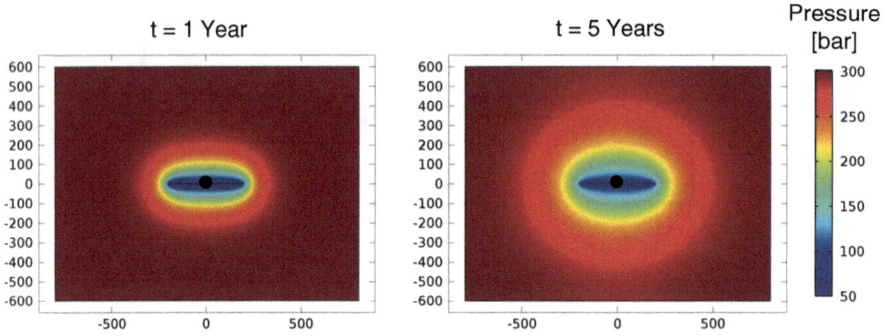


**Fig. 3** Fluid pressure distribution after one and five years

In Fig. 4 the maximum principal stress direction is plotted at the initial situation and after five years of production.

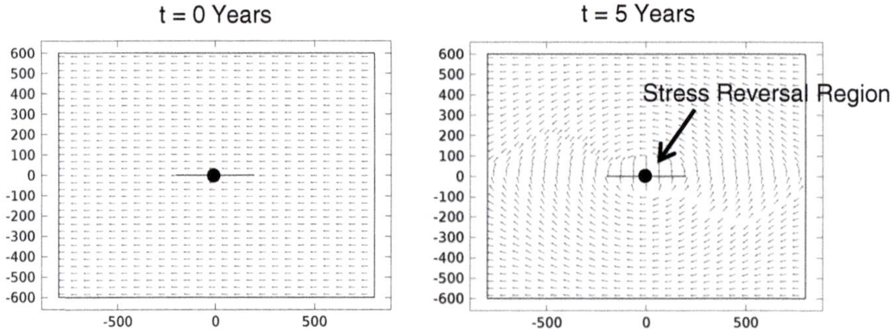


**Fig. 4** Direction of maximum principal stress initially and after five years

Initially the maximum principal stress is directed parallel to the x-axis because of the uniform definition of in-situ stresses. After five years the principal stress directions have changed. In the region around the fracture the direction of maximum principal stress is rotated by 90° due to the poroelastic effects. The stress lines bypass this region to balance the stress-state.

# 5 Re-fractured Vertical Well Model

For the investigation of a re-fracture treatment the simulation of the previous case was stopped after ten years and the stress-state used. At this point in time a re-fracture was added to the reservoir model. According to the size of the stress reversal region, this re-fracture propagates 88 m in a perpendicular direction to the initial fracture. At the isotropic point it starts to turn its direction and finally propagates parallel to the initial fracture. The direction of turn is thereby chosen randomly. The bottom hole flowing pressure of 50 bar is assigned to the re-fractured case as in the model before and the simulation continues for a further ten years.

## *5.1 Simulation Results*

The pressure distribution and maximum principle stress direction are presented in Fig. 5.

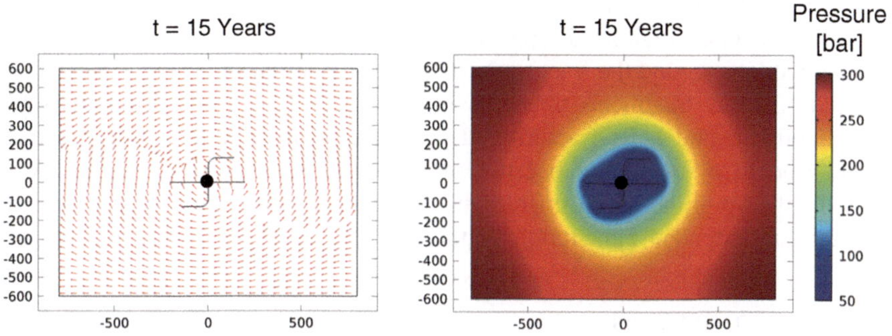


**Fig. 5** Pressure distribution and maximum principle stress direction after 15 years

The pressure distribution plotted indicates that the drainage area extends to the region around the re-fracture. The new drainage area remains shaped similar to an ellipse but the principle axis is rotated by about 30°. This behavior is reflected in the alteration of the stress-state. The maximum principle stress direction in the stress reversal region is rotated by additional 30°. Based on this behavior the orientation of a second re-fracture is again different. The second re-fracture would propagate parallel to the new maximum principle stress direction in the region which did not have a fracture until now. At the isotropic point it starts a 90° turn.

# 6 Horizontal Well Model

The geometry of the reservoir model was changed for the investigation of fractured horizontal wells. The model remains two-dimensional and represents a horizontal layer in the reservoir. The horizontal part of the well is added by a line parallel to the y-axis with a length of 800 m. Three transverse fractures with a half-length of 200 m are adjoined. The direction of propagation is related to the maximum horizontal stress which is parallel to the x-axis. The fractures are uniformly distributed with a spacing of 200 m. The total size of the reservoir model, selected using the Roussel and Sharma statement, is 1600 x 2000 m [10]. A simultaneous fracturing is assumed for the simulation implying all fractures start production at the same time. All other reservoir parameters and the boundary conditions remain unchanged.

## 6.1 Simulation Results

The behavior of pressure gradients reflects the change in the maximum principle stress directions. Initially the perpendicular flow to the fractures causes the occurrence of small stress reversal regions around each half-fracture (Fig. 7, left). The depletion of the regions between the fractures results in a change of flow regime after some months. As the drainage area spreads out to form an ellipse around the horizontal well, the maximum principle stress directions turn back and the stress reversal regions disappear. After five years the maximum principle stress directions around the fractures are equal to the initial state (Fig.7, right).

For re-fracture treatments three different positions have to be distinguished in this simplified model. Re-fracturing at the same position as the initial fractures would lead to a re-opening because of the unchanged stress directions. After attaining the tip of the initial fracture the direction of propagation should turn and therefore an extended penetration of the reservoir is limited (cf. Fig. 7, right).

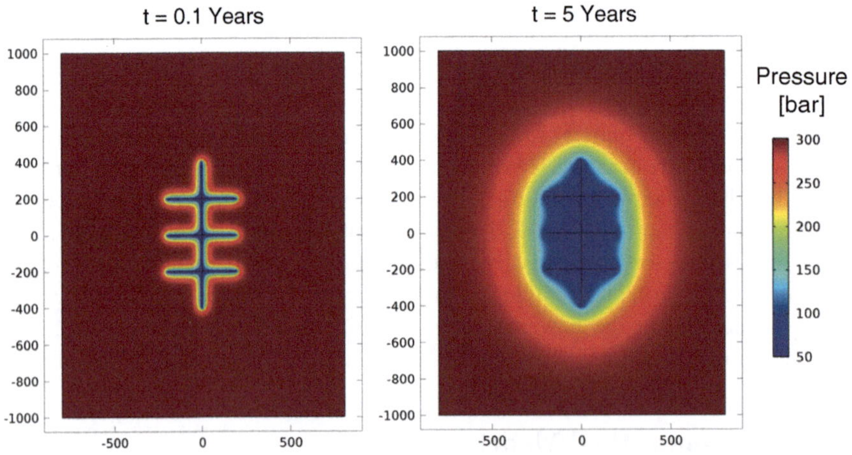


**Fig. 6** Pressure distribution after 0.1 (left) and 5 years (right)

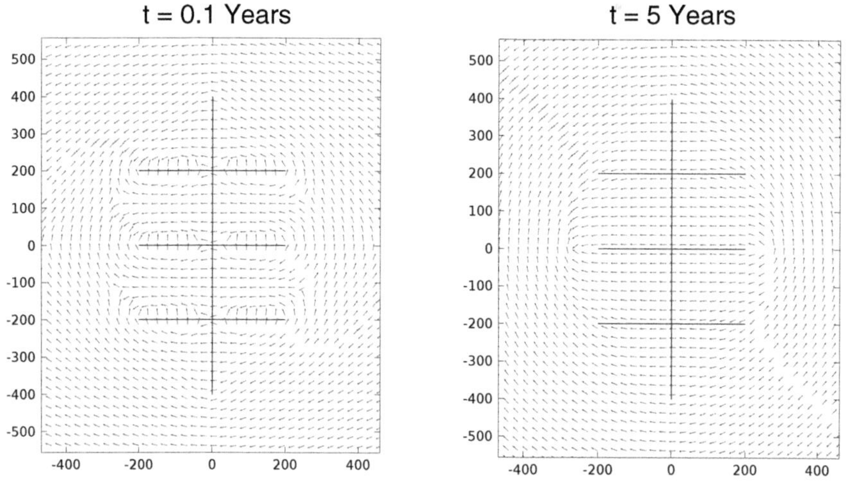


**Fig. 7** Maximum principle stress direction after 0.1 (left) and 5 years (right)

The second possibility is re-fracturing between the initial fractures. Thereby a new fracture would be created parallel to the initial fractures. However, any improvement in recovery is small because the regions between the fractures are already depleted. The last option is a re-fracture treatment at the origin or end-point of the horizontal well. The re-fracture would start to propagate perpendicular to the horizontal well. As the re-fracture spreads further the direction is slightly deflected towards the initial fractures because of the distortion of stress directions in this region. A slightly curved re-fracture would arise in this case.

# 7 Sensitivity Studies

The influences of different parameters to the dimension and timing of the stress reversal region are analyzed in this section. For the investigation the single fractured vertical well model is used and the base case model presented in section 4 is modified with different values for permeability, fracture half-length and horizontal stresses. For each case the depletion is simulated for 100 years and the distance from the well to the isotropic point is determined at several points in time. This distance is called $L_{sr}$ (cf. Fig. 1). Plots of this value versus time show the influence of each parameter.

## *7.1 Actual Value of Permeability*

The following plot shows the distance to the isotropic point as a function of time for three different reservoir permeabilities: 0.1 mD, 0.01 mD (base case) and 0.001 mD (Fig. 8).

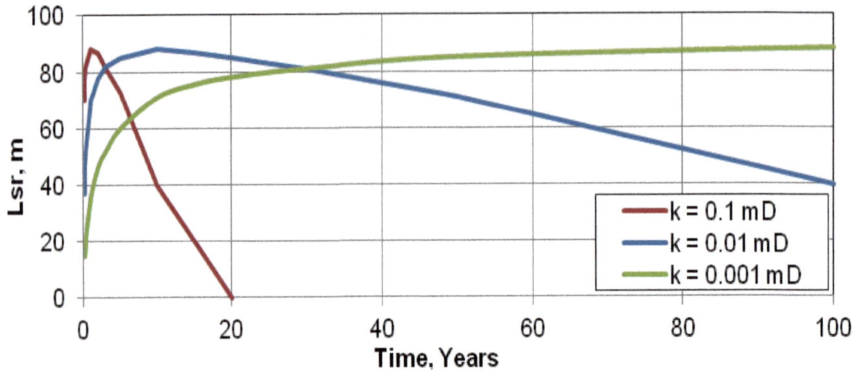


**Fig. 8** Influence of actual value of permeability

For the base case with permeability of 0.01 mD the distance to the isotropic point increases very quickly at the beginning. The maximum distance is attained after ten years with a value of 88 m. The subsequent decline is slow.

Observation of the study with ten times greater permeability shows similar behavior with a different timing. The maximum distance is equal, but the subsequent decline is faster and the stress reversal region disappears early.

The third study has ten times smaller permeability and the distance to the isotropic point increases much slower. The maximum value has not been achieved after100 years of production.

With respect to the stress reversal region the ideal times for a re-fracture treatment can be determined from Fig. 8 above. For the three analyzed cases the ideal times are after 2, 10 and 100 years.

## 7.2 *Permeability Anisotropy*

Two cases with anisotropic permeability are simulated and compared to the base case with k=0.01 mD. The first case has ten times greater permeability in the x-direction and the second case has ten times greater permeability in the y-direction. In Fig. 9 distances versus time to the isotropic point are plotted.

The green line represents an increased permeability in the x-direction. The curve developed is steeper than the base case at the beginning and a maximum distance of 102 m is achieved after 15 years. Therefore the maximum distance is increased and moved with time compared to the base case.

The red line represents an increased permeability in the y-direction. The maximum value is already achieved after 2.5 months. Compared to the base case this value is smaller with a value of 73 m. The subsequent decline is very fast and the stress reversal region almost disappears after two years.

The study indicates that a re-fracture treatment is very efficient for cases with greater permeability parallel to the initial fracture, whilst a greater permeability perpendicular to the initial fracture degrades the re-fracture treatment efficiency.

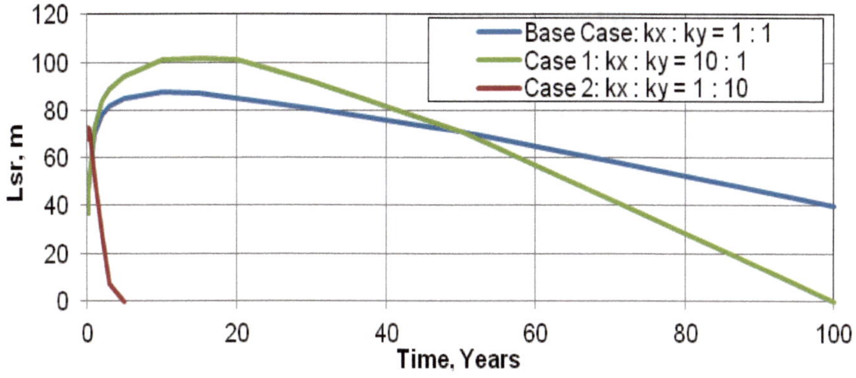


**Fig. 9** Influence of Anisotropic Permeability

## *7.3 Fracture Half-Length*

In Fig. 10 the distances versus time to the isotropic point are plotted for three different fracture half-lengths: 100 m, 200 m (base case) and 300 m.

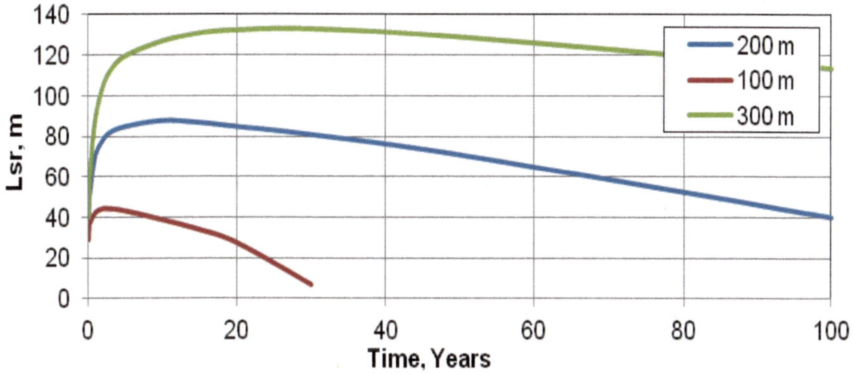


**Fig. 10** Influence of fracture half-length

For the case with a reduced fracture half-length the distance to the isotropic point is smaller and attains its maximum after two years with a value of 44.5 m.

The case with an increased fracture half-length shows a larger maximum distance to the isotropic point with a value of 133 m after 30 years.

A comparison of the three cases allows two conclusions: The maximum distance to the isotropic point increases in proportion to an increasing fracture half-length and the point in time, where the maximum value is reached, defers with an increasing fracture half-length.

### *7.4 Difference between Minimum and Maximum Horizontal Stress*

Two cases with different initial stress-states are compared to the base case in Fig. 11. Thereby the difference between minimum and maximum horizontal stress is defined with 0.1 MPa, 1 MPa (base case) and 2 MPa.

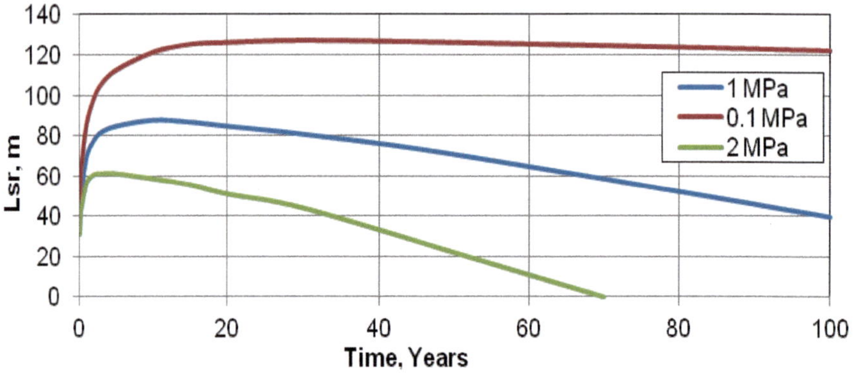


**Fig. 11** Influence of difference between minimum and maximum horizontal stress

The first case has a very small difference between the horizontal stresses. The distance to the isotropic point increases above 120 m. in the first years. The maximum is attained after 30 years and the subsequent decrease is very slow.

The second case has a higher difference between the horizontal stresses. Initially the distance to the isotropic point extends fast but its maximum is lower with a value of 60 m. The subsequent decline proceeds faster than in the base case and the stress reversal region disappears after 70 years.

It can be concluded that the occurrence of a stress reversal region strongly depends on the difference between the minimum and maximum horizontal stress. The maximum distance to the isotropic point decreases as the difference between the horizontal stresses increases.

## 8 Conclusions

- COMSOL Multiphysics enables the coupled simulation of fluid flow and rock mechanics.
- The simulation shows that an elliptically shaped stress reversal region arises, if the difference between minimum and maximum stress is small. The reservoir characteristics influence the size and time frame of this process.
- Based on the presented simplified model the optimum time for re-fracturing treatment can be predicted. Therefore the maximum extension of stress reversal region can be taken into account.

- The dimension of the stress reversal region initially extends quickly and after reaching its maximum it shrinks slowly.
- The actual value of the permeability influences the time, but not the maximum dimension of the development of the stress reversal region. The time frame of the stress reversal increases with decreasing permeability.
- Permeability anisotropy has a strong influence on both, the maximum dimension and the time development of the stress reversal region. A greater permeability parallel to the initial fracture enhances the efficiency of the re-fracture treatment.
- The maximum extension of the stress reversal region increases proportionally with an increasing fracture half-length.
- The initial difference between minimum and maximum horizontal stresses has a strong influence on the appearance of a stress reversal region. The size of the stress reversal region decreases with increasing stress difference.
- Re-fractured vertical wells show a further rotation of the maximum principle stress direction. Therefore the second re-fracture propagates in a new direction.
- The occurrence of stress reversal regions around fractured horizontal wells is very short-lived. Re-fractures would propagate into the same direction.

## 9 Nomenclature

| Symbol | Unit | Description |
|---|---|---|
| $\alpha$ | - | Biot's Coefficient |
| $\varepsilon$ | - | Strain |
| $\varepsilon_{vol}$ | - | Volumetric Strain |
| $\zeta$ | $m^3$ | Fluid content |
| $\sigma$ | Pa | Total Stress |
| $\sigma_{eff}$ | Pa | Effective Minimum Horizontal Stress |
| C | Pa | Elasticity Matrix |
| I | - | Unit Matrix |
| M | Pa | Biot Modulus |
| p | Pa | Pore Pressure |

## References

1. Elbel, J.L., Mack, M.G.: Refracturing: Observations and Theories. In: SPE 25464, Production Operations Symposium, Oklahoma City, USA, March 21-23 (1993)
2. Benedict, D.S., Miskimins, J.L.: Analysis of Reserve Recovery Potential from Hydraulic Fracture Reorientation in Tight Gas Lenticular Reservoirs. In: SPE 119355, SPE Hydraulic Fracturing Technology Conference, Woodlands, Texas, USA, January 19-21 (2009)

3. Weng, X., Siebrits, E.: Effect of Production induced Stress Field on Refracture Propagation and Pressure Response. In: SPE 106043, SPE Hydraulic Fracturing Technology Conference, College Station, Texas, USA, January 29-31 (2007)
4. Aghighi, M.A., Rahman, S.S., Rahman, M.M.: Effect of Formation Stress Distribution on Hydraulic Fracture Reorientation in Tight Gas Sands. In: SPE 122723, SPE Asia Pacific Oil and Gas Conference an Exhibition, Jakarta, Indonesia, August 4-6 (2009)
5. Li, P.: Theoretical Study on Reorientation Mechanism of Hydraulic Fractures. In: SPE 105724, Shanghai University (2007)
6. Siebrits, E., Elbel, J.L.: Parameters Affecting Azimuth and Length of a Secondary Fracture During a Refracture Treatment. In: SPE 48928, SPE Annual Technical Conference and Exhibition, New Orleans, USA, September 27-30 (1998)
7. Detournay, E., Cheng, A.H.D.: Fundamentals of Poroelasticity, Reprint of Chapter 5 in Comprehensive Rock Engineering: Principles, Practice and Projects. Analysis and Design Method, vol. II. Pergamon Press (1993)
8. COMSOL AB, Earth Science Module User's Guide, © COPYRIGHT (1998-2008)
9. Reuben, O.: Multiphase modeling of tissue growth in dynamic culture conditions. Thesis for the Degree of Doctor of Philosophy, University of Nottingham (2007)
10. Roussel, N.P., Sharma, M.M.: Role of Stress Reorientation in the Success of Refracture Treatments in Tight Gas Sands. In: SPE 124491, SPE Annual Technical Conference and Exhibition, Florence, Italy, September 19-22 (2010)

# Numerical Analysis of Polymer Micro-model Flooding Experiments

Jonas Wegner and Leonhard Ganzer

Institute of Petroleum Engineering, Clausthal University of Technology, Agricolastraße 10, 38678 Clausthal-Zellerfeld, Germany

**Abstract.** Adding polymers to the injection water leads to an increase in water viscosity together with a reduction in water permeability. As a result, the mobility ratio improves leading to a more efficient displacement process and a higher oil recovery factor. Different physical and chemical processes accompany the flow of aqueous polymer solutions in porous media resulting in the loss of polymer solution viscosity. The numerical simulation of polymer EOR in porous media relies on mathematical models that have already been proposed in literature. However, direct comparison between numerical simulations and experimental data from flood experiments is limited. Although important parameters can be obtained from core experiments, the local distribution of static and dynamic fluid transport parameters remains largely unknown. 2D micromodels such as silicon edged pore networks are an alternative as they provide visual access to the flooding process and enable a detailed quantification of the local distribution of relevant transport parameters. In this paper, a micromodel simulator that considers polymer related physical and chemical effects such as non-Newtonian rheology of the displacing phase, permeability reduction, retention and salinity effects is presented using comprehensive mathematical models. For the solution of transport system equations the software Matlab® by MathWorks, Inc. is used as it provides a flexible framework for implementing the underlying transport and auxiliary equations. A pattern generator code is used to define the 2D micromodel structure, generating predefined porosity and permeability averages with local heterogeneities. These generated patterns lead to the micromodels used in this work.

Results obtained from micromodel flooding experiments will be compared to results from numerical simulation in order to test the ability of existing mathematical models to reproduce dominating physico-chemical effects.

The work aims to provide and support selection criteria for optimizing polymer EOR by predicting polymer performance for a variety of critical input parameters such as polymer selection, polymer concentration, salinity and target viscosity for various average reservoir qualities (k, φ).

**Keywords:** micromodel, polymer flooding, simulation, enhanced oil recovery.

M.Z. Hou et al. (Eds.): *Clean Energy Systems in the Subsurface*, SSGG, pp. 131–142.
DOI: 10.1007/978-3-642-37849-2_11 

# 1 Introduction

The industry standards used to investigate polymer flow in porous media and thus to determine important parameters relevant to polymer EOR projects are core flood experiments. Although important parameters can be gathered from these experiments, the cores remain a “black box” to some extent. However, the distribution of static and dynamic flow properties throughout the flooding process contains valuable information relevant for the design of polymer EOR projects. Compared to cores, micromodels such as silicon edged pore-networks enable visual access to the flooding process. Existing mathematical models can be implemented into a simulator to reproduce different physico-chemical effects observed during experiments.

# 2 Micromodel Design and Experimental Set-Up

## *2.1 Micromodel Design – “Pattern Generator”*

Prior to building silicon edged micromodels, a digital representation of the desired 2D pore-structure in DXF file format is required. The “Pattern Generator” is a computer program capable of generating artificial pore-networks at a high level of complexity in the required format. Because of its extensibility and the large number of available modules, a hybrid framework of python and Matlab® [1] is used for the development of the “Pattern Generator”. Four different patterns have been created on a quadratic domain (20x20x0.02 mm) as depicted in Figure 1:

a.) A stochastic composition of circles with different radii based on a normal distribution and a fixed porosity.
b.) A pattern consisting of regular-spaced circles with quadrilaterals between them
c.) A pattern consisting of regular-spaced circles of the same size.
d.) A “dual porosity” pattern consisting of diamond-shaped collections representing the matrix and void space between them as well as fracture.

Porosity is determined as the fraction of the void area over the total area. The results are presented in Table 1. The porosity is between 18 % - 27 % for all structures which is representative of real porous media.

## *2.2 Experimental Set-Up*

A schematic diagram of the experimental flooding set-up used is presented in Figure 2 a).

Polymers are injected with a syringe pump via a multi-port valve into the micromodel holder and into the micromodel. A top view of a micromodel can be found in Figure 2 b). As depicted the micromodel consists of the pore structure, inlet and outlet channels and boreholes. The flooding process is visualized using a fluorescence microscope & camera.

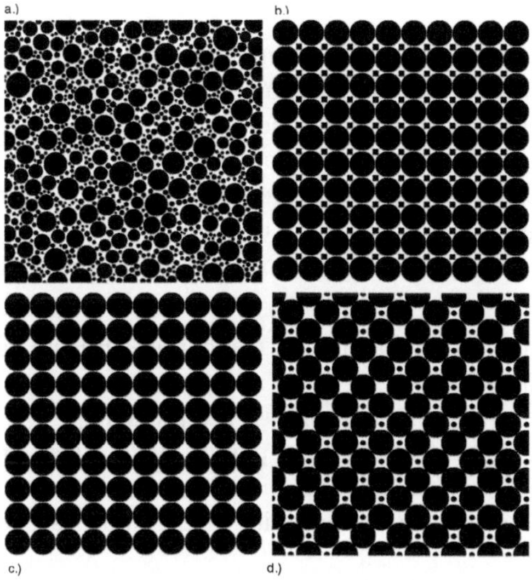


| Structure | Porosity |
|---|---|
| a.) Random | 26.8 % |
| b.) Rectangles and circles | 18.8 % |
| c.) Circles | 22.7 % |
| d.) Diamond | 19.7 % |

**Fig. 1** Four different example pore-networks (a–d) were generated with the "Pattern Generator". The original size of a) is 2x2 mm and (c-d) is 5x5 mm.

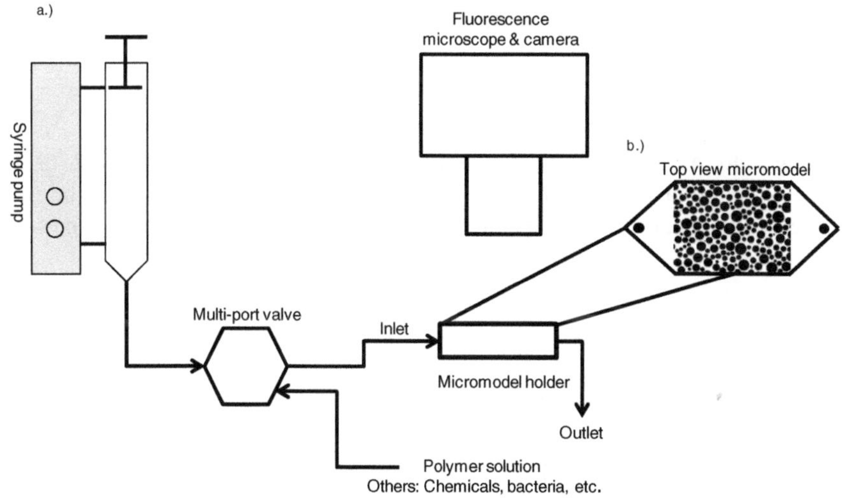


**Fig. 2** a) Schematic diagram of the experimental set-up and b) top view of the micromodel pore-network inlet/outlet channels and boreholes (According to Baumann (2006))

# 3 Micromodel Simulator

A two phase, four-component polymer EOR model is implemented into Matlab® [1] to simulate the displacement of oil by aqueous polymer solutions in micromodels. The oleic phase consists of a single component oil, while the aqueous phase contains the components water, polymer and salt. Although several complex physico-chemical processes influence the flow of aqueous polymer solutions in porous media under in-situ reservoir conditions, the following assumptions and modifications are madefor the micromodel experiments considered here:

a.) The model is two-dimensional, the properties of the porous structure are homogeneous, and gravity is neglected.
b.) Experiments are conducted at atmospheric temperature and pressure.
c.) Hence, the oelic and aqueous phases as well as the porous media edged onto the micromodel are assumed to be incompressible.
d.) Polymer adsorption reduces the relative permeability of the aqueous phase only.
e.) A Generalization of Darcy's law is applicable to multiphase flow.
f.) Multicomponent dispersion is neglected.
g.) Salt is not adsorbed onto the solid surface, but has an impact on the viscosity of the aqueous polymer solution.
h.) Salt, polymer and water are fully mixed.

## 3.1 Flow Equations

Taking into account the above assumptions, the simultaneous flow of two immiscible fluid phases in porous media can be described by the mass conservation equation for each phase. For the aqueous phase this is written as:

$$\frac{\partial(\phi S_a)}{\partial t} = \nabla \cdot \left( \frac{k_a}{\mu_a} \nabla p_a \right) + \tilde{q}_a \tag{1}$$

The mass conservation equation for the oelic phase is:

$$\frac{\partial(\phi S_o)}{\partial t} = \nabla \cdot \left( \frac{k_o}{\mu_o} \nabla p_o \right) + \tilde{q}_o, \tag{2}$$

where $p$ , $\mu$ , $S$ , $k$ , $\phi$ are the pressure, viscosity, saturation, effective permeability and porosity of the porous media. The subscripts $o$ and $a$ represent the oelic and aqueous phases, respectively. $\tilde{q}$ is the source/sink term.

The fact that the void space of the porous media is completely filled with the oelic and aqueous phase, leads to the following relation:

$$S_a + S_o = 1 \tag{3}$$

## 3.2 *Transport Equations*

Polymer and salt are converted with the bulk Darcy velocity of the aqueous phase. For both salt and polymer, a transport equation is required. The equation for polymer transport is written as:

$$\frac{\partial(\phi S_a C_p)}{\partial t} + \frac{\partial(A_v C_{pad})}{\partial t} = \nabla \cdot \left( C_p \frac{k_a}{\mu_a} \nabla p_a \right) + \tilde{q}_a C_{pw} \tag{4}$$

The equation for salt transport is written as:

$$\frac{\partial(\phi S_a C_s)}{\partial t} = \nabla \cdot \left( C_s \frac{k_a}{\mu_a} \nabla p_a \right) + \tilde{q}_a C_{sw}, \tag{5}$$

where $C$, $A_v$ are the component concentrations on a specific surface area of the micromoodel. The subscripts $p$, $s$ represent the polymer and salt components, respectively. $C_{pw}$ and $C_{sw}$ are the polymer and salt concentrations at the injection boundary. $C_{pad}$ is the mass of polymer adsorbed per unit mass of rock. Due to adsorption of polymer on the rock-fluid interface, physical parameters such as the rock permeability change. In addition to adsorption, non-Newtonian rheology and salinity effects have an impact on the aqueous phase viscosity. Therefore, additional constitutive equations are required to close and couple the system of equations (1) – (5).

## 3.3 *Cross Couplings*

### 3.3.1 Salinity and Concentration Effects

The viscosity of the polymer solution at "zero" shear rate depends on the salt as well as polymer concentration. A mathematical relation describing the dependence of the aqueous phase viscosity on polymer and salt concentration is the Flory-Huggins equation [3].

$$\mu_a^0 = \mu_w \cdot \left(1 + \left(a_1 C_p + a_2 C_p^2 + a_3 C_p^3\right) \cdot C_s^{Sp}\right) \tag{6}$$

where $a_1$, $a_2$, $a_3$, and $S_p$ are fitting constants. The units of the fitting constants are selected such that the items in the parenthesis become dimensionless.

### 3.3.2 Polymer Adsorption

Polymer adsorption causes a reduction of polymer concentration in the aqueous phase, hence, a reduction in aqueous phase viscosity. The degree of polymer adsorption depends on the type of polymer and rock, but in general it increases with higher polymer concentrations. The relation between the concentration of polymer in the aqueous phase and the concentration of polymer in the adsorbed state is described by the Langmuir isotherm [4].

$$\frac{C_{pad}}{C_{pad}^{\max}} = \frac{bC_p}{1+bC_p}, \tag{7}$$

Where $b$ is a Langmuir constant and $C_{pad}^{\max}$ is the maximum polymer concentration adsorbed to the rock. Adsorption can be considered reversible or irreversible. The Langmuir isotherm presented here is an equilibrium relation and cannot be used directly if the adsorption is irreversible. In such a case, the maximum adsorption must be traced for each time step.

$$C_{pad}^{\max} = \max\left\{(C_{pad})^1, (C_{pad})^2, \ldots, (C_{pad})^n\right\} \tag{8}$$

### 3.3.3 Permeability Reduction Factor

As a result of polymer adsorption on the rock surfaces, the permeability of the aqueous phase is reduced while the permeability of the oelic phase is unaltered. The permeability reduction factor representing the effect of polymer adsorption on the permeability of the aqueous phase is defined as $R_k = k_w / k_a$, where $k_w$ is the water permeability. Further the permeability reduction factor can be expressed as [5]:

$$R_k = 1 + (R_{RF} - 1)\frac{C_{pad}}{C_{pad}^{\max}}, \tag{9}$$

where $R_{RF}$ is the residual resistance factor. It represents the decrease in rock permeability when the maximum amount of polymer is adsorbed [1].

### 3.3.4 Non-Newtonian Rheology

The aqueous polymer solution behaves like a non-Newtonian fluid. Then, the following relation can be defined where shear rate and viscosity apply in the Carreau equation [6].

$$\mu_a - \mu_\infty = (\mu_{a0} - \mu_\infty)\cdot\left[1 + (\lambda\dot{\gamma}_{eq})^\alpha\right]^{(n-1)/\alpha} \tag{10}$$

Where, $\mu_\infty$ and $\mu_a$, are the water and aqueous phase viscosity, respectively. $\lambda$, $\alpha$, $n$ are polymer specific constants and $\dot{\gamma}_{eq}$ is the equivalent shear rate in the porous media. A relation to calculate the apparent shear rate exerted on polymers while flowing through porous media is given by [7]:

$$\dot{\gamma}_{eq} = C\left[\frac{(3n+1)}{4n}\right]^{n/(n-1)} \cdot \left[\frac{u_w}{\sqrt{\bar{k}k_{rw}S_w\phi / R_{RF}}}\right] \tag{11}$$

Where, $C$ is a constant, $\bar{k}$ is an average permeability and $u_w$ is the water velocity.

### 3.3.5 Saturation, Capillary Pressure and Relative Permeability

In order to solve the two-phase flow problem (Eqs. 1 to 3), three additional constitutive equations need to be specified. Due to the curvature and surface tension of the interface between the phases, the pressure in the wetting phase is less than in the non-wetting phase. The pressure difference is defined by the capillary pressure $p_c$, which is expressed as [8]:

$$p_c(S_a) = p_o - p_a \tag{13}$$

In this model, we used the following capillary pressure model [8]:

$$p_c = p_t \cdot S_n^{-\frac{1}{\varphi}} \quad , \tag{14}$$

where $\varphi$, $p_t$ are the Brooks-Corey coefficient and the capillary entry pressure and $S_n$ is the normalized water saturation defined as [8]:

$$S_n = \frac{S_a - S_{ra}}{1 - S_{ra} - S_{ro}} \quad , \tag{15}$$

$S_{ra}$ and $S_{ro}$ are the residual and immobile saturations of the aqueous and oelic phases. The effective permeabilities of water $k_{rw}$ and the oelic phase $k_{ro}$ are calculated using the Brooks-Corey correlation [8].

$$\begin{aligned} k_a &= \frac{k \cdot S_n^{\frac{2+3\varphi}{\varphi}}}{R_k} = \frac{k \cdot k_{rw}}{R_k} \\ k_o &= k \cdot (1 - S_n)^2 \left(1 - S_n^{\frac{2+\varphi}{\varphi}}\right) = k \cdot k_{ro} \end{aligned} \tag{16}$$

# 4 Model Parameterization and Fluid Properties

## 4.1 Micromodel Geometry

Figure 3 shows a microscopic image of the regular circled structure depicted in Figure 1 c) edged onto a silicon wafer. The size of the pore structure is 20 x 20 mm. Inlet/outlet boreholes appear black. Two fans are used to transport fluids from the inlet/outlet borehole to the porous structure. This approach flow concept was selected to establish linear flow conditions which are favorable for the Measurement of single phase flow properties of the structure. The porosity of the pore structure obtained from the pattern generator is 22.

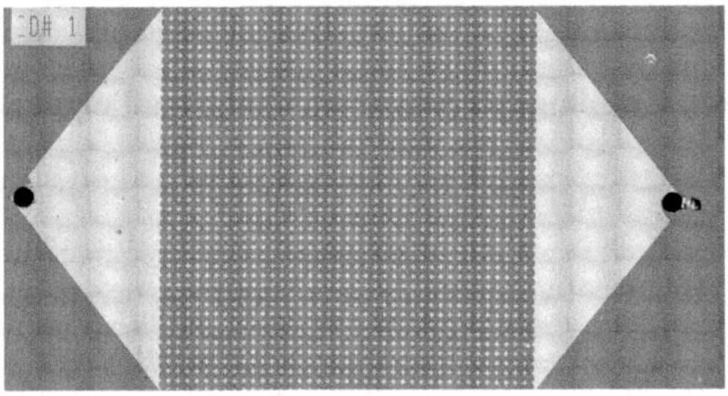

**Fig. 3** Top view of the micromodel used in the experiment

## 4.2 Micromodel Permeability

The permeability of the micromodel is determined by the following procedure. First, the micromodel is flooded with $CO_2$ to displace the remaining air. As all the $CO_2$ is dissolved in the water and 100 % water saturation is established, the pressure differential is measured at different injection rates. Based on Darcy's law the permeability can be calculated. The measurements show micromodel permeability of 480 mD which will be used in the numerical simulations.

## 4.3 Fluid Properties

The viscosity of the water with 1 % NaCl used at 25 °C is 1.1 cP and is constant for different shear rates. An aqueous polymer solution is a non-Newtonian fluid. Polymer X, with high temperature and salinity resistant characteristics has been used for the experiments.

The rheology was measured at 25 °C with a salt concentration of 1 % NaCl using the Kinexus pro Rheometer by Malvern [9]. Prior to the measurement the aqueous polymer solution was filtered through a 1.2 micron screen. The results are shown in Figure 4. Based on the bulk rheological measurements the following constants described previously $S_p$, $a_1$, $a_2$, $a_3$, $\lambda$, $\alpha$ and $n$ can be determined. All parameters used to populate the simulation model are summarized in Table 1. Where no experimental data is available, parameters are assumed.

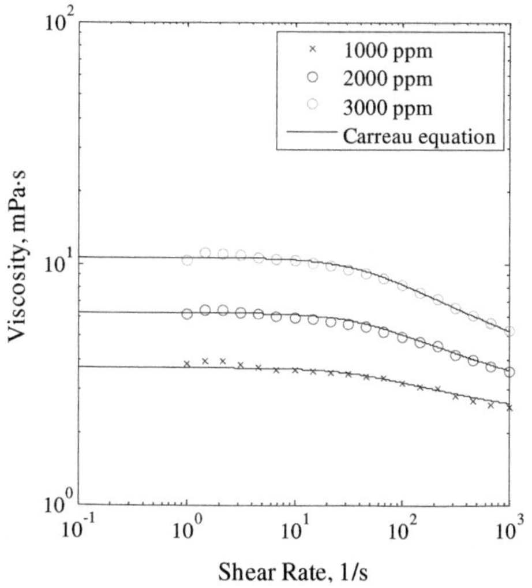


**Fig. 4** Rheology of the X polymer for different polymer concentrations, 1% NaCl and 25 °C

**Table 1** Physical parameters used as input in simulation

| Physical parameter | Value | Physical parameter | Value |
|---|---|---|---|
| Length, $x$ (cm) | 2 | Water viscosity, $\mu_w$ (mPa·s) | 1.1 |
| Width, $y$ (cm) | 2 | Specific surface area, $A_v$ (1/cm) | 160 |
| Height, $z$ (cm) | 20E-4 | Langmuir adsorption constant, b ($cm^3/mg$) | 10 |
| Gridblock size, $\Delta x$ , $\Delta y$ (cm) | 0.01 | Maximum adsorption, $C_{pad}^{max}$ ($mg/cm^2$) | 8.4E-4 |
| Porosity, $\phi$ | 0.23 | Permeability, k (mD) | 480 |
| Init. salt concentration, $C_s$ ($mg/cm^3$) | 10 | Shear rate exponent, n | 0.82 |
| Constant, $\lambda$ (s) | 1/25 | Constant, $C$ | 1 |
| Residual resistance factor, $R_{RF}$ | 2 | Constant, $\alpha$ | 2 |
| Injection Rate, Q ($cm^3/h$) | 0.0032 | Constant, $a_1, a_2, a_3$ (-) | 5, 0.4, 0 |
| Polymer inj. concentration, $Q_{pw}$ ($mg/cm^3$) | 1 (5 min)<br>0 (25 min) | Constant, $S_p$ | -0.414 |

# 5 Numerical Simulations

In order to generate results, the physical parameters presented in Table 1 are used in the simulator implemented into Matlab® [1]. Simulation results of the polymer concentration after 5 minutes are presented in Figure 5. In the case of heterogeneous and anisotropic permeability field, viscous fingering occurs as shown in Figure 5a. In the case presented here, the micromodel permeability is 480 mD and homogeneous and isotropic (Figure 5 b)), and a 1D representation of the results is sufficient.

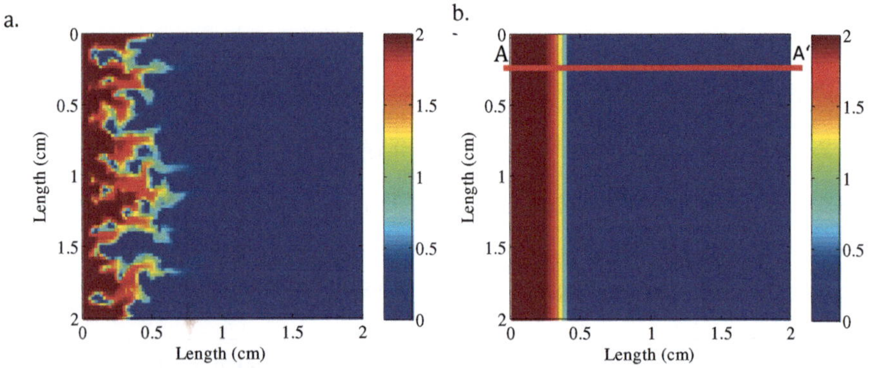


**Fig. 5** Polymer concentration in mg/cm³ after 5 minutes of polymer injection for a) a heterogeneous/anisotropic and b) homogeneous /isotropic case

The simulation results are presented in Figure 6 and are plotted along the cross section A-A' (Figure 5 b). During the first 5 minutes a polymer slug with a concentration of 2 mg/cm³ is injected into the micromodel (Figure 6 a.)). After 5 minutes the slug is displaced by water. The salt concentration is 10 mg/cm³ and constant. Due to adsorption the size of the polymer slug is reduced to a maximum concentration of 1.5 mg/cm³ after 25 minutes (Figure 6 a.)). As the polymer adsorption is assumed irreversible, no desorption occurs although polymer concentration decreases (Figure 6 b)).

Further, the permeability to the aqueous phase is reduced as can be seen in Figure 6 c) which is a result of polymer retention. Due to adsorption and shear thinning, the viscosity of the aqueous phase is slightly reduced (Figure 6 d)). The apparent shear rate depends on the level of adsorption and is between 7-10 1/s. Figure 6 d) depicts that in this case the viscosity alteration caused by shear thinning is small.

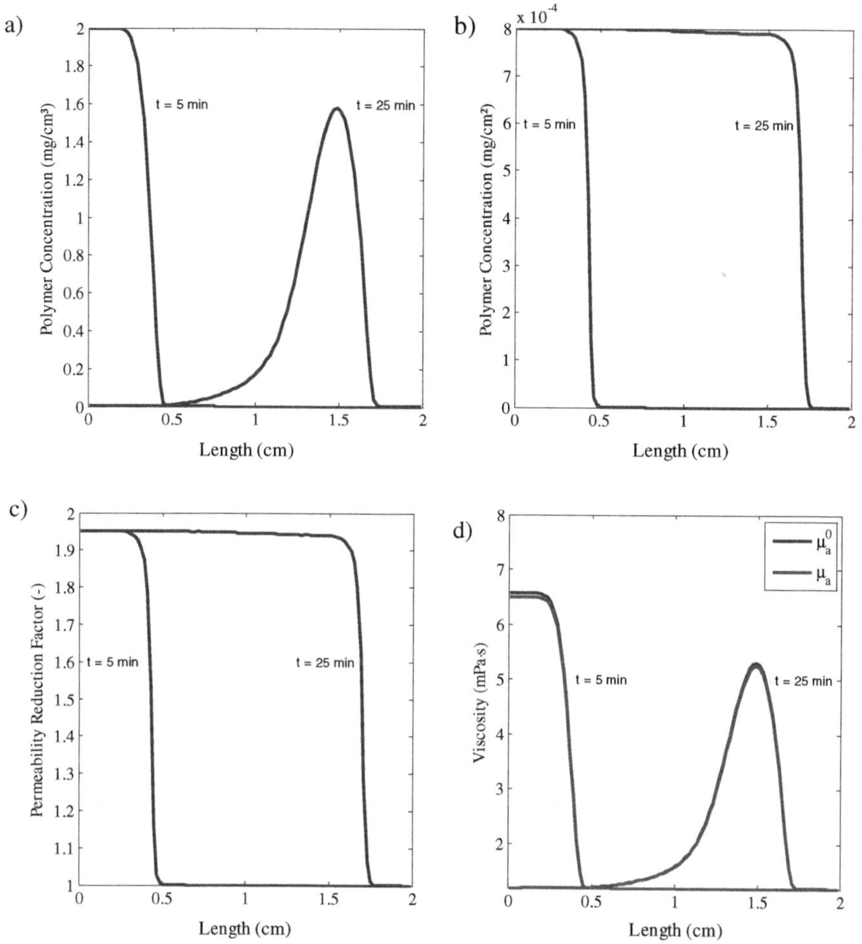


**Fig. 6** Simulation results of a) Polymer concentration, adsorbed polymer concentration, c) permeability reduction factor and d) aqueous phase viscosity with and without shear

# 6 Conclusions

(1) The implementation of a comprehensive mathematical polymer flood model has been presented. It considers different physico-chemical effects such as non-Newtonian rheology of the displacing phase, permeability reduction, retention and salinity effects by using comprehensive mathematical models.
(2) The micromodel properties such as porosity and permeability as well as fluid properties have been determined experimentally. Where parameters were not determined experimentally they are assumed.
(3) Polymer X with 1 % NaCl and a homogeneous and isotropic micomodel was used. Rheology was determined experimentally.

(4) The simulation results show, that the impact of shear on the apparent viscosity of the aqueous phase is small compared to the effect of polymer concentration. This is the result of the polymer type and low shear rates encountered. A shear rate between 7-10 1/s was simulated.

# 7 Outlook

Results obtained from the two phase micromodel flooding experiments will be compared with the results from numerical simulation in order to test the ability of existing mathematical models to reproduce dominating physico-chemical effects. Further, parameters such as adsorption isotherm, residual resistance factor, and parameters affecting apparent shear rate will be determined experimentally and will be used in simulation. This will establish a more detailed description of the micromodel properties.

# References

1. Matlab® The Language of Technical Computing by The MathWorks, Inc., `http://www.mathworks.de/`
2. Baumann, T., Niessner, R.: Micromodel study on reparationing phenomena of a strongly hydrophobic fluorophore at colloid/1-octanol interface. Water Resources Research 42, 12 (2006), doi:10.1029/2006WR004893
3. Flory, P.J.: Principles of Polymer Chemistry. Cornell University Press (1953)
4. Lakatos, I., Lakatos-Szabo, J., Toth, J.: Factors influencing polyacrylamide adsorption in porous media and their effect on flow behavior. Paper presented at the Symposium on Surface Phenomena in EOR, Stockholm (1989)
5. Islam, M., Farouq, S.: New scaling criteria for polymer emulsion and foam flooding experiments. J. Can. Petrol Technol. 28(4), 79–87 (1989)
6. Hirasaki, G.J., Pope, G.A.: Analysis of factors influencing mobility and adsorption in the flow of polymer solution through porous media. SPEJ, 337–346 (August 1974)
7. Carreau, P.J.: Rheological Equations from Molecular Network Theories. Trans. Soc. Rheol. 16(1), 99–127 (1972) (Ph.D. dissertation, University of Wisconsin, Madison, 1968)
8. Brooks, R., Corey, A.: Hydraulic Properties of Porous Media. In: Colorado State University Hydrology Paper, vol. 3, Colorado State University (1964)
9. © 2012- Malvern Instruments Ltd. a Spectris company (2012), `http://www.malvern.de`

# Recovery of the Geothermal Energy Stored in Abandoned Mines

Esmeralda Peralta Ramos and Gioia Falcone

Institute of Petroleum Engineering, Clausthal University of Technology

**Abstract.** Abandoned mines are already being used for various purposes, ranging from ultimate waste disposal to energy storage and the heating and cooling of spaces. Some examples of the energy storage systems in use include hydroelectric pumping storage, wind, and compressed air. These sites represent independent and sustainable sources of energy that can be exploited to added value to local communities after the cessation of mining activity.

After abandonment, mines usually flood due to the natural rise of the water table, or they can be artificially flooded. The relatively stable temperature of this flood water could be harnessed as a geothermal resource using heat pumps. For example, during winter, heat could be extracted from mine flood water and used for space heating, while in the summer the process could be reversed to provide space cooling.

The geothermal potential of mine flood water can be extracted via open or closed circulation systems, either through the shafts of the mine, or if still accessible, the galleries.

Although not widespread, the use of low enthalpy geothermal energy stored in abandoned mines for heating and cooling of buildings and industrial processes has already been implemented in Canada, Germany, USA and UK, with ongoing assessment projects in some communities in Europe.

The Harz region of Germany was an important mining zone from the Bronze Age until the 20$^{th}$ century, producing silver, lead, copper, and zinc. As most of the mines have been abandoned for at least 80 years, they could, in principle, be regarded as geothermal energy sources for the heating and cooling of community spaces. However, the quality of the mine water, the heat capacity and thermal conductivity of the rock, the total available length for water circulation, and the structural stability of the abandoned mines will have a significant impact on the technical feasibility of this concept.

This paper critically reviews implementation of this low enthalpy geothermal energy recovery strategy from the abandoned mines of the Harz region.

**Keywords:** geothermal energy, low enthalpy, abandoned mine, heat pump, Harz.

M.Z. Hou et al. (Eds.): *Clean Energy Systems in the Subsurface*, SSGG, pp. 143–155.
DOI: 10.1007/978-3-642-37849-2_12 

# 1 Introduction

Direct use of low-temperature geothermal resources includes district heating, greenhouses, fisheries and heating for industrial processes. Recovering the geothermal energy stored in abandoned underground mines represents an option for heating and cooling spaces. This concept is not new, as it was first implemented in Canada in the late 1980's. However, there are only a few existing and proposed installations around the world. Among the advantages associated with geothermal energy recovery from underground mines are the reduction of fossil fuels consumption, the decrease in $CO_2$ emissions, and the creation of jobs for the local communities beyond the natural life cycle of the traditional mining industry.

# 2 Background

## *2.1 Abandoned Underground Mines*

Typically, the mining process can last for many years, depending on the available technology and the abundance of the mineral resource. The primary reasons for ceasing mining activity are exhaustion of the deposits and economic and environmental factors, but they have also included war, accidents, diseases and listlessness [1]. Recently, abandoned mines have been used for various purposes, including ultimate waste disposal and energy storage (i.e. using hydroelectric pumping or compressed air).

From a geothermal stand point, the water stored in underground mines can be regarded as a heat source or a heat sink [2], and it represents an independent, sustainable source of energy. Of course, these resources can be exploited more efficiently if the geothermal recovery systems are planned and designed before the mine is abandoned.

### 2.1.1 Flooding of Abandoned Underground Mines

Underground mines tend to have a natural water influx, even during the mining process, and eventually they flood when they are abandoned. The geological structure and the type of mining activity are the factors that determine the inflow rate into the structure [1]. The quality and pH of the infiltrated water depend on the composition and reactivity of the rock to which the water is exposed and the time they remain in contact. When the mine is abandoned, the cavity left becomes the principal pathway for groundwater flow. As the mine water levels increase, the water flushes the sulphide oxidation products from the rock and the waste material, which increases the quantity of sulphates and metals in the water and raises its acidity. On the other hand, carbonates and other minerals contribute to the alkalinity of the water. [3]

Figure 1 shows an example of water flow through an abandoned mine with dewatering tunnels and galleries below them. The permeating water flows out through the deepest tunnel. The mine water temperature varies according to the in situ geothermal gradient.

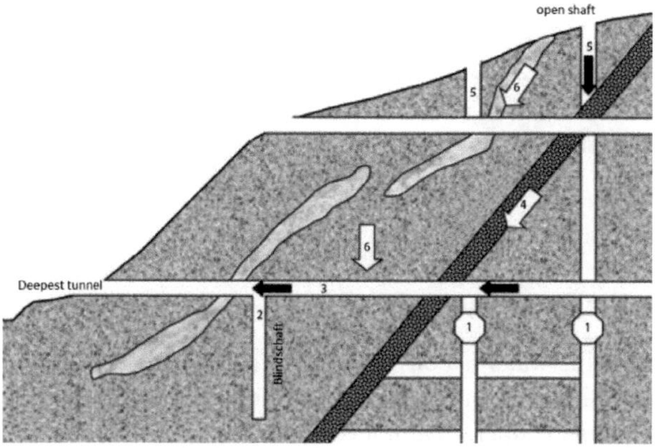


**Fig. 1** Example of mine water circulation (after [4])

### 2.1.2 Heat Pumps as Recovery Systems from Abandoned Mines

Conventionally, the geothermal energy stored in mine water can be recovered via heat pumps in combination with either open or close loops [2]. Heat pumps use a high-pressure refrigerant to capture and move thermal energy in the opposite direction to the spontaneous heat flow. The heat pump operates on a thermodynamic principle consisting of two isothermal and two isentropic processes. The full cycle is made up of four stages as shown in Figure 2: 1) a first heat exchanger (heater), where the heat of the mine water is captured to vaporise the refrigerant; 2) a compressor, which increases the pressure and temperature of the vapour; 3) a second heat exchanger (cooler), which releases the heat to the distribution system for space heating; 4) an expansion valve to decrease the pressure and temperature of the circulation fluid. By reversing the system, the pump can provide cooling. A heat pump is characterised by its coefficient of performance (COP), which is the ratio of the heating or cooling delivered over the equivalent electric energy input. A larger COP value indicates a lower consumption of energy to generate heat. Typical COP values for a heat pump are 3 to 4. [5]

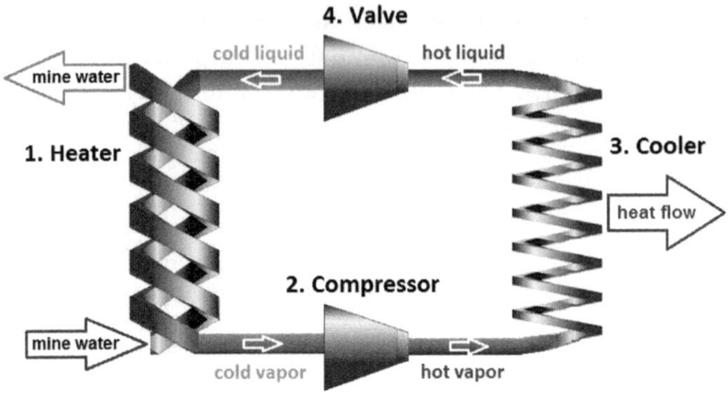


**Fig. 2** Heat pump working principle

### 2.1.3 Loop Configurations

The heat of the mine water is conveyed to the heat pumps by loops, which must be resistant to corrosion and must not obstruct the heat flow to the pump. Depending on whether the mine is operating, not operating but maintained, or closed, and also depending on drilling costs, water access and water quality, the geothermal potential can be extracted via open or closed circulation systems, through the shafts of the mine or, if still accessible, the galleries.

Some typical loop configurations will now be reviewed.

First of all, the harnessed water can be re-injected back into the mine, or simply disposed of at the surface. The re-injection can be achieved via the same shaft or well, or into a different shaft or well.

Secondly, the loop may be open or closed, depending on water quality and volume of water. In an open-loop, it is the mine water itself that flows, whereas in a closed-loop a circulation fluid is used, in combination with a heat exchanger.

Mine shafts can act as wells when still accessible. If they are not, and if the galleries have not been permanently isolated, wells can be drilled into mine galleries using directional drilling. Figures 3 to 5 summarise typical loop configurations used in the mining industry.

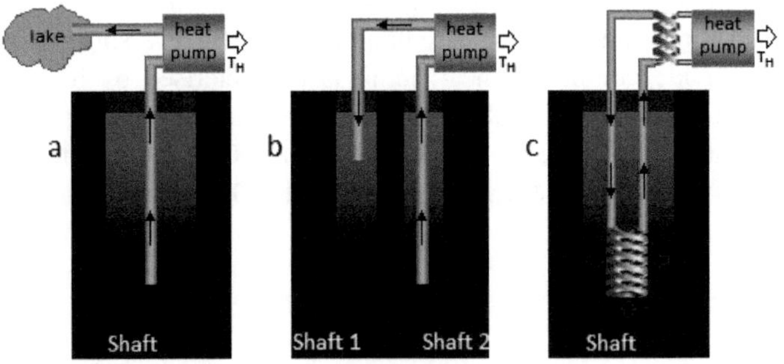


**Fig. 3** a. Open-loop, no recirculation, single shaft/well; b. Open-loop, double shaft/well; c. Closed-loop, single shaft/well with downhole heat exchanger

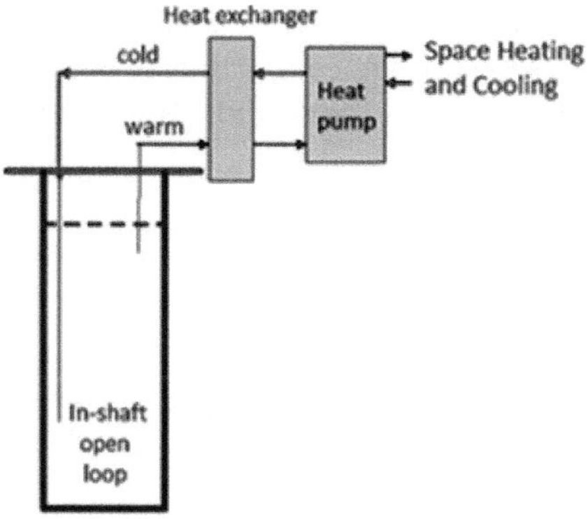


**Fig. 4** Open-loop, recirculation, single shaft/well [2]

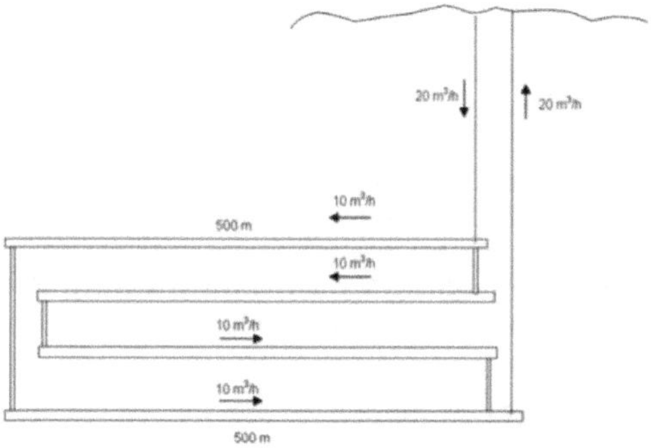


**Fig. 5** Open-loop, recirculation, double shaft/well, flow through galleries. Water is injected into a well and split into the two galleries. The split flows join again in the extraction well [6].

### 2.1.4 Modelling of Heat Extraction Processes

The system must be modelled in order to predict its long term behaviour. It is important to calculate the temperature increase of the fluid as it runs through an open-loop or a closed-loop, and to evaluate the geothermal capacity of the mine versus the size of the system and the extraction/circulation rate, in order to prevent water temperature decline. Modelling can also identify critical thermal shortcuts between a re-infiltration well and a source well [7]. Thermal and hydraulic parameters are the key to determining the efficiency of the heat recovery process [8]. Both numerical and analytical modelling techniques can be adopted for this purpose.

Rodríguez and Díaz [5] proposed an iterative calculation method for accessing the geothermal heat capacity of underground mine galleries, based on a thermal balance between rock and circulating water under the assumption of pseudo steady-state conditions.

Ghoreishi-Madiseh et. al. [9] proposed a transient, two-dimensional numerical approach where the heat flow from the rock mass to the mine cavities is controlled by conduction in the rock mass. A comparison of the performance of the two models for the geothermal system of Springhill (Nova Scotia, Canada) is shown in Figure 6.

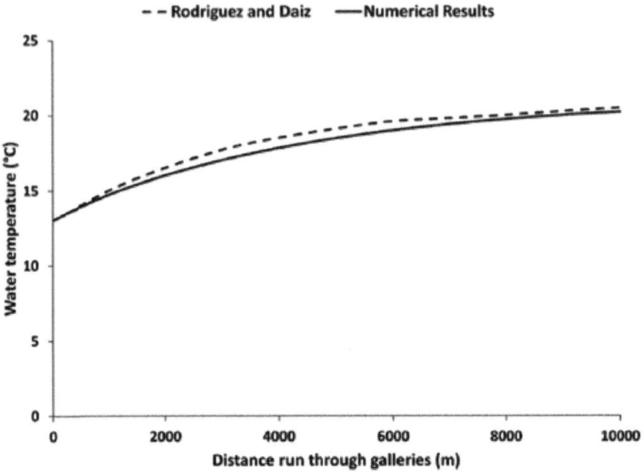


**Fig. 6** Comparison of Ghoreishi-Madiseh's numerical results vs. Rodríguez and Díaz for the geothermal system of Springhill (Nova Scotia, Canada) [9]

### 2.1.5 Operating Recovery Systems

The use of groundwater from abandoned mines for heating and cooling of buildings and industrial processes began in Canada in 1989 when the town of Springhill created an industrial park where companies could tap into the geothermal energy supply from the local abandoned coal mines [10]. There are other examples, in Germany, Wismut mine in Marienberg generates 690kW of heat capacity and a mine in Freiberg provides heating and cooling for a castle [8]. In the Netherlands, the Heerlen project was setup in 2005 by a diverse partnership of municipalities, building research institutes and housing associations, and received strong financial support from the EU. The power station consists of two sites in the centre and north of Heerlen. The proposed development included 350 dwellings, 3800m$^2$ of commercial buildings, 2500m$^2$ of public and cultural buildings, 11500m$^2$ of health care buildings and 2200m$^2$ of educational buildings [7]. Work began in March 2005, when two 825m wells were drilled at Heerlerheide to access trapped mine water at a temperature of 35°C. Three shallower wells were then drilled at Stadtpark Oranje Nassau, ranging in depth from 250m to 500m, to receive the Heerlerheide water after it has been used for heating [11]. Calculations suggest the concept could offer a 55% $CO_2$ reduction [12]. Table 1 presents some case studies that illustrate the versatility of the recovery concept.

**Table 1** Operating geothermal recovery systems [2, 13, 14]

| | Mine | | Heat Pump | | |
|---|---|---|---|---|---|
| Location | Output | Water Temp | Configuration | Quantity | Combined capacity |
| Folldal mine, Norway | Cu, Zn, S | ------- | Closed loop | 1 | 18 kW |
| Park Hills, Missouri, USA | Lead | 14°C | Open loop Re-injection | 9 | 113 kW |
| Shettleston, UK | Coal | 12°C | Open loop Re-injection | 2 | -------- |
| Lumphinnans, UK | Coal | 14.5°C | Open loop Re-injection | --- | -------- |

## 2.2 *Abandoned Mines in Harz*

### 2.2.1 Geological Structure in Harz

In northern Germany, the highest mountain chain is the Harz, covering approximately 2220km$^2$. Geologically, the region dates back to the Palaeozoic era, with subsequent orogeny. In spite of its small area, the diversity of the Harz geology is extraordinary, ranging from Silurian slates to Trias lacustrine limestone, plutonities and magmatic and ore deposits. The Harz can be divided into three geological areas: Upper Harz (up to 800m high), Middle Harz and Lower Harz (up to 400m high). The ore bodies found in the Harz have two different origins. In Goslar, during the Devonian period, hot metal-bearing thermal springs escaped to the sea floor and formed the vault bodies. On the other hand, a hydrothermal process took place in the Upper Harz [15].

### 2.2.2 Mining History

Mining activity in the Harz started approximately 3000 years ago; however, the discovery of a silver deposit near Goslar in 968 dramatically increased exploration and extraction across the mountain. The main products were lead, copper, iron, silver and, since the 19th century, zinc. The drainage system employed was made of tunnels and water column engines, including the Ernst-August tunnel, with a length of 26km, to transport the mine water by gravity from Lautenthal, Clausthal, Zellerfeld, Wildemann and Bad Grund to Gittelde. At the start of 1800, due to ore depletion, the prosperity of the Harz region declined and most of the mines were closed. With the advent of the Great Depression in 1930, almost all of the region's remaining mines were abandoned. Some of the mines were then used for electricity generation up to 1980, when the high operational costs triggered their permanent closure. Only the Samson mine in Sankt Andreasberg has continued to provide electricity generation. Other mines were converted into museums, like the Rammelsberg mine in Goslar, which has been declared a world heritage site by the UNESCO, along with the Samson pit and the Upper Harz water system used in the mining process [16, 17].

### 2.2.3 Selection of Abandoned Mines for Geothermal Energy Recovery

Seven mining districts can be identified in the Upper Harz: Altenau, Bad Grund, Clausthal-Zellerfeld, Lautenthal, Sankt Andreasberg and Wildemann. In addition, Goslar and Lower Harz have mines that are worthy of consideration as possible locations for the implementation of the previously described geothermal recovery concept. Seven criteria are proposed for screening these opportunities:

1. **Column height and temperature of the mine water.** There has to be sufficient water volume for the implementation of the loops and for the geothermal exchange. Temperature is a critical factor in determining the potential of the water as a geothermal energy source.
2. **Galleries.** The galleries, which are the main passages that connect the shaft to the stopes and possibly to other mines, can be kilometres in length and are where the main water volume is stored. Long galleries can also be regarded as efficient underground heat exchangers. However, where the galleries have already being closed following abandonment of the mine, there is accessible water at the shafts only, which may limit the thermal stability of the system.
3. **Old or newly abandoned mines.** The mines in the Harz region are classified as 'old' if they were mined before 1885 and 'new' if they were mined thereafter. The back filling of a mine differs, depending on its age. In the old mines, no particular standard was followed and the environmental impact of the filling material was not considered to be an important issue; for the more recent mines, concrete was employed.
4. **Shafts.** The shafts of decommissioned mines in the Harz region can be found either closed with a plug of concrete, or still open. If the shafts are closed, drilling operations may affect the stability of the abandoned mine.
5. **Structural stability.** It is important to determine if the underground structure can tolerate disturbances which may be caused by the implementation of the system.
6. **Mine water quality.** This is an important factor in determining not only the loop configuration, but also its materials (e.g. titanium for heat exchangers and pumps in the presence of high sodium chloride content). The probability of finding poor water quality is high in the case of old mines due to the back fill material that was used. The more recent mines may present a higher content of metal particles. Additionally, the pH of the mine water in both cases may be low due to the surrounding rock composition and chemical reactions with the water.
7. **Possible customers.** Site identification mapping is directly related to heating (and cooling) demand near the source.

In this paper, the mining district of Clausthal-Zellerfeld and the Rammelsberg mine in Goslar are screened using the criteria outlined above for possible implementation of the geothermal energy recovery concept.

### 2.2.4 Clausthal-Zellerfeld

The town of Clausthal-Zellerfeld has approximately 400 underground workings from which four ore bodies were mined.

1. The drainage level is recorded in the Ernst August tunnel with a mine water temperature between 24 and 26 °C. [17]. Table 2 lists the abandoned mines around the town that extend beyond the Ernst August tunnel's depth.

**Table 2** Working mines of Clausthal-Zellerfeld [16, 17]

| Ore body | | Shaft | Depth (m) | Drainage level (m) | Mining period | Old/New |
|---|---|---|---|---|---|---|
| Rosenhöfer | 1 | Ottiliae Schacht | 594 | 341 | 1868 – 1930 | New |
| | 2 | Silbersegener Richtschacht | 419 | 333 | 1817 – 1930 | New |
| | 5 | Alter Segener Schacht | 430 | 344 | 1591 – 1930 | |
| | 6 | Heilige Drei Königer Schacht | 370 | 344 | 1631 – 1817 | Old |
| | 9 | Unterer Rosenhöfer Schacht | 708 | 350 | 1649 – 1928 | |
| | 10 | Oberer Rosenhöfer Schacht | 400 | 343 | reported in 1588 | Old |
| | 11 | Braune Lilier Schacht | 430 | 344 | 1681 – 1818 | Old |
| | 13 | Johanneser Schacht<br>Thekla-Blindschacht | 663<br>242 | 392 | 1905 – 1930 | New |
| Burgstätter | 15a | Caroliner Schacht | 488 | 386 | 1711 – 1867 | |
| | 16 | Dorotheer Schacht | 576 | 386 | 1656 – 1886 | Old |
| | 18 | Königin Marien Schacht | 777 | 391 | 1856 – 1912 | New |
| | 20 | St. Elisabether Schacht | 551 | 360 | 1625 – 1885 | Old |
| | 28 | Anna Eleonorer Schacht | 728 | 360 | 1638 – 1908 | |
| | 30 | Herzog Georg Wilhelmer Schacht | 750 | 364 | 1644 – 1904 | |
| | 32 | Schacht Kaiser Wilhelm II (+ Blindschacht) | 947<br>390 | 357 | 1880 – 1930 | |
| | 42 | St. Lorenzer Schacht | 420 | 355 | 1550 – 1840 | Old |
| | 44 | Königin Charlotter Blindshacht | 524 | 371 | 1820 – 1870 | New |
| Zellerfeld | 56 | Rheinischweiner Schacht der Grube Ring und Silberschnur | 565 | 341 | 1558 – 1910 | |
| | 59 | Silberne Schreibfeder Schacht der Grube Regenbogen | 535 | 360 | 1561 – 1930 | |
| | 60 | Jungfrauer Schacht der Grube Regenbogen | 475 | 371 | 1560 – 1816 | Old |
| | 65 | Neuer Johanneser Schacht | 633 | 392 | 1926 – 1930 | New |
| Haus Herzberg | 73 | Neuer Haus Herzberger Schacht | 631 | 364 | 1925 – 1928 | New |

2. The galleries in the Clausthal-Zellerfeld mines have remained open, so the implementation of an open-loop double-well system can be evaluated. Operating systems report suction rates of 4 to 5 l/s, so as the water influx rate in the Ernst-August tunnel is around 10 l/s [17], the required heat pump supply could be achieved with minimal disturbance to the underground system.
3. Table 2 shows that 22 mines have been identified, which include 8 old, 7 new, and 7 that began mining activity before 1885 and continued until 1904, thus overlapping the old/new mine classification.
4. All the shafts of the abandoned mines in the town of Clausthal-Zellerfeld are closed.
5. Structural stability needs to be determined.
6. The presence of lead, sulphates oxidations contaminants and metal particles in the mine water is very likely. A chemical analysis of the water is essential to assess the feasibility of an open-loop system.
7. Figure 7, which shows the location of the mines lying beneath the town, suggests that it may be possible to implement the geothermal energy recovery concept. For this reason, possible customers range from residential houses to offices, commercial buildings and community structures.

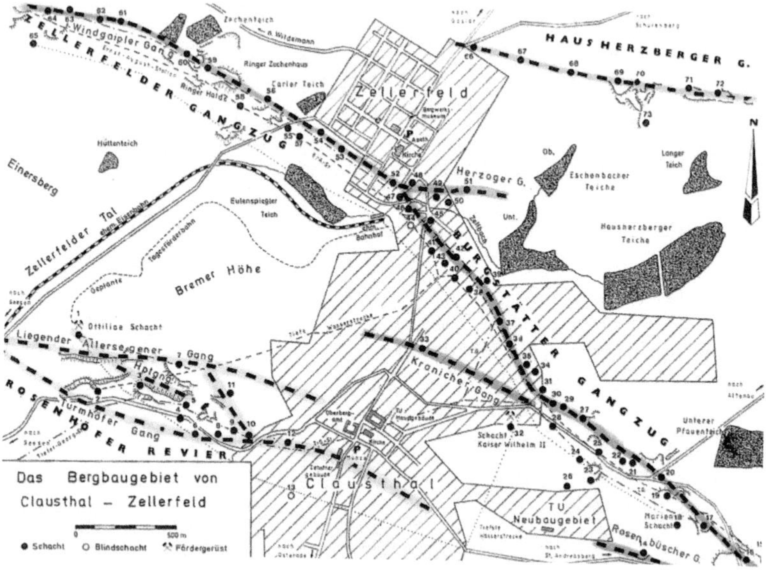


**Fig. 7** Location of the most important mines around Clausthal-Zellerfeld [after 16]

### 2.2.5 Rammelsberg Mine, Goslar

Mining activity in Rammelsberg lasted for over 10 centuries, leading to the extraction of approximately 400,000 tons of ore from two main bodies, with the following reported composition: 14% zinc, 6% lead, 1% copper, 120 grams/ton silver, about 1gram/ton gold and 20% barite [16].

1. The abandoned Rammelsberg mine has a total depth of 440 m, of which the lower 400 m are flooded. The estimated water temperature is of 20 - 22°C [17].
2. The galleries in the mine are not back filled.
3. Of the two ore deposits in the Rammelsberg mine, one is old and the other is new.
4. In 1988, due to the ore's exhaustion, the mine was closed and converted into a museum, maintaining the shaft open.
5. Structural stability to be determined.
6. A mine water analysis carried out in 1981 reported a pH of 2.5, 1mg/l lead, 1000mg/l zinc, 20mg/l cooper, 370mg/l iron, 6,500mg/l SO4. The water is sour because of the presence of pyrite in the ore body and surrounding rock. After treatment, carried out in the same year, the contaminants were significantly decreased and the pH increased to 8 [17]. Today, it is likely that increased quantities of metal and rock particle are again in the water; thus, a polyethylene closed-loop should be considered in this case.
7. The Rammelsberg museum, which is located directly above the abandoned mine, could be an obvious costumer. Additionally, there are residential buildings nearby.

## 3 Harz Mines vs. Currently Operating Systems

A high level comparison of the operating parameters of existing systems and those of the abandoned mines in the Harz Mountains was performed by assuming the water to be clean (specific heat capacity of 4.18 kJ/kg°C), and also assuming the same suction rate, applied work and temperature difference between the mine water and that supplied to the building as per the existing systems. Since the mine water temperatures in Clausthal-Zellerfeld and Goslar are greater than in the cases presented in Table 1, the performance of the heat pump would be better and the heat capacity for the same applied work on the system would also be higher. The data in Table 3 are obtained using the assumptions mentioned above, namely applied rate work of 52 kW and flow rate of 4 l/s.

**Table 3** Comparison between the Springhill system and the mine water of Harz

| | Location | Mine water | | COP | Heat capacity (kW) | $T_{provided}$ (°C) |
|---|---|---|---|---|---|---|
| | | $T_{suction}$ (°C) | $T_{discharge}$ (°C) | | | |
| Heating | Springhill | 18 | 13 | 3.6 | 188 | 25 |
| | Rammelsberg | 24 | 19 | 4.4 | 231 | 31 |
| | CLZ | 20 | 15 | 3.9 | 202 | 27 |
| Cooling | Springhill | 18 | 23 | 2.6 | 136 | 13 |
| | Rammelsberg | 24 | 32.8 | 3.8 | 199 | 19 |
| | CLZ | 20 | 26.3 | 3 | 157 | 15 |

However, the overall efficiency of a recovery loop depends on several factors, such as the properties of the fill back material, the working fluid and the surrounding rock, in addition to the flow rate and loop geometry. Therefore, even though the results of this high level comparison are encouraging, a detailed field investigation must be performed to thoroughly assess the feasibility of this concept in the Harz region.

## 4 Conclusions

The stable temperature of water in flooded abandoned mines is an independent and sustainable source of low enthalpy geothermal energy that can be exploited via heat pumps for the benefit of local communities.

Information gathered on abandoned mines in the Harz region suggests they could, in principle, become geothermal energy sources for the heating of community spaces.

The quality of the mine water, the heat capacity and thermal conductivity of the rock, the total available length for water circulation and the structural stability of abandoned mines have significant impact on the technical feasibility of this concept.

Detailed analyses are required to consider the installation of heat pumps, the economic feasibility, and the identification of target customers.

**Acknowledgements.** The authors would like to thank Dr. rer. nat. Claudia Pawellek for her valuable suggestions that helped improve this manuscript.

## References

1. Hartman, H.L., et al.: SME Mining Engineering Handbook. Cushing-Malloy Incorporation, Ann Abor (1996)
2. Hall, A., Scott, J.A., Shang, H.: Geothermal energy recovery from underground mines. Renewable and Sustainable Energy Reviews 15, 916–924 (2011)
3. http://www.gardguide.com/index.php/Chapter_4 (accessed April 26, 2012)
4. Ofner, C., Wieber, G.: Geothermische Potenziale gefluteter Bergwerke. Bundesamt für Bauwesen und Raumordnung Jahresmagazin, 72–77 (December 2008)
5. Cengel, Y.A., Boles, M.A.: Termodinámica. McGraw-Hill, México (2003)
6. Rodríguez, R., Díaz, M.: Analysis of the utilisation of mine galleries as geothermal heat exchangers by means a semi-empirical prediction method. Renewable Energy 34, 1716–1725 (2009)
7. Mine water Project co-financed by the European Union under the framework of INTERREG IIIB NWE programme. Mine water as a Renewable Energy Resource – An information guide based on the Minewater Project and the experiences at pilot locations in Midlothian and Heelen. PromoScence, (January 3, 2010), http://www.promoscene-database.eu/fileadmin/promoscene/downloads/Others/Minewater_Good_Practice_Guide.pdf

8. Kranz, K., Dillenardt, J.: Mine Water Utilisation for Geothermal Purposes in Freiberg, Germany: Determination of Hydrogeological and Thermophysical Rock Parameters. Mine Water Environ. 29, 68–76 (2010), doi:10.1007/s10230-009-0094-4
9. Ghoreishi Madiseh, S.A., Ghomshei, M.M., Hassani, F.P., Abbasy, F.: Sustainable heat extraction from abandoned mine tunnels: A numerical model. J. Renewable Sustainable Energy 4, 033102 (2012), doi:10.1063/1.4712055
10. Michel, F.A.: Utilisation of abandoned mine workings for thermal energy storage in Canada. Stockton College of New Jersey (2009), http://intraweb.stockton.edu/eyos/energy_studies/content/docs/effstock09/Session_11_1_Case%20studies_Overviews/105.pdf
11. European Commission, World's first mine-water power-station opens in Heerlen. European Commission – Environment – Eco-innovation Action Plan (February 15, 2009), http://ec.europa.eu/environment/ecoap/about-eco-innovation/good-practices/netherlands/328_en.htm
12. Whitbread-Abrutat, P., Coppin, N.: Renewables Revive Abandoned Mines. Wardell Armstrong International (April 13, 2012), http://www.renewableenergyworld.com/rea/news/article/2012/04/renewables-revive-abandoned-mines?page=2
13. Watzlaf, G., Ackman, T.: Underground mine water for heating and cooling using geothermal heat pump systems. Mine Water and Environment 25, 1–14 (2006)
14. Banks, D., Skarphagen, H., Wiltshire, R., Jessop, C.: Mine water as a resource: space heating and cooling via use of heat pumps. Land Contamination and Reclamation 11(2), 191–198 (2003)
15. http://www.showcaves.com/english/de/region/Harz.html (accessed June 16, 2012)
16. Ließmann, W.: Historischer Bergbau im Harz. Springer, Berlin (1997)
17. Documents in the Bergamt of Clausthal–Zellerfeld

# Carbon-dioxide for EOR in Upper Assam Basin

Subrata Borgohain Gogoi

Department of Petroleum Technology, Dibrugarh University, Assam, India

**Abstract.** This paper will deal with the injection of carbon-dioxide ($CO_2$) for the purpose of enhanced oil recovery (EOR) in mature and depleted oil and gas reservoirs of Upper Assam basin, India. $CO_2$ injection system is a procedure used to extract maximum oil from the reservoir. This system is performed through injecting natural gases like $CO_2$ into the oil wells. The main objective of the $CO_2$ injection is to stimulate the oil droplets that are inside the oil reservoir rock. Minimum Miscibility Pressure (MMP) is achieved by lowering the viscosity of the oil to make it flow easily to the surface. This paper will calculate the MMP at which miscible recovery takes place. This minimum dynamic miscibility pressure depends upon several factors, such as the composition of injected gas, reservoir temperature and pressure and the characteristics of the oil in place fluids.

**Keywords:** Carbon-dioxide, Upper Assam basin, Minimum Miscibility Pressure, factors.

## 1 Introduction

Petroleum is present in complex capillary networks in underground oil reservoirs. Production in most of the oil fields includes three stages namely, primary, secondary and tertiary or enhance oil recovery (EOR). EOR is defined as oil recovery by injection of fluids not normally present in reservoir which changes the intrinsic properties of the oil by using chemical thermal, Gas etc and excludes pressure maintenance or water flooding. The flow diagram in Figure 1 shows the type of EOR processes that are currently employed in the oil industry [1].

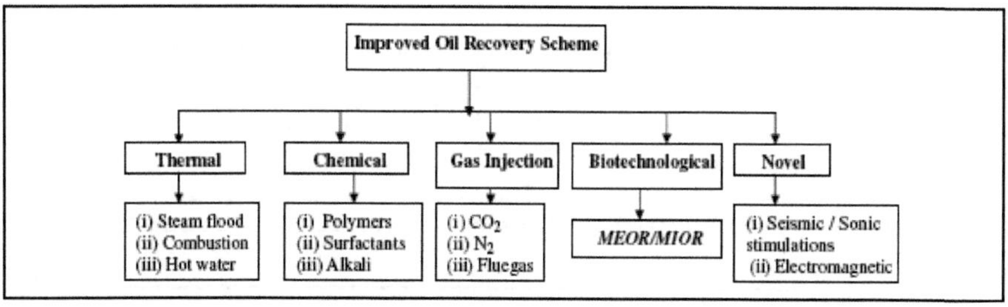


**Fig. 1** Flow sheet diagram showing the process steps in enhanced oil recovery (EOR)

M.Z. Hou et al. (Eds.): *Clean Energy Systems in the Subsurface*, SSGG, pp. 157–169.
DOI: 10.1007/978-3-642-37849-2_13 

Oil displacement by $CO_2$ injection depends on the phase behavior of the $CO_2$ and crude oil mixtures that are strongly dependent on reservoir temperature, pressure and crude oil composition. This method also plays an important role in the reduction of the green house effect representing how utilization of $CO_2$ can be combined with the rational use of natural resources and the saving of energy sources. These mechanisms range from oil swelling and viscosity reduction for the injection of immiscible fluids (at low pressures) to complete miscible displacement in high pressure applications.

The $CO_2$ miscible processes are first-contact miscible process or multiple-contact miscibility process. A volume of relatively pure $CO_2$ is injected to mobilize and displace residual oil. Through multiple contacts between the $CO_2$ and the oil phase, intermediate and higher-molecular-weight hydrocarbons are extracted into the $CO_2$ rich phase. Under proper conditions this $CO_2$-rich phase will reach a composition that is miscible with the original reservoir oil. From that point, miscible or near-miscible conditions exist at the displacing front interface [2, 3]. To achieve high sweep efficiency $CO_2$ and water are injected in alternate cycles.

In these applications, half to two-thirds of the injected $CO_2$ returns to the production well and is usually re-injected into the reservoir. $CO_2$ vaporizes or extracts hydrocarbon from the crude oil as heavy as the gasoline and the gas-oil fractions. Vaporization occurs at temperatures where the fluid at the displacement front is $CO_2$-rich gas, and extraction occurs at temperatures where the fluid at the displacement front is $CO_2$-rich liquid.

If reservoir pressure is considerably below MMP, large volumes of $CO_2$ will be needed to obtain the MMP. A semi-analytical method for predicting the MMP was later presented by Wang and Orr [4] who played an important role in the development and application of the analytical theory of gas injection processes [5]. The density of oil and water are similar in many reservoir conditions, but in reservoirs which are water flooded prior to $CO_2$ flooding, a density contrast may occur between water and $CO_2$, which may cause segregation.

## 2 Experiment

Figure 2 shows the experiments which will be carried out in this work.

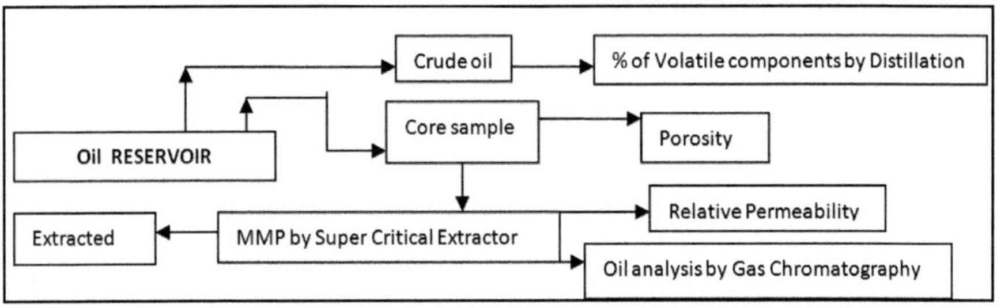


**Fig. 2** Layout of the Experiment

# 3 Materials

The core samples were collected from a depth 3929.8 – 3981.5 m, taken from the Oligocene sandstone reservoirs (Barail Formation) of Nahorkatiya in the Upper Assam basin, India. The porosities of the core samples were found to be 21.6%, 23.2% and 24.8% by the Helium Porosimeter, model no. TPI-219 made by Coretest systems. The average air permeability of the core sample was 2.7 mD measured by Temco's Gas Permeameter model no. RCHR-1, TEMCO INC. Tulsa, USA, using Darcy's equation. The core samples have an oil saturation of 33-40% after water flooding.

The crude oils obtained from Nahorkatiya oil fields have API gravities in the range of 15-58°API with an average of 30°API with significant wax content of 0.11-22%, low sulphur content and are generally moderately mature. Based on bulk geochemical characteristics the Assam oils are characterized to be from two major groups, (i) naphthenicaromatic and (ii) paraffinic. The former have 15-32° API gravity owing to their coal source while the latter group have 30-58° API. The crude oil for this study has an API gravity of 30° (sp. gr. = 0.88) and a viscosity of 10 cp. A deoiler by the trade name Catflo-T (Cationic Polyelectrolyte) was supplied by Thermax, Puna, India for the separation of the oil from the solvent.

If the temperature and pressure are both increased from standard temperature and pressure (STP) to be at or above the critical point for $CO_2$ it can adopt properties midway between a gas and a liquid. More specifically, it behaves as a supercritical fluid above its critical temperature (31.1 °C) and critical pressure (73.9 bar), expanding to fill its container like a gas but with a density like that of a liquid.

# 4 Methods

## *4.1 Porosity Calculation*

This method relies on the expansion of helium gas, and also measures the effective porosity of the rock. However, since helium is a slippery little molecule, it can penetrate pores which are much smaller. Hence, this method provides slightly higher porosity measurements on any given rock sample. This method uses the apparatus shown in Figure 3.

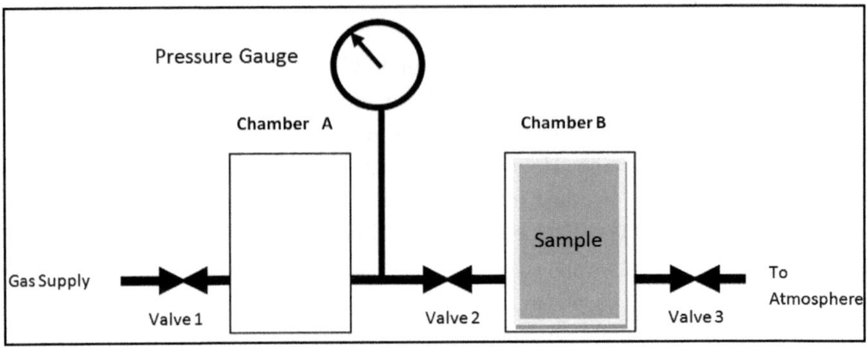


**Fig. 3** Porosity Experiment

Plugs from core samples were cleaned and the bulk volumes (VB) were obtained. In each case, valve 2 was closed and the helium gas was introduced into the chamber. A using valve 1 until a pressure Pi was reached. This pressure was approximately 6.89 bar (100 psi). The plug was inserted in chamber B, which was at atmospheric pressure and valve B was closed. Slowly valve 2 was opened to let the gas equilibrate through both the chambers. The gas penetrates into the pores of the rock sample. During this process the pressure was decreased to a new stable level, Pf. The drop of pressure depends on how much space there was in Chamber B, and that depends on how much of chamber B was occupied by solid rock particles. The measured gas pressures Pi, Patm and Pf alone were not sufficient to obtain the volume of the sample rock matrix. However, the system was calibrated by putting a range of metal cylinders of accurately known volume into chamber B and the experiment was repeated. When this was done, the calibration constant, and the pressures allowed the volume of the rock matrix (VM) to be obtained. The porosity was then calculated using the following formula.

Helium porosity, Φ calculated as:

$$\Phi = \frac{V_B - V_M}{V_B}$$

Where Bulk Volume = $V_B$

$$V_B = \frac{\pi}{4} d^2 l$$

$$V_B = \frac{\pi}{4} (2.4)^2 \times 3.4 = 61.56\text{cc}$$

**Table 1** Calculation of Porosity

| Serial no. | $V_B$ (cc) | $V_M$ (cc) | $\Phi = \frac{V_B - V_M}{V_B}$ (%) |
|---|---|---|---|
| 1 | 61.56 | 48.25 | 21.6 |
| 2 | 61.56 | 47.27 | 23.2 |
| 3 | 61.56 | 46.29 | 24.8 |

### *4.2 Distillation of Crude Oil (IP/55)*

100 ml of the crude oil sample was centrifuged in a closed container to remove the water. The sample was then filtered at atmospheric pressure and heated in a closed steel container fitted with a thermometer and a pressure gauge. The container was filled upto 70% capacity with the sample and heated till the temperature was 200°C. The container was allowed to cool and the water separated by decantation. 100 ml of the sample was distilled in a specified glass apparatus under standard conditions. The volume of the distillate obtained at each multiple of 25°C was recorded, up to a maximum of 300°C when the distillation was stopped. The volumes of distillates collected at 75°C, 100°C, 125°C etc were noted.

## 5 Minimum Miscibility Analysis by Super Critical Extractor

### *5.1 Preparation of the Sample*

Experimental analyses were performed on cylindrical core plug samples of 3.4 cm length and 2.4 cm diameter from depth of 3981.5 m, 3946.5 and 3929.8m. The samples were first placed in an oven for 24 hours in order to remove moisture. The dry and clean samples were weighed and placed in the saturation apparatus. Oil was introduced from one side of the flask, and the vacuum was created by a vacuum pump. The samples were kept for 48 hours in the same air tight flask to achieve proper saturation. After the saturation the excess liquid was removed and the sample was weighed. The weight of the saturation oil was calculated.

### *5.2 Test Procedure*

The core plug samples were transferred one by one into the extraction vessel. First supercritical $CO_2$ at 31°C and pressure of 74 bar was injected into the sample in the vertical position. The samples were 100% saturated with reservoir oil at 70°C, 60°C and 50°C respectively in the pressure difference range between the inlet and exit of the extraction vessel of 100 – 430 bar and a constant flow rate of 4L/min was maintained. This is an unsteady state operation at a constant flow rate. The extracted oil samples were collected at different pressures in the sampling vessels. The oil of the sampling vessels was separated by using a deoiler by the trade name Catflo-T (Cationic Polyelectrolyte). The separated oil samples were assessed both quantitatively and qualitatively. Quantitative assessment was done volumetrically, while quantitative assessment was done in the Gas Chromatograph (GC).

## 5.3 Equipment Operation

To remove the oil extracted from the core sample by supercritical fluid $CO_2$, a thermodynamic machine was required to extract the oil at high pressure. The design method of the Supercritical fluid extractor (SCFE) has been presented by different researchers [6-8]. In this research, SCFE can be used for separation and extraction of oil from plugs using SCFE. As it can be seen in Figure 4, this system can function in static and dynamic conditions. In this system two specially designed Transfer Vessels were used to provide system pressure using Nitrogen ($N_2$) gas. The possibility of establishing the flow of $CO_2$ gas in the machine in two different directions by fixing the existing valve in the machine, is a characteristic of the system. The extraction vessels were made of stainless steel like other parts of the system which can resist up to 689.5 bar (10,000 psi) pressure. The container has a side glass made of silicon material, and can withstand high pressures. The ability of the designed pump for rotating the supercritical fluid within the system is another unique feature of this machine in comparison to other devices. The mechanical part of this pump was designed and made manually. It has the ability to pump a two-phase fluid in a thermal range up to 120°C and with a flow rate of 4 L/min, regardless of the creation of Cavitations in the system, can be tuned by operator. This device uses an air bath system to provide temperature. The designed air bath was able to provide a uniform temperature of 120°C.

Extracted oil samples were collected at 100, 130, 160, 180, 200........430 bar pressures in the sampling vessels as in Figure 4. The results were interpreted graphically and the MMP was obtained. The composition of the oils extracted was determined by GC analysis. The extraction yield (Y) in supercritical carbon dioxide extraction (SC-$CO_2$) was calculated by equation [9]:

$$Y = \frac{m_{ext}}{m_{mat}} \times 100\%$$

where: $m_{ext}$ = mass of extract,
$m_{mat}$ = mass of raw material.

This review provides a detailed and updated discussion of applications of SC-$CO_2$ extraction in the EOR of crude oil from the underground porous medium.

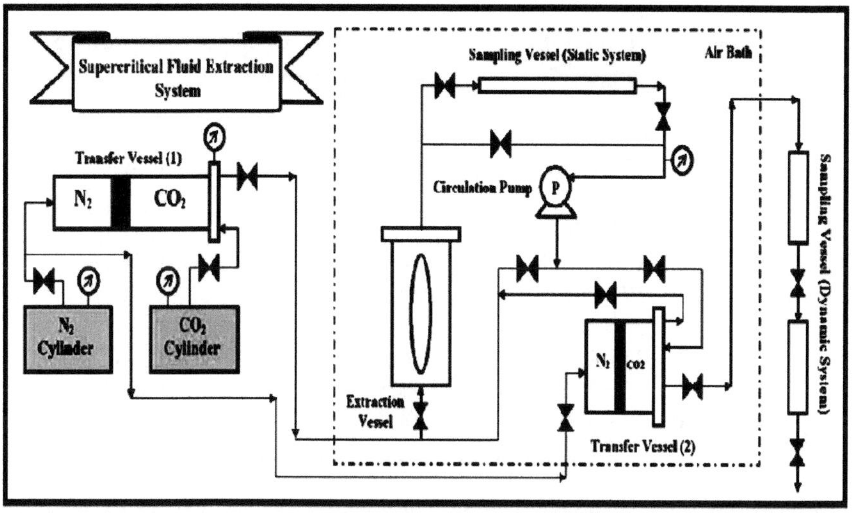


**Fig. 4** Schematic diagram of the Supercritical Fluid Extraction System (SCFE) for Extraction of oil from core samples with supercritical carbon dioxide

## *5.4 Gas Chromatography (GC)*

The GC analysis of oil samples was carried out in a *Thermo Fisher TRACE GC* coupled to a *DSQ, 2005.* In order to carry out GC analysis 0.05g of methylheptadecane was taken and diluted with 100ml of pentane reagent (grade 98%). Oil samples by SC-$CO_2$ collected from the SCE at a fixed temperature (50°C, 60°C, 70°C) and at 180 bar were diluted in test tubes so that the tubes were half filled with samples. The samples were injected into the heater zone, vaporized and transported by carrier gas helium into a packed column, which contained partitions for separating different samples. The column partitioned the components according to their boiling points, the eluted compounds were then carried by carrier gas into a detector, where their concentration was related to the area under the detector response-time curve. Individual peaks can be identified by comparing their retention times inside the column with those of known compounds previously analyzed under the same GC condition [10].

# 6 Results and Discussions

## *6.1 Distillation of Crude Oil (IP/55)*

The results predict the crude oil sample contains 44.6% distillates up to 300°C and are in Table 2.

**Table 2** Showing the distillates in crude oil upto 300$^0$C

| Serial no. | Temperature range ($^0$C) | Cumulative Distillate obtained (% vol) | Carbon no. |
|---|---|---|---|
| 1. | 0-75 | 3 | $C_1$-$C_4$ |
| 2. | 75-100 | 8 | $C_5$-$C_{12}$ |
| 3. | 100-125 | 13 | $C_5$-$C_{12}$ |
| 4. | 125-150 | 20 | $C_5$-$C_{12}$ |
| 5. | 150-175 | 24 | $C_5$-$C_{12}$ |
| 6. | 175-200 | 28.5 | $C_5$-$C_{12}$ |
| 7. | 200-225 | 33 | $C_{12}$-$C_{16}$ |
| 8. | 225-250 | 36.5 | $C_{12}$-$C_{16}$ |
| 9. | 250-275 | 43.5 | $C_{16}$-$C_{19}$ |
| 10. | 275-300 | 44.6 | $C_{16}$-$C_{19}$ |

It was important to know the amount of volatiles because when $CO_2$ is injected into an oil reservoir, volatiles or light hydrocarbons from the oil dissolve into $CO_2$. This occurs most readily when the $CO_2$ density is high i.e., when it is compressed. Below some minimum pressure, $CO_2$ and oil will no longer be miscible. As the reservoir temperature increases the density of $CO_2$ decreases or as the oil density increases due to the decrease of the light hydrocarbon fraction the minimum pressure needed to attain oil/$CO_2$ miscibility increases. For this reason, oil field operators must consider the pressure of a depleted oil reservoir when evaluating its suitability for $CO_2$ EOR. Low pressured reservoirs may need to be re-pressurized by injecting water.

## *6.2 Minimum Miscibility Analysis by Super Critical Extractor*

The results of the tests are shown in Figure 5. The graphs have irregular shapes, the peaks matching the highest rates of recovery were observed at 180 bar and at temperatures of 70°C, 60°C and 50°C. This MMP was far from the value indicated by three slim-tube results for oil that give an MMP of about 74 bar at 31°C [11-14]. The oil recovery rates in the pressure range from 100-430 bar were close at all three temperatures. The graphs for 60°C and 50°C were more similar. The graph at 70°C has more deviations, which do not coincide with the deviations in the curves at the other two temperatures. The oil displaced at all the three temperatures was highest at 180 bar. The increase in oil recovery rates in the range of 100-430 bar was temperature dependent. It was more gradual at 70°C and 50°C, while there was a sharp step at 160 bar for the 60°C curve. The visual observation of the oil samples shows that the viscosity decreases and flows easily as the pressure increases from 100 bar to 180 bar, but after crossing the180 bar pressure the oil becomes viscous and slowly ceases to flow. During the whole process of extraction 37.39% of oil was extracted at 70°C, 32.27% at 60°C and 25% at 50°C as shown in Table 3.

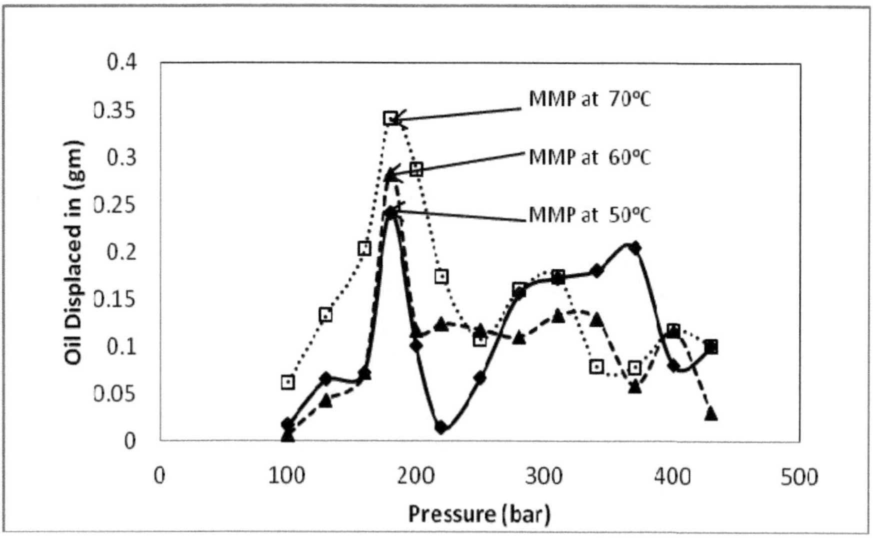


**Fig. 5** Amount of Oil displaced Vs saturation pressure at different temperatures

**Table 3** Showing the crude oil recovery from the SCE

| **Temp ($^0$C)** | **MMP (bar)** | **Total wt. of oil recove-red (gm)** | **Density of oil (gm/cc)** | **API gravity of oil ($^0$API)** | **Vol. of oil recovered (cc)** | **Saturated wt. of core sample (gm)** | **Dry wt. of sample (gm)** | **Wt. of oil saturated in core (gm)** | **% recovery** |
|---|---|---|---|---|---|---|---|---|---|
| 70 | 180 | 1.92 | 0.84 | 37.09 | 2.29 | 30.68 | 25.54 | 5.13 | 37.39 |
| 60 | 180 | 1.48 | 0.84 | 37.09 | 1.76 | 31.25 | 26.67 | 4.57 | 32.27 |
| 50 | 180 | 1.34 | 0.84 | 37.09 | 1.60 | 32.46 | 27.09 | 5.37 | 25.01 |

It was observed that the miscibility between $CO_2$ and crude oil requires restricted conditions of temperature and pressure rather than simply dissolving the $CO_2$ in the oil. Miscibility is attained when the IFT between $CO_2$ and oil is eliminated. At temperatures of 70°C, 60°C and 50°C the MMP was observed to be 180 bar, below which the interface will remain and it will be an immiscible flooding process. In addition $CO_2$ and crude oil, because of the difference in their properties and composition, will not become miscible at the first contact regardless of pressure. These materials will have what is called multiple contact miscibility i.e., $CO_2$ must repeatedly contact the oil. The concentration gradient from oil to $CO_2$ enables many hydrocarbon molecules specially C5-C30 to leave the oil and enter the $CO_2$. After several contacts many of these hydrocarbons will join the vapour ($CO_2$) phase, so that the vapour phase becomes miscible with crude oil. At this point the interface will disappear, capillary force will become zero and theoretically about 100% oil can be displaced from the part of the reservoir contacted by $CO_2$.

There is a difference between $CO_2$ dissolving in crude oil and $CO_2$ being miscible with crude oil. As pressure is applied to $CO_2$ – crude oil system, the crude oil will readily dissolve until the oil is saturated with $CO_2$ at the existing pressure and temperature. At that time both free $CO_2$ and $CO_2$ saturated oil will be present with an interface between the two phases. Dissolving the $CO_2$ in this manner will result in an expansion of the liquid phase and a reduction of the liquid viscosity. A $CO_2$ solution will take place regardless of the composition or API gravity of crude oil. It is obvious that the swelling of the oil will increase the oil saturation and therefore enhance the relative permeability of the rock to oil (Kro). In addition the residual oil saturation that will remain after $CO_2$ flooding will be swollen with a relatively large volume of $CO_2$. Both reduction of μo and Kro will facilitate flow of the swollen oil to the production well.

## 6.3 *Gas Chromatography*

To determine the hydrocarbon peak distribution in the oil samples obtained from SCE at the value of MMP (180 bar) and at temperatures of 70°C, 60°C and 50°C , a "standard" chromatography from the selected crude oil sample was run, and peaks from the sample were compared to the known peaks on the standard for identification. Figure 6 to Figure 8 shows the majority of n-alkane distributions for whole-oil gas chromatograms. A broad spectrum of n-alkanes ranging from n-C10 through n-C35 is observed. Modest odd-carbon preference in the n-C15 through to n-C19 range was also observed. The presence of isoprenoids pristane (C19H40) and phytane (C20H42) was also detected. In Figures 6 to 8 the heavy

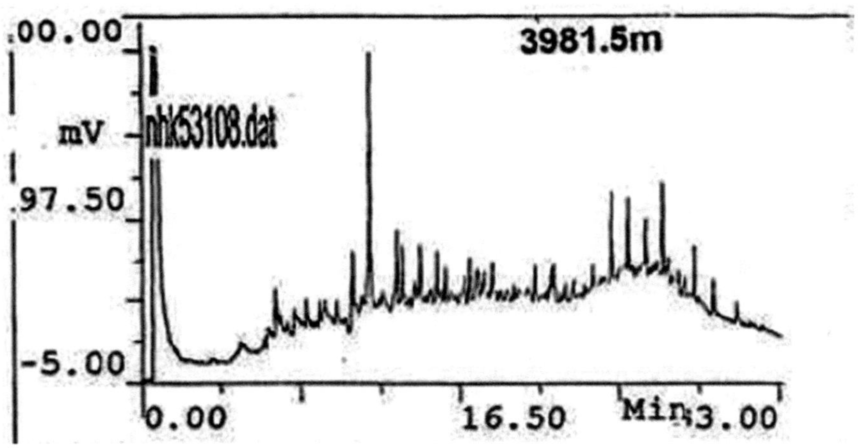


**Fig. 6** Shows the GC analysis of oil collected at 180 bar and temperature of 70$^0$C from the core sample

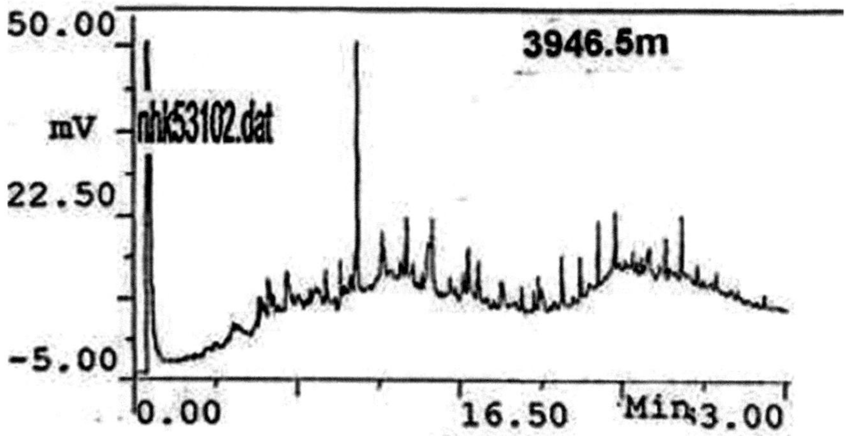


**Fig. 7** Shows the GC analysis of oil collected at 180 bar and temperature of $60^0C$ from the core sample

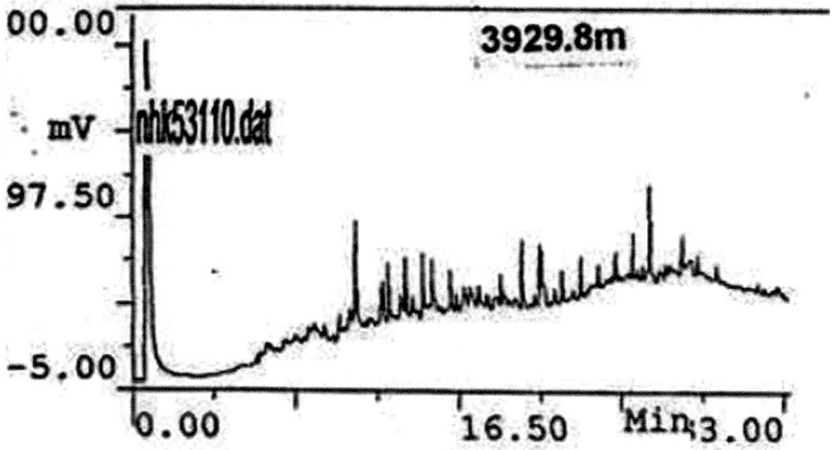


**Fig. 8** Shows the GC analysis of oil collected at 180 bar and temperature of $50^0C$ from the core sample

hydrocarbons from C18 to C20 can be seen which is a confirmation that $CO_2$ can displace heavy hydrocarbons by reducing the viscosity and improving the mobility ratio.

## 7 Conclusion

The percentage of lighter hydrocarbons from $C_1$-$C_{12}$ in the reservoir crude oil, as obtained from Table 2, was 8% and 44.6% of the distillates ($C_1$-$C_{19}$) up to 300°C.

When compressed $CO_2$ was injected into the plugs of the core at high density samples saturated with crude oil, became miscible with 44.6% distillates of the oil.

From visual observations it was clear that the viscosity of the extracted oil had reduced which indicates that $CO_2$ displaced the oil from the pores of the rock. As $CO_2$ dissolves in the oil the oil swells and reduces its viscosity which improves the efficiency of the displacement process.

From the GC analysis (Figures 6, 7 & 8) of the extracted oil samples, it was observed that a broad spectrum of n-alkanes ranging from n-$C_{10}$ through n-$C_{35}$ was present. Modest odd-carbon preference in the n-$C_{15}$ through n-$C_{19}$ range was observed and both isoprenoids pristane ($C_{19}H_{40}$) and phytane ($C_{20}H_{42}$) were also detected. The presence of heavy hydrocarbons from $C_{18}$ to $C_{20}$ predicts that $CO_2$ had displaced heavy hydrocarbons by reducing the viscosity and improving the mobility ratio.

The Nahorkatiya crude oil used for the study had an average API gravity of 30°. However, the extracted crude oil from the SCE had an average API gravity of 37° as shown in Table 3, which predicts that the oil became lean due to it swelling from the addition of lighter components

Results of Figure 5 predicts that during the whole process of extraction in the SCE 37.39% of oil was extracted at 700C, 32.27% at 60°C and 25% at 50°C at MMP of 180 bar in all the cases. This means that the higher the temperature of the plugs the better the oil recovery at MMP will be.

**Acknowledgements.** I would like to express my deepest gratitude to my supervisor, Professor Narendra Nath Dutta, who contributed immensely in carrying out the work. It was my privilege to be his scholar and to complete the work under his supervision.

## References

1. Sen, R.: Biotechnology in petroleum recovery: The microbial EOR. J. Progress in Energy and Combustion Science 34, 714–724 (2008)
2. Green, W., Willhite, G.: Enhanced Oil Recovery, Richardson, TX, USA. SPE Textbook Series, vol. 6 (1998)
3. Nedjad, M., Reza, A., Vafaie, S.M.: Determination of Minimum Miscibility Pressure by Analytical Method, Iran. J. Chem. Eng. 26, 11–17 (2007)
4. Wang, Y., Orr Jr., F.M.: Calculation of Minimum Miscibility Pressure. J. Pet. Sci. Eng. 27, 151 (2000)
5. Jessen, K., Standord, U., Stenby, E.H.: Fluid Characterization for Miscible EOR projects and $CO_2$ sequestration. In: SPE Paper 97192, SPE Annual Technical Conference and Exhibition, Dallas, Texas, October 9-12 (2005)
6. Rudyk, S., Spirov, P., Sogaard, E.: Extraction of Heavy Oil by Supercritical Carbon Dioxide. In: CHISA, Symposium on Supercritical Fluid Applications, C6.4, Serial number 0310 (2010)
7. Rudyk, S.N., Sogaard, E.G., Abbasi, W.A., Jorgensen, L.W.: Determination of minimum miscibility pressure in supercritical reactor using oil saturated sample. Chemical Engineering Transactions 17, 1531–1535 (2009a)

8. Rudyk, S.N., Sogaard, E.G., Abbasi, W.A., Jorgensen, L.W.: Determination of minimum miscibility pressure in supercritical reactor using oil-saturated sample. AIDIC Conference Series 9, pp. 253–260 (2009b)
9. Sovilj, M.N., Nikolovski, B.G., Spasojevic, M.D.: Critical Review of Supercritical fluid extraction of selected spice plant materials 30, 197–220 (2010)
10. Curvers, J., Van den Engel, P.: Gas Chromatographic method for simulated distillation up to a boiling point of $750^0$C using temperature-programmed injection and high temperature fused silica wide-bore columns. J. High Resolution Chromatography 12, 16–22 (1989)
11. Rahmatabadi, K.A.: Advances in Calculation of Minimum Miscibility Pressure. Dissertation for Doctor of Philosophy, The University of Texas at Austin (2011)
12. Creek, J.L., Sheffield, J.M.: Phase Behavior, Fluid Properties and Displacement Characteristics of Permian Basin Reservoir Fluid/$CO_2$ Systems. J. SPE Reservoir Engineering 8(1), 34–42 (1993)
13. Fong, W.S., Sheffield, J.M., Ehrlich, R., Emanuel, A.S.: Phase Behavior Modeling Techniques for Low-Temperature $CO_2$ Applied to McElroy and North Ward Estes Projects. In: Proceedings of SPE/DOE Enhanced Oil Recovery Symposium (1992)
14. Winzinger, R.: Design of a Major $CO_2$ Flood, North Ward Estes Field, Ward County, Texas. SPE Reservoir Engineering 6(1), 11–16 (1991)

# Performance Evaluation of Polymer Surfactant and Their Displacement Effects

Yikun Liu[1], Qingjun Deng[2], Gang Chen[2], and Shuang Liang[1]

[1] Key Laboratory of Education Ministry for Enhanced Oil Recovery, Northeast Petroleum University, 163318, Daqing China
[2] The First Oil Extraction Plant of Daqing Oilfield Limited Company, 163000, Daqing China

**Abstract.** Polymer surfactants have distinguished features such as salt resistant, high viscosity, high emulsification property and stability, with a displacement mechanism similar to ASP flood. It is one of the trendy chemical flooding technologies. This paper includes polymer surfactant physical-chemical properties experiments and a core displacement experiment. Research shows that the viscosity, viscosity stability and salt resistance performance of polymer surfactant are better than those of conventional polymers. When the polymer surfactant concentration reaches a certain value, an emulsification phenomenon is observed in the system. The study of the performance and displacement effect of polymer surfactant will be of great significance for future industrialized application in oil displacement.

**Keywords:** displacement effect, performance evaluation, Polymer surfactant.

## Preface

Polymer surfactant is a chemical agent developed by the Chemistry Institute of Academia Sinica. Despite its distinguished features such as being salt resistant, high viscosity, high emulsification property and stability, it is also of high technical sophistication, economic rationality and production management adaptability. It is the gordian technique for further enhanced oil recovery. Polymer surfactant flooding technology is currently at the stage of laboratory experiment and field test [1-3]. Further research is needed in the area of classifying polymer surfactant according to their types, physical and chemical properties and their displacement effects [4-6].This paper reports on an experiment to investigate the physical-chemical properties and core displacement effects of I type polymer surfactant. A comparison of polymer surfactant and conventional polymers will provide both a scientific and a theoretical basis for further enhanced oil recovery and industrialization application.

M.Z. Hou et al. (Eds.): *Clean Energy Systems in the Subsurface*, SSGG, pp. 171–176.
DOI: 10.1007/978-3-642-37849-2_14 

## Experiment Introduction

Quartz sand epoxy adhesive bond artificial cores, with a geometrical dimension of 30.0cm×4.5cm×4.5cm were used for the experiment. Gas logging permeability was about 2.0μm$^2$, the water for the experiment was from the Daqing Oilfield First Oil Extraction Plant sewage system. The oil used for the experiment has a viscosity of about 10mPa•s at 45°C and was from class 1 dewatered oil from the west block in the First Oil Extraction Plant, The polymer surfactant used was the I type developed by the Chemistry Institute of Academia Sinica, the basic physical-chemical parameters are shown in Table 1. The main experiment devices used are the RS-150 type rheometer and TX500 type interface tension instrument.

**Table 1** Basic performance of Polymer surfactant

| Solid content % | Hydrolysis degree mol% | Viscosity mPa.s | Insolubl e Wt% | Dissolution rate h | granularity % | |
|---|---|---|---|---|---|---|
| | | | | | ≤0.2m m | ≥1.0m m |
| 90.93 | 33.2 | 48.9 | 0.09 | <2 | 2.4 | 3.3 |

## 1 Evaluation of the Physico-chemical Properties

(1) Salt Resistance Property

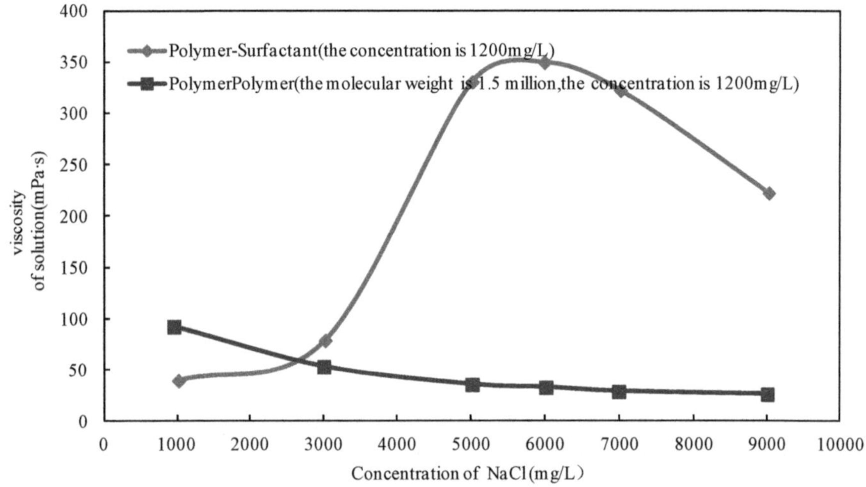


**Fig. 1** The viscosity and salinity relation curves of polymer surfactant and polymer in the same concentration

Figure 1 shows that, with increasing sodium chloride concentration, the viscosity of the conventional polymer solution declined while the viscosity of the polymer surfactant firstly increased to a certain maximum concentration and then began to decline The polymer surfactant had the advantage of being salt resistant under conditions of high salinity, which illustrated that the high salinity sewage prepared polymer surfactant solution was suitable for the actual application.

(2) Viscosity Property
Figure 2 showed that under the same concentration conditions, the viscosity of type I polymer surfactant solution was obviously higher than the conventional polymer solution. The polymer surfactant, whose increment of viscosity value was higher than that of the polymer solution as the concentration increased, showed a better viscosification power.

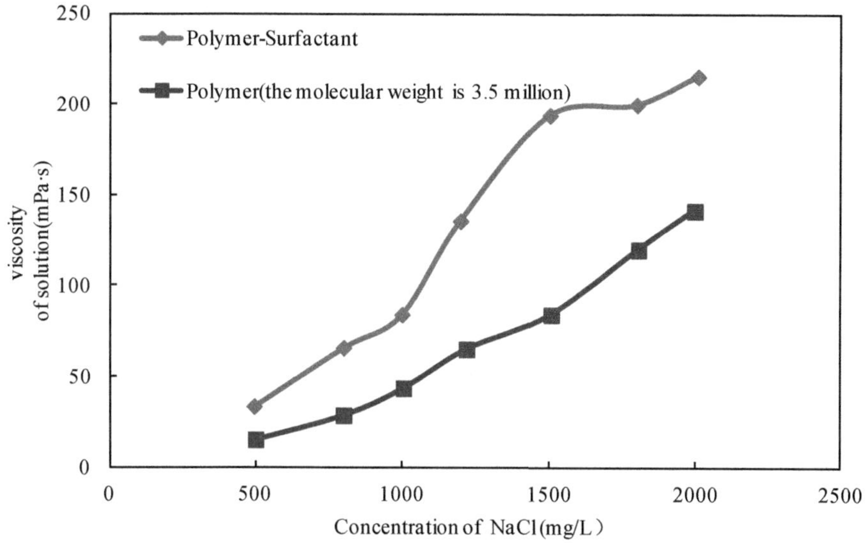


**Fig. 2** The concentration and viscosity relation curves of polymer surfactant and polymer

(3) Viscosity Stability
Table 2 shows that while the placing time was prolonged, the viscosity of polymer surfactant first increased and then decreased. The overall observation however, was that the viscosity of the polymer surfactant solution obviously increased and the higher the concentration, the larger the increment.

**Table 2** Sewage prepared polymer surfactant viscosity stability testing data table

| concentration \ Time (days) | 0 | 3 | 7 | 15 | 30 | 60 | 90 |
|---|---|---|---|---|---|---|---|
| 1500 (mg/L) | 251.7 | 201.7 | >1000 | 599 | 596 | 517 | 477 |
| 2000 (mg/L) | 297.6 | >1000 | >1000 | >1000 | >1000 | >1000 | >1000 |

(4) Emulsification Property
Different concentrations of polymer surfactant solution were prepared using on-site sewage mixed with crude oil on a 1:1 ratio and then placed in a 45°C incubator for observation of the emulsification behavior. After inspection it was observed that the type I polymer surfactant showed an emulsification phenomenon when the concentration reached 800 mg/L. The emulsion was primarily a reverse emulsion type which could not be separated after having been vibrated for 16 days.

## 2 Analysis of Core Displacement Effect

**Table 3** The core displacement experiment results

| Chemical agent types | Preparation water | Viscosity (mPa.s) | Core physical property | | | Displacement effect | | |
|---|---|---|---|---|---|---|---|---|
| | | | Kg (mD) | Φ (%) | So (%) | Water flooding (%) | Enhanced recovery (%) | Final enhanced rate (%) |
| polymer surfactant | sewage | 104.5 | 554 | 26.2 | 68.9 | 41.94 | 19.89 | 61.83 |
| polymer | sewage | 29.2 | 546 | 26.8 | 73 | 44.03 | 14.0 | 58.03 |

Table 3 shows the core displacement experiment results with a polymer surfactant concentration of 1000mg/L). The polymer solution with a molecular weight of 15 million was produced by Daqing Refining and Chemical Company. It was observed from table 3 that the displacement effect of the polymer surfactant solution was obviously better than that of the conventional polymer solution. A water flooding scheme was implemented using a polymer surfactant displacement, during which an enhanced oil recovery of 19.89% was achieved. After water flooding further polymer displacement enhanced recovery by a further 14%. Table 4 shows that with an increase in polymer surfactant concentration, the displacement effect was also improved.

**Table 4** The artificial homogeneous core displacement experiment results for different concentrations of polymer surfactant

| Scheme | Permeability (mD) | Porosity (%) | Oil saturation (%) | Water displacement recovery (%) | Improved value of Polymer surfactant flooding (%) | Total recovery (%) |
|---|---|---|---|---|---|---|
| 300mg/L | 1154 | 23.5 | 74.8 | 47.6 | 8.8 | 56.4 |
| 0.6PV | 1158 | 24.8 | 73.9 | 47.5 | 9.2 | 56.7 |
| 500mg/L | 1186 | 25 | 74.9 | 47.7 | 10 | 57.7 |
| 0.6PV | 1184 | 24.6 | 74.3 | 47.5 | 9.8 | 57.3 |
| 800mg/L | 1066 | 21.6 | 74 | 47.5 | 11.7 | 59.2 |
| 0.6PV | 1056 | 22.8 | 75.6 | 47.6 | 12 | 59.6 |
| 1000mg/L | 1107 | 21.6 | 71.9 | 47 | 15.9 | 62.9 |
| 0.6PV | 1123 | 22.6 | 73.5 | 47.3 | 16.2 | 63.5 |

## 3 Conclusions

Polymer surfactants have the advantage of a higher viscosifying property and better viscosity stability. While the placing time was prolonged, the viscosity of polymer surfactant increased. With increasing salinity, the viscosity of conventional polymer solution declined. The viscosity of the polymer surfactant first increased reaching a certain maximum concentration, and then began to decline, but the solution viscosity remained at a high value. Upon reaching a certain concentration, polymer surfactant concentration exhibited the emulsification phenomenon. The emulsion was primarily of the reverse emulsion type.

(2) The displacement effect of polymer surfactant solution was obviously better than the middle-to-high molecular weight polymer solution in the same concentration and as the polymer surfactant concentration increased, the displacement effect was also better. Further studies on the performance and displacement effect of polymer surfactant will have a great significance for industrialized application in future oil displacement.

**Acknowledgement.** This work is financially supported by the National Natural Science Foundation of China serial number 51074035. This work is also financially supported by the Major Projects of National Science and Technology of China serial numbers 2011ZX05010-002-05 and 2011ZX05052-12.

## References


1. Wang, D.: Develop new theory and technique of tertiary production to ensure continuous and stable development of Daqing oilfield. Petroleum Geology & Oilfield Development in Daqing 20(3), 1–7 (2001) (in Chinese)
2. Niu, J.: Practices and understanding of polymer flooding enhanced oil recovery technique in Daqing oilfield. Petroleum Geology & Oilfield Development in Daqing 23(5), 91–93 (2004) (in Chinese)
3. Wang, J.: Reservoir engineering method research and application. Petroleum Industry Press (September 1998 edition) (in Chinese)
4. Guo, W., Cheng, J., Liao, G.: Present Situation and Direction of Tertiary Recovery Technique in the Future in Daqing Oilfield. Petroleum Geology & Oilfield Development in Daqing 21(3), 21–24 (2002) (in Chinese)
5. Wang, K., Zhao, L., Shao, J., Gao, H.: Experimental Studies on the Performance and Oil Flooding Effects of Polymer Surfactant Solution. Petroleum Geology & Oilfield Development in Daqing 29(2), 1–10 (2010) (in Chinese)
6. Wang, Q., Ji, B., Sui, J., Guo, W., Ji, B.: Practice and Knowledge of Tertiary Recovery Technique in Daqing Oilfield. Petroleum Geology & Oilfield Development in Daqing 20(2), 11–17 (2001) (in Chinese)

# Production of Coal Bed Methane in Germany

Oliver Langefeld and Amit Agasty

Department of Underground Mining Methods and Machinery, Institute of Mining, Clausthal University of Technology, 38678 Clausthal-Zellerfeld, Germany

**Abstract.** The political situation in Germany and the EU in general, has provided an opportunity for energy sources like CBM. The Decision of the Federal Government to move from the traditional Nuclear Power to renewable and sustainable sources has led to the implementation of various energy sources including Geothermal and Power generation from mine gas. This does not only ensure energy security for the country but also mitigates climate change by keeping the greenhouse emissions within the EU norms and standards. Methane, the major component of mine gas, is a 20 times more potent greenhouse gas than $CO_2$. Extraction of gas from coal seams for power generation and heat supply thus reduces the emissions significantly. Abandoned Mine Methane (AMM) and Coal Mine Methane (CMM) projects are now prevalent in several sites in North Rhine-Westphalia and Saarland. As mining companies start to wind-up their activities in the country more AMM projects will start up. Along with this there is a need to invest in the development of virgin Coal Bed Methane projects to secure an abundant energy supply for the future. The Münsterland Basin has good potential for this. This paper discusses the state-of-the-art of degasification technology and the present production scenario for coal seam methane.

**Keywords:** Coal Mine Methane (CMM), Abandoned Mine Methane (AMM), Coal Bed Methane (CBM), Germany, Block Heat and Power Plants (Blockheizkraftwerke BHKW).

## 1 Introduction

According to phase III of the European Union Emission Trading Scheme (EU ETS) the annual allowable $CO_2$-Emmisions in the entire EU should be reduced by 21% of the present 2.1 billion tons by 2020 [1]. In addition, the decision made by the Federal Government of Germany in 2011 to accelerate the phasing out of the Nuclear Power on one hand necessitates securing the Energy supply in Germany and on the other hand gives a competitive advantage to the development of renewable energy sources. This leads to the "Renewable Energy Act"

M.Z. Hou et al. (Eds.): *Clean Energy Systems in the Subsurface*, SSGG, pp. 177–185.
DOI: 10.1007/978-3-642-37849-2_15 

(Erneuerbare-Energien-Gesetz, EEG), established in April 2000. The 2004 version of the EEG regulates compensation for power generation from mine gas, in cases where mine gas is produced from active or abandoned mines [2, 3].

The German mining laws (Bundesberggesetz) permit the extraction/production of mine gas from coal seams, a mineral resource. Methane, the major component of mine gas, is a 20 times more potent greenhouse gas than $CO_2$, and so its use for power production can contribute tremendously to the protection of the climate, while at the same time providing a sustainable source of energy. In the German state of North Rhine-Westphalia, where most of the coal seams are located, the government supports and recognizes the importance of mine gas as a future energy resource ("Landesinitiative Zukunftsenergien" 2001).

Different technologies are available to recover Coal Seam Methane, depending on the stage of recovery. This paper discusses the aspects related to these recovery stages, a state-of-the-art coal seam methane drainage technology and the current situation in Germany with regard to the production of mine gas.

## 2 Coal Seam Methane and Coal Seams in Germany

Mine gas consists mainly of methane, with its content varying from about 20% to over 90 % during production, depending on the drainage stage and method used. Methane found in coal seams was formed at the same time as coal. The methane content of a virgin seam depends on the depth of the seam, age and rank of the coal. Figure 1 shows an idealized concept of gas content increasing with coal rank [4]. Additionally Table 1 gives the mean methane content as a function of depth [5].

The German coal seams are concentrated in three main geographical regions. The largest of which is the Ruhr-Region in western Germany. Second is the state of Saarland and third is the Ibbenbüren-Region. In the east Ruhr region the seams are at a depth of about 1000 m with a vitrinite reflection of 1.7 %, accounting for about 88 % of the carbon. However, the west Ruhr region also contains Anthracite with a vitrinite reflection of 2.3% at same stratigraphic level, but different geothermal gradient and variable overburden. In principle, all types of coal can be found in the Ruhr region. Generally, coal seams are flat, with a slight dip towards the north-east. In the south of the Ruhr-region the seam is exposed to the surface. The overburden thickness increases towards the north, as depicted in Figure 3.

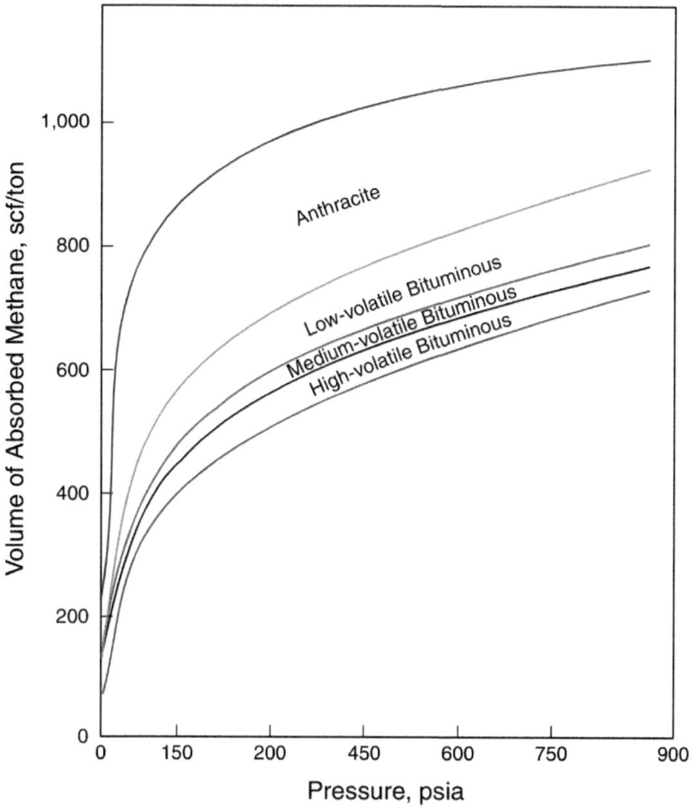


**Fig. 1** Gas content of a maturing coal seam [4]

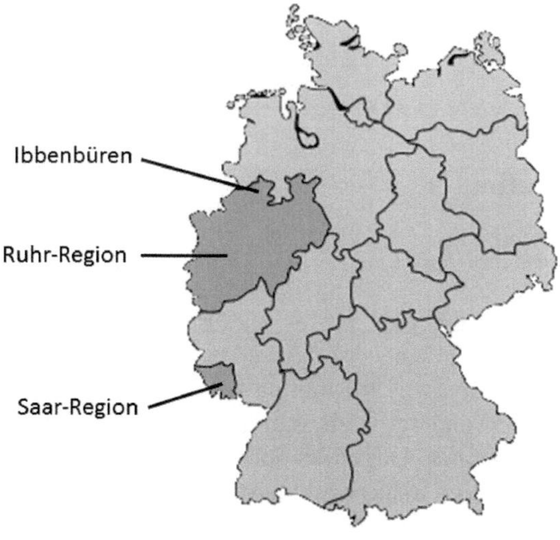


**Fig. 2** Coal Mining Regions in Germany

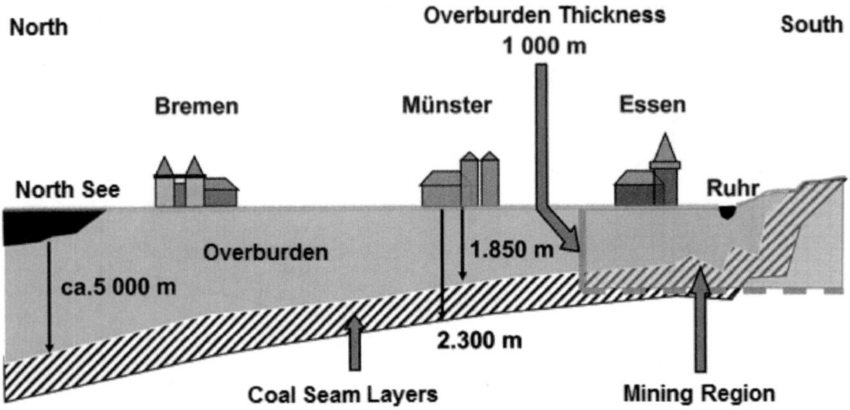


**Fig. 3** Dipping of the coal seam from the Ruhr-region to the North [2]

The total German coal bed methane reserve is about 3 trillion cubic meters, but the recoverable portion is restricted by geo-technical constraints, mostly the low permeability.

**Table 1** Methane content in coal [5]

| Depth Interval (meters) | Mean methane content (cubic meters per ton of coal) |
|---|---|
| 100 | 0.02 |
| 500 | 0.99 |
| 1000 | 3.73 |
| 1500 | 4.89 |
| 2000 | 7.09 |

# 3 Degasification

## *3.1 Coal Mine Methane*

In mining terminology the methane-air mixture is called "Fire Damp", and is highly explosive within a range of 5 to 15% of methane concentration in the mixture. This is the reason why the mining laws regulate the maximum allowable methane concentration underground, and appropriate dilution measures must be taken by the mine operator. Degasification can be achieved either by supplying enough fresh air ventilation to the working stations or by various possible methane drainage methods.

Ventilation of the underground workings serves two purposes, first to supply enough fresh air to the workforce and the equipment and second to dilute the contaminants, including poisonous or explosive gases, dust, heat etc. Fresh air is drawn into the mine by intake shafts or inclines and channeled through the intake airways to direct the required quantity to the working face. From here the polluted air is taken out of the mine by draining through the exhaust galleries and up the cast shaft. The methane content in the exhaust air is generally very low, usually less than 1 %. This is called Ventilation Air Methane (VAM).

In the case of a very gassy mine, methane drainage, either in the front or behind the working face is necessary for providing a safe environment underground. There are several different systems used in the world for draining methane from a gassy mine. When the seams are not at a great depth, drilling boreholes from the surface is a likely option (Figure 4). In Germany, however, where the coal seams lie at great depths, boreholes from another seam or in-seam are more common. These systems involve drilling boreholes, either from the surface or inside the mine to drain the methane from the coal seam sandwiched between the overlying and underlying strata so that the ventilation is not overloaded. Three different drainage options, however, exist for shallow seams: Surface pre-mining boreholes, horizontal pre-mining boreholes and post-mining (gob) boreholes. The first two options for degasification fall under the stage of methane drainage from active mines. Methane recovered during the mining activities, i.e., when the coal is in the process of being extracted, is known as Coal Mine Methane (CMM). In figure 4 different options of degasification are schematically represented, (1) horizontal pre-mining, (2) surface pre-mining (3) gob well and (4) VAM.

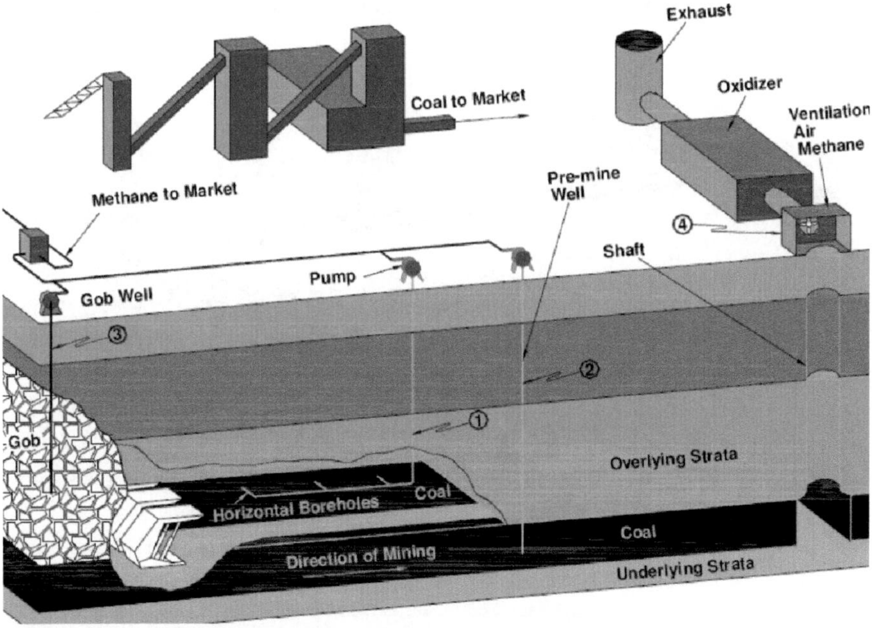


**Fig. 4** Types of Methane Degasification from an active mine [5]

In case of longwall mining 50 to 200 m into the coal seam from the longwall face, the over lying beds start to subside as the overburden pressure starts building up [6], this caused by the weight of the strata above that was previously supported by the coal extracted (now gob). The stress at the longwall front can be 4 to 5 times higher than the in situ stress. As a result of the high pressure the elasticity of the coal seam is exceeded and this results in the release of absorbed gases. The density of the mine gas released in this way is about 0.72 kg/m$^3$. As a consequence this gas tends to settle near the roof of the galleries resulting in the phenomenon of "Methane-Layering".

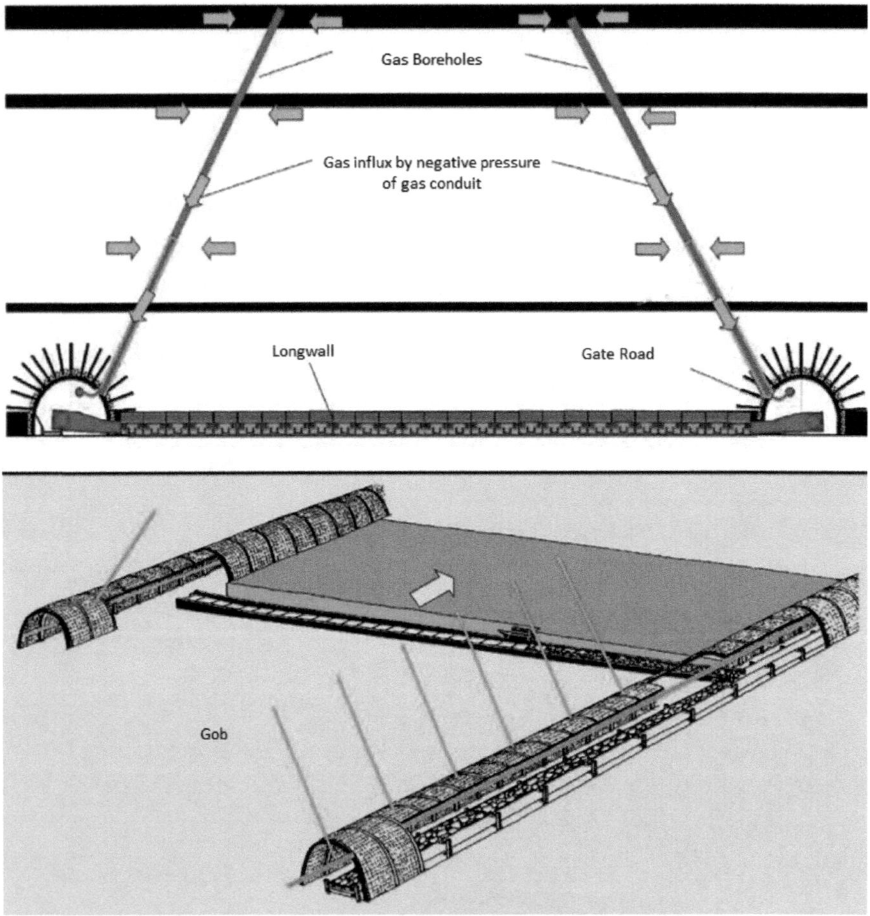


**Fig. 5** Horizontal pre-mining Methane Drainage in Ruhr-Region RAG [7]

### 3.1.1 Block Heat and Power Plants (Blockheizkraftwerk, BHKW)

BHKW is a modular unit used for generating power and heat. It works on the power-heat coupling and vis-à-vis principle. The mine gas recovered is compressed and supplied to a BHKW unit. In the state of North Rhine-Westphalia (NRW) multiple decentralized units are in operation while in Saarland this is achieved with a mine gas network of about 110 km [1]. A BHKW unit with a capacity of 1.35 MW prevents about 50,000 t $CO_2$-emissions annually. A STEAG Grubengas-Holding NRW GmbH (60 %) and RWE Power AG together own the Mingas-Power GmbH. Mingas-Power GmbH is involved in power generation from the active mine in NRW. It generates a total of 39 BHKW units from 15 sites, and its annual power production in 2012 was 191.1 GWh from 28 BHKW and a heat supply of 40.5 GWh.

## *3.2 Coal Bed Methane*

This is the methane recovered from virgin coal seams. These coal seams may be mined in the future but this is largely dependent upon geological factors such as coal depth and quality. As the coal seam is not yet loose the gas stays adsorbed and thus corresponds to higher methane content. Methane is recovered from unmined seams for two purposes, first when it is necessary to drain the gas before mining as a safety precaution against explosions during future mine operations, and second for energy production regardless of whether the coal is going to be excavated. CBM is recovered by means of surface boreholes and fracking techniques, which aim at increasing the permeability of the strata and hence, gas desorption. The coal seam needs to be dewatered in order to reduce pressure and release the methane from its adsorbed state. The horizontal drilling option is chosen as suitable in several articles, as it can recover higher volumes of methane in comparison to vertical wells. Its efficiency, however, remains low and is dependent on the length of the borehole through the seam.

Coal seams at extreme depths cannot be excavated because of safety, technological and economic reasons and in this case methane extraction is limited to purposes of energy generation. The methane recovery from this method may be over 95 %, as a result of not being exposed to the ventilation air as in case of CMM.

According to the 2006 estimations of CBM, a total global resource of 143 trillion cubic meters exists, of which only a small fraction has so far been recovered. This is as a result of lack of government incentives for its recovery. As mentioned above, compensation by the German government for power generation from mine methane is limited to mine gas recovered from active and abandoned mines. Russia, Canada, China, Australia and USA top the list of estimated resource bases in the world. The deeper coal seam towards the north, as represented in Figure 3, presents an opportunity for Virgin CBM recovery, since coal extraction at such extreme depths is likely to be impossible.

### *3.3 Abandoned Mine Methane*

When considering methane recovery from abandoned mines UK, USA and Germany have taken the lead in technology and project development. The primary aim for such projects is generally energy production but recently their significance in attempts towards greenhouse gas emissions control has been appreciated. The abandoned mines generally get flooded with water. In cases where the shafts are sealed off and the sealing standard is good, the volume of methane will be good. Another option is to leave the mines vented, such that the ventilation shafts are not sealed and fresh air flows through the mine allowing methane to escape into the atmosphere. As expected, abandoned mines that were sealed off present the highest potential for methane recovery, although their economic advantage in recovery lies within the first couple of years post-abandonment. The research in the Ruhr-region pointed out that 10-30 % of gas from the initial pre mining period remains in the abandoned mines. Recovery of the remaining gas is possible either by using the previous methane drainage system or else boreholes need to be drilled to access the gas reserve. The latter option poses a greater risk, as the borehole needs to be drilled in broken, fractured and subsided ground, which may result in the borehole collapsing.

The STEAG Grubengas-Holding NRW GmbH, Essen (74.8 %), the Green Gas Germany GmbH, Meerbusch and the Lambda Gesellschaft für Gastechnik mbH, Herten (12.6 % each) own the Minegas GmbH. Mine gas GmbH has so far produced 58 BHKW units of mine gas from 16 sites as of the end of 2012. The year's power production was 385.9 GWh from 48 units with a total heat supply of 4.8 GWh. Table 2 presents the cumulative supply of energy from mine gas in Germany, an initiative from the independent mining organizations.

**Table 2** Cumulative Mine gas use in Germany up to the year 2012 [8]

| Drainage Stage | Power Production (kWh) | Heat Supply (kWh) | Avoided $CO_2$-Equivalent (t $CO_2$) |
|---|---|---|---|
| AMM | 550,693,681 | 31,777,692 | 2,417,821 |
| CMM | 257,484,558 | 76,300,992 | 1,412,594 |
| Total | 808,178,239 | 108,078,684 | 3,830,415 |

## 4 Conclusion

The BHKW unit's capacity has increased over the years to 130 MW. As the mining operations in Germany start to wind up the CMM projects will need to be converted into AMM projects. This needs proper planning for mine closures. Moreover, since such reserves are limited it will be necessary to take up CBM projects to recover mine gas from deeper coal seams.

## References


1. Gesamtverband Steinkohle e.V.: Steinkohle und Umwelt aus dem Jahresbericht des Gesamtverbands Steinkohle e.V., pp. 44–53 (2012)
2. Minke, A., Beßelmann, F.J.: Gewinnung und Verwertung von Grubengas aus der Karbonlagerstätte in Nordrhein-Westfalen. 7. Altbergbau – Kolloquium, Freiberg (2007)
3. Bundesgesetz, Gesetz für den Vorrang Erneuerbarer Energien (Erneuerbare-Energien-Gesetz - EEG). Teil 3 §26 (retrieved February 18, 2013)
4. Kim, A.G.: Estimating Methane Content of Bituminous Coalbeds from Adsorptive Data. U.S. Bureau of Mines, Report of Investigation, RI 8245 (1977)
5. U.S. EPA, Identifying Opportunities for Methane Recovery at U.S. Coal Mines: Profiles of Selected Gassy Underground Coal Mines 2002 - 2006, EPA -430-K-04-003, pp. 2–5 (January 2009)
6. Reuther, E.-U.: Lehrbuch der Bergbaukunde. Auflage, Band, vol. 11, pp. 578–628. Verlag Glückauf GmbH, Essen (1989)
7. Junker, M., Barabasch, U., Thyrock, K.: RAG Mining Solutions. Arbeitskreis Grubengas Exportaktivitäten. Energie Agentur NRW (February 2010)
8. Weiß, E.G.: Personal communication on 21 January 2013 (2013)

# Numerical Study of $CO_2$-Injection Borehole Integrity with Consideration of Thermo-mechanical Effects

Michael Zhengmeng Hou[1,2], Zhi Qiao[3], Xuan Luo[4], and Qingyuan Wang[2]

[1] Institute of Petroleum Engineering,
Clausthal University of Technology,
Clausthal-Zellerfeld, Germany
[2] Sino-German Energy Research Center,
Sichuan University, Chengdu, China
[3] Herrenknecht AG, Schwanau,Germanny
[4] Energie-Forschungszentrum Niedersachsen,
Goslar, Germany

**Abstract.** Underground storage of $CO_2$ is one efficient solution to reduce $CO_2$ emissions and to slowdown the greenhouse effect. The objective of the CLEAN project is to investigate the feasibility of $CO_2$ storage in a depleted oil and gas reservoir. For this purpose, one part of the onshore-gas field "Altmark gas field" has been provided for the research. The objective of this paper is to verify the integrity of the research borehole during cooler $CO_2$-injection and to validate the closure of the injection induced cracks in cement due to salt creep by using numerical simulator FLAC3D. After numerical simulation, the critical value of the decreasing temperature in different segments was indicated. It was also proved that, the creep behavior of salt rocks can indeed close the cracks in the cement which leads to the recovery of borehole integrity.

**Keywords:** Altmark gas field, Borehole integrity, CCS, Borehole Sealing.

## 1 Introduction

$CO_2$ emission is one of the most dominant reasons for the greenhouse effect. Different technologies have been developed to reduce the $CO_2$ emissions. Underground storage of $CO_2$ is one such efficient solution. The second largest European on-shore gas field Altmark has been provided by Gaz De France-Suez Germany Ltd. for the research. The injecting borehole is constructed with casing and cement, which is located at a depth between ca. 2600 m and 3500 m surrounded by rock salt and sandstone. After 30 years of gas production, the temperature field in the casing, cement and rocks has reached equilibrium. When the cooler liquid $CO_2$ is injected through the casing into the reservoir, the

M.Z. Hou et al. (Eds.): *Clean Energy Systems in the Subsurface*, SSGG, pp. 187–196.
DOI: 10.1007/978-3-642-37849-2_16 

temperature field will be disturbed and the equivalent stress increased, which will induce shear failure of the steel and damage the borehole integrity. Moreover, the temperature interruption will cause different shrinkage of the steel (casing), cement and rocks, there may be cracks building in the casing and cement, especially at the contact areas between casing and cement or between cement and rock mass. However such cracks might be closed due to the creep of salt rock.

## 2 Numerical Simulation

Wellbore ITE1, through which gas was produced for 30 years, was constructed in 1984. As shown in Figure 1 the reservoir sandstone layer and bedding salt rock layer were simulated with FLAC3D from -2790 m to -3470 m. The sandstone section is located from -3310 m to -3470 m, while salt rock is located from -2790 m to -3310 m. The parameters for the wellbore construction and mechanical and thermal properties are listed in Tables 1, 2 and 3.

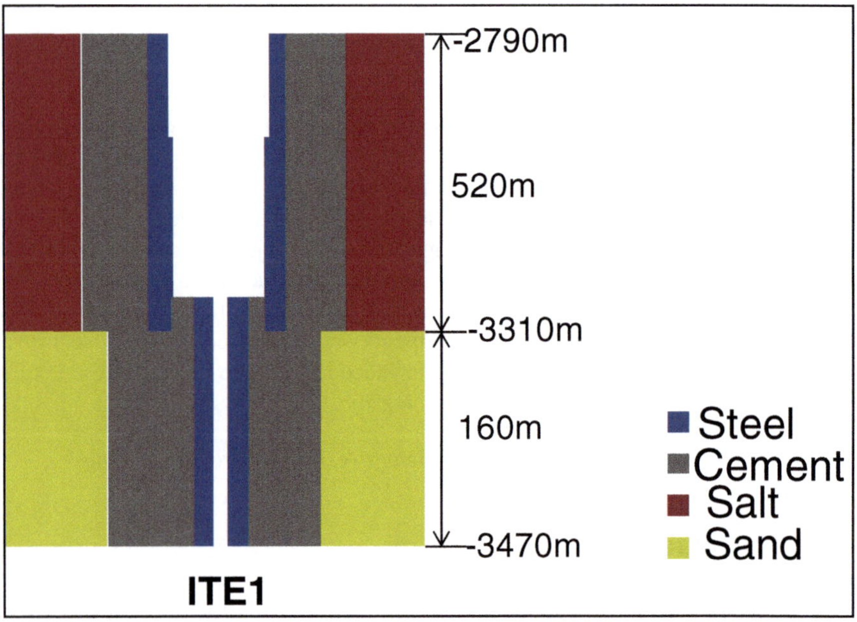


**Fig. 1** Construction of wellbore ITE1

Before $CO_2$-injection the wellbore was used as a production borehole for about 30 years. Therefore, the production phase must be simulated first and the injection phase should be simulated based on the results of the production phase. Because of limited hardware only the three segments shown in Figure 2 were numerically simulated. Detailed geometry is presented in Figure 3. Moreover, because of the radial symmetry of the wellbore only three 5 ° sectors were simulated. The salt rock, sandstone, steel and cement were characterized with LUBBY2, Strain-Softening, von-Mises and Mohr-Coulomb's constitutive model respectively.

According to the project partner's study, the temperature inside the casing will decrease by about 22 °C within the first day of $CO_2$-injection. If the injection rate is constant, the temperature field should stay constant. However, there may be perturbation during the injection, which will cause an immediate decreasing temperature [2]. Therefore, two concepts for the decreasing temperature were simulated: one is immediate decrease by more than 22 °C in the inner surface of steel pipe (temperature boundary condition) on the first day; the other one is the stepwise decrease of 22 °C in the inner surface of steel pipe (temperature boundary condition) on the first day followed by an immediate decrease of 1-10 °C (22+x °C). The vertical primary stress is equal to the integration of density along the depth. The two horizontal stress components are identical and equal $k$ multiplied by the vertical stress ($k$ is assumed to be 0.5). By analyzing the numerical simulations, the critical value of the interruption in the temperature due to the two decreasing concepts is likely to maintain the borehole integrity.

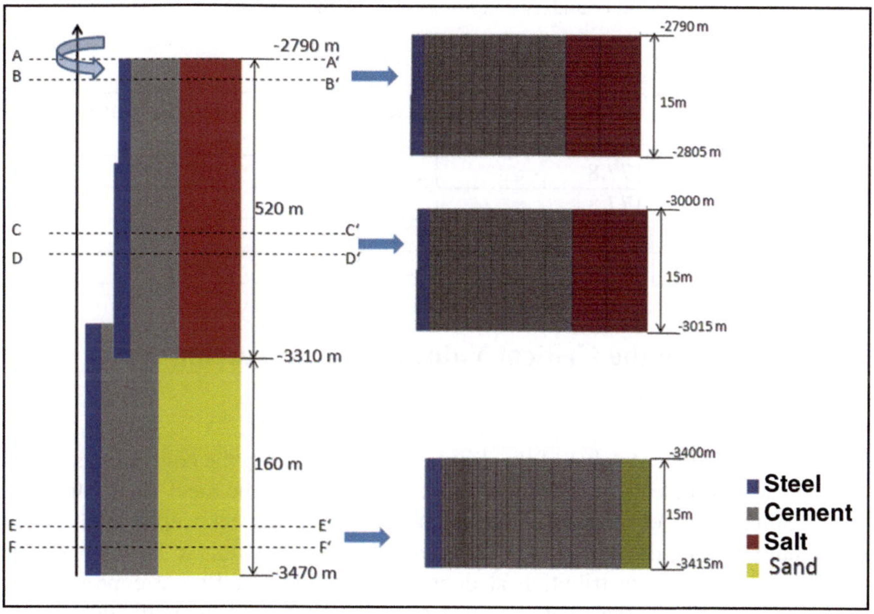


**Fig. 2** Three segments for simulation (ordered from the top down: segment 1, 2, 3)

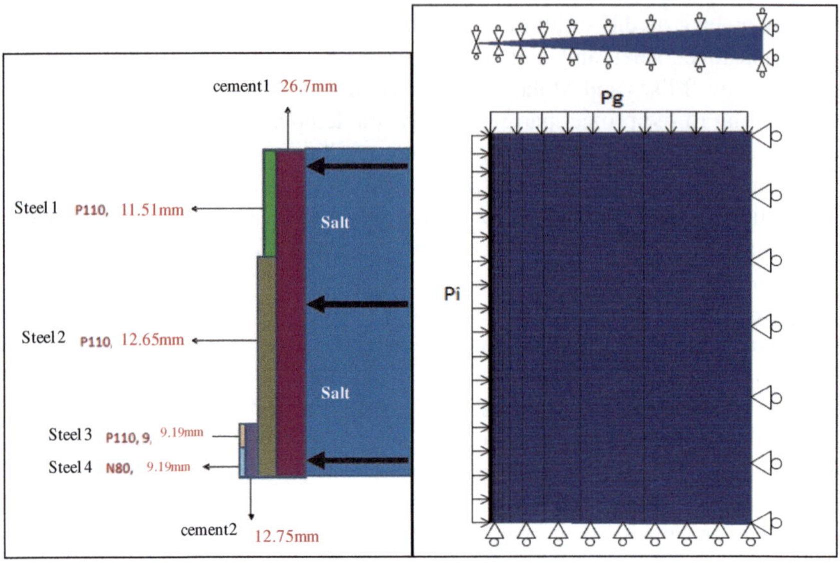


**Fig. 3** Detail geometry and boundary conditions of the simulated model

**Table 1** Geometrical parameters of borehole construction

| | outer diameter [mm] | range [m] | width [mm] |
|---|---|---|---|
| Casing string | 177.8 | 2770-2900 | 11.51 |
| Casing string | 177.8 | 2900-3310 | 12.65 |
| Liner | 127 | 3250-3275 | 9.19 |
| Liner | 127 | 3275-3470 | 9.19 |

## 3 Indicating the Critical Value of the Decreasing Temperature

To indicate the critical value of the decreasing temperature, the equivalent stresses around the borehole were recorded and compared with the steel shear strength. After simulation of the two temperature decreasing concepts, the stress distributions at the three segments were recorded. As an example Figure 4 shows the equivalent stress distribution at depth 2800 m under different conditions. It shows that, the more the temperature decreases, the bigger the equivalent stress ($\sigma_1$-$\sigma_3$) is. When the immediate decreasing temperature reaches 29 °C, the equivalent stress exceeds the steel shear strength (shown with a red dash), which means a shear failure. The stress distributions were recorded for all three segments, so that the critical value of the decreasing temperature for each of the three segments can be detected. Figure 5 shows the shear failure for segment 1

**Table 2** Mechanical parameters of steel, sandstone, rock salt and cement

| parameter | symbol | unit | steel | Sandstone | rock salt | cement |
|---|---|---|---|---|---|---|
| compression modulus | K | [MPa] | 1.75e5 | 5.376e3 | - | 4.7619 E3 |
| Shear modulus | G | [MPa] | 8.0769e4 | 3.702e3 | - | 4.3478E3 |
| Young's modulus | E | [MPa] | 2.1e5 | 9032 | 23363 | 1E5 |
| inter friction angle | $\phi$ | [°] | - | 37.5 | - | 29 |
| density | $\rho$ | [kg/m$^3$] | 7900 | 2260 | - | 2400 |
| Poisson value | $\nu$ | [-] | 0.3 | 0.22 | 0.3 | 0.15 |
| strength | F | [MPa] | 861.5(P110*)/655(N80*) | 3.69** | - | - |
| cohesion | C | [MPa] | - | 17.28 | - | - |
| Kelvin-shear module | $\overline{G}_k^*$ | | - | | 1.4e04 | - |

* P110 and N80 are the two different types of steel; the value is the compression strength of steel** tensile strength

**Table 3** Thermal parameters for steel, cement and rock salt

| | Temperature [°C] | heat capacity [J/kg·K] | thermal [[W/mK] | thermal expansion coefficient [K$^{-1}$] |
|---|---|---|---|---|
| steel | - | 480 | 48 | 1.2e-4 |
| cement | - | 1880 | 0.87 | 1.6e-4 |
| rock salt | 25 | 826 | 5.51 | 4.2e-5 |
| | 50 | 867 | 5.1 | |
| | 100 | 876 | 4.26 | |
| | 180 | 890 | 3.33 | |

immediately after a decreasing temperature of 30 °C. For the second decreasing temperature concept (step by step plus an immediate decrease of 22+x °C), the critical value is higher than that of the first concept for segment 1, and reaches 32 °C. The equivalent stresses are shown in Figure 6.

It is found that, in the 3 segments there were already 3 failures after the first day of injection (temperature decreasing by 22 °C). Therefore, the simulation of the second decreasing concept was not performed for segment 3. The critical value of two decreasing temperature concepts in segment 1 and 2 are listed in Table 4.

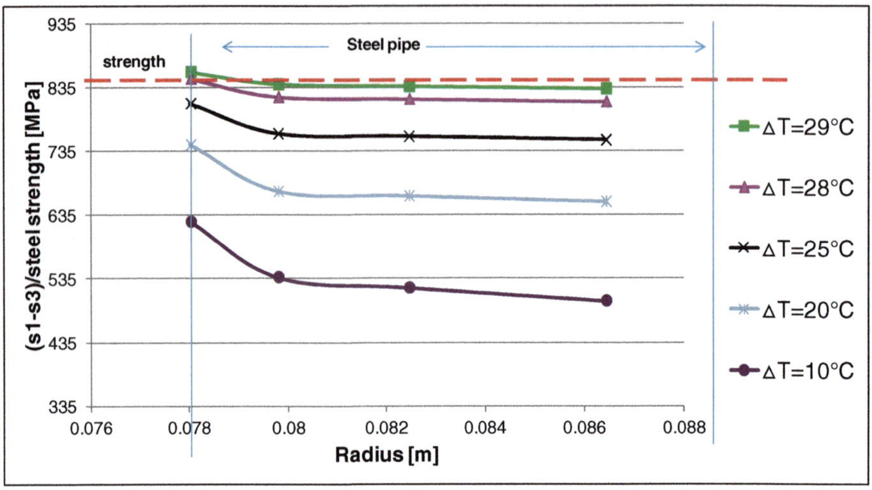


**Fig. 4** Equivalent stresses after different decreasing temperatures in the steel pipe (immediately)

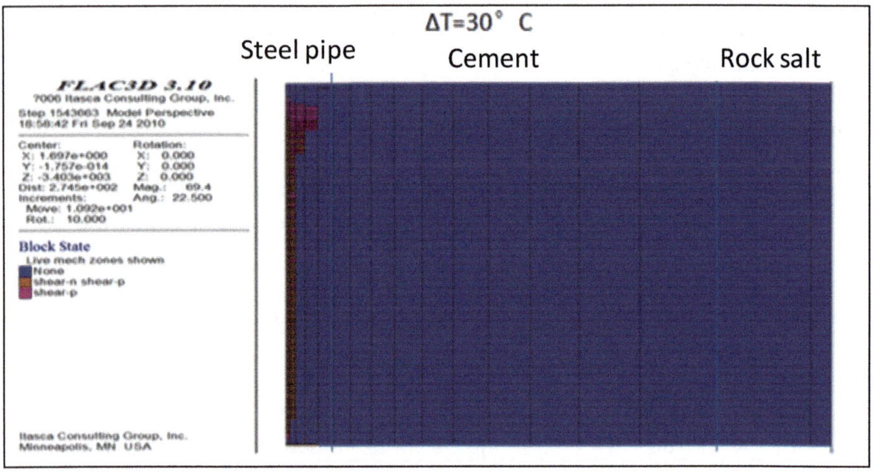


**Fig. 5** Shear failures in segment 1 after decreasing temperature of 30 °C

# 4 Analyzing the Influence of Salt Creep Behavior on Borehole Integrity

Another objective of this paper is to the find the reason for the closure of the $CO_2$ injection induced cracks. During the decreasing temperature process, the steel and cement shrink differently and induce cracks in the casing and cement, especially at

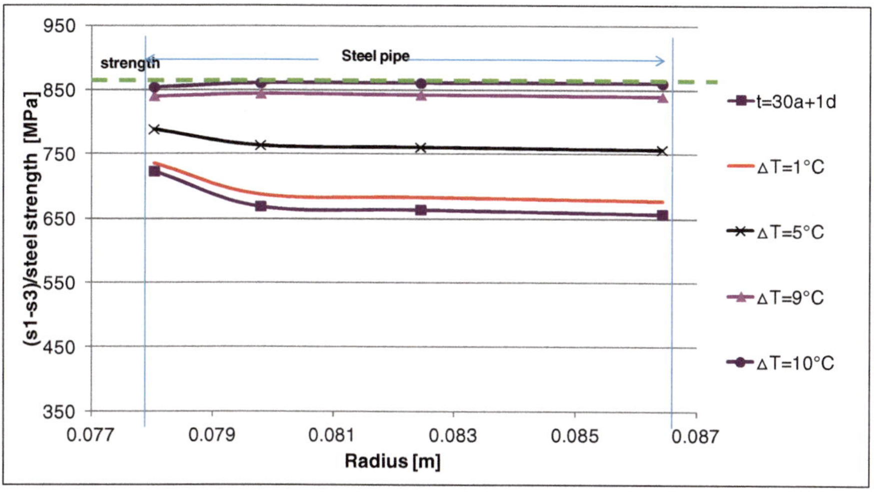


**Fig. 6** Equivalent stresses after one day step by step decreasing plus further immediate decreasing in the steel pipe (segment 1)

**Table 4** Critical value of decreasing temperature

| | step by step +immediately | Immediately |
|---|---|---|
| Segment 1 | 32 °C | 28 °C |
| Segment 2 | 30 °C | 31 °C |

the contact areas between casing and cement or between cement and rock mass. After the temperature decreases the salt rocks creep and push the cement in the steel direction. Therefore, it is possible that the cracks are closed due to the pushing effect. This question can be answered by analyzing the displacement in salt rock, cement and steel. Segment 1 was analyzed to find the reason for the closure of the cracks. After detecting the critical temperature, it is known that the integrity is damaged when the value of the decreasing temperature reaches 29 °C in segment1. Therefore, the displacement caused by the immediate decreasing temperature of 29 °C was analyzed (see Fig. 7). Generally, the nearer the point is to the steel, the bigger the displacement is. However, Figure 7 shows that the displacements of points 1, 2, 3 are different from each other. Point 2 has the greatest displacement, which is an anomaly (the displacement in the borehole direction is negative). This phenomenon can be explained by the creep behavior of salt rock.

The displacement of point 2 consists of two components, displacements induced by cooling and creep. To better understand the effect of creep behavior, another simulation using an elastic model is performed for comparison. The displacement differences of the two simulations are shown with Figure 8. The difference between point 1 and 2 means the displacement of point 1 minus the displacement of point 2. Therefore, if the displacement of point 2 is bigger than point 1, the displacement different between point 1 and 2 should be positive, because the displacement in borehole direction is negative. That means the bigger the different is, the greater the creep behavior is.

Comparing the difference between point 1 and 2, it is found that the displacement difference decreases at the beginning when using the visco-plastic model, which means that point 1 moves further than point 2 at the beginning because of the cooling effect. Then the increase difference slows, which means that point 2 moves further than point 1 because of the creep behavior. Compared with the result when using an elastic model for steel, sand and rock salt, the difference between point 1 and 2 is smaller than using the visco-plastic model while the difference between point 2 and 3 is bigger which indicates that point 2 moves more in the direction of the borehole so the cracks could be closed by this movement. The situation of the step by step decreasing temperature is different. The displacement of the situation (22+10 °C) in segment 1 is shown in Figures 9 & 10. It shows that all three points move fast into the borehole on the first day due to the shrinkage of the steel. Then the three points move slowly into the borehole until the fifth day. As in the first concept, the displacement of point 2 is bigger than point 1, which proves that the cracks can also be closed by the creep behavior of rock salt under the second decreasing temperature concept.

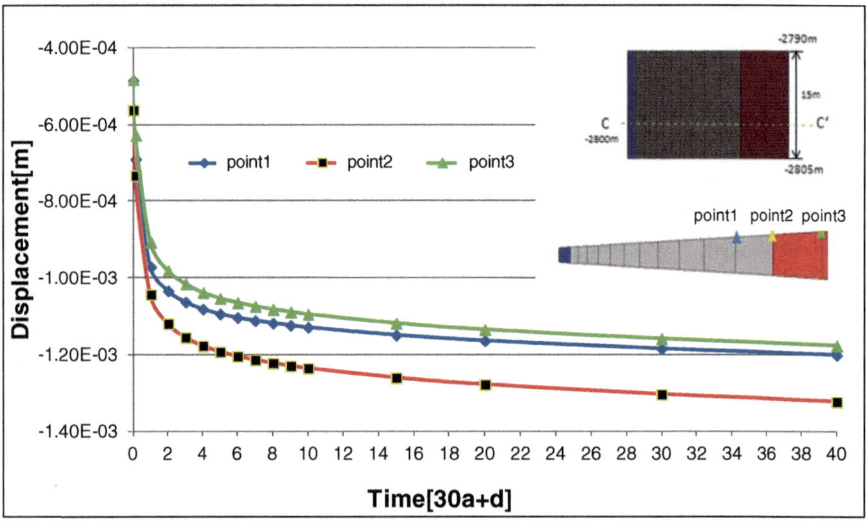


**Fig. 7** Radial displacement vs. time (Segment 1, ΔT=29 °C, imediately)

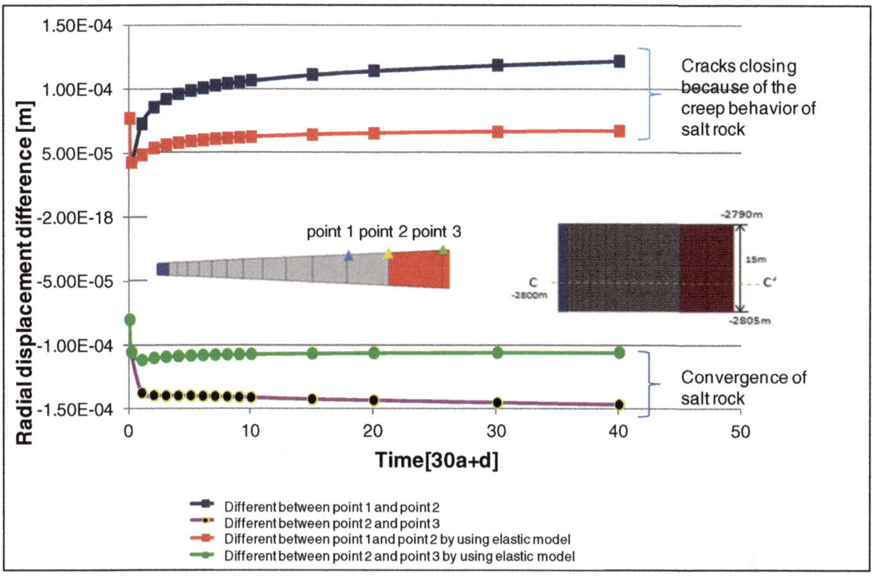


**Fig. 8** Radial displacement difference vs time (Segment 1, ΔT=29 °C, immediately)

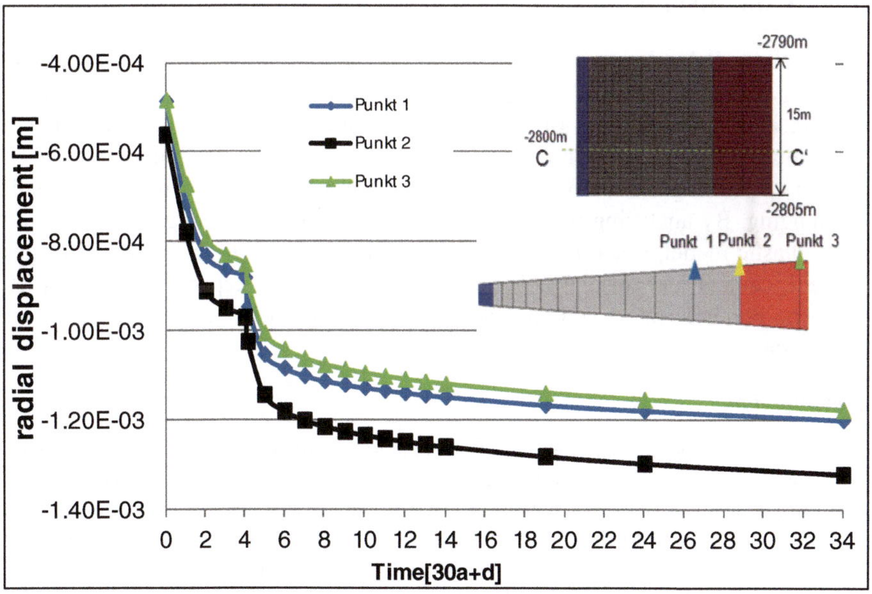


**Fig. 9** Radial displacement vs. time (ITE1, ΔT=22+10 °C)

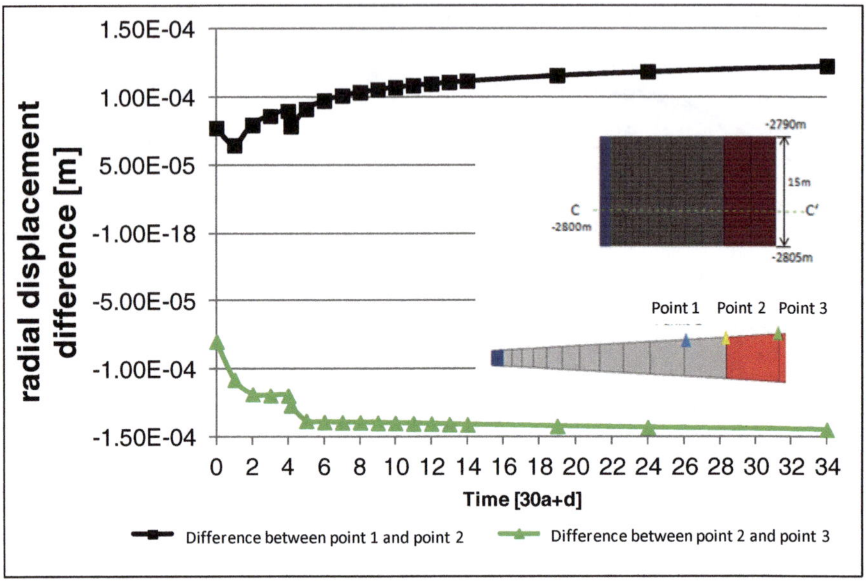


**Fig. 10** Radial displacement vs. time (Segment 1, ΔT=22+10 °C)

## 5 Conclusions

The numerical simulation of the $CO_2$ injection process verified that the borehole integrity can be maintained if the injection induced decreasing temperature is not too big. The critical value of the decreasing temperature is indicated under a different concept for a different segment by using the numerical simulation, even though controlling the temperature of steel during or after $CO_2$ injection is still challenging. By analyzing the displacement after simulation using the elastic and visco-plastic model, it is found that cracks are induced due to different shrinkages of steel, cement and rock, which can be closed by the creep movement of rock salt. The borehole integrity is recovered by the creep behavior of rock salt after several days.

## References


1. Hou, Z., Wundram, L., Gou, Y.: Entwicklung, Erprobung und Modellierung eines Langzeit-Bohrungsverschlusskonzeptes unter $CO_2$-Einwirkungen; Untersuchung zu Primärspannungszuständen, geomechanischen und hydromechanisch gekoppelten Eigenschaften von Reservoirgesteinen und Cap-Rocks in der Altmark; Numerische Untersuchungen zu THM-gekoppelten Prozessen im Reservoir Nahfeld. Abschlussbericht zum BMBF-Projekt (03G0704Q), Energie-Forschungszentrum Niedersachsen an der TU Clausthal (2012)
2. DBI, Concept suggestion for the $CO_2$-injection starting of the borehole Aaz 144 of DBI at GDFSUEZ within the project CLEAN, TV III. P II process simulation. DBI-gastechonogie institute GmbH Freiberg (2009) (not published)

# Numerical Simulation of $CO_2$ Leakage through Abandoned Wells during $CO_2$ Underground Storage

Mingxing Bai[1,2] and Kurt M. Reinicke[2]

[1] Northeast Petroleum University,
163318, Daqing, China
[2] Institute of Petroleum Engineering,
Clausthal University of Technology,
38678, Clausthal-Zellerfeld, Germany

**Abstract.** The storage of $CO_2$ in depleted oil and gas reservoirs, coal seams or saline aquifers is an important means of mitigating greenhouse effect in the environment. In $CO_2$ underground storage projects, well integrity is very important to ensure a safe execution of projects and storage of $CO_2$. It is a prerequisite for such projects and must be demonstrated for all wells affected by the injected $CO_2$ including abandoned wells. As part of one comprehensive methodology proposed by the authors, the leakage of $CO_2$ from the storage reservoir into the atmosphere or overlying aquifers over a certain time frame has been simulated. By building a model consisting of the critical system components, e.g., storage reservoir, casing-cement-rock composite system, injected $CO_2$ etc, according to a detailed study of features, events and processes (FEPs) which affect well integrity, a simulation is conducted for a time frame of 1000 years without consideration of geochemical influences. For the scenarios simulated, results show that $CO_2$ leakage rate is very small except for a reservoir under high pressure and poor quality cement-sheaths, which can lead to leakage rates in excess of the maximum allowable value.

**Keywords:** $CO_2$ underground storage, well integrity, abandoned well, leakage rate.

## 1 Introduction

The high energy production in the world results in a correspondingly high emission of $CO_2$ which is one of the greenhouse gases. To mitigate the negative influences on the environment, one possibility is to store $CO_2$ in the underground, e.g., depleted oil and gas reservoirs, saline aquifers or coal seams. It is usually referred to as $CO_2$ Sequestration and specifically as Carbon Capture and Storage (CCS). One important prerequisite for a CCS project is to ensure a long-term and

M.Z. Hou et al. (Eds.): *Clean Energy Systems in the Subsurface*, SSGG, pp. 197–210.
DOI: 10.1007/978-3-642-37849-2_17 

safe containment of $CO_2$ without adversely impacting the environment and the human health during both the injection and post-injection phase. The period for $CO_2$ sequestration projects is typically 10 to 50 years for the operation phase and 100 to 10,000 years for the post-injection phase (Gérard et al. 2006, Duval 2004) with a most likely value of 1000 years, recommended by IPCC (2005) and Bouc et al. (2007). Usually there are large numbers of wells in the field, some of which are abandoned and plugged. The integrity of wellbores is a risk for the long-term security of geological storage facilities, because $CO_2$ is known to cause severe corrosion in oil and gas production and transportation facilities, in particular if not designed for $CO_2$ service. Therefore it is of paramount importance to verify the well conditions and evaluate well integrity prior to the commencement of $CO_2$ injection.

The background of this work is linked to the $CO_2$ Large–scale Enhanced Gas Recovery project, conducted on the Altmark Natural Gas Field (CLEAN) in Germany. The Altmark natural gas field is the second largest onshore gas field in Europe and it covers an area of more than 1000 $km^2$. The natural gas is contained in the geological Rotliegend formation at a depth of approx. 3500 m. There is a geological barrier mainly consisting of a Zechstein salt layer above the sandstone formation. This salt layer can provide a natural barrier against $CO_2$ and formation fluid migration, and provide a unique abandonment method using the creep effect of salt. During the lifetime of the pilot project, nearly 100,000 tons of $CO_2$ are to be injected into the hydraulically and structurally isolated Altensalzwedel subfield at a depth of more than 3000 m.

Evaluation of actual well integrity is relatively simple for accessible wells, since they can be surveyed to directly assess the conditions of the wells. The evaluation of the integrity of abandoned wells or the evaluation of well integrity development over longer time frames can only be done indirectly and requires relevant information on geology and wells, and a comprehensive understanding of relevant THMC (thermal–hydraulic–mechanical–chemical) processes affecting well integrity. A comprehensive methodology to evaluate abandoned well integrity starting from a detailed FEP (Features, Events and Processes) analysis has been proposed in Reinicke et al. (2011a and 2011b). As part of this method, the leakage rate of $CO_2$ from reservoir to the atmosphere and overlying aquifer over long timeframe will be simulated by building a model consisting of the critical system components, e.g., storage reservoir, casing-cement-rock composite system, injected $CO_2$ etc, according to a detailed study of FEPs which affect well integrity. A simulation is conducted for a timeframe of 1000 years without considering geochemical influences.

## 2 Problem Description

A scenario is achieved by a combination of Features, Events and Processes (FEPs) which are all static and dynamic activities influencing well integrity. By using some semi-quantitative tools, e.g., interaction matrix, risk matrix etc, the most critical FEPs and system components are recognized. The developed scenarios will be evaluated by using simulation models. For the simulation, the radial models shown in Figure 1 are employed to simulate the leakage into (a) the atmosphere and (b) an overlying aquifer. In steps of increasing complexity, single-phase and two-phase simulations will be carried out. A Schlumberger Black Oil simulator E100 will be used for the simulations including chemical processes, more advanced simulators are planned for use in the future.

In the simulations, the reservoir is modeled as a closed tank containing 100,000 ton (approx. $5.55\times10^7$ $Sm^3$) of $CO_2$. Representative reservoir properties of interest are assigned to the model, i.e. a depth of 3,400 m, a temperature of 125 $^0$C, a porosity of 15 %, an initial water saturation of 20 %, and a permeability of 100 mD. Two values were assumed for the reservoir pressure. The first pressure of 50 bar reflects the current pressure while the second pressure of 450 bar resembles initial conditions. For this model, potential leakages were simulated for 1000 years assuming three different cement permeabilities: permeabilities representing good cement, poor cement, and defected cement (e.g., cracked cement, debonded cement, or cement with channels).

A three dimensional radial model was built with 10 grids in the radial direction, 4 grids in the azimuth direction and 39 grids in the vertical direction. The top view and side view of the model are shown in Figure 2. The basic parameters for system geometry and reservoir parameters, used in the simulation work, are listed in Table 1. They are representative of the area of interest for the CLEAN project.

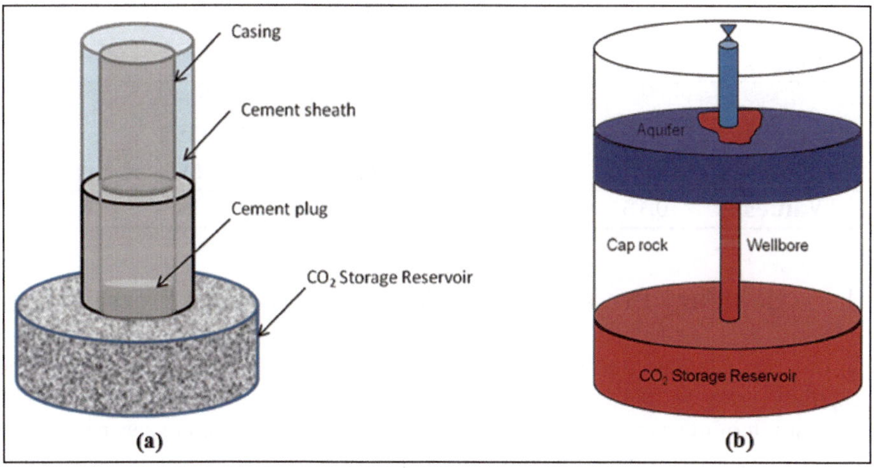


**Fig. 1** Radial model used to simulate (a) $CO_2$ leakage to the atmosphere and (b) an overlying aquifer

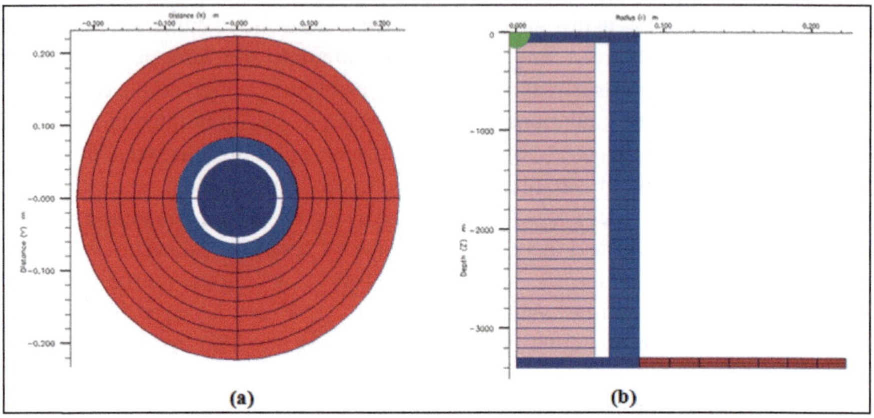


**Fig. 2** Well model for $CO_2$ leakage through cement sheath (a: top view; b: side view. In this figure, the red color represents formation, light blue represents cement sheath, deep blue represents cement plug, pink represents mud between plugs, and white represents casing)

**Table 1** Basic data for the simulation

| **Parameters** | depth, m | casing inner radius, m | casing outer radius, m | cement sheath thickness, m | gross res. thickness, m |
|---|---|---|---|---|---|
| **Values** | 3400 | 0.053 | 0.063 | 0.02 | 140 |
| **Parameters** | porosity, fraction | res. permeability, mD | initial res. $S_w$, fraction | brine density , kg/m$^3$ | brine viscosity, cp |
| **Values** | 0.15 | 100 | 0.2 | 1196 | 0.37 |

## 3 Capillary Pressure and Relative Permeability Correlations

In order to simulate the liquid and gas two-phase flow through the cement, the capillary pressure and relative permeability curves play a paramount role. The acquisition of the capillary pressure function is usually based on experimental data or in its absence by empirical correlations. The two models most often used are the Brooks and Corey correlation and the van Genuchten correlation (Brooks and Corey 1964 and 1966, van Genuchten 1980, Burdine 1953, Mualem 1976). The relationship of capillary pressure in terms of water saturation as well as the relative permeability equation for Brooks and Corey correlation is:

$$S_e = \frac{S_l - S_{lr}}{1 - S_{lr} - S_{gr}} = \left(\frac{p_d}{p_c}\right)^{\lambda} \quad \text{for } p_c \geq p_d \tag{1}$$

$$k_{rw} = S_e^{(2+3\lambda)/\lambda} \tag{2}$$

$$k_{rnw} = (1 - S_e)^2 (1 - S_e^{(2+\lambda)/\lambda}) \tag{3}$$

where Se is the effective water saturation, Pd is the displacement pressure or threshold pressure and $\lambda$ is pore size distribution index which is a material constant. The parameter $\lambda$ typically ranges from 0.5 for a wide range of pore sizes to 5 for a uniform pore size rock (Flett et al. 2004). In the Brooks & Corey equation, if $\lambda$=2, we can use the standard Corey correlation.

The relationship of capillary pressure in terms of water saturation and the relative permeability equation for the van Genuchten correlation is:

$$S_e = \frac{S_l - S_{lr}}{1 - S_{lr} - S_{gr}} = \left[1 + (\alpha p_c)^n\right] \quad \text{for pc} \geq 0 \tag{4}$$

$$k_{rw} = S_e^{\varepsilon}\left[1 - (1 - S_e^{1/m})^m\right]^2 \tag{5}$$

$$k_{rnw} = (1 - S_e)^{\gamma}\left[1 - S_e^{1/m}\right]^{2m} \tag{6}$$

In the equation, $\alpha$, $\varepsilon$, $\gamma$, n and m are Van Genuchten parameters.

After a detailed evaluation of empirical relative permeability correlations, in this work two different models are selected. For a good cement sheath whose permeability is assumed to be lower than 0.01 mD, the van Genuchten correlation is used. In the equation, $\varepsilon = 0.5$, $\gamma = 0.5$, m = 0.44 according to Mainguy and Coussy (2001). For poor and highly defected cement, the Brooks and Corey correlation is used. The relevant parameters in the equation are from a DGMK project which has been conducted by the Institute of Petroleum Engineering at Clausthal University of Technology. In the project the capillary pressure was measured in the lab for four types of rocks with different porosities, based on which the relative permeability is obtained.

## 4 Numerical Model Development

For the modeling process, the following assumptions have been made. Altogether these assumptions should lead to conservative results in the opinion of the author.

- The model consists of the most important system components: part of a reservoir, composite system, cement plug, CO2 and reservoir fluid (brine).
- Caprock above the formation is considered impermeable to the flow for CO2 and water. Leakage from the reservoir via the wellbore is the only way for CO2 to escape.
- The initial conditions in the reservoir are homogeneous.
- The reservoir is penetrated by only one abandoned well.
- No ageing processes (casing corrosion and cement carbonation) are considered.

The simulation processes cover two steps: single-phase flow (only $CO_2$) and two-phase flow ($CO_2$ and formation water) from the reservoir up to the surface or overlying aquifer which has a depth of 500 m.

During the life time of the wellbore, many processes could result in a mechanical failure of the casing-cement-rock composite system, leading to defects, e.g., gas channel, micro-annulus and micro-cracks, etc. The estimation of cement permeability is based on a detailed mechanical integrity evaluation. The resulting defects can be used to estimate the permeability based on empirical equations shown in Equation 7 and 8 (Huerta and Checkai 2009, Etiope and Martinelli 2001, Carey et al. 2009, Tran Viet 2012). The permeabilities are average permeabilities across the whole annulus area (A). Since the mechanical model is not included in this paper, three different types of cement are simulated. Each cement is assigned a different permeability value, for demonstration purposes. Table 2 lists the different cement permeabilities selected for the simulation.

For:

- Micro-annulus between cement and casing or cracks of cement sheath

$$kA = \frac{WB^3}{12} \tag{7}$$

- Gas channels

$$kA = \frac{\pi R^4}{8} \tag{8}$$

where A is annulus area, W is inner casing/cement sheath circumference for micro-annulus or cement sheath thickness for micro-cracks, B is the separation between casing and cement sheath for micro-annulus or aperture of the crack for micro-cracks, R is radius of the channel.

**Table 2** Different cement qualities and effective permeabilities

| Cement Quality | Permeability (mD) | Micro-annulus (mm) | Cracks (mm) | Gas channel (mm) |
|---|---|---|---|---|
| Good cement | 0.01 | - | - | - |
| Poor cement | 10 | 0.014 | 0.038 | 0.124 |
| Defected cement | 1000 | 0.065 | 0.177 | 0.392 |

## 5 Results and Discussion

For single-phase flow, the cement and the formation are assumed to be saturated with $CO_2$. The results are shown in Table 3 for an initial reservoir pressure $P_R$=50 bar and in Table 4 for $P_R$=450 bar. For the two-phase simulations it has been assumed that $CO_2$ is not soluble in water due to high salinity in water. The results for flow to the atmosphere are shown in Table 5 ($P_R$=450 bar). For two-phase flow into an overlying aquifer, the wellhead is shut off and $CO_2$ cannot escape to the atmosphere. However it will flow into the near ground aquifer. The brine content in the closed aquifer is assumed to be $1\times10^6$ $Sm^3$. The depth is 500 m. The results are shown in Table 6.

**Table 3** Simulation results for different cement qualities in the case of $P_R$=50 bar (Single phase)

| Cement quality | Permeability (mD) | Leakage volume (%) | Reservoir pressure drop (%) | Peak leakage rate ($Sm^3$/day) |
|---|---|---|---|---|
| Good cement | 0.01 | 0 | 0 | 0.000004 |
| Poor cement | 10 | 0.005 | 0.004 | 0.0074 |
| Defected cement | 1000 | 0.5 | 0.4 | 0.742 |

**Table 4** Simulation results for different cement qualities in the case of $P_R$=450 bar (Single phase)

| Cement quality | Permeability (mD) | Leakage volume (%) | Reservoir pressure drop (%) | Peak leakage rate (Sm³/day) |
|---|---|---|---|---|
| Good cement | 0.01 | 0.001 | 0.04 | 0.0006 |
| Poor cement | 10 | 0.226 | 1 | 0.343 |
| Defected cement | 1000 | 16.8 | 30 | 34.45 |

**Table 5** Simulation results for two phase flow to atmosphere in the case of $P_R$=450 bar

| Cement quality | Permeability (mD) | Leakage volume (%) | Reservoir pressure drop (%) | Peak leakage rate (Sm³/day) |
|---|---|---|---|---|
| Good cement | 0.01 | 0 | 0.016 | 0 |
| Poor cement | 10 | 0.003 | 0.03 | 0.00883 |
| Defected cement | 1000 | 4.39 | 11.79 | 7.597 |

**Table 6** Simulation results for two phase flow into an overlying aquifer in the case of $P_R$=450 bar

| Cement quality | Permeability (mD) | Leakage volume (%) | Reservoir pressure drop (%) | Peak leakage rate (Sm³/day) |
|---|---|---|---|---|
| Good cement | 0.01 | 0 | 0.016 | 0 |
| Poor cement | 10 | 0.004 | 0.03 | 0.006 |
| Defected cement | 1000 | 2.212 | 6.15 | 3.363 |

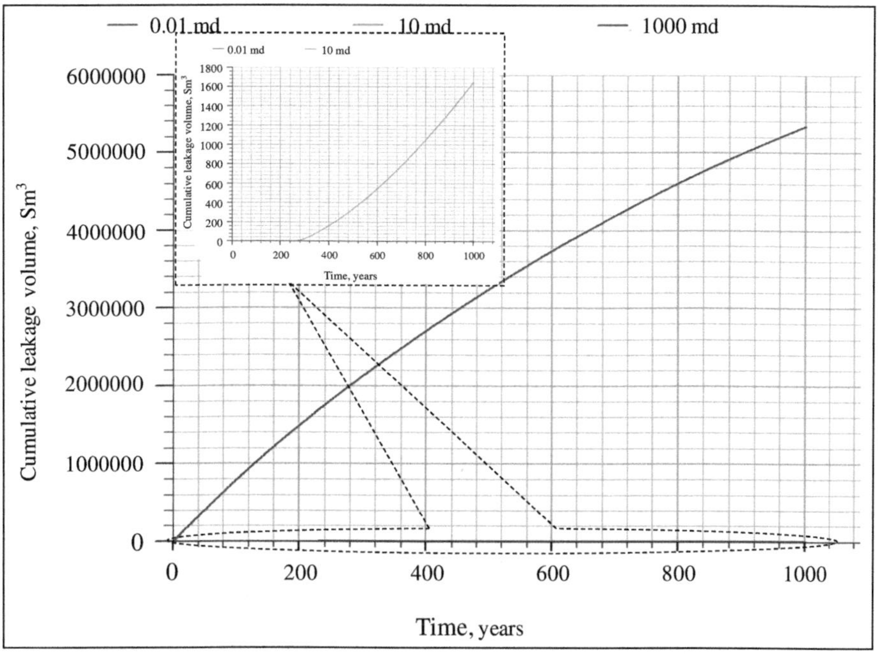


**Fig. 3** Cumulative leakage to the atmosphere for two-phase flow ($P_R$ =450 bar)

Figure 3 and Figure 5 show the cumulative leakage for two-phase flow under an initial reservoir pressure of 450 bar into the atmosphere and the overlying aquifer, respectively. Figure 4 and Figure 6 show the initial gas saturation as well as the development in 1000 yrs for three different types of cement in the case of two-phase flow into the atmosphere and the aquifer, respectively. In this figure, the blue represents the original gas saturation of 0% in the cement sheath, and the red represents gas saturation of 100%.

Based on all the results, several observations can be made:

- For severely depleted reservoirs the leakage rate from the reservoir to the biosphere is small, even for relatively high permeability of the cement sheath, e.g., a significantly damaged cement sheath.
- In a more realistic modeling of the two-phase flow of CO2 and water, there is a significant reduction in the leakage rate in comparison to a single-phase modeling because of the two-phase flow characteristics and the capillary pressure. The capillary pressure prevents the CO2 from displacing water.

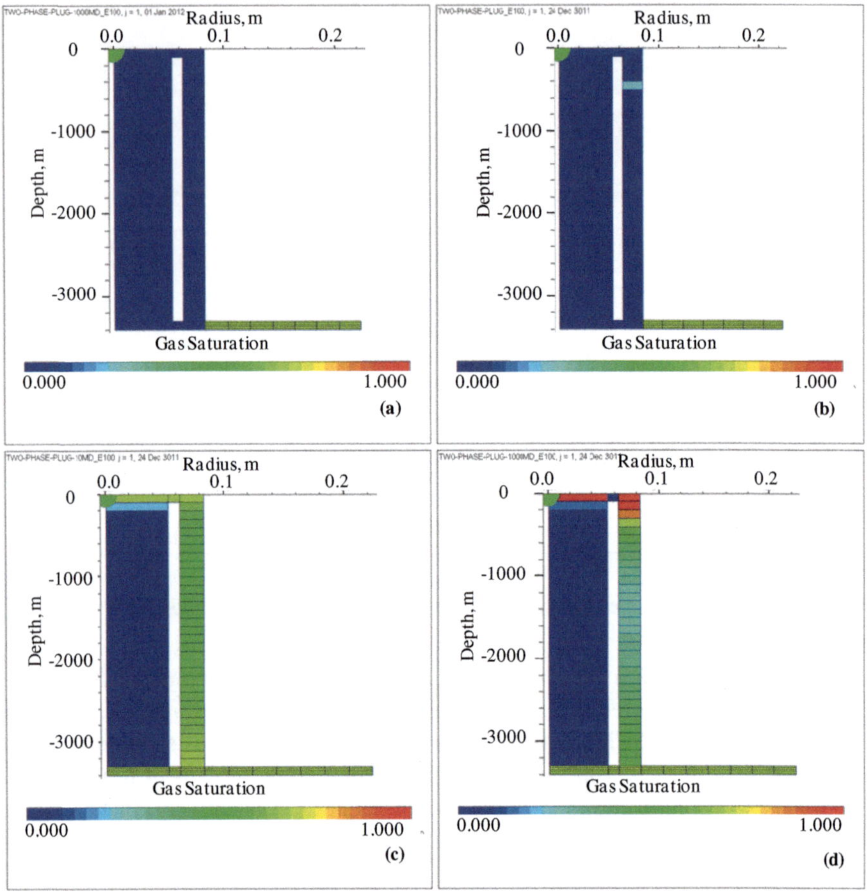


**Fig. 4** Gas saturation when t=0(a) and t=1000 yrs for k=0.01 mD(b), 10 mD(c) and 1000 mD(d) in the case of two-phase flow to the atmosphere

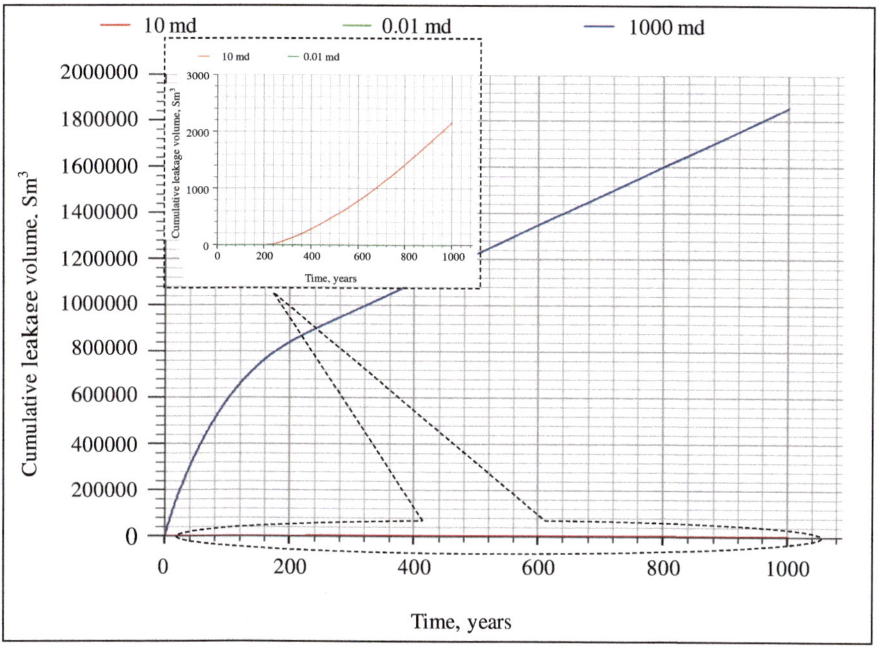


**Fig. 5** Cumulative leakage into the aquifer for two-phase flow (PR =450 bar)

- For the higher initial reservoir pressure of 450 bar, i.e. hydrostatic pressure conditions, the leakage rate is much higher.
- Among the scenarios, only when assuming a very pessimistic, worst-case scenario of a single-phase flow under the influence of a hydrostatic reservoir pressure towards the atmosphere (the permeability of the cement is 1000 mD) can the leakage rate reach a value which is above the minimum detectable limit of 50 kg/day (ca. 26.3 Sm3/day) according to SMRI (1996). This leakage rate will lead to a cumulative leakage per year which is above the maximum allowable value that is 0.01%, of the total cumulative stored volume according to Hepple and Benson (2002).
- Because of the aquifer backpressure, the leakage rate and volume into the aquifer are smaller than into the atmosphere under the same conditions.

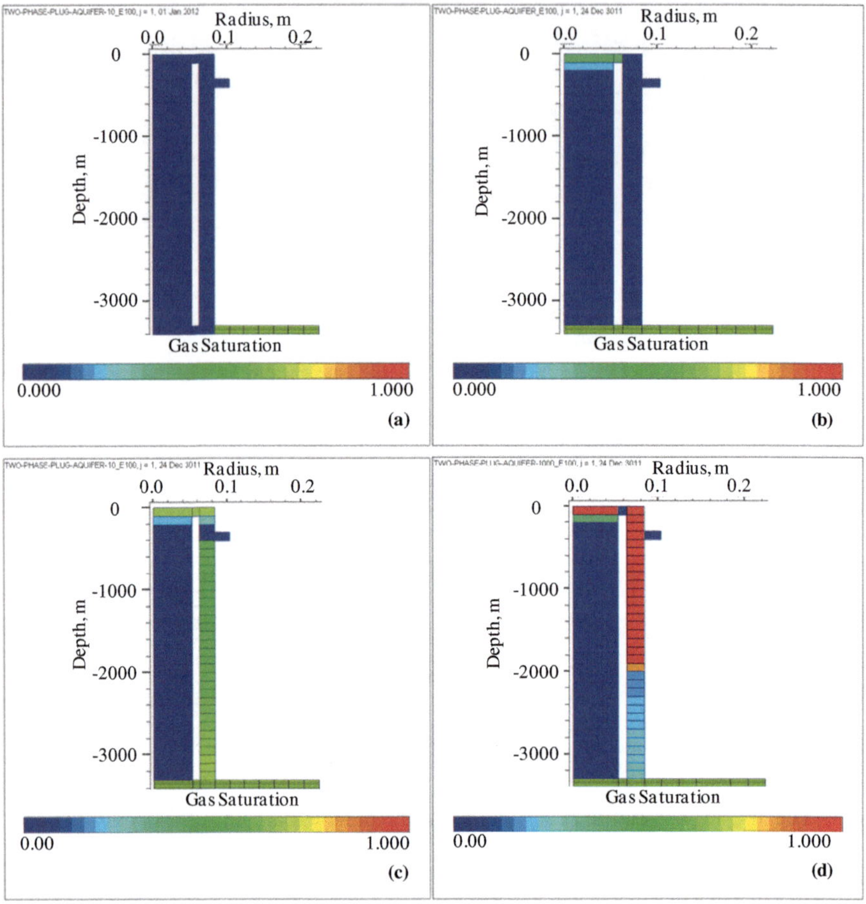


**Fig. 6** Gas saturation when t=0(a) and t=1000 yrs for k=0.01 mD(b), 10 mD(c) and 1000 mD(d) in the case of two-phase flow into the overlying aquifer

## 6 Conclusions and Recommendations

The leakage of $CO_2$ through an abandoned well has been simulated in this work. The model focuses on a one-phase flow of $CO_2$ and a two-phase flow of $CO_2$ and formation water, neglecting well cement degradation. For this, the commonly used relative permeability correlations have been reviewed. The simulation results show that for severely depleted reservoirs the leakage rate from the reservoir to the biosphere is small, even for relatively high permeability of the cement sheath, e.g., significantly damaged cement sheath. Among the scenarios, only when assuming a very pessimistic, worst-case scenario of a single-phase flow under the influence of a hydrostatic reservoir pressure towards the atmosphere (the permeability of cement is 1000 mD), can the leakage rate reach a value above the minimum detectable limit proposed by researchers. Cement based materials are reactive porous media. When exposed to an acidic environment, some dissolution/precipitation processes can occur and lead to modification of mechanical and transport properties. The coupled geo-chemical and geo-mechanical effects on cement properties should also be included into a future model with the help of other types of software.

**Acknowledgement.** This paper is part of the joint research project CLEAN, funded by the German Federal Ministry of Education and Research (BMBF) and the Germany Research Council (DFG). The authors would like to thank GDF SUEZ E&P DEUTSCHLAND GMBH for the research collaboration and the permission to publish this paper.

## References

1. Bouc, O., Quisel, N., Le Gouevec, J.: Risk and safety evaluation for $CO_2$ geological storage. In: Geotechnologien Science Report on 1. French-German Symposium on Geological Storage of $CO_2$, June 21-22. Geoforschungszentrum Potsdam (2007)
2. Brooks, R.H., Corey, A.T.: Hydraulic properties of porous media. Hydrology paper No. 3, Fort Collins, CO, Colorado State University (1964)
3. Brooks, R.H., Corey, A.T.: Properties of porous media affecting fluid flow. Journal of Irrigation and Drainage Division, Proceedings of the American Society of Civil Engineers (ASCE) 92(IR2), 61–68 (1966)
4. Burdine, N.T.: Relative permeability calculations from pore size distribution data. Transactions of the American Institute of Mining and Metallurgical Engineers 198, 71–78 (1953)
5. Carey, J.W., Svec, R., Grigg, R., Zhang, J., Crow, W.: Experimental investigation of wellbore integrity and $CO_2$–brine flow along the casing–cement microannulus. International J. of Greenhouse Gas Control 2010, 272–282 (2010)
6. Duval, P.P.: Sustainable future for hydrocarbons. IFP Contribution to TOG 2004 Tarablous, Libya (2004)
7. Etiope, G., Martinelli, G.: Migration of carrier and trace gases in the geosphere: an overview. Physics of the Earth and Planetary Interiors 129, 185–204 (2001)
8. Flett, M., Gurton, R., Taggart, I.: The function of gas-water relative permeability hysteresis in the sequestration of Carbon Dioxide in saline formations. In: SPE 88485 presented at the SPE Asia Pacific Oil and Gas Conference and Exhibition Held in Perth, Australia, October 18-20 (2004)
9. Gérard, B., Frenette, R., Augé, L., Desroches, J., Barlet, V., Jammes, L.: Well integrity in CO2 environments performance & risk – technologies. Schlumberger Presentation at Berkeley (March 21, 2006)
10. Hepple, R.P., Benson, S.M.: Implications of surface seepage on the effectiveness of geologic storage of carbon dioxide as a climate change mitigation strategy. In: Proceedings of the Sixth International Greenhouse Gas Technologies Conference, Kyoto, Japan, October 1-5 (2002)
11. Huerta, N., Checkai, D.A.: Utilizing sustained casing pressure analog to provide parameters to study $CO_2$ leakage rates along a wellbore. In: SPE International Conference on $CO_2$ Capture, Storage, and Utilization, San Diego, California, November 2-4 (2009)

12. IPCC, IPCC Special Report on Carbon Dioxide Capture and Storage, p. 442. Cambridge University Press, Cambridge (2005)
13. Mainguy, M., Coussy, O.: Role of air pressure in drying of weakly permeable materials. Journal of Engineering Mechanics 127, 582–592 (2001)
14. Mualem, Y.: A new model for predicting the hydraulic conductivity of unsaturated porous media. Water Resources Research 12(3), 513–522 (1976)
15. Reinicke, K.M., Weichmann, M.J., Teodoriu, C., Bai, M., Fichter, C., Krebs, R., Weinlich, F.: Research network well integrity-Project Well Condition. To published in GEOTECHNOLOGIEN (2011a)
16. Reinicke, K.M., Bai, M., Fichter, C., Weichmann, M.J. Weinlich, F.H.: Leckagerisiko von Bohrungen unter Einfluss von $CO_2$: Von den FEPs bis zu den Szenarien. In: DGMK/ÖGEW-Frühjahrstagung 2011, Celle, April 11-12 (2011b)
17. Tran Viet, T.: Methodische Entwicklung eines praktikablen Konzeptes zur Bewertung der Dichtheit verfüllter Bohrungen. PhD Dissertation, Clausthal University of Technology (2012)
18. Van Genuchten, M.T.: A closed-form equation for predicting the hydraulic conductivity of unsaturated soils. Soil Science Society of America Journal 44(5), 892–898 (1980)

# A Study on Shear Characteristics of a Smooth Rock Surface under Different Thermal and Mechanical Conditions

Taehyun Kim, Nan Zhang, and Seokwon Jeon

Department of Energy Resources Engineering,
Seoul National University, Korea

**Abstract.** High-level radioactive waste repositories are designed to ensure long-term stability, and it is necessary to consider different effects at depth in various aspects. Many uncertainties are involved in characterizing rock mass deformation and failure in the designing process. The interactions between fluid flow, high temperature and stress disturbance must be considered. In this study, multi–stage triaxial compression tests were conducted to investigate the shear characteristics of a saw-cut surface of crystalline rock under different thermal-mechanical conditions. Granitic rock was taken from KURT (KAERI Underground Research Tunnel) for the tests. The artificial shear surface was cut, using diamond saw, at 28° from the direction of loading. The thermal-mechanical testing condition was decided by considering the actual condition at the vicinity of a canister. Three different confining pressures, 5, 10, 15 MPa, were applied during the test at two different temperatures 20° C and 80° C. The shearing behavior of a saw-cut surface was analyzed based on the Mohr-Coulomb failure criterion. From the experimental results, the shear characteristic of the granitic rock was observed to be sensitive to confining pressure but not to temperature. Numerical analysis was conducted using COMSOL to simulate the test, providing a good agreement with the experimental results.

**Keywords:** Multi-stage triaxial compression test, T-M coupled process, shear behavior, crystalline rock.

## 1 Introduction

A high-level radioactive waste disposal facility is usually designed to be stable for an extended period of time as long as 10,000 years. To ensure the mechanical, hydrological and chemical stability of the facility, long-term performance is assessed with respect to many different variables and their coupled effects. It is well-known that the international cooperative project DECOVALEX accumulated a lot of information and knowledge of coupled processes in fractured rock masses. In addition, many independent studies were carried out to investigate the T-H-M-C coupled effects in numerical and experimental methods. They include research

M.Z. Hou et al. (Eds.): *Clean Energy Systems in the Subsurface*, SSGG, pp. 211–219.
DOI: 10.1007/978-3-642-37849-2_18 

works by Lockner et al. (1982), Elliott and Brown (1988), Yoon and Jeon (2004), Wanne and Young (2008) and many more.

However, few studies have been carried out on the shearing behavior of a shear surface of crystalline rock at an elevated temperature. In this study, a series of laboratory scale experiments were carried out to investigate the shearing behavior of a smooth surface of crystalline rock under triaxial compression with the simulated stress condition at depth and temperature condition in the vicinity of a canister for high-level radioactive waste. The test results were analyzed using Mohr-Coulomb criterion. In addition, numerical analysis was carried out using COMSOL Multiphysics to simulate the test.

# 2 Experimental Study

## *2.1 Sample Preparation*

Granitic rock cores obtained during site investigation at KURT (KAERI Underground Research Tunnel) were taken as test specimens. The specimens used in the tests were from a depth of 70 - 80 m. The size of the specimen for the triaxial compression test was approximately 47.7 mm in diameter and 100 mm in height. To make a smooth shear surface, the specimens were cut by a diamond saw and the angle of the saw-cut surface, θ, was 28° from the axis of the specimen as shown in Fig. 1. Table 1 presents the specification of a typical specimen.

## *2.2 Experimental Method*

Multi-stage triaxial test can generate a full failure envelope using a single specimen (Kovari et al. 1983). In this study, the multi-stage triaxial compression tests were performed under various confining pressure and temperature conditions.

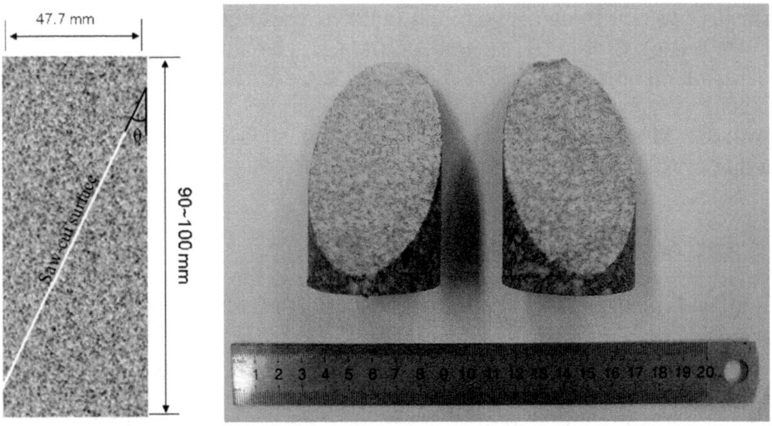


(a) (b)

**Fig. 1** Configuration of the test specimen: (a) cross-sectional view of the specimen, and (b) picture of the specimen with saw-cut surface

**Table 1** Specification of the specimens

| Testing condition | Sample ID | Diameter (mm) | Height (mm) | Inclined angle θ (°) |
|---|---|---|---|---|
| M | A_M_1 | 47.7 | 96.5 | 28 |
| | A_M_2 | 47.7 | 97.8 | 28 |
| T-M | A_TM_1 | 47.7 | 101.6 | 28 |
| | A_TM_2 | 47.7 | 86.7 | 28 |

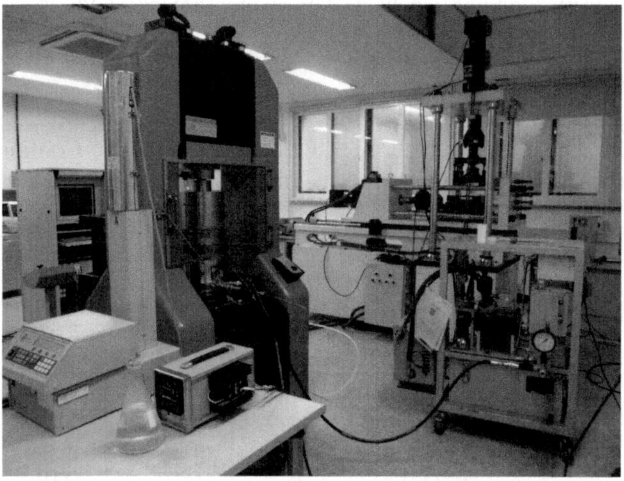

**Fig. 2** View of experimental set-up of multistage triaxial compression test under elevated temperature condition

To apply heat to the specimen during the test, the triaxial chamber was filled with oil and heated by a thermal jacket. The temperature was controlled at an accuracy of ±5°C. Pre-heating was required to make the temperature even throughout the chamber. The axial deformation of the specimen was not allowed during the pre-heating stage, which caused stress build-up in the specimen. Fig. 2 shows the view of the experimental set-up for the multi-stage triaxial compression test.

## 2.3 *Testing Conditions*

After preliminary study, variable testing conditions were decided as presented in Table 2. The temperature was set at 20°C and 80°C for the test. The rock temperature around the canister was reported to reach up to 80°C according to

Kwon and Cho (2009). The heating rate was set at 3°C per minute to prevent thermal shock to the specimen. In order to reduce the thermal gradient inside the specimen, the peak temperature was maintained for 90 minutes. Different confining pressures, 5, 10, 15 MPa, were applied to the specimen using a servo-controlled hydraulic unit with the accuracy of 0.1 MPa.

## 2.4 *Results and Discussion*

As the test specimen approached the peak axial stress, sliding occurred along the saw-cut surface creating small amount of gouge. Peak shear strength and friction angle of the shear surface were investigated using Mohr-Coulomb failure criterion. Results of the experiments are summarized in Table 3. Fig. 3 shows the relationship between the normal stress and shear stress acting on the shear surface.

**Table 2** Variable conditions for the multi-stage triaxial compression test under various mechanical and thermal conditions

| Temperature of specimen (°C) | Heating rate (°C/min) | Loading rate (mm/min) | Confining pressure (MPa) |
|---|---|---|---|
| 20, 80 | ≤ 3.0 | 3.0 | 5, 10, 15 |

**Table 3** Results of multi-stage triaxial compression test on saw-cut specimens under the room temperature (20° C) and the elevated temperature conditions (80°C)

| Testing condition | Sample ID | Confining pressure, $\sigma_3$ (MPa) | Peak axial stress, $\sigma_1$ (MPa) | Normal stress, $\sigma_n$ (MPa) | Shear stress, $\tau$ (MPa) | Friction angle (°) |
|---|---|---|---|---|---|---|
| M | A_M_1 | 5 | 10.28 | 6.16 | 2.19 | 30.47 |
| | | 10 | 23.96 | 13.08 | 5.78 | |
| | | 15 | 40.89 | 20.71 | 10.73 | |
| | A_M_2 | 5 | 10.73 | 6.26 | 2.38 | 28.13 |
| | | 10 | 25.24 | 13.36 | 6.32 | |
| | | 15 | 38.75 | 20.23 | 9.85 | |
| T-M | A_TM_1 | 5 | 13.13 | 6.79 | 3.37 | 29.36 |
| | | 10 | 29.65 | 14.33 | 8.15 | |
| | | 15 | 42.42 | 21.04 | 11.37 | |
| | A_TM_2 | 5 | 9.46 | 5.98 | 1.85 | 26.38 |
| | | 10 | 25.55 | 13.43 | 6.44 | |
| | | 15 | 35.50 | 19.52 | 8.50 | |

The friction angle does not show much difference between room temperature (20°C) and the elevated temperature (80°C). The peak shear stress increased with the increase of confining pressure, but it did not show much variation with the increase of temperature. According to Lockner et al. (1982), the frictional property of sandstone was independent of temperature at 150°C and 240°C, the friction angle of granite was also independent of the temperature in the range of 80°C in this study.

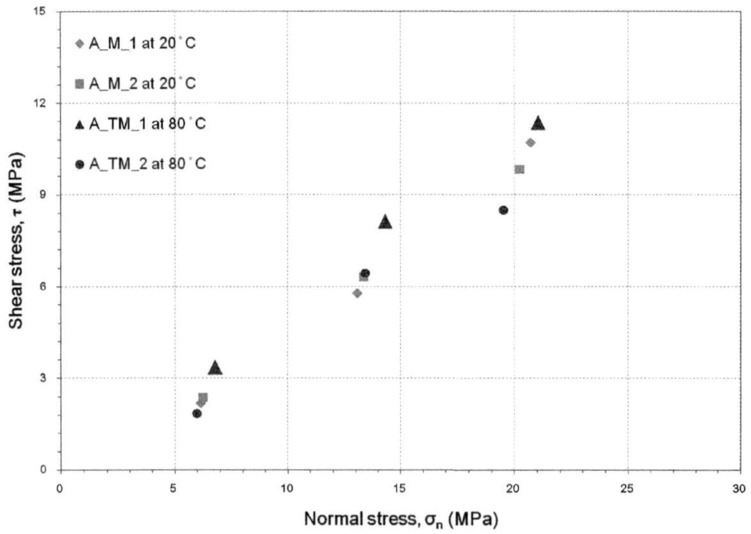


**Fig. 3** Relations between normal stress and shear stress acting on shear surface under the room temperature (20°C) and the elevated temperature (80°C) conditions

# 3 Numerical Simulation

## *3.1 Model Introduction*

The numerical analysis was carried out using COMSOL Multiphysics, which provides a powerful and flexible modeling environment for solving partial differential equations. The numerical simulation for multi-stage triaxial compression test on a saw-cut specimen was carried out under a plane strain condition in the two-dimensional model.

The built-in Structural Mechanics Module and Heat Transfer Module of COMSOL Multiphysics are used to simulate the thermal-mechanical coupled behavior. The saw-cut surface was set as a boundary contact pair and frictional characteristics were modeled based on the Coulomb friction criterion (COMSOL AB. 2008).

The numerical model has the same dimension as the test specimen. The linear quadrilateral element was used and 14,880 quadrilateral meshes were generated as shown in Fig. 4. Rock was assumed to be a homogeneous and isotropic material. The input parameters used in the numerical simulation are listed in Table 4.

For the simulation at the elevated temperature condition, the coefficient of thermal expansion of granite specimen was assumed to be $10 \times 10^{-6}$/°C (Kim et al. 2011). The model expands with the increase in temperature, causing thermal strain in the material. The thermal stress is induced by the thermal expansion of the model.

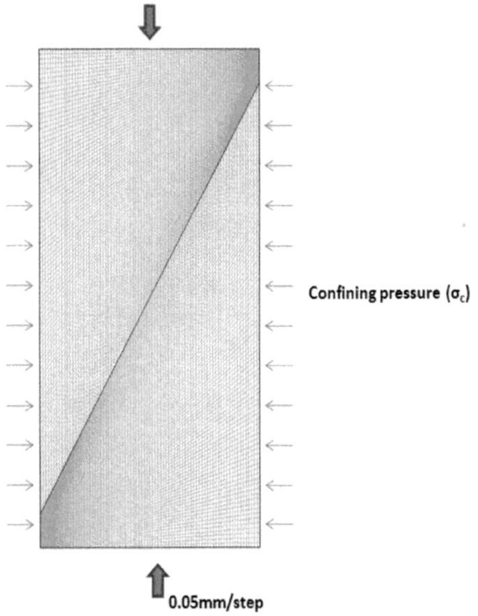


**Fig. 4** Two-dimensional model setup and boundary condition

**Table 4** Input parameters used in the numerical simulation

| Parameter | Value |
|---|---|
| Young's modulus (GPa) | 60 |
| Poisson's ratio | 0.20 |
| Density (kg/m$^3$) | 2660 |
| Coefficient of thermal expansion ($10^{-6}$/°C) | 10 |

## 3.2 Results and Discussion

Fig. 5 shows the deformation and maximum principal stress distribution in the specimen during the multi-stage triaxial compression test at room temperature. Sliding occurred along the saw-cut surface and the results of the numerical analysis are summarized in Table 5.

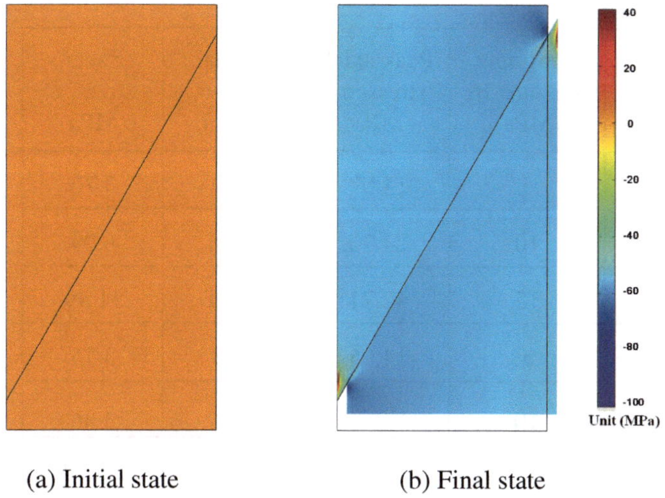


**Fig. 5** Deformation presented in the numerical model with maximum principal stress distribution under the room temperature (20° C) condition

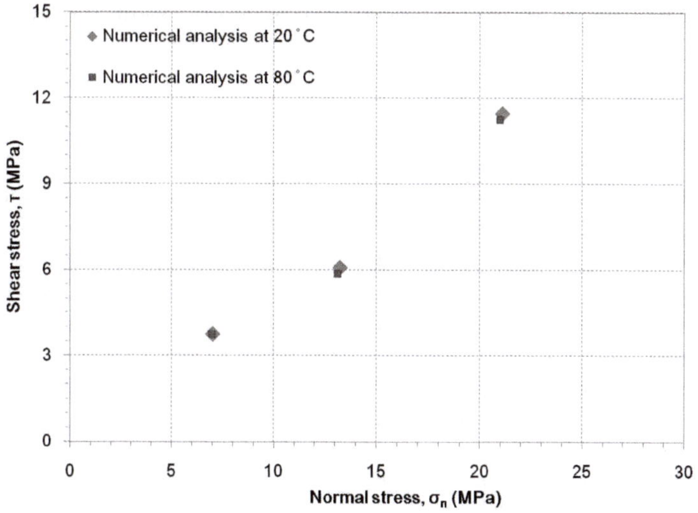


**Fig. 6** Relations between normal stress and shear stress acting on shear surface at room temperature (20°C) and elevated temperature (80°C)

The shear behavior under multi-stage triaxial compression conditions and the friction angle of the numerical analysis coincided with those observed in the laboratory tests. Fig. 6 shows the relationship between the normal and the shear stress acting on the shear surface under different temperature conditions.

**Table 5** Results of numerical simulation

| Testing condition | Confining pressure, $\sigma_3$ (MPa) | Peak axial stress, $\sigma_1$ (MPa) | Normal stress, $\sigma_n$ (MPa) | Shear stress, $\tau$ (MPa) | Friction angle (°) |
|---|---|---|---|---|---|
| Room temperature (20°C) | 5 | 14.07 | 7.00 | 3.76 | 28.97 |
| | 10 | 24.70 | 13.24 | 6.09 | |
| | 15 | 42.71 | 21.11 | 11.49 | |
| Elevated temperature (80°C) | 5 | 14.05 | 7.00 | 3.75 | 28.53 |
| | 10 | 24.15 | 13.12 | 5.86 | |
| | 15 | 42.16 | 20.99 | 11.26 | |

The shear stress increased with the increase of normal stress however the friction angle and shear stress did not vary with the increase of temperature. From the analysis, the thermal stress due to the change in temperature was too small to cause a noticeable change in the shearing characteristics of the specimen.

## 4 Conclusions

A multi-stage triaxial compression test was carried out for a granitic rock specimen having a saw-cut shear surface. The confining pressure and temperature were varied in the test the values of which were decided considering the actual condition in the vicinity of the canister. The results were analyzed by Mohr-Coulomb failure criterion. Peak shear stress varied with the increase in confining stress however the peak shear stress and friction angle of the granitic rock did not vary much at the elevated temperature of 80°C in the present study.

In order to simulate the test, a series of biaxial compression tests were carried out using COMSOL Multiphysics, providing a good agreement with the experimental results.

In the present study, the thermal effect on shearing behavior of the crystalline rock was investigated. Based on the results, the shearing behavior under added hydraulic conditions is to be studied in the future.

## References

1. COMSOL AB, COMSOL Multiphysics version 3.5, User's guide, p. 604 (2008)
2. Elliott, G.M., Brown, E.T.: Laboratory measurement of the thermo-hydro-mechanical properties of rock. Quarterly J. Eng. Geology 21, 299–314 (1988)

3. Kim, T., Lee, C.S., Jeon, S.: A numerical study on shear characteristics of jointed rock under Thermo-mechanical coupled condition. In: International Conference on Analysis of Discontinuous Deformation, Hawaii, USA (2011)
4. Kovari, K., Tisa, A., Einstein, H.H., Franklin, J.A.: Suggested methods for determining the strength of rock materials in triaxial compression. Int. J. Rock Mech. and Mining Sci. & Geomech. Abstr. 20, 285–290 (1983) (revised version)
5. Kwon, S.K., Cho, W.J.: A sensitivity analysis of design parameters of an underground radioactive waste repository using a backpropagation neural network. Tunnelling and Underground Space Technology 19(3), 203–212 (2009)
6. Lockner, D.A., Summers, R., Moore, D., Byerlee, J.D.: Laboratory measurements of reservoir rock from the Geysers geothermal field, California. Int. J. Rock Mech. and Mining Sci. & Geomech. Abstr. 19, 65–80 (1982)
7. Wanne, T.S., Young, R.P.: Bonded-particle modeling of thermally fractured granite. Int. J. Rock Mech. and Mining Sci. 45(5), 789–799 (2008)
8. Yoon, J., Jeon, S.: Experimental verification of a pts mode II for rock. Int. J. Rock Mech. and Mining Sci. 41(3), 8–13 (2004)

# Hydro-Mechanical Coupled Discrete Element Modeling of Geothermal Reservoir Stimulation and Induced Seismicity

Jeoung Seok Yoon, Arno Zang, and Ove Stephansson

GFZ German Research Centre for Geosciences, 14473 Potsdam, Germany

**Abstract.** The injection of fluid underground, the use of $CO_2$, water or waste for storage, or disposal purposes or both, results in a stress field change, the creation of fractures and the reactivation of pre-existing faults and joints. Sometimes these man-made processes are associated with seismicity, e.g. seismicity of a local magnitude 3.4 occurred in Basel Switzerland enhanced geothermal system (EGS) site. Such phenomena have lead to the development of numerical tools that are able to simulate fluid injection into underground reservoirs and predict induced seismicity. Appropriate measures for mitigating larger magnitude events (LME) can then be established after the reliability of the numerical tools is validated. In this context, this paper introduces hydro-mechanical coupled discrete element fracture network models developed for simulating hydraulic fracturing and induced seismicity. Particle Flow Code 2D is used in which the hydro-mechanical coupling routine is implemented, plus the seismicity computation routine. A fractured granitic reservoir with dimension of 2 km x 2 km is constructed using laboratory and field data from Soultz-sous-Forêts European Hot Dry Rock project. For mechanical and hydraulic data of the pre-existing fractures, measured data from Forsmark Sweden was adopted for planning the construction of the final repository for spent nuclear fuels. Hydraulic fracturing of intact reservoirs (without fractures) and fractured reservoirs is performed by means of a fluid injection at the model centre under two different scenarios: 1) a one day injection with a monotonic bottom hole pressure (BHP) increase followed by 1.5 days of shut-in where BHP decays non-linearly, 2) a one day injection with a cyclic BHP increase and the same BHP decays. Simulation results are examined in terms of: 1) fracture propagation pattern, 2) magnitude of induced events, 3) potential of LME in post shut-in, 4) influence of different injection schemes on the fracturing pattern and magnitude of induced events. The final objective of this paper is to examine if the presented modeling approach is capable of capturing the field observations and providing a good understanding of the key issues in the development of EGS, in particular for soft stimulation strategies.

**Keywords:** Discrete element model, Hydro-mechanical coupling, Geothermal reservoir stimulation, Induced seismicity.

M.Z. Hou et al. (Eds.): *Clean Energy Systems in the Subsurface*, SSGG, pp. 221–231.
DOI: 10.1007/978-3-642-37849-2_19 

# 1 Introduction

It is noted in Soultz-sous-Forêts, The Geysers, and other crystalline sites, that the largest magnitude induced events tend to occur on the fringes and outside the main cloud of induced events during stimulation. Moreover, large magnitude events are often observed after the shut-in of EGS injection operations making such events still more difficult to control (Majer et al. 2007). The need for better understanding of the processes underlying the occurrence of post-shut-in seismicity has become an important issue. The development and use of suitable coupled reservoir fluid flow, geomechanical simulation codes will be a great help in this respect, and advances are being made in this area (Hazzard et al. 2002, Kohl & Mégel 2005 and Ghassemi et al. 2007) (Majer et al. 2007).

This paper investigates reservoir stimulation by means of hydraulic fracturing and induced seismicity using a discrete element based numerical model. Commercial Particle Flow Code 2D (*PFC2D*) is used to handle the mechanical process. On the *PFC2D*, a modified hydro-mechanical coupling routine is adopted to simulate fluid flow in a porous medium and flow driven failure of rock mass and pre-existing joints (smooth joint contact model) in Mode I (tension) and Mode II (shear). The seismicity computing algorithm taken from Hazzard and Young (2002, 2004) is modified to compute the magnitude of Mode I and Mode II failures of pre-existing joints. The aim of this paper is: 1) to simulate reservoir stimulation by means of fluid injection and hydraulic fracturing, 2) to test the modeling results if they show similarity to field observations, 3) to investigate if the modeling approach has the potential to be applied in the soft stimulation strategies of the EGS industry.

# 2 Methodology

## 2.1 *Particle Flow Code 2D (PFC2D)*

*PFC2D* is a two-dimensional distinct element geomechanical modeling software. The reservoir rock material is modeled as an aggregate of circular particles bonded at their contacting points with finite thickness of cementing around the contact with pre-specified Mohr-Coulomb strength parameters (Itasca 2012 – enhanced parallel bond model). Under the applied load, the bonds can break either in Mode I (tensile) or Mode II (shear). The calculation cycle in *PFC2D* is a time stepping algorithm that requires repeated application of the law of motion applied to each particle and a linear (non-linear) force displacement law applied to each contact. For more detail, refer to Potyondy and Cundall (2004).

## 2.2 *Hydro-Mechanical Coupling*

A hydro-mechanical (HM) coupling scheme is used which enables fluid flow in the bonded particle assembly and fluid pressure (flow volume) driven breakages of bonds in Mode I and Mode II, in which poro-elasto-plasticity is modeled indirectly. HM coupling in a PFC framework was first introduced by Kamp et al. (1999), Cundall (unpublished technical note, 2000) and Konietzky et al. (2001). HM coupling was later modified by Hazzard et al. (2002). Yoon (2007) and Yoon et al. (2008) and was applied to the simulation of drained biaxial compressive tests and Mode II fracture toughness tests of saturated rock model. Yoon and Jeon (2009) applied this method and simulated blast load driven fracture nucleation and propagation at multiple blastholes.

Fluid flow is simulated by assuming that each particle bonded contact is a flow channel and these channels connect up small pores that can store pressure, hence making a continuous pore-network model (Fig. 1 inset).

Cubic law (Table 1) is used assuming the flow is laminar between two smooth parallel plates. Hydraulic aperture, *e*, changes as a function of normal stress, $\sigma_n$. This paper adopted, as the target simulation model, a granite reservoir, a laboratory derived *e* vs. $\sigma_n$ relation from the through investigations of the granitic rock mass at Forsmark Sweden where construction of a final repository for spent nuclear fuels is planned (SKB 2010).

Fluid pressure ($P_f$, Fig. 1 inset) is calculated from the fluid bulk modulus, volume of pore space and flow volume and converted to a force term by multiplying the length *d* (Fig. 1 inset) and unit thickness of 1 m in an out-of-plane direction. The resulting force term is applied to particle A from which the law of motion computes particle velocity and displacement which subsequently changes stress states at the surrounding contacts which in turn changes the hydraulic aperture and thereby the flow field.

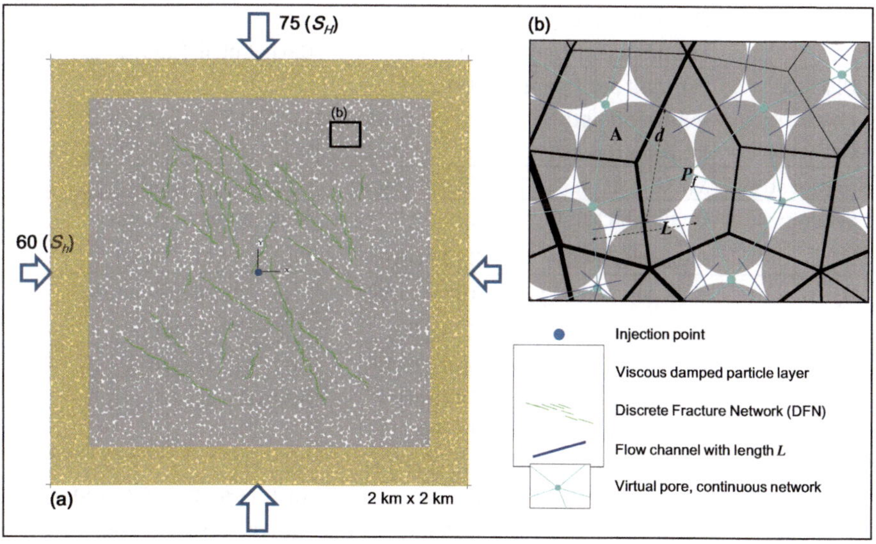


**Fig. 1** Fractured reservoir model 2 km x 2 km in size with viscous damped layer near boundaries and subjected to differential in-situ stresses ($S_H$ = 75 MPa, $S_h$ = 60 Mpa)

## 2.3 Computation of Seismic Parameters from a Bond Breakage

Each bond breakage in the bonded particle reservoir model is assumed to be a fracture process associated with seismic energy radiation. *PFC2D* runs in dynamic mode with low levels of numerical damping where a realistic level of energy attenuation in rock is simulated. In this modeling, on a bond breakage by Mode I or Mode II, part of the accumulated strain energy at the broken bond is released to the surrounding area in a form of a seismic wave. A numerical technique for

**Table 1** Model parameters used to describe Soultz-sous-Forêts reservoir characteristics

| Category | Property | Value (Unit) | Reference/Remarks |
|---|---|---|---|
| Intact rock | Density | 2600 (kg/m$^3$) | |
| | Tensile strength, $T_0$ | 11.5±1.5 (MPa) | 10% UCS (Valley & Evans 2006) |
| | Cohesion, $c$ | 30±5 (Mpa) | |
| | Friction angle, $\varphi$ | 52 (deg.) | Aue granite (Yoon et al. 2012) |
| | Young's modulus, $E$ | 60 (Gpa) | Valley & Evans (2006) |
| | Friction coefficient, $\mu$ | 0.9 (-) | Cornet et al. (2007) |
| Discrete fracture | Normal stiffness, $k_n$ | 200 (Gpa/m) | SKB (2010) |
| | Shear stiffness, $k_s$ | 50 (Gpa/m) | SKB (2010) |
| | Tensile strength, $T_{0,f}$ | 0 (Mpa) | |
| | Cohesion, $c_f$ | 0.5 (Mpa) | |
| | Friction angle, $\varphi_f$ | 30 (deg.) | |
| | Dilation angle, $\psi_f$ | 3 (deg.) | |
| | Friction coefficient, $\mu_f$ | 0.9 (-) | Cornet et al. (2007) |
| Discrete Fracture Network | No. joint set / per set | 3/10 (-) | |
| | Joint set length | (s1) 150-200 (m), (s2) 0-150 (m), (s3) 100-200 (m) | |
| | Joint set dip direction | (s1) S10°-30°E, (s2) N10°-30°E, (s3) S40°-60°E | |
| In-situ stress | $S_H/S_h$ | 75/60 (Mpa) | Cornet et al. (2007) |
| Seismic | Quality factor, Q | 140 (-) | Michelet et al. (2005) |
| | S-wave velocity | 3500 (m/s) | Cuenot et al. (2008) |
| Hydraulic | Bulk conductivity, $K$ | $10^{-12}$ (m/s) | Fresh granite (Bear 1972) |
| | Flow rule equation | Cubic law | $Q=(e^3\Delta P)/(12\mu L)$ |
| | Fracture aperture, $e_0$ | 650 ($\mu$m) | Aperture at zero $\sigma_n$ |
| | Fracture aperture, $e_{\text{inf}}$ | 50 ($\mu$m) | Aperture at $\infty$ $\sigma_n$ |
| | Fluid viscosity, $\mu$ | 1 (cP) | |
| | Fluid bulk modulus | 2.2 (Gpa) | |
| | $e$ vs. $\sigma_n$ relation | $e=e_{\text{inf}}+(e_0-e_{\text{inf}})*\exp(-0.15*\sigma_n)$ | SKB (2010) |
| | Formation pressure, $p_f$ | 0 (Mpa) | Initially dry reservoir |
| FBP | Min/Max. estimates | 80/85 (Mpa) | $p^{HF}-p_f=(3S_h-S_H+T_0)/1.44$ |

calculating this seismic source information in *PFC2D* has been proposed by Hazzard and Young (2002, 2004) and modified in this paper so as to consider seismic wave propagation and compute focal mechanism and resulting magnitudes from Mode I (tensile) and Mode II (shear) failures at pre-existing joints, i.e. at smooth joint contacts. More details of its use in rock modeling can be found in Hazzard and Young (2002, 2004) and Yoon et al. (2012).

## 3 Model Description

The reservoir model is 2 km x 2 km in size (Fig. 1a). Diameters of particles to pack the given volume are in a range between 20-30 m, resulting in ca. 7500 particles. The model contains three set of joints with their strength and stiffness properties defined in Table 1 (s1, s2, s3). After the model is constructed and DFN is embedded, the model is compressed using the servo-controlled routine to achieve in-situ stresses of $S_H$ = 75 Mpa and $S_h$ = 60 Mpa, assuming that the model section is located at the5 km target depth reservoir, using the stress-depth relation at Soultz from Cornet et al. (2007). Along boundaries, ca. 150 m thick a region (shaded region, Fig. 1a) is assigned with high viscous damping properties in both normal and shear stress to model energy absorption. This concept is taken in order to exclude any side effects on bond breakages coming from the kinetic seismic wave emanating from an induced bond breakage propagating outward and bouncing back when the wave is reflected from the boundary. The onset of Mode I and Mode II fractures of intact rock (enhanced parallel bond model) and pre-existing joint (smooth joint model) are governed by Mohr-Coulomb failure criterion where the parameters, e.g. tensile strength, cohesion and friction angle, are listed in Table 1. Friction coefficients ($\mu$, $\mu_f$) govern slip (sliding) behavior at the nbounded contacts but with non-zero normal stress, by limiting the shear force, $F_s$=$\mu F_n$. Friction angle ($\varphi_f$) and dilation angle ($\psi_f$) of the discrete fracture are responsible for an increase in normal force resulting from the shear displacement during slip.

## 4 Modeling of Reservoir Stimulation

### *4.1 Intact Reservoir with Monotonic BHP Increase*

An intact reservoir, i.e. without discrete fracture network (DFN), is stimulated by increasing the bottomhole pressure (BPH) at the model center monotonically for 1 day and decreasing it for 1.5 days in a form of negative exponential decaying function. Injection is fluid pressure controlled, whereas injection at EGS operation is often carried out by controlling the injection rate, e.g. liter per second. Numerical procedure for the flow rate controlled injection is under testing. Stimulation of the intact reservoir is performed in order to examine the following issues: 1) to check if induced fractures are aligned parallel to the maximum in-situ stress, 2) to check if BHP at the onset of first fracturing matches with the Fracture Breakdown Pressure (FBP) estimation (Haimson-Fairhurst equation (Zang & Stephansson 2010), see Table 1). Fig. 2a shows BHP, a number of induced events during a time interval dt = 60 sec. (thus occurrence rate) and moment magnitude ($M_w$) of induced events vs. time for stimulation of the intact reservoir with monotonic BHP increase. The three largest magnitude events are represented by stars, which average 0.5. Fig. 3a and 3b show distribution of the induced cracks that occurred before and after shut-in, respectively. Black and red dots represent rock failure in Mode I and II, respectively. Two events occurred at the very beginning near the model boundary and are excluded from the analysis (Fig. 1a, (o)).

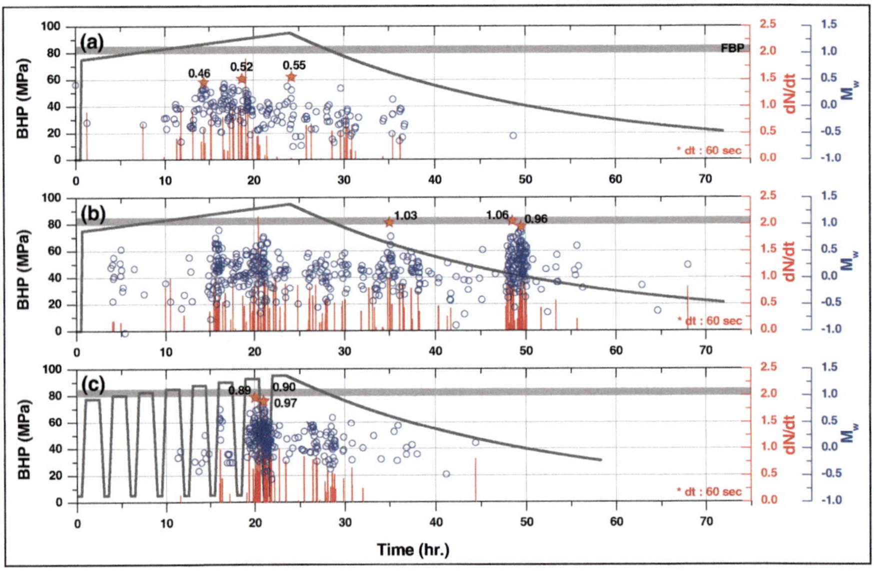


**Fig. 2** Bottom hole pressure (BHP), event occurrence rate (dN/dt, dt = 60 sec.) and moment magnitude ($M_W$) vs. time (hr.) for (a) intact reservoir, (b) fractured reservoir with monotonic BHP increase and (c) fractured reservoir with cyclic BHP increase. Shaded bar represents the estimated range of fracture breakdown pressure. Red stars indicate LME.

## 4.2 *Fractured Reservoir with Monotonic BHP Increase*

The fractured reservoir, i.e. with DFN, is stimulated by monotonic BHP increase and non-linear decay the same as before. Fig. 2b shows BHP, the number of induced events during the time interval, and moment magnitude ($M_w$ ~0.5) of induced events vs. time (hr.). The three largest magnitude events are represented by red stars, which have an average of 1. Compared to the intact reservoir where the length of potential fractures are in a range between 20-30 m (particle diameters), the presence of discrete fractures (with length scale of up to 200 m, see Table 1) results in more than double the amount of maximum moment magnitudes ($M_w$ ~1.0). Consequently, the magnitude-frequency distribution is different for the intact and fractured reservoirs, e.g. the total number of cumulative seismic events is 200 (intact) and 500 (fractured reservoir), respectively. Fig. 3c and 3d show the distribution of induced cracks that occurred before and after shut-in, respectively. In addition to Mode I and II failures of intact rock denoted by black and red dots, respectively, blue and pink dots represent Mode I and Mode II failure of pre-existing joints of whose locations and orientations are denoted by green lines. The magnitude of LME is double compared to those induced in the intact reservoir.

### *4.3 Fractured Reservoir with Cyclic BHP Increase*

The fractured reservoir, i.e. with DFN, is stimulated with a cyclic increase of BHP from 75 MPa to 95 MPa and down to 5 MPa, in 8 cycles, and then decays non-linearly the same as before. Fig. 2c shows BHP, the occurrence rate of induced events (dN/dt) and the magnitude of induced events vs. time. The three largest magnitude events are represented by stars, which have an average of 0.9 and appear prior to shut-in. Fig. 3e and 3f show the distribution of the induced cracks that occurred before and after shut-in, respectively.

## 5 Observations

From the results, the following observations are made and brief explanations are provided. 1) The overall pattern of fracture propagation is parallel to the maximum in-situ stress direction ($S_H$) in the case of the intact reservoir, and sub-parallel in the case of the fractured reservoir. Due to the heterogeneitic nature of the discrete element model (random particle arrangement), fracture propagation patterns are asymmetric. 2) The overall pattern of hydraulic fracture propagation deviates slightly from the maximum in-situ stress direction ($S_H$). It is observed that the hydraulic fracture propagates conjugate at the pre-existing dominating DFN which is oriented NW-SE. 3) BHP at the onset of fracturing near the injection point matches well with the estimated FBP range (shaded horizontal bar in Fig. 2). 4) Post-shut-in induced events occurred in the outer region of the pre-shut-in, induced events cloud and their magnitudes are relatively higher (see Fig. 4).

This result matches well with field observations, especially in Soultz EGS (Michelet et al. 2004) and Basel EGS (Bachmann et al. 2012). 5) In the case of the fractured reservoir with a monotonic BHP increase, relatively larger magnitude events occurred long after the shut-in (see Fig. 2b post-shut-in LME and Fig. 4b stars). 6) The potential for the occurrence of post-shut-in LME is higher in case of a monotonic BHP increase in the fractured reservoir. 7) In the fractured reservoir with a cyclic BHP increase, the fracturing pattern becomes relatively isotropic even though the reservoir is subjected to a high differential stress state.

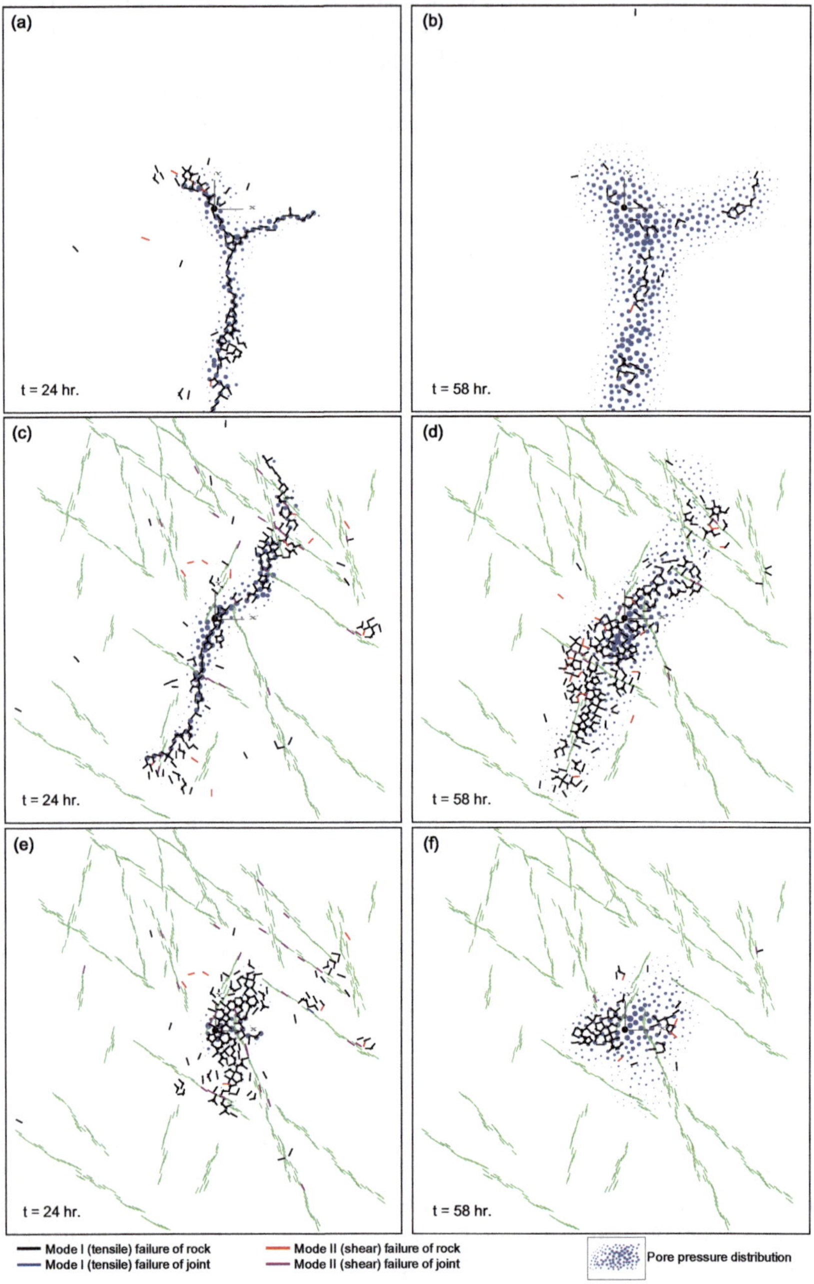


**Fig. 3** Distribution of fluid injection induced cracks during pre-shut-in (left) and post-shut-in (right) periods in (a, b) the intact reservoir model with monotonic BHP increase; (c, d) the fractured reservoir model with monotonic BHP increase; (e, f) the fractured reservoir model with cyclic BHP increase

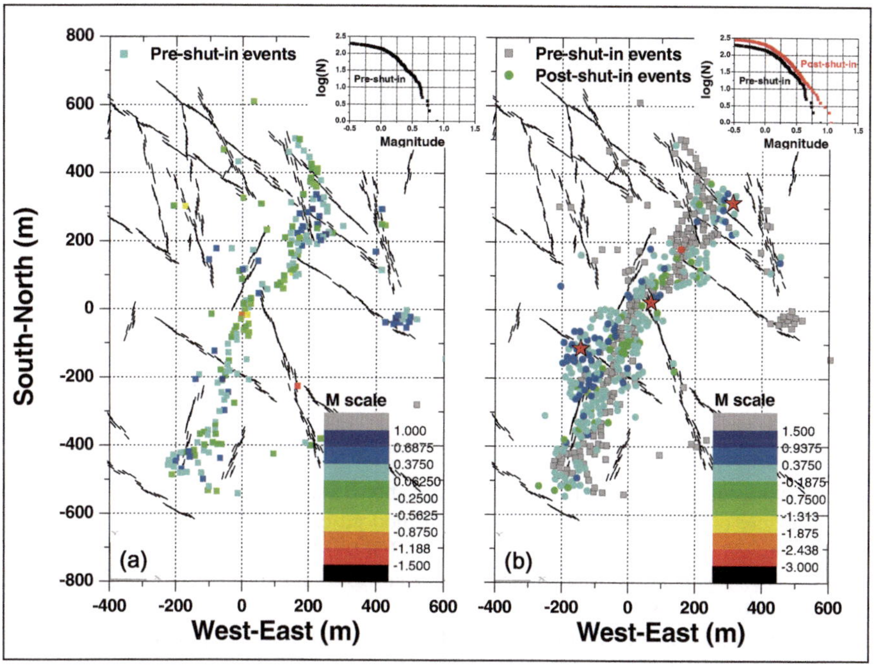


**Fig. 4** Distribution of (a) pre-shut-in and (b) post-shut-in induced events plotted with their magnitude scale in the fractured reservoir with a monotonic BHP increase. Coordinates of the injection point are (0,0). Color scales for the magnitude used are different in (a) and (b). In (b), the three largest events are denoted by red stars. Inset shows the magnitude-frequency relation of induced seismicity.

## 6 Conclusions

The following conclusions can be drawn from simulating geothermal reservoir stimulation and associated induced seismic events by hydro-mechanical coupled, discrete element fracture network models. 1) The onset of hydraulic fracture matches with the fracture breakdown pressure in the classical Haimson-Fairhurst theory. 2) The orientation of hydraulic fracture is tilted (30-45°) with respect to the maximum compressive stress azimuth in the reservoir. This is due to (a) heterogeneity in the intact reservoir (random particle arrangement) and (b) orientation distribution of pre-existing joints in the fractured reservoir. 3) The maximum moment magnitude of seismic events in the intact reservoir ($M_w$ ~0.5; diameters of particles in range 20-30 m) is one half of the magnitude observed in the fractured reservoir ($M_w$ ~1.0; pre-existing joints length $\leq$ 200 m). 4) Post-shut-in seismic events occur at the outer region of the pre-shut-in event cloud and their magnitudes are relatively higher in the case of the fractured reservoir which accords with field observations in Soultz and Basel. However, this trend is not observed in the case of the fractured reservoir stimulated with cyclic injection.

Cyclic BHP loading is one way to produce an isotropic event cloud in an isotropic stressed reservoir. One explanation can be that the reduction of BHP from 80 to 5 MPa implies fluid production, which then leads to a stress field change around the borehole. It consequently results in fracture propagation in a reoriented direction leading to an isotropic fracture pattern.

**Acknowledgements.** The original hydro-mechanical coupling scheme is provided by Dr. J. Hazzard (Itasca) to whom the authors are grateful. This work was supported by the European Union funded project GEISER (Geothermal Engineering Integrating Mitigation of Induced Seismicity in Reservoirs, FP7-ENERGY-2009-1, Grant agreement no.: 241321-2). In this respect, we would like to thank Ernst Huenges and David Bruhn (GFZ, 4.1 Reservoir Technology).

## References

1. Bachmann, C.E., Wiemer, S., Boertz-Allmann, B.P., Woessner, J.: Influence of pore-pressure on the event-size distribution of induced earthquakes. Geophysical Research Letters 39, L09302 (2012), doi:10.1029/2012GL051480
2. Bear, J.: Dynamics of fluids in porous media. Dover Publications (1972) ISBN 0-486-65675-6
3. Cornet, F.H., Bérard, T., Bourouis, S.: How close to failure is a granitic rock mass at a 5 km depth? Int. J. Rock Mech. Min. Sci. 44, 47–66 (2007)
4. Cuenot, N., Dorbath, C., Dorbath, L.: Analysis of the Microseismicity Induced by Fluid Injections at the EGS Site of Soultz-sous-Forêts (Alsace, France): Implications for the Characteriyation of the Geothermal Reservoir Properties. Pure and Applied Geophysics 165, 797–828 (2008)
5. Ghassemi, A., Tarasovs, S., Cheng, A.H.D.: A 3-D study of the effects of thermomechanical loads on fracture slip in enhanced geothermal reservoirs. Int. J. Rock Mech. Min. Sci. 44, 1132–1148 (2007)
6. Hazzard, J.F., Young, R.P.: Moment tensors and micromechanical models. Tectonophysics 356, 181–197 (2002)
7. Hazzard, J.F., Young, R.P.: Dynamic modeling of induced seismicity. Int. J. Rock Mech. Min. Sci. 41, 1365–1376 (2004)
8. Hazzard, J.F., Young, R.P., Oates, S.J.: Numerical modeling of seismicity induced by fluid injection in a fractured reservoir. In: Proceedings of the 5th North Americal Rock Mechanics Symposium on Mining and Tunnel Innovation and Opportunity, Toronto, Canada, pp. 1023–1030 (2002)
9. Itasca, Technical Memorandum – 5.0 Parallel Bond Enhancement (2012)
10. Kohl, T., Mégel, T.: Coupled hydro-mechanical modeling of the GPK3 reservoir stimulation at the European EGS site Soultz-sous-Forêts. In: Proceedings of 39th Workshop on Geothermal Reservoir Engineering, pp. 320–327. Stanford University, Stanford (2005)
11. Konietzky, H., te Kamp, L., Blümling, P., Mayor, J.C.: Micro-mechanical analysis of excavation disturbed zones around tunnels. In: Proceedings of 10th International Conference on Computer Methods and Advances in Geomechanics, vol. 1, pp. 543–546. AA Balkema, Tucson (2001)

12. Majer, E.L., Baria, R., Stark, M., Oates, S., Bommer, J., Smith, B., Asanuma, H.: Induced seismicity associated with Enhanced Geothermal Systems. Geothermics 36, 185–222 (2007)
13. Michelet, S., Baria, R., Baumgaertner, J., Gérard, A., Oates, S., Hettkamp, T., Teza, D.: Seismic source parameter evaluation and its importance in the development of a HDR/EGS system. In: Proceedings of the 29th Workshop on Geothermal Reservoir Engineering, January 23-28. Stanford University, Stanford (2004)
14. Michelet, S., Toksöz, N.: Characterizing the mechanics of fracturing from earthquake source parameter and multiplet analysis: Application to the Soultz-sous-Forêts Hot Dry Rock site. Earth Resource Laboratory Industry Consortia Annual Report 2005-2 (2005), http://hdl.handle.net/172.1/67880
15. Potyondy, D.O., Cundall, P.A.: A bonded-particle model for rock. Int. J. Rock Mech. Min. Sci. 41, 1329–1364 (2004)
16. SKB Svensk Kärnbränslehantering AB, THM-issues in repository rock–Thermal, mechanical, thermo-mechanical and hydro-mechanical evolution of the rock at the Forsmark and Laxemar sites. Techanical Report TR-10-23, pp. 38–43 (2010)
17. te Kamp, L., Konietzky, H., Blümling, P., Pineau, F., Ando, K., Mayor, J.C.: Micromechanical backanalysis of lab tests on rock. In: Proceedings of VII International NUMOG Symposium (Numerical Models in Geomechanics) Graz, September 1-3. AA Balkema, Rottermam (1999)
18. Valley, B., Evans, K.F.: Strength and elastic properties of the Soultz granite. In: EHDRA Scientific Conference, June 15-16 (2006)
19. Yoon, J.: Hydro-mechanical coupling of shear-induced rock fracturing by bonded particle modeling. Doctoral dissertation. Seoul National University, Korea (2007)
20. Yoon, J., Jeon, S., Stephansson, O., Dresen, G.: Experimental analysis of confined shear fracturing of saturated granite and numerical analysis by 2D hydro-mechanical coupled bonded-particle modeling. In: Proceedings of 42nd US Rock Mechanics Symposium and 2nd US-Canada Rock Mechanics Symposium, San Francisco, USA, June 29-July 2 (2008)
21. Yoon, J., Jeon, S.: Use of a modified particle-based method in simulating blast-induced rock fracture. In: Proceedings of the 9th International Symposium on Rock Fragmentation by Blasting – FRAGBLAST 2009, Granada, Spain, September 13-17, pp. 371–380 (2009)
22. Yoon, J.S., Zang, A., Stephansson, O.: Simulating fracture and friction of Aue granite under confined asymmetric compressive test using clumped particle model. Int. J. Rock Mech. Min. Sci. 49, 68–83 (2012)
23. Zang, A., Stephansson, O.: Stress Field of the Earth's Crust, p. 143. Springer Science + Business Media BV, Dordrecht (2010)

# Simulation of $CO_2$-Water-Rock Interaction Processes-Mineral Scaling Problems in Saline Formations

Hejuan Liu[1], Michael Zhengmeng Hou[1,2], Yang Gou[1], and Patrick Were[1]

[1] Lower Saxony Energy Research Centre, Goslar, Germany
[2] Institute of Petroleum Engineering, Clausthal University of Technology, Agricola str. 10, 38678 Clausthal-Zellerfeld, Germany

**Abstract.** Many problems may occur in the process of $CO_2$ injection which usually lasts for many years. Injection efficiency depends on many factors, such as temperature (T), well bottomhole pressure (P), well heading pressure, injection rate, reservoir properties (porosity, permeability, wettability) and scales of some minerals. The scaling problem in the wellbore and near-well region in reservoir (usually a few meters away from the borehole), may have a large negative effect on the reservoir properties such as a decrease in porosity and permeability near the well borehole region. As a result, the amount of $CO_2$ injected will be restricted. The purpose of this study is to predict mineral scales formation and distribution that happen in near-well reservoir using the simulation method. In this paper, TOUGH2 and TOUGHREACT software are used, and a 1D model has been set up. In this benchmark simulation of scaling problems, T and P are chosen to be at 100°C and 4MPa respectively. Simulation results show that pressure and gas saturation of the reservoir had been changed greatly after $CO_2$ injection. Illite and calcite may be the main mineral scales in the near-well region. At different time after the injection of $CO_2$, there are different changes of mineral types and mineral volume, illustrating that at the beginning of the injection period, the precipitated minerals are illite, oligoclase and calcite, with time, ankerite, smectite-Ca and dawsonite will precipitate. In order to control the scale problems and alleviate destruction of the reservoir and petroleum equipment, chelants (such as EDTA, DTPA) should be used.

**Keywords:** scaling problem, injection efficiency, chemical reaction, reservoir prevention.

## Introduction

Mineral scale formation is a big problem in industrial aqueous systems (oilfield & geothermal, cooling water system, municipal projects and water reuse), causing reductions in the production and damage to the equipment (Elguedri 1999). For

M.Z. Hou et al. (Eds.): *Clean Energy Systems in the Subsurface*, SSGG, pp. 233–248.
DOI: 10.1007/978-3-642-37849-2_20 

geothermal and CCS projects, precipitation of mineral scale can occur either in the formation pores in the vicinity of injection and production wells or in tubing used in $CO_2$-water-rock interaction (CWRI). Scale deposits have very large effects on reservoir porosity, permeability, injectivity, safety and heat transfer characteristics (Williams et al. 1999). During the supercritical $CO_2$ injection phase, mineral scale could lead to a decrease in injectivity.

Scaling problems caused by $CO_2$-water-rock interaction have been reported by many projects, such as Nag-67 in the Tiwi field and the wells in the Soultz project's Hot Dry Rock research (Charleston 1968, Carrel 1987, Hardy et al. 1992, Xu et al. 2004a, 2004b, Garcia et al. 2006, Taheri et al. 2011).

$CO_2$-water-rock interaction is the main reason for scaling problems. Many researchers have done experiments to study $CO_2$-water-rock interaction mechanisms under different conditions. Ross et al. (2003) carried out a series of experiments using sandstone with 20% carbonate cements, dolomite and oolites at 6.8MPa, 13.8MPa, under 20°C and 80°C conditions, showing that the permeability of these three kinds of rocks varied, and among them, the permeability of the lower porosity oolites changed remarkably. Slices of the rocks showed that there was selectivity in mineral dissolution and the carbonate minerals next to the pore throat dissolute first, which magnify the rock permeability. Besides dissolution of carbonate minerals, new carbonate cements will generate in the pores of the rock during $CO_2$ transport along the pores under particular conditions such as ankerite ($CaFe(CO_3)_2$), siderite ($FeCO_3$), dawsonite($NaAlCO_3(OH)_2$) etc. Robert et al (2005) studied the $CO_2$-brine-rock interactions at elevated temperature and pressure. Abdulrazag et al (2009) have studied the supercritical $CO_2$-brine-carbonate minerals. Many other experimental and simulation methods have also been used to study the $CO_2$-water-rock interaction mechanisms (Edmunds et al. 1982, Xu et al. 2001, Ueda et al. 2005, Bertier et al. 2006, Pang et al. 2008, 2012, Fan et al. 2009, Fischer et al. 2010, Zhu et al. 2011).

In this paper, the mineral scales formation in the reservoir will be simulated using a benchmark study. In addition to understanding the fundamentals of mineral scale formation, the pressure of the reservoir, gas saturation changes, the type of mineral scales and their distribution in various places at different time scales are simulated by Tough2 and ToughReact software. Also scale prevention measures are introduced.

## 1 The Scale Formation in the Reservoir

The process of scale formation occurs in a series of steps (Duggirala 2002), mainly caused by the mixing effect of different fluids in an aqueous environment (Mclin et al. 2006, Fig1.), such as oilfield & geothermal, cooling water, municipal projects and water reuse (BJΦRNSTAD et al. 2006).

Faced with scale damage to the tubing and reservoir, treatments should be taken to handle the scale problem. For different reservoirs, the types of the scale deposits are different according to the different chemical components in the environment, and the corresponding treatments should be different as well (Charleston 1968, Carrel 1987, Hardy et al. 1992, Taheri et al. 2011).

The $CO_2$ injected into the reservoir will interact with the rock and formation water, which will cause complex mineral dissolution and deposition. Reservoir physical properties (porosity, permeability and wettability) would be changed due to $CO_2$-rock-formation water interaction. On one hand, the mineralization of $CO_2$ is very useful considering long-term $CO_2$ storage safety; on the other hand, the mineral scales in the well borehole and near-well region of the reservoir, whose formation and distribution should be studied, will affect the oil or natural gas production, decreasing the efficiency of $CO_2$-EOR and $CO_2$-EGR. Based on favorable initial porosity changes of about 15% (due to the dissolution of calcite), the combination of $CO_2$ co-injection with other mitigation strategies might alleviate some of the wellbore scale and formation plugging problem in the near-well region.

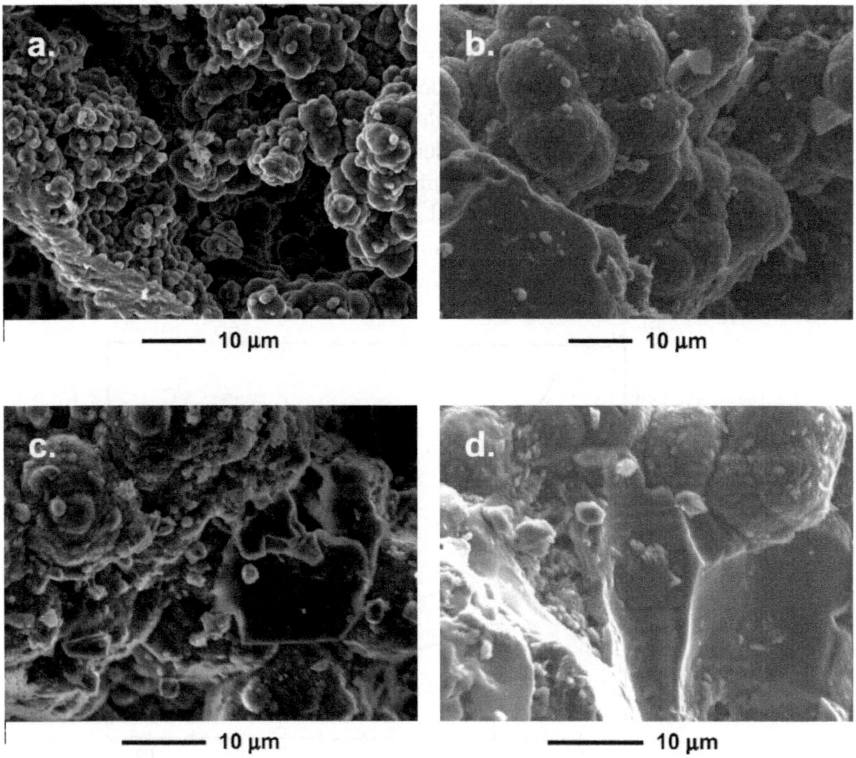


**Fig. 1** (a)-(d) SEM images of the scale from some samples from the well at different depth (Mclin et al. 2006)

## *1.1 The Mineral Assemblages of Different Reservoirs*

In different kinds of reservoirs, such as sandstone reservoirs and carbonate reservoirs, there are different mineral compositions. For sandstone reservoirs, the clastic particles are mainly quartz, K-feldspar, alkali-feldspar, plagioclase, rock debris, clay and sundry. The compositions of cement (possible mineral scale) are mainly quartz, calcite, chlorite, albite, illite, anhydrite and so on.

Carbonate reservoirs are mainly composed of calcite, dolomite, aragonite, ankerite, magnesite, siderite etc.

## *1.2 The Factors Affecting the Formation of Mineral Scales*

Scales can occur when the content of scaling ions is higher than the critical concentration. The water containing $CO_2$ reacts with the carbonate minerals (such as calcite and dolomite), and then $Ca^{2+}$ and $Mg^{2+}$ can be released into the water.

Temperature, pressure, $CO_2$ injectivity, $CO_2$ injection pressure ($pCO_2$) can all affect the deposition of the carbonate cement in the reservoir (Fig. 2). Supersaturation and accelerated kinetics are the driving forces that accelerate the formation of scale, and the kinetics induced scale formation including temperature shocks, intense mechanical and hydrodynamic shear forces, optimum pH conditions, and sudden changes in pressure (Rhudy et al. 1993).

With increased temperature, the solubility of $CO_2$ and $CaCO_3$ decreases, and calcite deposits easily, but the pressure has the opposite effect on the formation of $CaCO_3$. According to the pressure gradients of strata, it can be seen that with increased depth, the trend in reservoir scaling problems decreased.

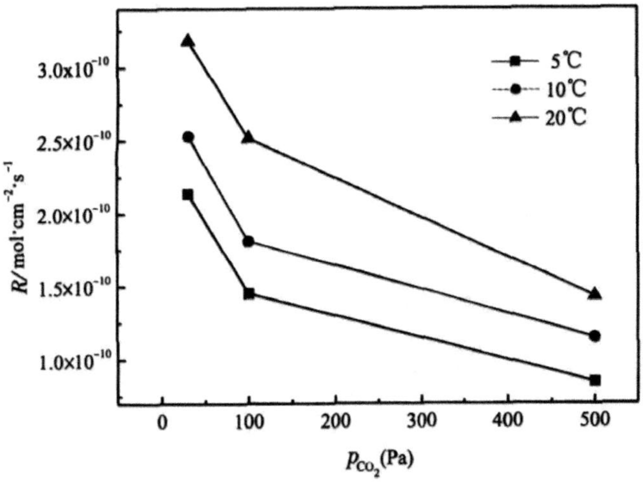


**Fig. 2** The deposition rate of calcite under different temperature condition (Liu et al. 2002)

$$CO_2 + H_2O \leftrightarrow H_2CO_3 \leftrightarrow HCO_3^- + H^+ \leftrightarrow 2H^+ + CO_3^{2-} \tag{1}$$

$$Ca^{2+} + 2HCO_3^- \leftrightarrow CaCO_3 \downarrow + H_2O + CO_2 \tag{2}$$

$$Ca^{2+} + CO_3^{2-} \leftrightarrow CaCO_3 \downarrow \tag{3}$$

Equation (1) shows that with increased $pCO_2$, pH decreases and the dissolution of the carbonates will be strengthened; from equations (2) and (3), it can be seen that, with decreased $pCO_2$, the reaction tends to the positive direction and calcite will deposit, illustrating that the drop of $CO_2$ pressure will cause the precipitation of calcite in the well borehole and the near-well region.

The morphology of scale is dependent on many factors including ion type in the water, concentrations, mixing ratio and other factors (Rhudy et al. 1993). For example, the formation of calcite scale depends on the concentration of $Ca^{2+}$ and $CO_3^{2-}$. There are four stages of calcite formation according to the concentration of $CO_3^{2-}$, from the least likely to the most likely, according to the concentration of $CO_3^{2-}$ and the increase of the solute pH (Fig 3). At the source of $CO_3^{2-}$, a few scales are from the second grade ionization of $H_2CO_3$ which is minor and can be negligible, and the majority is from the transformation of $HCO_3^-$ under alkaline conditions:

$$HCO_3^- + OH^- \rightarrow CO_3^{2-} + H_2O \tag{4}$$

Some chemical reaction equations (4.1, 4.2, 4.3, 4.4, 4.5, and 4.6) show that, with increased $pCO_2$, the reaction will move towards the positive direction resulting in dawsonite. The reactants can be albite, kaolinite, K-keldspar etc.

$$Al(OH)_3 + Na^+ + CO_2 \leftrightarrow NaAlCO_3(OH)_2 + H^+ \tag{4.1}$$

$$NaAlSi_3O_8 + H_2O + CO_2 \leftrightarrow NaAlCO_3(OH)_2 + 3SiO_2 \tag{4.2}$$

$$CaAl_2Si_2O_8 + 2Na^+ + 2CO_2 + 3H_2O \leftrightarrow 2NaAlCO_3(OH)_2 + 3CaCO_3 + 2SiO_2 + 2H^+ \tag{4.3}$$

$$3Na_{0.33}Al_{2.33}Si_{3.67}O_{12}H_2O + CO_2 + 4H_2O \leftrightarrow NaAlCO_3(OH)_2 + 3Al_2Si_2O_5(OH)_4 + 5SiO_2 \tag{4.4}$$

$$NaAl_3Si_3O_{10}(OH)_2 + H_2O + CO_2 \leftrightarrow NaAlCO_3(OH)_2 + Al_2Si_2O_5(OH)_4 + SiO_2 \tag{4.5}$$

$$Al_2Si_2O_5(OH)_4 + H_2O + 2CO_2 + 2Na^+ \leftrightarrow NaAlCO_3(OH)_2 + 2SiO_2 + 2H^+ \tag{4.6}$$

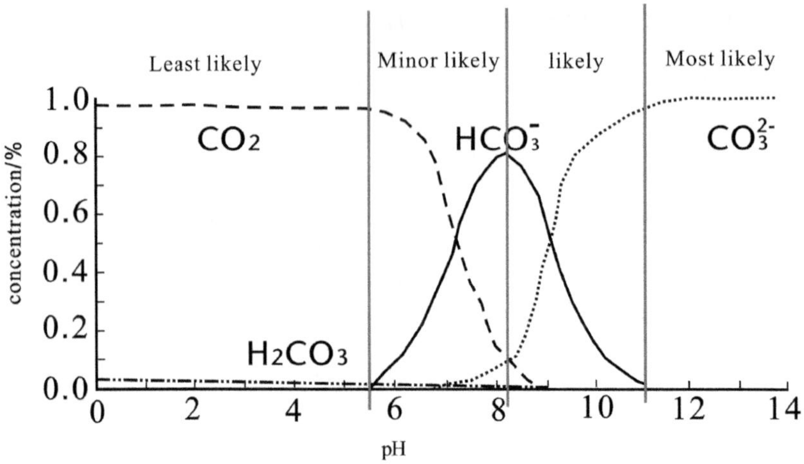


**Fig. 3** The relationship between pH and the concentration of $CO_2$, $H_2CO_3$, $HCO_3^-$ , $CO_3^{2-}$ in solute (Cheng et al. 2003)

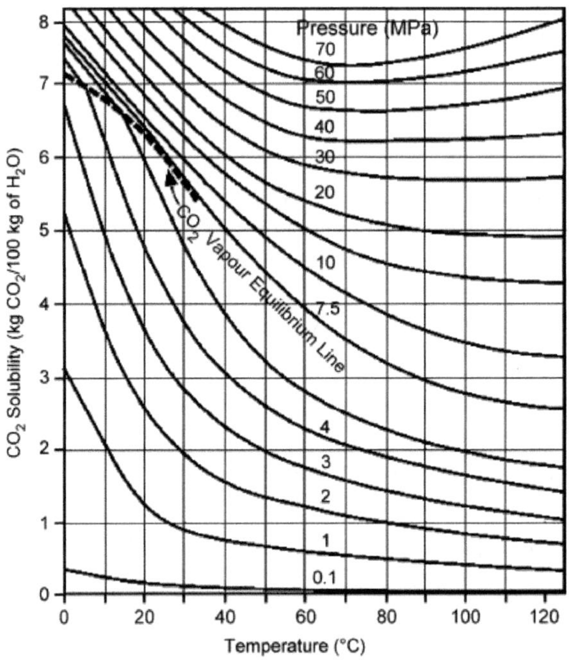


**Fig. 4** Solubility of $CO_2$ in water (Kohl & Nielsen 1997)

## 2 The Prediction of Scale Distribution in Saline Formation

In this paper, a benchmark simulation model has been set up. The pressure is 4MPa, and reservoir temperature is 100°C. Under these conditions, $CO_2$ solubility in water is 1.5kg/100kg $H_2O$ (Kohl and Nielsen 1997, Fig. 4). Before the simulation started, there was a negligible amount of $CO_3^{2-}$ in the formation water in near-well region according to the pH of 4.7-6.1. The minerals and aqueous species content used in this simulation are based on those from the wells in the Norddeutsch basin. Thus there is no scaling effect before the injection of $CO_2$, and before the simulation starts.

The process of scaling in the vicinity of the wellbore has been simulated by TOUGHREACT developed by Lawrence Berkely National laboratory, and the ECO2N model has been chosen.

## 2.1 Geological Model

The geologic formation is assumed to be infinite-acting and homogeneous with a thickness of 20m. A simple 1D radial flow model is set up, consisting of 50 radial blocks representing a distance of 2000m with logarithmically increasing radii (Fig. 5). The initial porosity and permeability of the reservoir rock are 10% and 8e-14 $m^2$ respectively, and the T and P are 100°C and 4MPa.

The injection rate of supercritical $CO_2$ is 3.17kg/s, which equals 100k tons a year. The simulation process lasts for 10 years and the injection period is 1 year. The hydrogeological parameters are shown in Table 1.

## 2.2 Geochemical Data

The types and initial abundances of primary minerals are determined from the reported data of S1 well drilling cores with a depth of 3430-3450m. Secondary minerals were also determined from field observations (Table 2).

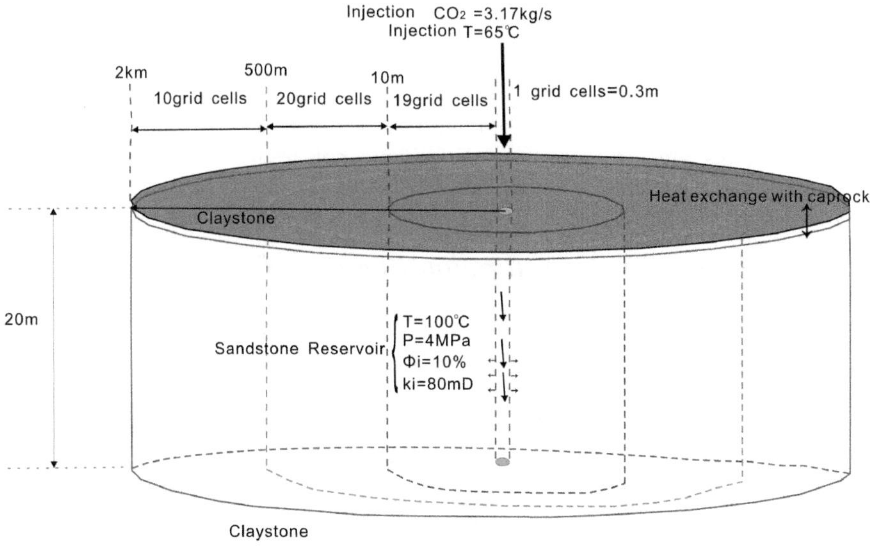


**Fig. 5** Simplified conceptual model for the injection and production well

**Table 1** Hydrogeologic parameters for the radial fluid flow problem

| | |
|---|---|
| Compressibility | $4.5\times10^{-10}$ Pa$^{-1}$ |
| Salinity | 0.06 (mass fraction) |
| Initial Gas saturation | 0 |
| $CO_2$ injection rate | 3.17 kg/s |
| ***relative permeability model*** | |
| Liuquid (van Genuchten, 1980) $k_{rl}=\sqrt{S^*}\left\{1-\left(1-\left[S^*\right]^{1/m}\right)^m\right\}^2$ | $S^*=(S_l-S_{lr})/(1-S_{lr})$ |
| irreducible water saturation | $S_{lr}=0.30$ |
| Exponent | m=0.457 |
| **gas(Corey,1954):** | |
| $k_{rg}=\left(1-\hat{S}\right)^2/\left(1-\hat{S}^2\right)$ | $\hat{S}=\frac{(S_l-S_{lr})}{(S_l-S_{lr}-S_{gr})}$ |
| irreducible gas saturation | $S_{gr}=0.05$ |
| ***Capillary pressure:*** van Genuchten (1980) $P_{cap}=-P_0\left(\left[S^*\right]^{-1/m}-1\right)^{1-m}$ | $S^*=(S_l-S_{lr})/(1-S_{lr})$ |
| irreducible water saturation | $S_{lr}=0$ |
| Exponent(m) | m = 0. 457 |
| strength coefficient | $P_0$ = 19. 61 kPa |

Anhydrite is assumed to react at equilibrium because of its rapid reaction rate. Other minerals are set to react under kinetic constrains. Thermodynamic and kinetic data is taken from Xu et al. (2004a, 2004b).

## 2.3 *Pressure and Gas Saturation Changes after the Injection of $CO_2$*

At the beginning of injection period (36.5days), the reservoir pressure increases quickly, especially for the region around the borehole (< 200m). But for regions far from the injection point, the pressure decreases sharply, and for the area in between (>2km), the pressure has only a tiny variation; after 1 year, the reservoir pressure increases to 9MPa near the injection point, and to 7MPa at 3km away from the injection point; 10 years after the initial injection, the reservoir pressure had remained at 7MPa (Fig. 6).

**Table 2** list of minerals and aqueous species considered in the simulations

| Mineral | surface area($cm^2/g$) | mineral initial volume fraction (%) | $k_{25}$ ($moles/m^2/s$) | Ea (KJ/mol) | n | Primary aqueous components |
|---|---|---|---|---|---|---|
| **Primary:** | | | | | | $H_2O$<br>$H^+$<br>$Ca^{2+}$<br>$Mg^{2+}$<br>$Na^+$<br>$Cl^-$<br>$SiO_2(aq)$<br>$Fe^{2+}$<br>$HCO_3^-$<br>$SO_4^{2-}$<br>$K^+$<br>$AlO_2^-$ |
| Quartz | 9.8 | 69.2 | 1.2589e-14 | 87.50 | 0 | |
| K-feldspar | 9.8 | 3.6 | 1.0000e-12 | 57.78 | 0 | |
| | | | 3.5481e-10 | 51.83 | 0.4 | |
| Oligoclase | 9.8 | 4.4 | 1.3323e-10 | 69.12 | 1 | |
| Muscovite | 9.8 | 1.2 | 1.9180e-14 | 60.26 | 1 | |
| Illite | 151.6 | 2.0 | 1.0000e-13 | 62.76 | 0 | |
| | | | 4.3652e-12 | 62.76 | 0.17 | |
| Calcite | 9.8 | 3.8 | 1.0000e-13 | 62.76 | 0 | |
| | | | 6.9183e-2 | 18.98 | 1 | |
| Anhydrite | equilibrium | 13.6 | | | | |
| Chlorite | 9.8 | 0.2 | 2.5119e-12 | 62.76 | 0 | |
| Hematite | 12.9 | 1.6 | 2.5120e-15 | 66.2 | 0 | |
| | | | 4.0740e-10 | 66.2 | 1 | |
| Low-albite | 9.8 | 0.6 | 2.7540e-13 | 69.8 | 0 | |
| | | | 6.9180e-11 | 65.0 | 0.457 | |
| **Secondary:** | | | | | | |
| Magnesite | 9.8 | 0 | 4.5710e-10 | 23.5 | 0 | |
| | | | 4.1690e-7 | 14.4 | 1 | |
| Dolomite | 9.8 | 0 | 2.9510e-8 | 52.2 | 0 | |
| | | | 6.4570e-4 | 36.1 | 0.5 | |
| Siderite | 9.8 | 0 | 1.2600e-9 | 62.76 | 0 | |
| | | | 6.4570e-4 | 36.1 | 0.5 | |
| Ankerite | 9.8 | 0 | 1.2600e-9 | 62.76 | 0 | |
| | | | 6.4570e-4 | 36.1 | 0.5 | |
| Dawsonite | 9.8 | 0 | 1.2600e-9 | 62.76 | 0 | |
| | | | 6.4570e-4 | 36.1 | 0.5 | |
| Ca-smectite | 151.6 | 0 | 1.6600e-13 | 35.0 | 0 | |
| | | | 1.0470e-11 | 23.6 | 0.34 | |
| Kaolinite | 151.6 | 0 | 6.9180e-14 | 22.2 | 0 | |
| | | | 4.8980e-12 | 65.9 | 0.777 | |

From the gas saturation changes diagram (Fig. 7), it can be observed that the gas saturation first increases, and then returns to the original level of the reservoir.

Fig. 8 gives the distribution of $CO_2$, the affected distance is only about 200m, and the decreasing trend shows the movement of $CO_2$ in the saline formation.

Fig. 9 shows the minor changes in the porosity of sandstone, illustrating the dissolution and scale deposits after $CO_2$ injection.

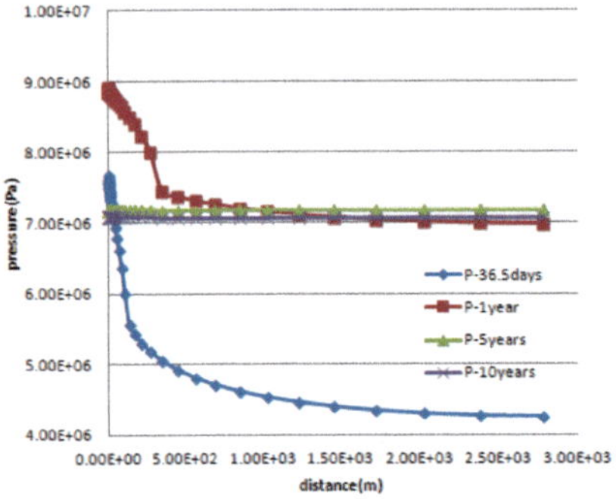


**Fig. 6** Pressure changes at different time scale after injection of $CO_2$

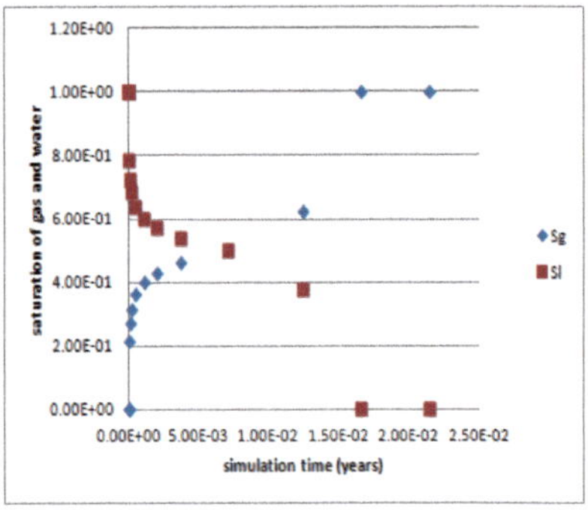


**Fig. 7** Saturation changes of liquid and gas phase at 4m far away from injection point

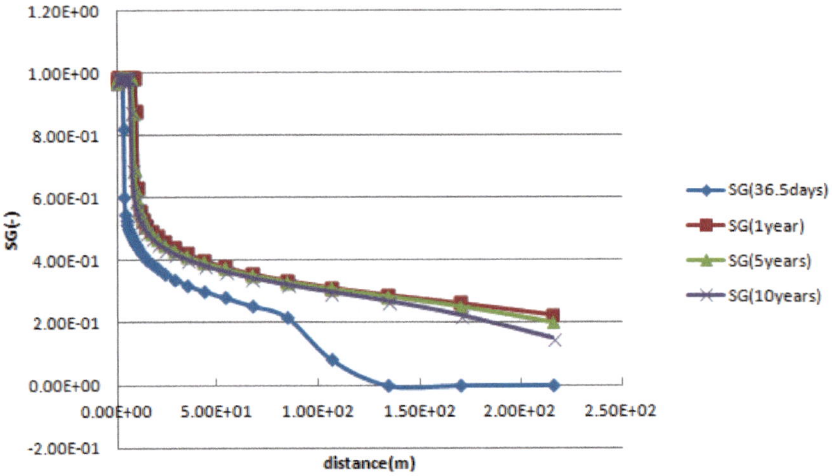


**Fig. 8** Gas saturation changes at all distances of different time

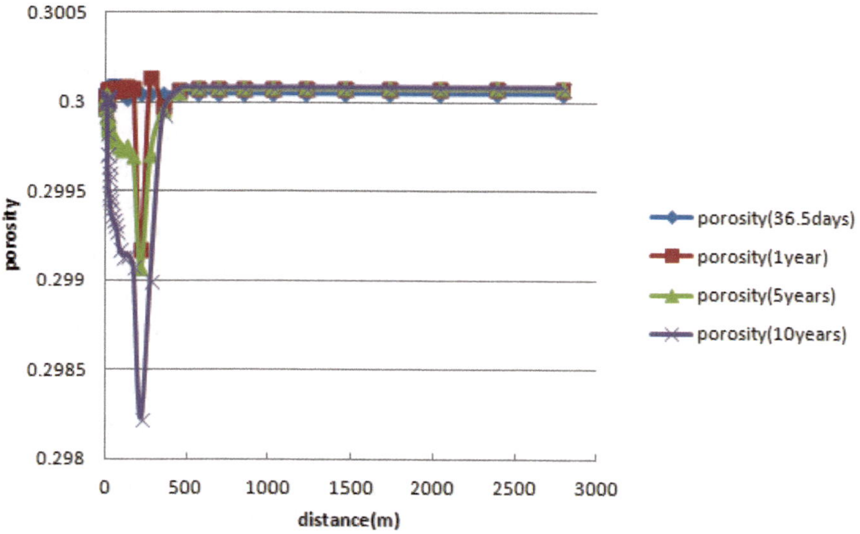


**Fig. 9** Porosity changes of the reservoir at all distances of different time

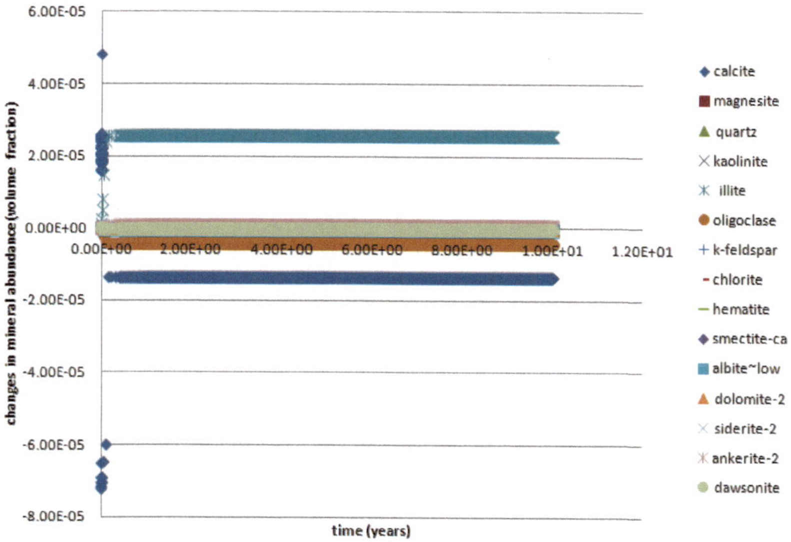


**Fig. 10** All the changes in mineral abundance at distance of about 4m from the injection point (In the volume fractions, positive values indicate precipitation and negative values dissolution)

Fig. 10 illustrates the changes in the abundance of minerals at a distance of about 4m from the injection point, illite is the only mineral precipitated, and calcite has a trend towards total dissolution, while near the injection point, calcite changes strongly and can both dissolve and precipitate. Therefore, illite and calcite may be the main mineral scales in near-well region of the reservoir.

Fig. 11 shows that, at the beginning of the $CO_2$ injection period (36.5days), calcite, chlorite and albite dissolve, especially at the near-well region of about 100m in diameter, just as expected. The dissolution of albite is the predominant phenomenon, which will affect the whole reservoir for a distance of about 3km. The precipitated minerals are illite, oligoclase and calcite. Near the well borehole, illite and oligoclase are the main mineral scales, but calcite dissolves in the near-well region but precipitates far from the well borehole.

At the end of the injection period (1 year), the chemical reaction will focus at a point within a distance of about 500m. On one hand, illite and ankerite will precipitate, on the other hand, chlorite and calcite will dissolve.

5years after the injection of $CO_2$, dawsonite, illite, smectite-ca and ankerite will deposit, and at the same time, chlorite and calcite will dissolve.

After 10 years, the trend of mineral changes is similar to that of 5 years. But the mineral variation degree is different.

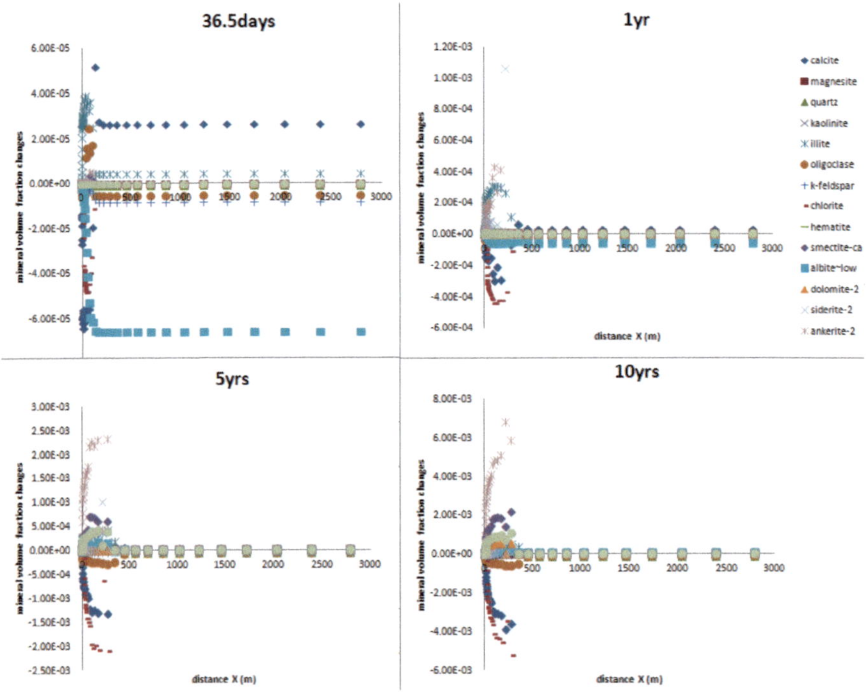


**Fig. 11** The changes in mineral abundance at different time scale (in volume fraction, positive values indicate precipitation and negative dissolution)

# 3 Scaling Problem Controlling

## *3.1 The Corrosion Problem Control in Equipment*

Drilling equipment, such as tubing, may be corroded during the injection of $CO_2$ processes, thus, after the injection of $CO_2$, some other chemical material should be injected in order to decrease the corrosion rate of the tubing. For petroleum engineering, some commercial scale dissolvers will be used, for example, formate can be used to protect the tubing and inhibit the scale near wellbore and thus increase the injection efficiency (Rhudy et al. 1993, Williams 1999, Taheri et al. 2011). Because the liquid after ionization of formate is basic, it can neutralize a little $CO_2$, thus inhibiting the corrosion of tubing (Li et al. 2011).Therefore, a $CO_2$-formate-$CO_2$ injection style can be selected for the CCS project.

## *3.2 The Problem of Scaling Control in the Near-Well Region*

The scale formed in the reservoir near the wellbore region can decrease the injection efficiency of water or gas and the hydrocarbon production efficiency, inhibiting the gas and oil flow towards the well. Therefore, measures should be carried out to avoid scaling in the near-wellbore reservoir (Baraka et al. 2012).

Scale can be ameliorated by chemical, physical or biological methods. These can be divided into three main categories: those that affect solubility, those that alter the growth mechanism of crystals, and those that change the potential of a surface to foul (Jitka et al. 2004). Generally, chelants (such as EDTA, DTPA) can be used to control scale formation. Chelants are molecules that sequester soluble cationic scaling species (e.g. $Ca^{2+}$) in order to prevent their reaction with anions (such as $CO_3^{2-}$ and $SO_4^{2-}$). Polymeric antiscalants with higher efficiency can also be used to control particle growth, agglomeration, and morphology (Fig. 12), thus significantly impeding the formation of thick, adhesive scale deposits in the well borehole and near-well reservoir.

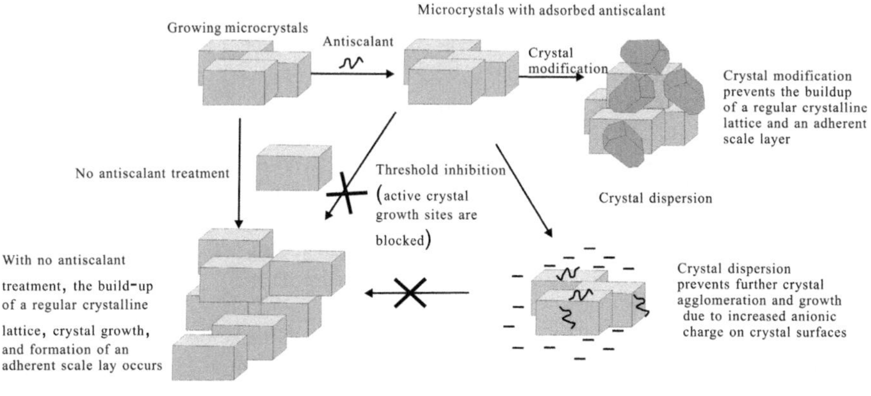


**Fig. 12** The ways antiscalant impede the formation of scale (Duggirala 2002)

# 4 Conclusions

(1) After the injection of supercritical $CO_2$ into saline formation reservoir, pressure will change with time and distance, however, with different characteristics at different time scales.
(2) The injection of $CO_2$ can cause the dissolution of the original minerals and the precipitation of new minerals (scales), causing minor changes of porosity in the reservoir not far from the injection well and a redistribution of the system.
(3) At the beginning of $CO_2$ injection, illite and oligoclase are the main scaling minerals near the well borehole. Calcite dissolves near the injection point and precipitate far away from the injection well. After some time , chlorite and calcite will only dissolve, oligoclase will dissolve later than chlorite and calcite, dawsonite, illite, smectite-Ca and ankerite will deposit last (5 and 10 yrs after the $CO_2$ injection) ;
(4) Chelants and polymeric antiscalants can be used to control particle growth, agglomeration, and morphology, thus significantly impeding the formation of thick, adhesive scale deposits in the well borehole and the neighboring region of the reservoir.

## References

1. Duggirala, P.Y.: Formation of calcium carbonate scale and control strategies in continuous digesters. Research Associate, Pulp and Paper Research, Nalco Company, Naperville (2002)
2. Rosenbauer, R.J., Koksalan, T., Palandri, J.L.: Experimental investigation of $CO_2$–brine–rock interactions at elevated temperature and pressure. Implications for $CO_2$ sequestration in deep-saline aquifers. Fuel Processing Technology 86, 1581–1597 (2005)
3. Abdulrazag, Y.Z., Shedid, A.S., Reyadh, A.A.: Investigation of supercritical carbon dioxide, aspheltenic crude oil, and formation brine interactions in carbonate formations. J. Petr. Sci. Engine 69(1-2), 63–70 (2009)
4. Ross, G.D., Todd, A.C., Tweedie, J.A.: The effect of simulated $CO_2$ flooding on the permeability of reservoir rocks. In: Proc. 3rd European Symp. Enhanced Oil Recovery, pp. 351–366 (1981)
5. Zhu, Z., Li, M., Lin, M., Peng, B., Sun, L.: Review of the CO2-water-rock interaction in reservoir. Bulletin of Mineralogy, Petrology and Geochemistry 30(1), 104–112 (2011) (in Chinese)
6. Kohl, A.L., Nielsen, R.B.: Gas Purification. Gulf Publishing Company, Houston (1997)
7. Cheng, Y., Pan, W., Sun, S.: Study on infection of fouling of oil deposit caused by $CO_2$ injection. Corrosion and Protection in Petrochemical Industry 20(6), 21–23 (2003) (in Chinese)
8. Xu, T., Sonnenthal, E., Spycher, N., Pruess, K.: A Simulation Program for Non-isothermal Multiphase Reactive Geochemical Transport in Variably Saturated Geologic Media. TOUGHREACT User's Guide (2004a)

9. Xu, T., Pruess, K.: Numerical Simulation of Injectivity Effects of Mineral Scaling and Clay Swelling in a Fractured Geothermal Reservoir (2004b)
10. Li, J., Zhao, C., Lü, E., et al.: Characteristics of formate and organic salt base fluid. Drilling Fluid & Completion Fluid 28(4), 72–77 (2011)
11. Liu, Z., Wolfgang, D.: Physicoshemical mechanisms of rate-determining of calcite deposition and their implications for paleo-environmental reconstruction. Cars Sinica 21(1), 252–257 (2002) (in Chinese with English abstract)
12. MacAdam, J., Parsons, S.A.: Calcium carbonate scale formation and control. Reviews in Environmental Science and Bio/Technology 3, 159–169 (2004)
13. Xu, T., Apps, J.A., Pruess, K.: Analysis of mineral trapping for $CO_2$ disposal in deep aquifers. Lawrence Berkeley National Laboratory (2001)
14. Pang, Z., Yang, F., Duan, Z., Huang, T.: Gas-water-rock interactions in sandstone reservoirs. Implications for enhance re-Injection into geothermal reservoirs and $CO_2$ geological sequestration. Presented at the Workshop for Decision Makers on Direct Heating Use of Geothermal Resources in Asia, organized by UNU-GTP, TBLRREM and TBGMED, Tianjin, China, May 11-18 (2008)
15. Bertier, P., Swennen, R., Laenen, B., Lagrou, D., Dreesen, R.: Experimental identification of $CO_2$-water-rock interactions caused by sequestration of $CO_2$ in Westphalian and Buntsandstein sandstones of the Campine Basin (NE-Belgium). Journal of Geochemical Exploration 89, 10–14 (2006)
16. Edmunds, W.M., Bath, A.H., Miles, D.L.: Hydrochemical Evolution of the East Midlands Triassic Sandstone Aquifer, England. Geochimica et Cosmochimica Acta 46, 2069–2081 (1982)
17. Ueda, A., Kato, K., Ohsumi, T., Yajima, T., Ito, H., Kaieda, H., Metcalfe, R., Takase, H.: Experimental Studies of $CO_2$-Rock Interaction at Elevated Temperatures under Hydrothermal Conditions. Geochemical Journal 39, 417–425 (2005)
18. Fischer, S., Liebscher, A., Wandrey, M., The $CO_2$SINK Group: $CO_2$-Brine-Rock Interaction—First Results of Long-Term Exposure Experiments at in Situ P-T Conditions of the Ketzin $CO_2$ Reservoir. Chemie. der Erde 70(s3), 155–164 (2010)
19. Fan, M., Hu, K., Jiang, X., Liu, W., Zhang, J., Chen, H.: Effect of Acid Fluid on Carbonate Reservoir Reconstruction. Geochimica 38(1), 20–26 (2009)
20. Pang, F., Mcpherson, B.J.O.L., Lichtner, P.C., et al.: Numerical Simulations of Interactions between Supercritical $CO_2$ and High-Temperature Fractured Rocks in Enhanced Geothermal Systems. In: Proceedings of 37th Workshop on Geothermal Reservoir Engineering, Stanford, California, January 30-February (2012)
21. Williams, G.D., Collins, I.R., SPE, BP Amoco Exploration et al.: Enhancing Mineral Scale Dissolution in the Near-Wellbore Region SPE No.56774 (1999)
22. Garcia, A.V., Thomsen, K., Stenby, E.H.: Prediction of mineral scale formation in geothermal and oilfield operations using the extended UNIQUAC model Part 2 Carbonate-scaling minerals. Geothermics 35(3), 239–284 (2006)
23. Charleston, J.: Scale Removal in the Virden, Manitoba Area. In: SPE 2160 presented at the 43rd Annual Fall Meeting, Houston, TX, September 29-October 2 (1968)
24. Carrel, K.D.: The Occurrence, Prevention and Treatment of Sulfate Scale in Shell Expro. In: SPE 16538 presented at Offshore Europe 1987, Aberdeen, Scotland, September 8-11 (1987)
25. Hardy, J.A., Barthorpe, R.T., Plummer, M.A., Rhudy, J.S.: Control of Scaling in the South Brae Field. In: OTC 7858 presented at the 1992 Offshore Technology Conference, Houston, TX, May 4-7 (1992)

26. Rhudy, J.S., Marathon Oil Co.: Removal of Mineral Scale from reservoir Core by scale dissolver. In: SPE 25161, presented at the SPE International Symposium on Oilfield Chemistry, New Orleans, LA, USA, March 2-5 (1993)
27. Taheri, A., Zahedzadeh, M., Masoudi, R., et al.: Simulation and experimental studies of mineral scale formation effects on performance of Sirri-C oil field under water injection. Iran J. Chem. Chem. Eng. 30(3), 9–24 (2011)
28. Baraka-Lokmane, S., Hurtevent, C., Ohanessian, J.L., et al.: Prediction of Mineral Scaling in a MEG Loop System of a gas production offshore. In: SPE 155124, presented at the SPE International Conference and Exhibition on Oilfield scale, Aberdeen, UK, May 30-31 (2012)
29. Bjφrnstad, T., Stamatakis, E.: Mechanisms of mineral scaling in oil and geothermal wells studied in laboratory experiments by nuclear techniques. Czechoslovak Journal of Physics 56(suppl. D), D405–D416 (2006)
30. Elguedri, M.: Assessment of scaling and corrosion problems in the Kebili geothermal field, Tunisia. Geothermal Training Programme, Reports (1) (1999)
31. McLin, K.S., Moore, J.N., Hulen, J., Bowman, J.R., Berard, B.: Mineral Characterization of Scale Deposits in Injection Wells. Coso and Salton Sea Geothermal Fields, CA. In: Proceedings of Thirty-First Workshop on Geothermal Reservoir Engineering, January 30-February 1. Stanford University, Stanford (2006)

# Simulation of Hydraulic-Mechanical (HM) Coupling Geo-processes at Ketzin $CO_2$ Sequestration Site Using the Code TOUGHMP-FLAC3D

Qun Wang[2], Michael Zhengmeng Hou[1,2], and Patrick Were[2]

[1] Institute of Petroleum Engineering, TU Clausthal, Clausthal-Zellerfeld, Germany
[2] Energy Research centre of Lower Saxony, Goslar, Germany

**Abstract.** $CO_2$ injection into saline aquifer triggers some coupled geo-processes including hydraulic-mechanical effects. In this study, hydraulic-mechanical (HM) coupling effects were investigated by means of the simulation code TOUGHMP-FLAC3D, and the results were compared with those obtained using the TOUGH2-MP code. Results of the measured and simulated parametric data give a relatively good match. The evolution of the pore pressure and formation stresses in the reservoir and caprock are measured and analyzed during $CO_2$ injection. Simulation results show that at the given optimal $CO_2$ injection rate, the HM coupling effects neither affect the integrity of the caprock nor reactivate the existing closed fault.

**Keywords:** HM coupling, $CO_2$ injection, Cap rock stability, Stress regime, Permeability.

## 1 Introduction

Deep saline aquifers have received special attention lately because they are the most promising option for geological storage of $CO_2$. Comparative storage volumes for several geological storage options are provided in Table 1, which clearly shows that the volume that can be stored in deep saline aquifers greatly exceeds the capacity of other geological options. Bounding seal integrity in deep saline aquifers is likely to be governed by the presence of pre-existing discontinuities and the induced buoyancy driven flow of $CO_2$ through those discontinuities. Concerns about leakage due to hydraulic fracturing or shearing of caprocks are minor because the facility is designed in such a way that pressure build-up is minimized through the injection of $CO_2$ in sweet spots of large permeability, leading to minimal geomechanical impact on caprock integrity. This is the case at the Ketzin $CO_2$ sequestration sites, where bottomhole pressures due to $CO_2$ injection have changed very little. In deep saline aquifers, it is likely that

M.Z. Hou et al. (Eds.): *Clean Energy Systems in the Subsurface*, SSGG, pp. 249–264.
DOI: 10.1007/978-3-642-37849-2_21 

the only wellbores present will be those related to the storage of $CO_2$, which are boreholes for investigation, injection, and monitoring. These wells will therefore be drilled and completed with state of the art techniques to guarantee their integrity.

**Table 1** Estimative of the volumes that can be sequestered for different formations (Holloway et al. 1996)

| Sequestration option | Worldwide capacity* |
|---|---|
| Ocean | 1000's GtC |
| Deep saline formation | 100's-1000's GtC |
| Depleted oil and gas reservoirs | 100's GtC |
| Coal seams | 10's-100'sGtC |

*worldwide total anthropogenic carbon emissions are around 7GtC per year, (1GtC= 1 billion metric tons of Carbon equivalent)

$CO_2$ SINK/MAN was the first on-shore storage project for $CO_2$ injection into a deep saline aquifer in Europe. The project has an overall objective to understand the mechanism and extent to which greenhouse gases, particularly $CO_2$, can be permanently and safely sequestered in geological formations. The ultimate deliverable of the project is a credible evaluation of the permanent containment of injected $CO_2$ determined by long-term predictive simulations and formal risk analysis techniques. It the hope of all parties involved that this assessment will be helpful to the regulatory government bodies to answer questions regarding the security of large volume $CO_2$ storage in deep saline aquifers in Europe and other areas of geological similarities worldwide, where permanent $CO_2$ storage is contemplated.

Since the commencement of the $CO_2$SINK/MAN project in 2004, a substantial amount of research and investigation into the properties of seals has been conducted in different perspectives. For such large-scale geological storage projects, however, $CO_2$ injection will alter the in-situ characteristics of caprocks, possibly degrading its sealing properties. The in-situ effective stresses, for example, will change, resulting in consolidation or swelling and deformation. Such changes will affect the hydraulic integrity of caprock because the permeability of geomaterials is strongly dependent on its mechanical behavior. Knowledge of how those properties evolve under downhole injection conditions is weak. Therefore, to make geological sequestration of $CO_2$ an environmentally acceptable practice it is necessary to understand how bounding seals will react under pre- and post-operational conditions, with a focus on the hydraulic integrity over both the short-term (injection stage) and long-term (hundreds of years after shutting down the injection wells). Injection of $CO_2$ will give rise to a variety of coupled physical and chemical processes that will affect the bounding seals and their surroundings. Among the most notable are changes to the stress field due to pore pressure and temperature changes, immiscible displacement of one of the phases by $CO_2$, partial dissolution of $CO_2$ into the aqueous phase, and chemical interactions between the $CO_2$, the aquifer and caprock. Therefore, permeability is affected by these changes, resulting in changes in the hydraulic integrity of caprocks during the life time of the project. It is expected that such changes will be most prominent during the short-term. One primary concern in the short-term is the stability of the caprock seals, how they will react to a rapidly changing environment, and how their geomechanical, geochemical, and hydrogeological properties will be affected. In the long-term, hydrostatigraphic and/or geochemical trapping will be the two basic mechanisms that will keep $CO_2$ from reaching portable water sources and /or the biosphere. Since the $CO_2$ plume covers a very large area, the heterogeneities in the reservoir and its caprock will play a key role in $CO_2$ migration, pressure build-up, geochemical reactions, and potential leakage flow paths. When the main mechanism is hydrostratigraphic trapping, the movement of the $CO_2$ plume and the continuity and stability of the bounding seals will determine its performance. On the other hand, when geochemical trapping is the main mechanism, the chemical reactions between the plume and the rock will control its performance. Besides experimental research, monitoring and numerical simulations have been carried out using a variety of computer simulators (Kempka et al. 2010, Nimtz et al. 2010), including Eclipse 100, Eclipse 300, and MUFIE-UG user-developed software. However, none of these simulations could take into account the hydraulic-mechanical coupled effects caused by $CO_2$ sequestration. It is therefore the main objective of this paper is to study the HM coupling effects caused by $CO_2$ injection.

*Location and Geological Settings*

The Ketzin $CO_2$ sequestration site is near the town of Ketzin, Germany, which is about 40 km northwest of Berlin. The Ketzin geological formation is well investigated by boreholes as it has been used as seasonal natural gas storage in lower Jurassic strata at around 280 m deep since 1960 (Martens et al. 2012). In the northern part of the survey area, approximately 1.5 km from an injection well, a fault-and-graben system was imaged, however this was not detected by 2D vintage data, but attention has been paid to well and production data gathering (Lueth et al. 2011).

Ketzin district is a multilayered geological system, besides the upper layer, there are 2 layers of permeable saline aquifer. Each of them is overlain by caprock.

The target saline aquifer for $CO_2$ storage at Ketzin is the Stuttgart Formation of Triassic age, which has a thickness of 80 m and is located between 630 m to 710 m deep. Stuttgart Formation is of fluvial origin and exhibits a heterogeneous lithology (Förster et al. 2008); the storage formation consists of sandstone and siltstone interbedded with mudstone. The upper part of Stuttgart formation is a 14-m-thick, fine-grained, cross-bedded channel facies sandstone underlain by a thin layer siltstone and overlain by a thin mud stone. Sandstones in this formation are fine-grained to medium-grained and well-sorted. The channel-facies sandstone between depths of 630-650 m is the primary target reservoir for $CO_2$ storage. The initial reservoir conditions of the injection zone are 33 °C and 6.2 MPa. The average calculated reservoir porosity is about 26 % (Förster et al. 2008), and the reservoir permeability calculated based on the thickness of the permeable zones of the formation lies between 40 mD and 80 mD (Schilling et al. 2005).

The cap rock of the $CO_2$ storage formation is the Weser Formation, which consists mostly of mudstone (Triassic claystone), clayey siltstone and anhydrite and has a cumulative thickness of about 210 m (Kempka et al. 2010). The average porosity is 8 % and the average gas-permeability is 0.010 mD (Förster et al. 2008).

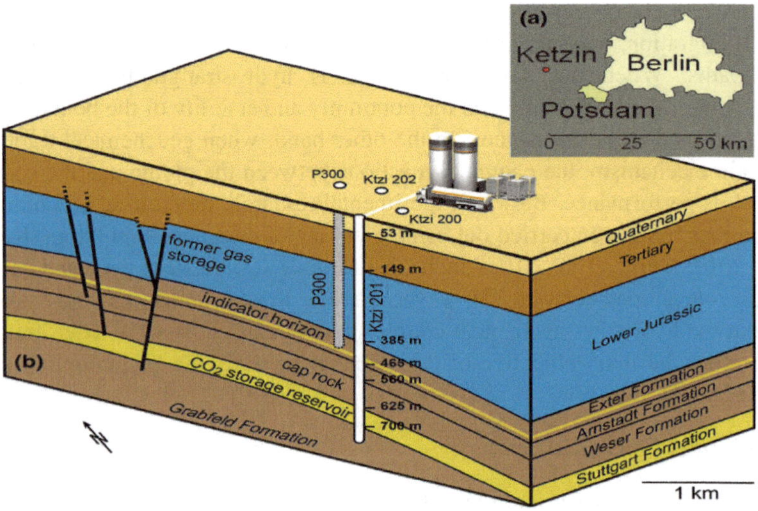


**Fig. 1** Geological formation in Ketzin district and injection wells (Martens et al. 2012)

*$CO_2$ Injection*

Well $CO_2$ Ktzi 201/2007, drilled in 2007, abbreviated as Ktzi 201 is used for the injection of $CO_2$, while the other two wells $CO_2$ Ktzi 200/2007(Ktzi 200) and $CO_2$ Ktzi 202/2007 (Ktzi 202) serve as observation wells for observing the injection and migration of $CO_2$. Here, food grade carbon dioxide ($\geq$ 99.9 % purity) has been injected into the sandstone layer of the Stuttgart Formation since June 2008 (Würdemann et al., 2009). Liquid $CO_2$ is delivered by road tankers to Ketzin and $CO_2$ and injected into the reservoir in a supercritical state. Since the start of the $CO_2$ injection on June 30, 2008, the injection facility has been operating safely and reliably. By the end of September 2011, approximately 53,000 tones of $CO_2$ had been injected via the well Ktzi 201 (Fig. 2). The injection facility at Ketzin is designed to allow for injection rates ranging from 0 to 78 tons per day.

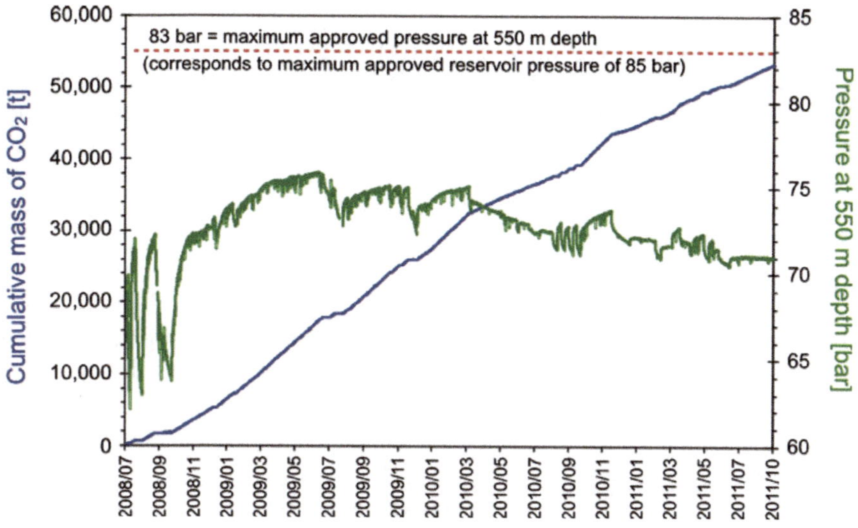


**Fig. 2** Evolution of downhole pressure at mass of injected $CO_2$ (Martens et al. 2012)

# 2 Numerical Simulation and Modeling

## 2.1 *Applied Numerical Simulator*

Two numerical codes were applied to simulate $CO_2$ sequestration in this study, namely TOUGH2-MP and TOUGHMP-FLAC3D. Initially, bottomhole pressures were simulated using TOUGHMP-FLAC3D by implementing the stress-dependent permeability model into the code. For comparison purposes another simulator, TOUGH2-MP, which simulates only the hydraulic phase of $CO_2$ propagation, was engaged to perform a parallel simulation. $CO_2$ plume distribution, pore pressure distribution as well as stress evolution were observed during $CO_2$ injection and caprock integrity has been assessed based on stress evolution.

*TOUGH2-MP Code*

TOUGH2-MP is a massively parallel (MP) version of the TOUGH2 code, designed for computationally efficient parallel simulation of isothermal and non-isothermal flows of multicomponent, multiphase fluids in one, two and three-dimensional porous and fractured media (Zhang et al. 2008). ECO2N module was designed in 2005 for applications to geological sequestration of $CO_2$ in saline aquifers. It includes a comprehensive description of the thermodynamics and thermo-physical properties of $H_2O$-NaCl-$CO_2$ mixtures that reproduce fluid properties largely within experimental error for the temperature, pressure, and salinity conditions of interest (10 °C≤T≤110 °C, P≤600 bar; salinity up to full halite saturation).

*TOUGHMP-FLAC3D Code*

The simulator used in this work, TOUGHMP-FLAC3D code, was developed by Gou (2011) based on the THOUGH-FLAC code (Rutqvist et al. 2002). The hydraulic-mechanical coupling is realized by implementing a stress-dependent permeability model and a stress-dependent porosity model into the code. The hydraulic phase, namely the pore pressure is calculated in TOUGH2-MP, the pore pressure is transferred to FLAC3D to calculate the mechanical change and then the porosity and permeability are updated using the new stress.

*Stress-Dependent Permeability Model*

The stress-dependent permeability model is developed based on the data of north German basin (Albrecht et al. 2010). Permeability is found to depend on the effective mean stress:

$$k = k_0 \exp\left(-m \frac{\sigma_m^{'}}{\varphi_0 (180\sigma_m^{'} + 6000)}\right) \quad (1)$$

The validated permeability model is written into TOUGHMP-FLAC3D code through Fortran 90.

## 2.2 Modeling

*Model Setting Up*

An axial symmetric quarter model is set up to take into account the three dimensional aspects of the injection through a vertical well. The model is a 4 x 4 km square area with a height of 800 m. Corresponding to the real formations, the model includes 7 groups, which are upper layer (sandstone), caprock1(claystone), aquifer1(sandstone), caprock2 (claystone), aquifer2 (sandstone), baselayer1 (claystone) and baselayer2 (mudstone). The model contains 39×39×37=56277 elements.

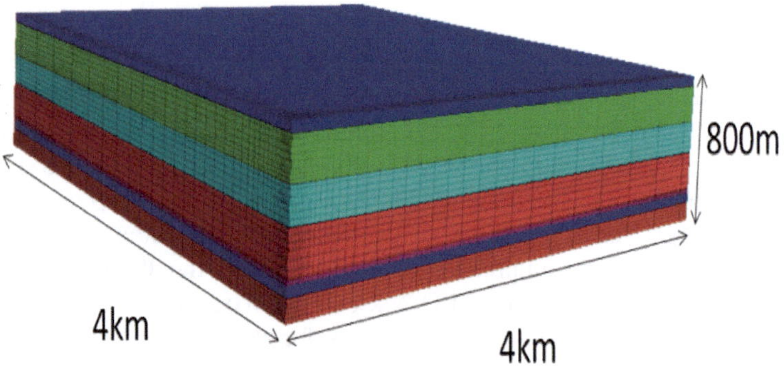


**Fig. 3** 3D model of the Ketzin formation

Mechanical and hydraulic properties used for the simulation are derived from literature from the Ketzin project, whereas those which are not available in literature are assumed to be realistic. The properties of the simulation model are presented in Table 2.

**Table 2** Mechanical properties of simulation model

| Layer | D (m) | Rock | K (GPa) | G (GPa) | f (°) | C (MPa) | T (MPa) | α | Φ (%) | $k_h$ ($m^2$) | m |
|---|---|---|---|---|---|---|---|---|---|---|---|
| U1 | 0-50 | Sa | 9.5 | 9.4 | 30 | 11.9 | 3 | 0.645 | 0.1 | 3E-15 | 38 |
| C1 | 50-250 | Cl | 12.88 | 9 | 22 | 26 | 3 | 0.358 | 0.08 | 1E-17 | 38 |
| A1 | 250-420 | Sa | 9.5 | 9.4 | 29 | 11.9 | 3 | 0.645 | 0.26 | 4E-14 | 38 |
| C2 | 420-630 | Cl | 12.88 | 9 | 22 | 26 | 3 | 0.358 | 0.08 | 1E-17 | 38 |
| A2 | 630-650 | Sa | 9.5 | 9.4 | 29 | 11.9 | 3 | 0.645 | 0.26 | 4E-14 | 38 |
| C3 | 650-700 | Cl | 12.88 | 9 | 22 | 26 | 3 | 0.358 | 0.08 | 1E-17 | 38 |
| B | 700-800 | Mu | 12.88 | 9 | 22 | 26 | 3 | 0.358 | 0.08 | 6E-20 | 38 |

Note: U-Upper layer, C1-Caprock1, A1-Aquifer1, C2-Caprock2, A2-Aquifer2, B-Baselayer, D-Depth, K-Bulk modulus, G-Shear modulus, C-Cohesion, T-tension, Sa-sandstone, Cl-Claystone, Mu-Mudstone.

## 2.3 *Initial and Boundary Conditions*

*Hydraulic Static Pressure*

In an undisturbed system, the hydrostatic pressure in a fully saturated condition is assumed to be the pressure gradient multiplied by the depth. The hydrostatic gradient is 0.98 bar/m.

*Permeability Anisotropy*

The saline aquifer permeability adopted here are calculated based on the thickness of the permeable zones of the formation, which lies between 40 mD and 80 mD (Schilling et al. 2005). The permeability anisotropy in a saline aquifer was determined as $k_v / k_v = 1/3$ (Kempka et al. 2010).

*Temperature Gradient*

The temperature of the formation layers is expressed with a geothermal gradient. In the Ketzin area, a geothermal gradient of 0.035°C/m is assumed and the ground temperature is 10 °C (Nimtz et al. 2010).

*Aquifer Salinity*

The content of the total dissolved solids increases stepwise with depth, from 0.2-1.4 g/L in Quaternary formations and 47-50 g/L in Jurassic formations to 250-321 g/L in Triassic formations (Foester et al. 2006). For simplicity, the salinity of the whole model is set at 20 %, which is the average value of the reservoir formation.

*Initial Stress Conditions*

The initial stress conditions in the Ketzin area are not available. However, the Groß Schönebeck which is about 70 km north east of Ketzin area has a stress regime $\sigma_h = 0.56\sigma_z, \sigma_H = 0.78\sigma_z$ (Moeck et al. 2009). Combined with world stress map (GFZ), the stress regime in the Ketzin area is conservatively assumed to be: $\sigma_h = \sigma_H = 0.8\sigma_z$.

**Table 3** Relative permeability and capillary pressure parameters

| Layers | Corey $S_{lr}$ | Corey $S_{gr}$ | $\lambda$ | $S_{lr}$ | $P_0$ | $P_{max}$ | $S_{ls}$ |
|---|---|---|---|---|---|---|---|
| U1 | 0.15 | 0.05 | 0.457 | 0 | 1.61E-04 | 5.00E+07 | 0.99 |
| C1 | 0.15 | 0.05 | 0.457 | 0 | 1.61E-04 | 5.00E+07 | 0.99 |
| A1 | 0.15 | 0.05 | 0.457 | 0 | 5.03E-05 | 5.00E+07 | 0.99 |
| C2 | 0.15 | 0.05 | 0.457 | 0 | 1.61E-04 | 5.00E+07 | 0.99 |
| A2 | 0.15 | 0.05 | 0.457 | 0 | 5.03E-05 | 5.00E+07 | 0.99 |
| C3 | 0.15 | 0.05 | 0.457 | 0 | 1.61E-04 | 5.00E+07 | 0.99 |
| B | 0.15 | 0.05 | 0.457 | 0 | 1.61E-04 | 5.00E+07 | 0.99 |

# 3 Simulation Results and Discussion

## *3.1 Bottom-Hole Pressure History Match*

The injection data is retrieved during the monitoring operations at the Ketzin site. Monitored pressure is the pressure measured at 550 m depth plus an additional 2 bars. TOUGH2-MP and TOUGHMP-FLAC3D simulated bottom-hole pressure as well as the monitored pressure are presented in Figure 4. Both the bottom-hole pressures simulated from TOUGH2-MP and TOUGHMP-FLAC3D are higher than the monitored pressure, which might be due to the conservative parameters, such as permeability, the formation factor. Bottom-hole pressure simulated by TOUGH2-MP is slightly higher than that of TOUGHMP-FLAC3D. A reasonable explanation is that the increasing pore pressure gives rise to an enlargement of the porosity, and further an increase in permeability, which in turnprevents the pore pressure increasing.

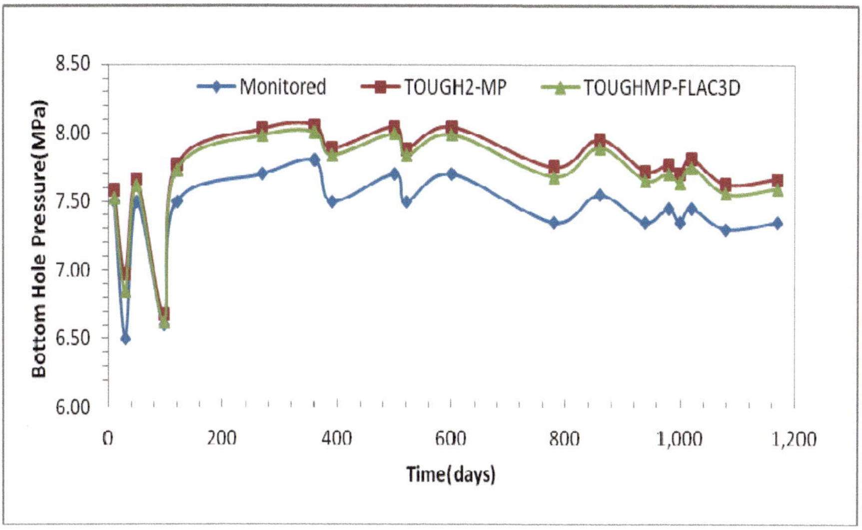


**Fig. 4** Bottom-hole pressure match

Permeability and porosity evolution in the injection zone during $CO_2$ injection are presented in Fig. 5. Permeability and porosity show distinct pressure dependence on effective stress and bottom-hole pressure accordingly.

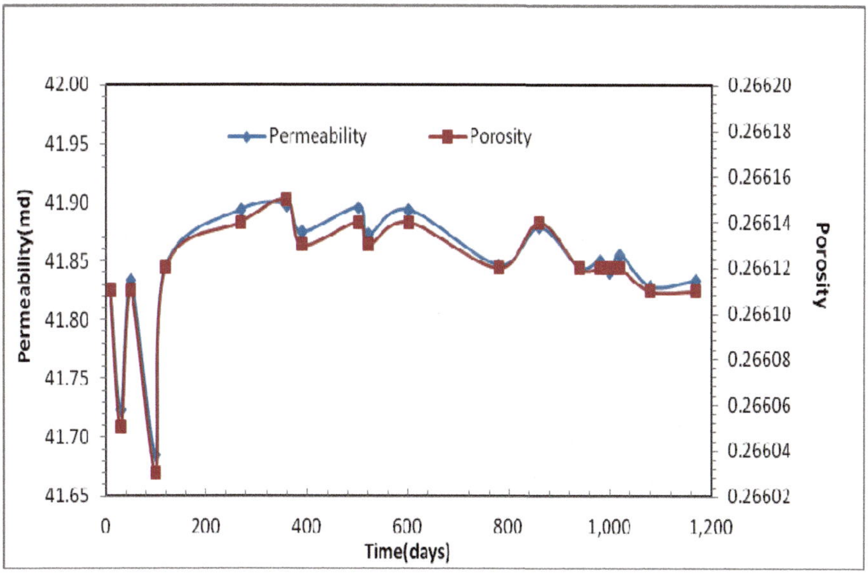


**Fig. 5** Permeability and porosity development of the injection zone during $CO_2$ injection

## 3.2 Stress Alteration

Stress alteration caused by pore pressure alteration has been studied during $CO_2$ injection (Rutqvist et al. 2008, Gou 2011). The stress response indicates that the in situ stress field does not remain constant during fluid injection, but evolve with time and space. The stress path, namely the pore pressure and horizontal stress variation ratio, $\beta = \Delta\sigma_{H/h} / \Delta p$, varies between 0.5 and 1.2 (Addis et al. 1998). The analytical solution is given based on linear poroelastisity, on the assumption that a reservoir will behave under uniaxial strain conditions.

$$\sigma_{H/h} = \alpha \frac{1-2\upsilon}{1-\nu} \Delta p \tag{2}$$

where $\alpha$ is Biot's coefficient, which is usually assumed to be 1, but for sandstone, is lower than 1.

**Table 4** Comparison of the total horizontal stress change calculated from both the analytical solution and the simulation

| | Pore presure (MPa) | Poission ratio | Biot coefficient | Caculated value (MPa) | Simulated value (MPa) |
|---|---|---|---|---|---|
| Bottom of caprock | 1.464 | 0.214 | 0.358 | 0.328 | 0.380 |
| Top of reservoir | 1.482 | 0.129 | 0.645 | 0.7457 | 0.815 |

The comparison of the total horizontal stress change calculated from the analytical solution and the simulation is presented in Table 4. The simulated value is smaller than the analyzed value. A possible reason is that vertical displacement of the reservoir is hampered by the side band, which further causes the limitation of the horizontal strain and horizontal stress change.

## 3.3 Performance Analysis

*$CO_2$ Plume Evolution*

Figure 6 shows the $CO_2$ plume evolution with injection time. $CO_2$ first propagates upwards at the beginning of injection, which is due to fact that the density of $CO_2$ is smaller than saline water, and afterwards $CO_2$ begins to spread in a horizontal direction. The $CO_2$ plume reaches about 28 m each side of the well in a horizontal direction and a height of 45 m in 10 days; 60 m in the horizontal direction and a height of 50 m in 120 days; after injecting for 360 days, it increases to 130 m in the horizontal direction and a height of 60 m; and at the end of injection, the value increases to 300 m in the horizontal direction and the height stays at 60 m, which is far from the fault zone. The $CO_2$ arrival time in Ktzi 200 is 30 days, and 300 days in Ktzi 202 compared with 21 days in Ktzi 200 and 270 days in Ktzi 202 from the monitored data, which is also longer than the simulation results of Kempka et al. (2010), the possible reason is the conservative parameters chosen, like lower permeability; and formation factor.

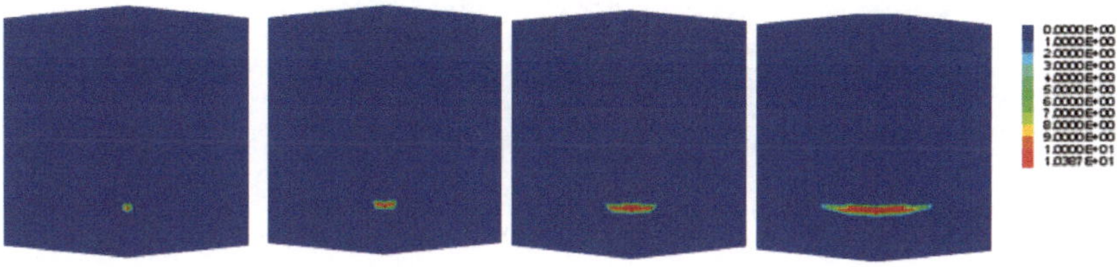

**Fig. 6** CO2 plume evolution with time

*Pore Pressure Alteration Distribution*

Due to the different injection velocities, pore pressure in the reservoir fluctuates with $CO_2$ injection and the bottom-hole pressure alteration reaches a maximum value of about 1.58 MPa, after 360 days of injection. Pore pressure change stays at a value of 1.27 MPa.

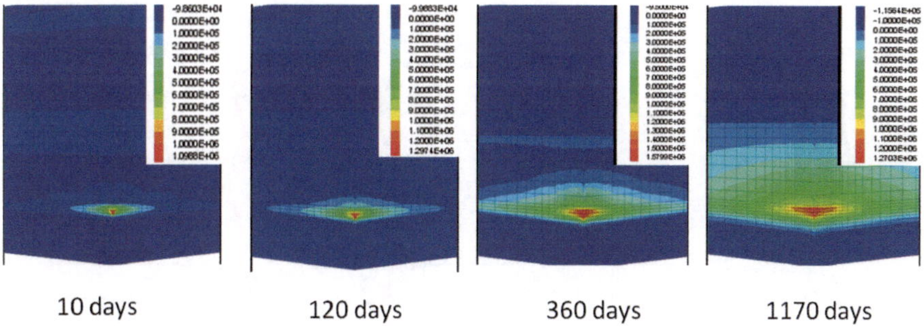


**Fig. 7** Pore pressure alteration distribution with time, unit (Pa)

*Induced Ground Movement*

One target of HM simulation is to investigate the ground movement caused by the $CO_2$ injection. Ground movement during $CO_2$ injection has been studied by many researchers. Rutqvist et al. (2009) calculated the surface uplift in In-Salah $CO_2$ injection site and they found the maximum uplift at the top of injection zone and an attenuated uplift at the ground surface, which has been proven correct with the monitored data. The simulation of $CO_2$ geological storage in the Dogger carbonates of the Paris Basin by Vidal-Gilbert et al. (2010) showed similar results. Figure 7 presents the vertical displacement of the model after 360 days of injection. A maximum vertical displacement of approximately 1 mm is observed at the top of reservoir and 0.5 mm at the surface.

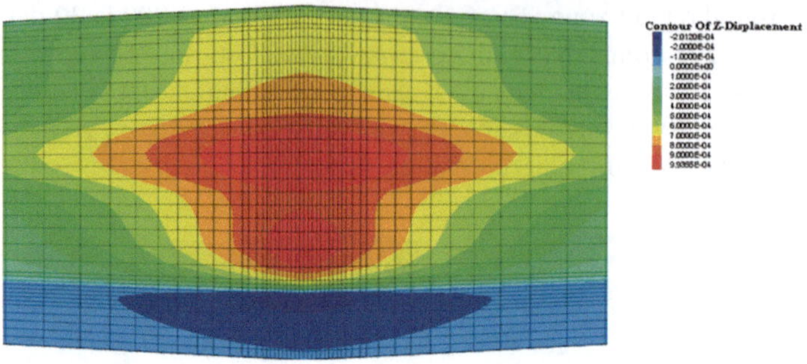


**Fig. 8** Vertical displacement of model after 360days of injection, unit, m

*Profile Analysis*

Figure 9 shows that the bottom-hole pressure reaches a maximum value after injecting for 360 days. The pore pressure, total stress and effective stress of the profile through the injection well are analyzed. It shows clearly that there is a large pore pressure build-up in the reservoir which gives rise to a large total horizontal stress change and a small total vertical stress change. While the effective vertical stress makes a large alteration in the reservoir the effective horizontal stress only a slight change.

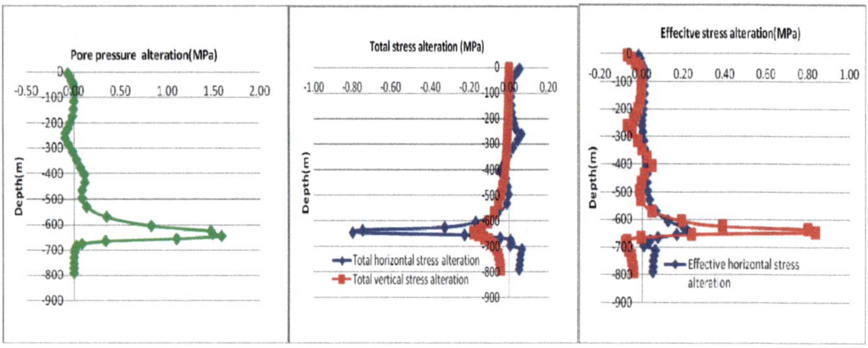


**Fig. 9** Pore pressure, total stress, effective stress alteration of profile through the injection point after injection for 360 day

*Point Analysis*

Two zones, the bottom of caprock and the top of the reservoir over the injection zone were chosen for analysis. Figure 10 presents the pore pressure, total stress and effective stress evolution at the top of the reservoir. The pore pressure evolution has a similar trend to the bottom-hole pressure. The total vertical stress varies slightly with the $CO_2$ injection, the total horizontal stress fluctuates with pore pressure change. The effective vertical stress fluctuates with the pore pressure change, while the effective horizontal stress varies slightly. The pore pressure, total stress and effective stress evolution at bottom of caprock has a similar trend to that at the bottom of the reservoir. Abnormal behavior of the total vertical stress and the effective stress are noticed at the beginning of the $CO_2$ injection, in which the total vertical stress shows a great change. A possible reason could be an unstable injection rate.

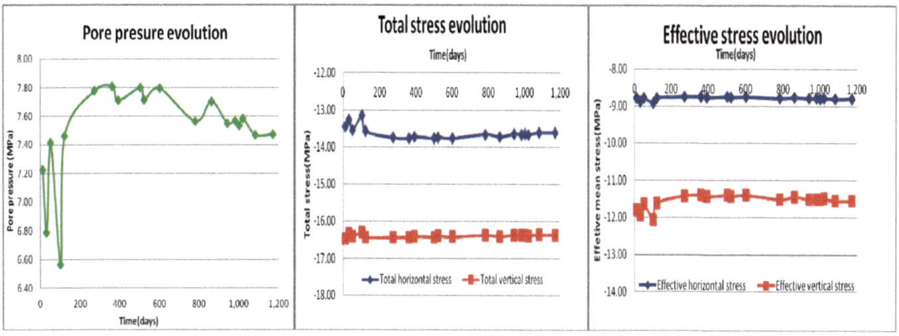


**Fig. 10** Pore pressure, total stress, effective stress evolution at the top of reservoir

*Caprock Stability Analysis*

The seal capability of the caprock is of vital importance in $CO_2$ sequestration and caprock stability is the most often used criterion for caprock capability. The Mohr-Coulomb yield criterion; and the Drucker-Prager yield function are adopted to analyze caprock stability (Rutqvist et al. 2008, Vilarrasa et al. 2010). The stability analysis is based on the Mohr-Coulomb yield criterion and is analyzed with a $\sigma_1^{'} - \sigma_3^{'}$ diagram. The stress path is analyzed at the bottom of the caprock in the given stress regime in Figure 11. A distinct change in the stress path is observed

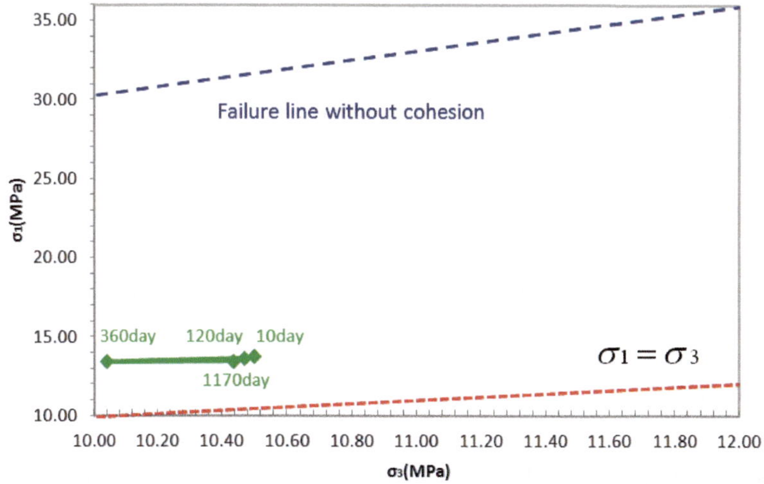


**Fig. 11** Stress path evolution at the bottom of caprock during $CO_2$ injection in three stress regimes

after 360days' injection, when the bottomhole pressure reaches the maximum value, however, the movement is far beyond the failure regime, which means there is no shear failure in the analyzed zone. The stress path moves back as the pore pressure decreases.

## 4 Conclusions

1. The Bottom-hole pressure evolution simulated by TOUGHMP-FLAC3D simulator yields provides a better fit with the trend of the measured values than that simulated by TOUGH2-MP, which means the mechanical effects caused by $CO_2$ injection should not be neglected.
2. The $CO_2$ injection rate used in this simulation, prevented the $CO_2$ from breaking through the fault to cause any leakage, and the caprock stayed stable during the $CO_2$ injection.
3. Since the permeability (40 mD) used in the simulations is the lower limit for the measured values, a conservative stress state was assumed. Therefore, the simulation results are more conservative when compared to the real case.

## Nomenclature

$k_0$ - Initial permeability, md

$\varphi_0$ - Initial porosity

$\sigma'_m$ - Effective mean stress, MPa

$m$ - Exponent

$S_{lr}$ - Corey irreducible liquid saturation

$S_{gr}$ - Corey irreducible gas saturation

$\lambda$ - Van Genuchten exponent

$P_0$ - Van Genuchten's air-entry pressure (MPa)

$P_{max}$ -Van Genuchten parameter

$S_{ls}$ -Liquid saturation

**Acknowledgements.** The authors would like to thank Mr Yang Gou for his technical support.

## References

1. Albrecht, D., Ganzer, L., Pusch, G., Reitenbach, V.: Bestimmung von In-situ Fließenschaften von Tight Gas Reservoirgesteinen unter Porendruck mittels Korrelationen aus Labormessungen. DGMK/ÖGEW-Frühjahrstagung (2010)
2. Addis, M.A.: The stress-depletion response of reservoirs. In: SPE 38720, pp. 55–65 (1997)

3. Förster, A., Norden, B., Zinck-Jørgensen, K., Frykman, P., Kulenkampff, J., Spangenberg, E., Erzinger, J., Zimmer, M., Kopp, J., Borm, G., Juhlin, C., Cosma, C., Hurter, S.: Baseline characterization of the $CO_2$SINK geological storage site at Ketzin, Germany. Environmental Geosciences 13, 145–161 (2006)
4. Förster, A., Giese, R., Juhlin, C., Norden, B., Springer, A., $CO_2$SINK GROUP: The geology of $CO_2$SINK site: from regional scale to a laboratory scale. Energy Procedia (2008)
5. Holloway, S., et al.: The Underground Disposal of Carbon Dioxide: Summary Report. JOULE II Project No. CT92-0031, British Geological Survey (1996)
6. Gou, Y.: Untersuchungen zu THM-gekoppelten Reaktionen der Speicherformationen und des Deckgebirges bei $CO_2$-EGR mit dem integrierten Code TOUGH2-FLAC3D. Master Thesis, TU Clausthal (2011)
7. Kempka, T., Kühn, M., Class, H., Frykman, P., Nielsen, C.M., Probst, P.: Modelling of $CO_2$ arrival time at Ketzin. Part I International Journal of Greenhouse Gas Control 4, 1007–1015 (2010)
8. Lueth, S., Bergmann, P., Cosma, C., Enescu, N., Giese, R., Götz, J., Ivanova, A., Christopher, J., Kashubin, A., Yang, C., Zhang, F.: Time-lapse seismic surface and down-hole measurements for monitoring $CO_2$ storage in $CO_2$SINK project (Ketzin, Germany). Energy Procedia 4, 3435–3442 (2011)
9. Martens, S., Kempka, T., Liebscher, A., Lüth, S., Möller, F., Myrttinen, A., Norde, B., Schmidt-Hattenberger, C., Zimmer, M., Kühn, M., The Ketzin Group: Europe's longest-operating on-shore $CO_2$ storage site at ketzin Germany: a progress report after three years of injection. Environmental Earth Sciences (2012), doi:10.1007/s12665-012-1672-5
10. Nimtz, M., Klatt, M., Wiese, B., Kuehn, M., Krautz, H.J.: Modelling of the $CO_2$ process and transport chain in CCS systems-Examination of transport and storage processes. Chemie der Erde (Geochemistry) 70, 185–192 (2010)
11. Rutqvist, J., Wu, Y.S., Tsang, C.F., Bodvarsson, G.: A modeling approach for analysiss of coupled multiphase fluid flow heat transfer, and deformation in fractured porous rock. International Journal of Rock Mechanics and Mining Sciences 39, 429–442 (2002)
12. Rutqvist, J., Birkholzer, J.T., Tsang, C.T.: Coupled reservoir geomechanical analysis of the potential for tensile and shear failure associated with $CO_2$ injection in multilayered reservoir-caprock systems. International Journal of Rock Mechanics and Mining Sciences 45, 132–143 (2008)
13. Rutqvist, J., Vasco, D.W., Myer, L.: Coupled reservoir-geomechanical analysis of $CO_2$ injection and ground deformations at In Salah, Algeria. International Journal of Greenhouse Gas Control (2009), doi:10.1016/j.ijggc. 2009.10.017
14. Schilling, F., Borm, G., Würdemann, H., Möller, F., Kühn, M.: $CO_2$ Sink Group Status report on the first European on-shore $CO_2$ storage site at Ketzin (Germany). Energy Procedia 1, 2029–2035 (2009)
15. Vidal-Gilbert, S., Nauroy, J., Brosse, E.: 3D geomechanical modeling for $CO_2$ geological storage in the Dogger carbonates of the Paris Basin. International Journal of Greenhouse Gas Control 3, 288–299 (2009)
16. Vilarrasa, V., Olivella, S., Bolster, D., Carrera, J.: Coupling Hydromechanical modeling of $CO_2$ sequestration in deep saline aquifers. International Jounal of Greenhouse Gas Control 4, 910–919 (2010)

17. Würdemann, H., Möller, F., Kühn, M., Heidug, W., Christensen, N.P., Borm, G., Schilling, F.R., $CO_2$SINK Group: $CO_2$SINK-From site characterization and risk assessment to monitoring and verification: One year of operational experience with the field laboratory for $CO_2$ storage at Ketzin, Germany. International Journal of Greenhouse Gas Control 4, 938–951 (2010)
18. Zhang, K., Wu, Y.S., Pruess, K.: User's Guide for TOUGH2-MP-A Massively Parallel Version of the TOUGH2 Code. Earth Sciences Division, Lawrence Berkeley National Laboratory (2008)
19. Moeck, I., Schandelmeier, H., Holl, H.G.: The stress regime in Rotliegend reservoir of the Northeast German Basin. International Journal of Earth Sciences 98, 1643–1654 (2009)

# The Application of the Strength Reduction Method in Site Stability Analysis of CCS

Wei Yuan, Xiaochun Li, and Bing Bai

State Key Laboratory of Geomechanics and Geotechnical Engineering,
Institute of Rock and Soil Mechanics,
Chinese Academy of Sciences,
Wuhan, Hubei 430071, China

**Abstract.** The factor of safety (FOS) as a technical index, especially adopted in slope engineering, is widely used in the stability evaluation of geotechnical engineering. It is popularly accepted because of its conciseness and straight forward physical meaning. However, very little analysis of this concept has been found for a large scale site involving regional structural geology. In this paper, FOS was introduced into the stability analysis of the site during $CO_2$ injection. To get this FOS, the strength reduction technique was used. In most existing engineering projects, shear failure is the most usual mode in rock mass, but the tensile failure may also occur in certain areas which means that tensile failure should also be taken into consideration in strength reduction methods. Hence a tensile-shear strength reduction method which claims that the tensile strength, friction angle and cohesion should satisfy an inequality during the strength reduction process is put forward. This new strength reduction method was applied to the geological sequestration of $CO_2$ to acquire the FOS of the stratum during the $CO_2$ injection process. What makes this new method unique is that it can be presented with an evolution history of FOS during the whole injection process.

**Keywords:** Strength Reduction Method, Tensile Failure, Site Stability Analysis, $CO_2$ injection, Safety Factor.

**FOS.** the factor of safety
**SRM.** strength reduction method

## 1 Introduction

The factor of safety (FOS) is a kind of technical index used for evaluating the safety of project objectives and it is widely used in geotechnical engineering particularly for slope engineering due to its conciseness and straight forward physical meaning. Zienkiewicz [1] (1975) was the first to suggest using a strength reduction method (SRM) to calculate the FOS of slopes, which is the proportion of

M.Z. Hou et al. (Eds.): *Clean Energy Systems in the Subsurface*, SSGG, pp. 265–280.
DOI: 10.1007/978-3-642-37849-2_22 

actual shearing resistance strength of rock mass and the shearing strength after a reduction in critical status. The FOS value calculated by Griffith [2], et al. from the U.S. in 1999 by using a finite element strength reduction method is very close to that obtained through the traditional limit equilibrium method, thus promoting the popularity of research on SRM. The shear failure mode of rock mass is generally considered instead of the tensile failure mode in the previous research [3-5]. Although the tensile failure mode of materials is considered in the calculation process, the following two methods are generally used: (1) Shearing resistance strength parameters instead of the parameter of tensile strength are reduced [6, 7] (2) The same reduction coefficient is applied to both shearing resistance strength parameters and tensile strength parameter [8, 9]. It is self-evident that the physical significance of the above method for considering tensile strength is not explicit.

Geological storage of $CO_2$ is a coupled thermal-hydrologic-mechanical process. Pore pressure caused by $CO_2$ seepage will exert interference to the initial stress field, which possibly causes excessive surface uplift and a whole or partial instability of the areas. In particular, the risks of fault slide and dislocation are increased in the fault area. Therefore, the evaluation and research on the regional stability of the geological storage of $CO_2$ have a significant scientific and practical value. Research on the mechanical stability of cap rocks and faults is currently focused on the mutual influence of the stress and seepage fields under multi-fields coupling, but research on a concise and practical stratum failure criterion and an evaluation index of safety in the geological storage of $CO_2$ is rarely carried out. For instance, Jurgen E. Streit and Richard R. Hillis had analyzed the influence rules of pore pressure on the stress field of the fault area during $CO_2$ injection into stratum [10]; J. Rutqvist, J. Birkhozer, et al. had analyzed the relationship between the maximum injection pressure and the fault sliding direction [11] and had also discussed the sliding mechanism of faults in the case of $CO_2$ sequestration by the continuous shear-slip analysis method and discrete shear-slip analysis method [12]; J. Rutqvist, Y.S Wu, et al. had compared the shear failure mode of the cap rock and $CO_2$ leakage path under circumstances with or without faults [13]; Joseph P, et al. used NUFT code to study mechanical deformation of stratum rock mass due to the injection of $CO_2$ by taking the Salah Project as an example [14].

Therefore, this article is intended to establish the relationship between shearing resistance strength parameters and the tensile strength parameter from the actual phenomenon of the project and then to carry out a coupling calculation of the fluid migration and subsurface mechanical process during the process of $CO_2$ injection by taking advantage of the TOUGH-FLAC joint program. Besides, the strength reduction method which considers tensile failure is used in such coupling calculations and the concept of FOS is expected to be adopted in evaluating the site stability during the injection of $CO_2$ into stratum.

## 2 Coupling Simulation Methods of FLAC and TOUGH

In 2002, Jonny Rutqvist suggested the coupling calculation by linking TOUGH and FLAC and had successfully applied it in the multi-fields coupling analysis of $CO_2$ geological sealing [19, 22]. On the basis of the framework idea of the TOUGH-FLAC joint simulation, a component-based programming design approach is used in this article, with regards to TOUGH, FLAC and the joint programs designed here are relatively independent components for compilation, linking and operating separately. The communication between them is carried out through disk files; data transfer direction and order in every Newtonian iteration and the overall calculation steps are controlled through the DOS batch processing order. The calculation process of each Newtonian iteration is shown in Figure 1.

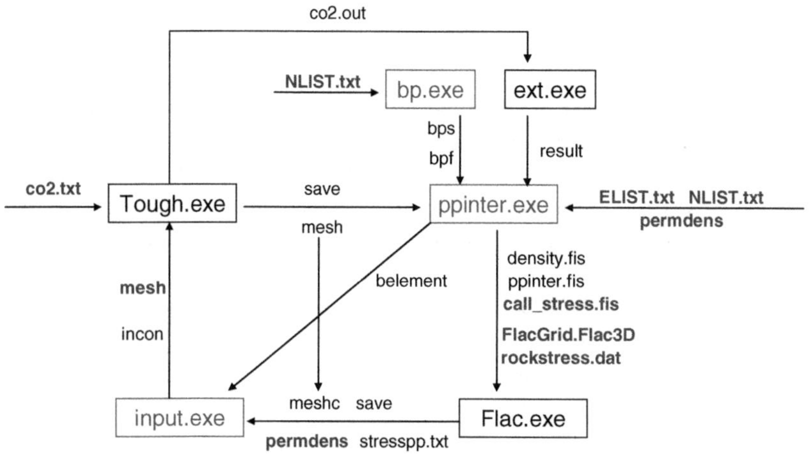


**Fig. 1** Calculation process in every Newtonian iteration of every time step

**Table 1** Function of Major pre-processing Documents

| File name | Function | Remarks |
|---|---|---|
| NLIST.txt | Nodes file of mechanical calculation | |
| ELIST.txt | Elements file of mechanical calculation | |
| Mesh | Elements file of seepage calculation | Updated after each Newtonian iteration |
| permdens | Material parameters file | |
| Call_stress.fis | Extracting total stress and pore pressure of every element | Shared for FLAC in each Newtonian iteration |
| CO2.txt | Input file of seepage calculation | |
| Rockstress.dat | Input file of mechanical calculation | |

**Table 2** Function of Joint Programs

| Name | Function |
|---|---|
| bp.exe | Generating the first and second boundary condition |
| ppinter.exe | Inserting the pore pressure of the element central point of each Newtonian iteration to the element corner points reversely, calculating the saturated density of each element, transforming the numbering of TOUGH elements to relate to the numbering of FLAC elements |
| input.exe | Calculating the correction factor of the permeability and capillary pressure and porosity of every element; transforming the numbering of FLAC elements to relate to the numbering of TOUGH elements. |

The blue fonts in Fig.1 are documents to be prepared before calculation while black fonts are temporary documents generated in each Newtonian iteration process. Black boxes represent the executable program of TOUGH and FLAC while red boxes represent joint programs designed in the course of this study. See Table 1 and Table 2 for the major pre-processing documents and joint programs.

The node of every TOUGH element is in the center of the element, but the nodes of every FLAC element are at the corner points of the elements. Therefore, FLAC cannot be used for mechanical calculations until the data in the center of elements is interpolated to the corner points. As shown in Fig. 2, suppose the central position coordinates of four adjacent elements are ①$(x_1, y_1)$, ②$(x_2, y_2)$, ③ $(x_3, y_3)$ and ④ $(x_4, y_4)$ , the pore pressure on different nodes are $P_1, P_2, P_3, P_4$,the coordinate of the corner point ⑤ connecting such four elements is $(x, y)$ and the pore pressure to be calculated according to equation (1) is P,

$$P = \sum_{i=1}^{4} P_i \cdot N_i \tag{1}$$

Where, $N_i$ is calculated as follows:

$$\begin{aligned} N_1 &= \frac{(x-x_2)\cdot(y-y_3)}{(x_1-x_2)\cdot(y_1-y_3)} \quad N_2 = \frac{(x-x_1)\cdot(y-y_4)}{(x_2-x_1)\cdot(y_2-y_4)} \\ N_3 &= \frac{(x-x_4)\cdot(y-y_1)}{(x_3-x_4)\cdot(y_3-y_1)} \quad N_4 = \frac{(x-x_3)\cdot(y-y_2)}{(x_4-x_3)\cdot(y_4-y_2)} \end{aligned} \tag{2}$$

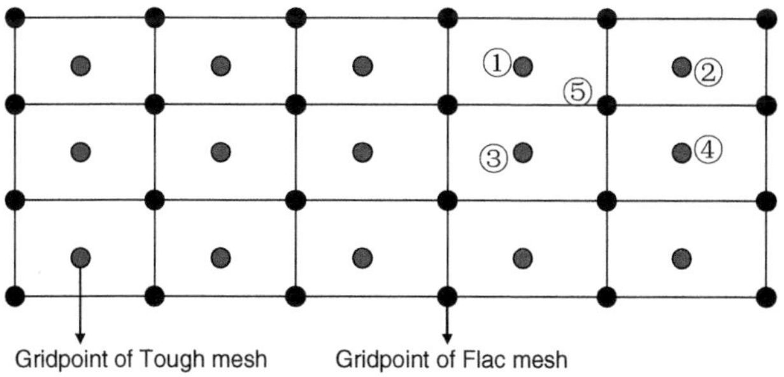


**Fig. 2** Grid Nodes of TOUGH and FLAC

The injection of $CO_2$ into stratum will result in the change of rock-mass density. The new density is calculated according to the following equation:

$$\rho = \rho_d + n \cdot (s_l \cdot \rho_l + s_g \cdot \rho_g + s_s \cdot \rho_s) \tag{3}$$

Where, $\rho_d$ stands for the dry density of the rock mass, $n$ for porosity, $s_l$, $s_g$ and $s_s$ for the saturation of liquid , gas and solid phases respectively while $\rho_l$, $\rho_g$ and $\rho_s$ stands for their densities respectively.

The change of stress field will influence seepage and this is evident by a change of porosity, permeability and capillary pressure. Porosity $\phi$ is a function of average effective stress $\sigma'_M$, as shown in equation (4) [20]:

$$\phi = (\phi_0 - \phi_r) \cdot \exp(5 \cdot 10^{-8} \cdot \sigma'_M) + \phi_r \tag{4}$$

Where $\phi_0$ is the initial porosity when no stress is applied, $\phi_r$ is the residual porosity at high stress and $\sigma'_M$ is defined from the principal stress as:

$$\sigma'_M = \frac{1}{3}(\sigma'_1 + \sigma'_2 + \sigma'_3) \tag{5}$$

The permeability $k$ is correlated to the porosity according to the following exponential function [20]:

$$k = k_0 \cdot \exp[22.2(\phi/\phi_0 - 1)] \tag{6}$$

Where $k_0$ stands for the zero stress permeability.

Except for the two coupling equations (Equations 4 and 6), the capillary pressure $P_c$ is changed according to the change of porosity and permeability, as shown in the equation below [21]:

$$P_c = P_{c0}(s_l)\frac{\sqrt{k_0/\phi_0}}{\sqrt{k/\phi}} \tag{7}$$

Where $P_{c0}(s_l)$ is controlled by the Van Genuchten function.

Therefore, porosity, permeability and capillary pressure are all directly or indirectly affected by the average effective stress [22].

## 3 Strength Reduction Method Considering Tensile Failure

The tensile crack is extended to a certain depth vertically or approximately vertically to the natural slope or the artificial slope due to excavation. In particular, when the slope is under critical status, the depth and the distribution scope of the tensile cracks are much larger [15-17]. Therefore, such actual engineering phenomena will be taken into consideration when the strength reduction method is used to obtain the FOS.

The Mohr-Coulomb Criteria has been widely used in geotechnical engineering but it can only describe the shear failure mode even though both shear failure and tensile failure always coexist in real projects. Thus, to remedy the gap in the Mohr-Coulomb Criteria, a much simpler and practical shear-tensile compound failure criterion was proposed [18], which means the Mohr-Coulomb. Criterion is adopted when the normal stress on the failure surface is compressive stress and the tensile strength failure criterion is adopted when the normal stress on the failure surface is tensile stress, as shown in Fig. 3. A demonstration of the major stress space is shown in equation (8).

$$\begin{cases} F^t = \sigma_3 - f_t = 0 \\ F^s = \sigma_1 - \dfrac{1+\sin\varphi}{1-\sin\varphi}\cdot\sigma_3 - \dfrac{2\cdot c\cdot\cos\varphi}{1-\sin\varphi} = 0 \end{cases} \tag{8}$$

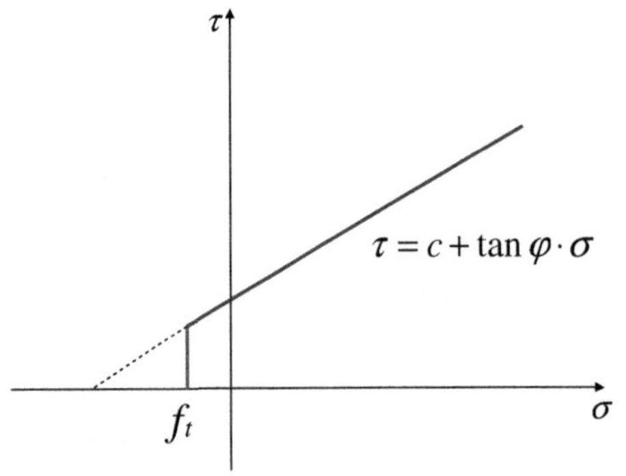


**Fig. 3** Shear-Tensile Compound Failure Criterion

Suppose one potential sliding surface of the slope under critical status is composed of a tensile failure section (Segment AB) and a shear failure section (Segment BC), the tensile failure section is vertical and Point $B$ is the inflection point of the tensile failure section and the shear failure section, the depth of Point B is supposed to be $Z_0$, as shown in Fig. 4.

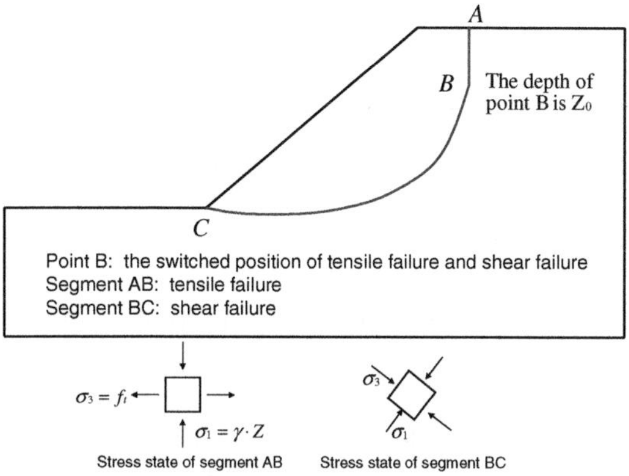


**Fig. 4** Schematic Diagram of Potential Sliding Surface

Suppose the cohesion of the rock mass is $c$, the fiction angle is $\varphi$, the tensile strength is $f_t$ and the weight is $\gamma$, the principal stress will be negative in tensile stress and positive in compression stress. The principal stress of Segment AB is represented in equation (9) and the direction is illustrated in Fig. 4.

$$\begin{cases} \sigma_1 = \gamma \cdot Z \\ \sigma_3 = -f_t \end{cases} \tag{9}$$

Obviously, the maximum principal stress in Segment AB becomes larger with the increase in depth and the minimum principal stress is always equal to - $f_t$. At this time, however, the inclined straight line does not make a tangent with the circle of stress. The circle of stress will be at a tangent with the vertical and inclined straight lines when the depth reaches point B. Different points in Segment BC are all in a shear failure state. The inclined straight line makes a tangent with the circles of stress and the direction of principal stress changes according to the change of position. The demonstration of the stress status of the whole potential sliding surface in $\sigma - \tau$ coordinate axis is shown in Fig. 5.

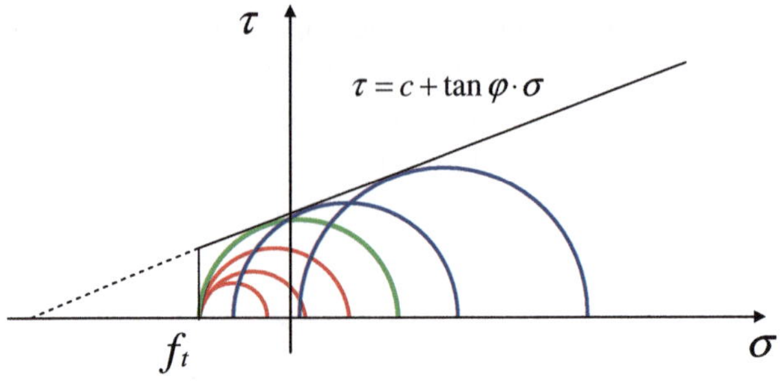


**Fig. 5** Stress Status of the Potential Sliding Surface

The following equation can be obtained by substituting the stress status of Point B in equation (8):

$$Z_0 = \left( -\frac{1+\sin\varphi}{1-\sin\varphi} \cdot f_t - \frac{2 \cdot c \cdot \cos\varphi}{1-\sin\varphi} \right) \Big/ \gamma \tag{10}$$

Obviously, the depth of Point B is larger than 0. Suppose $Z_0 > 0$, the following can be obtained:

$$f_t < \frac{2 \cdot c \cdot \cos\varphi}{1+\sin\varphi} \tag{11}$$

Therefore, on the basis of the tensile-shear compound failure criterion, to satisfy the tensile crack at a certain depth at the top of the slope, $c$, $\varphi$ and $f_t$ of rock mass must have a relation (11). After referring to nearly one hundred different tensile and shearing resistance strength parameters of different rock mass, it was found out that all the parameters above satisfy equation (11). Therefore, when SRM is used for calculating the FOS, and required to represent the existence of a tensile crack, both the tensile strength parameter and shearing resistance strength parameters need to satisfy equation (11) as long as every reduction calculation step is carried out. It has been demonstrated that when the FOS is greater than 1.0, the first kind of strength reduction method considering tensile failure cannot ensure the establishment of equation (11). When the FOS is less than 1.0, the second kind of strength reduction method considering tensile failure cannot ensure the establishment of equation (11). This paper does not include this demonstration in details owing to the limited space here. To ensure the constant establishment of equation (11) during the reduction process, the constants of equation (12) can be established in each reduction process:

$$\frac{2 \cdot c_i \cdot \cos \varphi_i}{1+\sin \varphi_i} = \beta \cdot f_{ti} \quad (12)$$

where $f_{ti}$ · $c_i$ and $\varphi_i$ correspond to the tensile strength, cohesion and angle of friction after the reduction in the step $i$; $\beta$ for constant, therefore $\beta$ is given as:

$$\beta = \frac{2 \cdot c \cdot \cos \varphi}{1+\sin \varphi} \cdot \frac{1}{f_t} \quad (13)$$

Where $f_t$, $c$ and $\varphi$ are the real strength parameters of the rock mass.

When the reduction coefficient is intended to be infinite, $c_i \to 0$ and $\varphi_i \to 0$. It can be determined from equation (12) that $f_t \to 0$. This implies a rock without shearing resistance strength and tensile strength, i.e., similar to water. Thus, the rationale behind equation (12) can be seen to some extent. In fact, what equation (12) establishes is the relationship between the tensile and shearing resistance strength of rock mass but the necessary test and the mechanical foundation are lacking. Therefore, equation (12) aims at making the critical status acquired through SRM more accessible to the actual situation of the project. From another perspective, although equation (11) is obtained by deducing from one specific project objective, it complies with the common mechanical property of the rock mass. As a result, when SRM is used in other projects for stability analysis, it is suggested that the strength parameters of the rock mass during the reduction process will comply with such inequality.

## 4 Illustrative Examples

To take advantage of the concept of FOS and SRM to evaluate the stratum stability during the process of $CO_2$ injection, the following concept model is established: No displacement constraint is set at the top of the model, horizontal displacement constraint is set at the right and left hand sides while the whole displacement constraint is set at the bottom. Suppose that the pressure at the top of model is 1 atm (0.1MPa) with a temperature of 10°C and the temperature at the bottom of model is 85°C. Before commencing $CO_2$ injection, the stress field, pore pressure field and temperature field should be initialized. After initialization is finished, the temperature near the injection well is about 51.5°C, the initial pore pressure is about 16.385 MPa and the initial stresses are; $\sigma_h = 44.783\ MPa$ and $\sigma_v = 14.928\ MPa$. See Fig. 6 for the geometric dimension and stratum partition. See Table 3 for the corresponding material parameters. It is necessary to note that the model size and the hydraulic parameters of the example refer to information in the literature [23], and the mechanical parameters of the four strata take examples from the 'Manual of the Rock Mechanics Parameters', which was published by Water Resources and Electric Power Press in China.

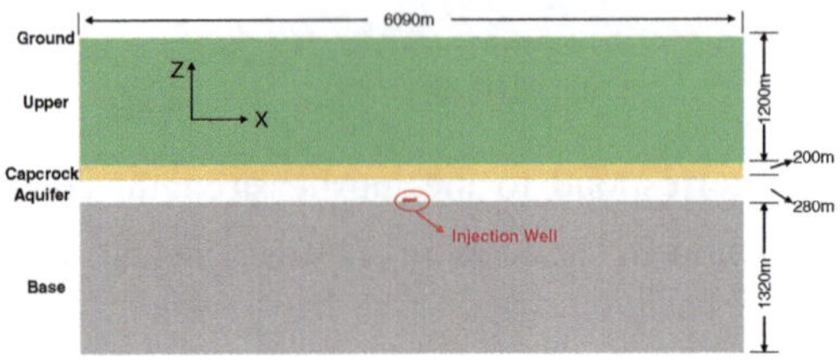


**Fig. 6** Stratum Partition Map of Vertical Section

**Table 3** Material property

| Property | Upper | Caprock | Aquifer | Base |
|---|---|---|---|---|
| Young's modulus, E(Gpa) | 5 | 5 | 5 | 5 |
| Poisson's ratio, $\upsilon$(-) | 0.25 | 0.25 | 0.25 | 0.25 |
| Cohesion, c (MPa) | 1.98 | 2.45 | 3.08 | 5.30 |
| Friction, $\varphi$ (°) | 48 | 50 | 52 | 57 |
| Tensile strength, $f_t$ (MPa) | 1.254 | 1.457 | 1.920 | 2.539 |
| Dry rock density, $\rho_d$ (Kg/m³) | 2600 | 2600 | 2600 | 2600 |
| Zero stress porosity, $\phi_0$ (-) | 0.20 | 0.10 | 0.30 | 0.10 |
| Residual porosity, $\phi_r$ (-) | 0.19 | 0.09 | 0.28 | 0.09 |
| Zero stress permeability, $k_0$ (-) | 5.0E-15 | 1.0E-16 | 1.0E-12 | 5.0E-15 |
| Corey's irreducible gas saturation, $S_{rg}$ (-) | 0.05 | 0.05 | 0.05 | 0.05 |
| Corey's irreducible liquid saturation, $S_{rl}$ | 0.3 | 0.3 | 0.3 | 0.3 |
| van Genuchten's air-entry pressure, $P_0$ (KPa) | 196 | 3100 | 19.6 | 3100 |
| van Genuchten's exponent, m | 0.457 | 0.457 | 0.457 | 0.457 |

Compressed $CO_2$ is continually injected at a constant rate of 0.2 kg/s per meter (per meter normal to the two-dimensional model) and injection is stopped after continuous injection for 7.93 years. A stability analysis of the stratum is carried out for three time points, i.e., t=6.34 year, t=7.93 year and t=9.51 year and the FOS of different time points is calculated accordingly by SRM.

① t=7.93 year

At first, the coupling calculation is executed without the reduction of the strength of the rock mass. See Fig. 7 and Fig. 8 for $CO_2$ saturation and the pore pressure echogram after 7.93 years of injection. See Fig. 9 for the echogram of displacement along Direction Z. The pore pressure reaches 40MPa at the bottom of the cap rock (Z=1400m) and the pore pressure above the 1400m elevation is less. $CO_2$ saturation at the bottom of the cap rock is larger than that in other parts of the rock and $CO_2$ is mainly distributed at the bottom of the cap rock. It can also be concluded from the Fig.8 that the $CO_2$ has spread over 1500m under the cap and has penetrated upwards into the cap rock by about 40m.The integrity of the cap rock has played a direct and obvious role in sealing $CO_2$. The displacement along the positive direction of Z near the bottom of cap rock is about 400mm and the trend of upward displacement on the ground surface is not obvious. The increase in pore pressure causes a reduction of average effective stress with the injection of $CO_2$.The area at the bottom of the cap rock is the position of the greatest change of pore pressure and average effective stress. Therefore, the change in effective stresses at the bottom of cap rock resulting in a poro-elastic

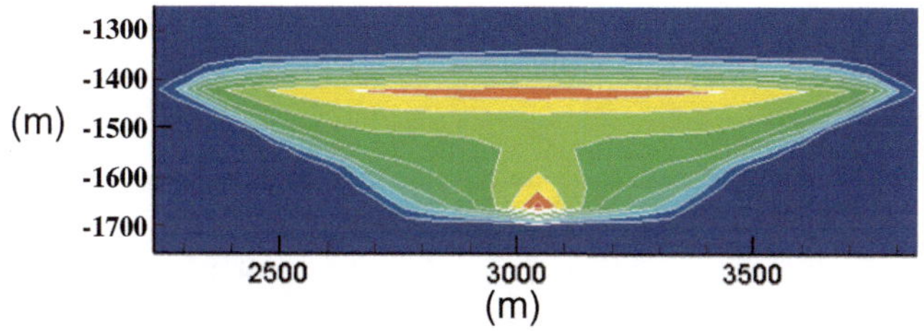


**Fig. 7** Distribution Map of $CO_2$ saturation (t=7.93 years)

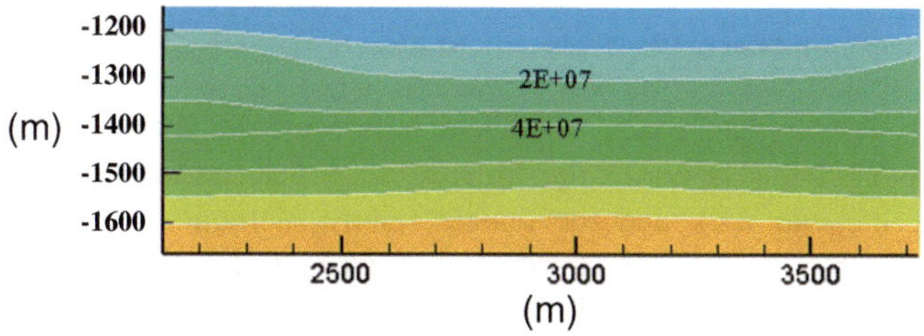


**Fig. 8** Distribution Map of Pore Pressure (t=7.93 years)

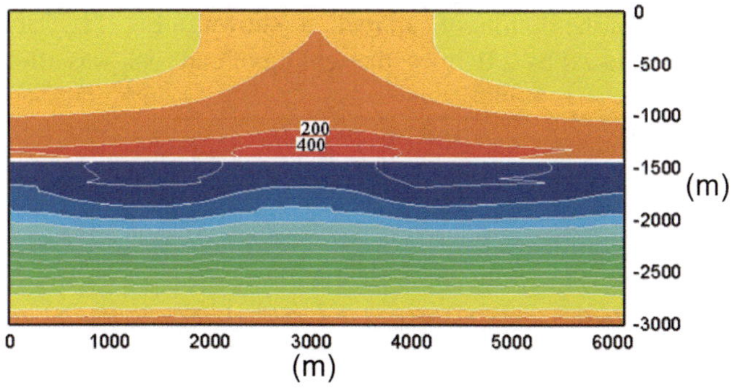


**Fig. 9** Distribution Map of Displacement along Direction Z (t=7.93 years)

expansion of the rock mass is most apparent. In addition, pore pressure supplies lift the force at the bottom of the cap rock. This causes the cap rock to lift up towards the ground surface. However, it is found through calculation that no plastic zone exists in the whole calculation model. Therefore, the stratum at t=7.93 year satisfies the requirements of mechanical stability.

According to the reduction principle presented in the previous section, a stability analysis with the strength parameters reduced is implemented. When the reduction coefficient K is equal to 1.6, a shear failure area started to appear at the bottom of the cap rock and some shear failure zones will have also been in existence in the upper rock mass, as shown in Fig. 10. With the effect of multi-directional stresses, the normal stress on the shear failure surface is pressure stress which does not affected the leak proof properties of the cap rock. Meanwhile, the shear failure scope of the whole area is very small. It is therefore considered that the stability of the stratum is within the safety limit when the strength has been reduced by 1.6 times.

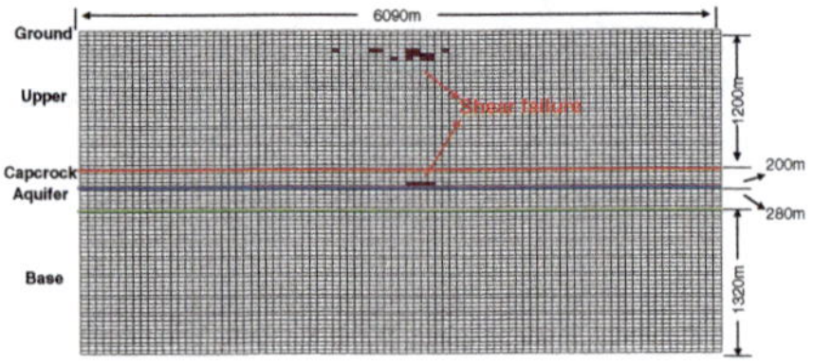


**Fig. 10** Distribution Map of Plastic Zone (K=1.6)

When K is increased to 2.0, the shear failure area at a certain depth from the surface has expanded accordingly. Meanwhile, tensile failure has occurred at the bottom of the cap rock and the tensile failure elements are on both sides of the shear failure elements. The horizontal effective stress of some elements at this portion would have surpassed the tensile strength of the stratum. In addition, one reverse bending to a certain degree may occur on both sides of the bulged portion at the bottom of the cap rock, causing a tensile stressed area. When the tensile stress surpasses the tensile strength, tensile failure may also occur. Thus, tensile cracks first appear at the bottom of cap rock, as shown in Fig. 11. Therefore, when the strength was reduced by 2.0 times, the sealing of cap rock was affected.

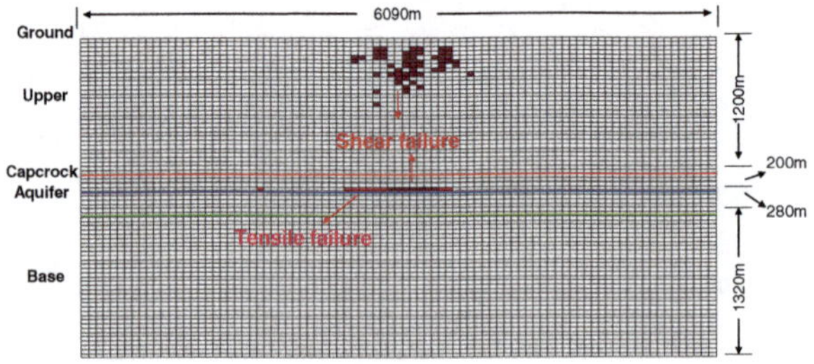


**Fig. 11** Distribution Map of Plastic Zone (K=2.0)

When K is increased to 2.8, the shear failure area near the surface and the tensile failure area at the bottom of the cap rock have expanded further. Obviously, the shear failure area within the area of 450m vertical depth and 1000m horizontal width is a wedge-shaped groove, as shown in Fig. 12. When the strength is reduced by 2.8 times, the displacement of the cap rock towards the ground surface has increased due to the reduction of the shearing resistance strength. It exerts a larger extrusion function on the overlapping rock mass than without the reduction of the strength parameters, a shear failure area has occurred within the scope of the free face and the ground surface has become unstable. Meanwhile, with the increase of in the reduction coefficient, the tensile strength has been reduced and the tensile failure reduced. Therefore the area has increased. As a result, the sealing of the cap rock is seriously damaged.

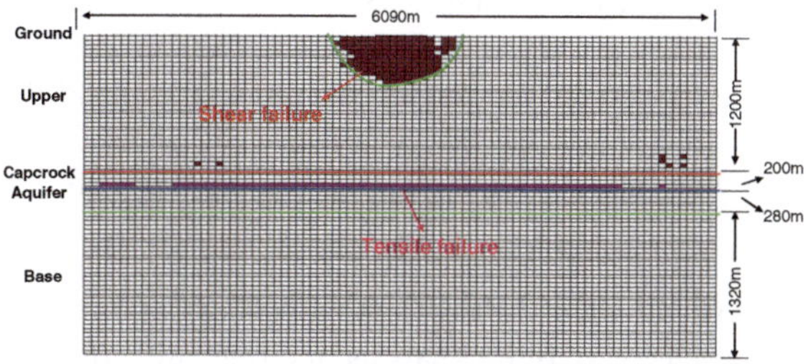


**Fig. 12** Distribution Map of Plastic Zone (K=2.8)

It can be seen from the above analysis that if the tensile failure area occurring in the cap rock is taken as the criterion of critical status, then the FOS (t=7.93 year) can be regarded as 2.0 and if the connectivity of local plastic zones is taken as the criterion of critical status, then the FOS (t=7.93 year) can be regarded as 2.8.

② t=6.34 year and t=9.51 year

Firstly, coupling calculations are executed without the reduction of the strength of rock mass at t=6.34 years and t=9.51 years, respectively. It is obvious to find that the scope of the $CO_2$ distribution and the pore pressure at the bottom of cap rock at 6.34 years are less than those at t=7.93 years. Therefore, both the cap rock sealing and mechanical stability at this time point are much better than those at t=7.93 years. In addition, the horizontal distribution scope of $CO_2$ at t=9.51 years reaches about 1700 m and the pore pressure at the bottom of cap rock is about 30 MPa. It is illustrated that, as time goes on, $CO_2$ spreads along the horizontal direction of the cap rock, reducing the clustering at the bottom, and thus the pore pressure has decreased.

By gradually increasing the reduction coefficient, when the tensile failure zones start to occur in the caprock, the reduction coefficient K is equal to 2.9 at t=6.34 years, however K is equal to 3.1 at t=9.51 years. In addition, when the reduction coefficient K reaches 3.7, the shear failure zones having through-going status occurs at an approximate depth of 500m at t=6.34 years, and when the reduction coefficient K reaches 4.0, the shear failure zones having through-going status occur at the depth of approximately 250m. Distribution maps of the plastic zones at t=6.34 years and t=9.51 years are ignored owing to space reasons.

In conclusion, the FOS is smallest at the moment the $CO_2$ injection is stopped (t=7.93 years). With the continuous injection of $CO_2$, the pore pressure keeps increasing and the safety risks to the cap rock and stratum also become higher. When the injection of $CO_2$ is stopped, $CO_2$ clustered at the bottom of the cap rock will spread horizontally causing a decrease in pore pressure. As a result, the safety risk of cap rock and stratum will reduce and the FOS will increase. Thus, the FOS could be used as one index to evaluate the safety.

## 5 Conclusions and Discussions

A SRM including tensile failure is proposed in this article and the concept of the FOS is introduced to the site stability evaluation of geological storage of $CO_2$. Finally, the following main results were obtained from the stability analysis through SRM:

(1) To make the critical status relate to the actual situation as much as possible when SRM is used for stability analysis, the relevancy of tensile and shearing resistance strength parameters should be considered and the condition stipulated by equation (11) should be established and be met in each reduction process. Although equation (11) is deduced from slope engineering, it could reflect the common mechanical characteristics of rock mass. Thus, it is also appropriate for the stability analysis of other engineering objects through SRM.

(2) The change of average effective stress and pore pressure at the bottom of cap rock is the largest and the stress status here is the most complicated. Due to the lifting force of pore pressure and the poro-elastic expansion of the rock mass, the upward bulging of the cap rock extrudes the overlaying rock mass, which results in the vertical displacement. Therefore, the safety risk here is also the largest.

(3) With the increase of the strength reduction coefficient, the change in the failure area entails a gradual process: the shear failure area occurs firstly at the bottom of cap rock, followed by the tensile failure area at the bottom of cap rock and the shear failure area in the overlaying rock mass. With a further increase in the reduction coefficient, the through-going shear failure area occurs due to the surface becoming locally unstable. Meanwhile, the tensile failure area at the bottom of cap rock enlarges further.

(4) The FOS will vary according to the different criteria of the critical status. If no tensile crack is allowed to appear in the cap rock, the FOS of the three different time points in the previous analysis will correspond to 2.9, 2.0 and 3.1 respectively. If the connecting of the plastic zones is considered as the

criterion of critical status, the FOS of the three different time points in the previous analysis will correspond to 3.7, 2.8 and 4.0 respectively.

(5) The FOS at the moment when the $CO_2$ injection is stopped (t=7.93 years) is the smallest and the FOS before and after this time point are larger. The FOS could be considered as one index to evaluate the site stability during the injection process of $CO_2$.

Some worthy conclusions have been obtained through the above analysis, but the followings need to be further researched and discussed:

(1) The relationship between shearing resistance and tensile strength parameters established in equation (12) is only to make the critical status acquired through SRM more accessible to the actual situation but corresponding tests and the mechanical basis are currently lacking and the physical significance of $\beta$ is not explicit.

(2) The selection of a proper criterion of the critical status needs further discussion. Different criteria result in different FOS and the two criteria selected in this article are only on in preliminary discussion.

(3) The selection of the threshold value of the FOS needs further discussion. That is to say, the value of the FOS for the project objective could be recognized as being under safety status. For instance, in slope engineering, the slope can be recognized as being safe when the FOS is greater than 1.0.

**Acknowledgements.** The authors gratefully acknowledge the support of the safety system of the project on deep rock masses (the project of the innovation team in the Chinese Academy of Science). Project No. Y011091C05.

## References

1. Zienkiewiez, O.C., Humpheson, C., Lewis, R.W.: Associated and non-associated visco-Plasticity and Plasticity in soil mechanics. Geostechnique 25(4), 671–689 (1975)
2. Griffiths, D.V., Lane, P.A.: Slope stability analysis by finite element. Geotechnique 49(3), 387–403 (1999)
3. Matsui, T., San, K.C.: Finite element slope stability analysis by shear strength reduction technique. Soil and Foundations 32(1), 59–70 (1992)
4. Wang, D., Zhang, L., Xu, J.: Seismic stability safety evaluation of gravity dam with shear strength reduction method. Water Science and Engineering 2(2), 52–60 (2009)
5. Cheng, Y.M., Lansivaara, T., Wei, W.B.: Two-dimensional slope stability analysis by limit equilibrium and strength reduction methods. Computers and Geotechnics 34, 137–150 (2007)
6. Lin, H., Cao, P., Gong, F.: Directly searching method for slip plane and its influential factors. J. Cent. South Univ. Technol. 16, 131–135 (2009)
7. Dai, Z., Liu, Z., Liu, C.: Numerical analysis of soil slope stability considering tension and shear failures. Chinese Journal of Rock Mechanics and Engineering 27(2), 375–382 (2008) (in Chinese)

8. Alejano, L.R., Alonsl, E.: Application of the shear and tensile strength reduction technique to obtain factors of safety of toppling and footwall rock slopes. Impact of Human Activity on the Geological Environment, 7–13 (2005)
9. Lai, J., Fang, Y.: Stability analysis of rock slope with extroversion joint considering tensile strength. Journal of Logistical Engineering University 27(6), 7–12 (2011) (in Chinese)
10. Streit, J.E.: Estimating fault stability and sustainable fluid pressures for underground storage of $CO_2$ in porous rock. Energy 29, 1445–1456 (2004)
11. Rutqvist, J., Birkholzer, J.T., Tsang, C.F.: Coupled reservoir-geomechanical analysis of the potential for tensile and shear failure associated with $CO_2$ injection in multilayered reservoir-caprock systems. International Journal of Rock Mechanics &Mining Sciences (2007)
12. Rutqvist, J., Birkholzer, J., Cappa, F.: Shear-slip anlysis in multiphase fluid-flow reservoir engineering applications using TOUGH-FLAC. In: Proceedings of Tough Symposium (2006)
13. Rutqvist, J., Wu, Y.S., Tsang, C.F.: A modeling approach for analysis of coupled multiphase fluid flow, heat transfer, and deformation in fractured porous rock. International Journal of Rock Mechanics & Mining Sciences 39, 429–442 (2002)
14. Morrisa, J.P., Hao, Y., Foxall, W.: A study of injection-induced mechanical deformation at the In Salah $CO_2$ storage project. International Journal of Greenhouse Gas Control 5, 270–280 (2011)
15. Kogure, T., Aoki, H., Maekadl, A.: Effect of the development of notches and tension crack on instability of limestone coastal cliffs in the Ryukyus, Japan. Geomorphology 80, 236–244 (2006)
16. Brown, I., Hittinger, M., Goodman, R.: Finite element study of the Nevis Bluff rock slope failure. Rock Mechanics 12, 231–245 (1980)
17. Kroeger, E.B.: Analysis of plane failures in compound slopes. International Journal of Surface Mining, Reclamation and Enviroment 14, 215–222 (2000)
18. Paul, B.: Modification of the Coulomb-Mohr theory of fracture. J. Appl. Mech. 28, 259–268 (1961)
19. Rutqvist, J.: Status of the TOUGH-FLAC simulator and recent applications related to coupled fluid and crustal deformation. Computers & Geosciences 37, 739–750 (2011)
20. Davis, J.P., Davis, D.K.: Stress-dependent permeability: characterization and modeling. Society of Petroleum Engineers, SPE Paper no. 56813 (1999)
21. Leverett, M.C.: Capillary behavior in porous media. Trans. AIME 142, 341–358 (1941)
22. Rutqvist, J., Tsang, C.F.: Tough-Flac: A numerical simulator for analysis of coupled T-H-M processes in fractured and porous geological media under multi-phase flow conditions. In: Proceedings TOUGH Symposium (2003)
23. Rutqvist, J., Tsang, C.T.: A study of caprock hydrymechanical changes associated with $CO_2$-injection into a brine formation. Environmental Geology 42, 296–305 (2002)

# Different Methods to Evaluate Strength from Compression Tests for Rock Salt

Yu Bian[1,2], Jianfeng Liu[1,2,3,*], Guosheng Ding[4], Liang Chen[5], Zhide Wu[4], Lina Ran[4], Michael Zhengmeng Hou[6], and Yang Gou[6]

[1] State Key Laboratory of Hydraulics and Mountain River Engineering, Sichuan University, Chengdu 610065, China
[2] State Key Laboratory of Geohazard Prevention and Geoenvironment Protection, Chengdu University of Technology, Chengdu 610059, China
[3] Key Lab. of Energy Engineering, Safety and Mechanics on Disasters (Sichuan Univ.), Ministry of Education, Chengdu, 610065, China
[4] Langfang Branch, Research Institute of Petroleum Exploration and Development, Langfang 065007, China
[5] Beijing Research Institute of Uranium Geology, Beijing 100029, China
[6] Institute of Petroleum Engineering, Technical University of Clausthal, Clausthal-Zellerfeld 38678, Germany

**Abstract.** As a special soft rock, rock salt can exhibit a large deformation under triaxial stress conditions. In this paper, the author presents triaxial compressive tests which were carried out under different confining pressures on rock salt samples taken from Anning natural gas storage cavern, located in Yunnan Province, China. The axial stress of rock salt was analyzed by four methods based on initial cross-sectional area, corrected cross-sectional area according to engineering strain, logarithmic strain and volume deformation. Results show that the peak stress and strain determined were different for each method. The peak stress and strain determined by the method of initial cross-sectional area were the largest, while those obtained by the method of corrected cross-sectional area according to logarithmic strain were the smallest. The maximum difference is 36.15%. Based on the results of this study, it is recommended that the logarithmic strain method be used to determine the axial peak stress and strain of rock salt in triaxial stress condition.

**Keywords:** rock salt, triaxial compressive stress, cross-sectional area, axial stress, axial strain.

---

[*] Corresponding author.

M.Z. Hou et al. (Eds.): *Clean Energy Systems in the Subsurface*, SSGG, pp. 281–291.
DOI: 10.1007/978-3-642-37849-2_23 

# 1 Introduction

Rock salt is widely considered an ideal medium for oil/gas storage and radioactive waste disposal because of its small porosity, low permeability, large plasticity and self sealing capacity. Many rock salt energy storage caverns have already been built in Europe, USA, Canada etc. China built the first natural gas storage cavern in Asia in 2007 located in Jintan, Jiangsu Provence. However, the study of physical-mechanical behavior of rock salt in China has been done later than in other countries.

Based on the extensive studies carried out, numerous experimental and theoretical findings on salt rock have been obtained throughout the world [1-3]. With the construction and operation of the Jintan gas storage cavern, more attention has been given to rock salt in China. Yang and Zhou [4-6] characterized the variation of volumetric deformation of rock salt based on creep tests, and proposed a fractional order creep model that can describe the creep behavior of rock salt. Xie [7] studied the fractal mechanisms of 3D acoustic emission distribution during the full failure process of rock salt for uniaxial compression tests and indirect tensile tests. The results revealed that the rock failure corresponds to an accumulation of energy released, and a decreasing tendency of fractal dimensions [7]. Based on experimental investigations on the tensile characteristics of rock salt with a different testing method, Liu [8] thought that the direct tensile test is more reliable and precise to determine the tensile strength of rock salt. Furthermore, many other researchers [9-14] have studied the characteristics of rock salt from the perspective of strength, deformation and damage. However, little attention has been given to the method of determining the triaxial compressive strength.

As a special soft rock, rock salt can show a large plastic deformation before the appearance of surface failure, especially under high confining pressure. In this case, if the initial cross-sectional area is used in the determination of compressive strength, the results cannot represent the real bearing capacity of rock salt, and a misleading conclusion may be achieved. In this paper, four methods were employed for the comparative analysis of the deformation characteristics of rock salt under triaxial stress conditions. Based on the analysis of experimental data, a relatively reliable method to determine the compressive strength is proposed.

# 2 Experiment Setup and Testing Methods

## *2.1 Experiment Setup*

All tests were carried out using MTS815 Flex Test GT rock test equipment. The controlled range of the principal parameters is listed as follows: axial load: 0~4600 kN; axial displacement (LVDT): 0~100 mm; confining pressure and pore pressure: 0~140 MPa.

### 2.2 Sampling and Testing Methods

Rock salt samples were taken at the depth of 500 m in Anning, Yunnan Province. Due to the various thicknesses of rock layers, the NaCl content of rock salt from different depths is different. The samples in this paper have an average NaCl content of 79%. Specimen preparations were carried out on a dry lathe following the ***Standard for Tests Method of Engineering Rock Mass*** (GB/T50266-99). The cylindrical specimens prepared were 90mm in diameter and 180mm high. The test temperature was room temperature (25°C). The confining pressures were 5, 10, 15, 20 MPa, respectively, and the loading rate of confining pressure was 3 MPa/min. The axial loading started after the confining pressure reached its target value. The axial loading was a displacement control, with a loading rate 1.0 mm/min.

## 3 Calculation Methods for Triaxial Stress

Rock salt usually exhibits a large axial and lateral deformation under triaxial stress conditions. The deformed sample has a barrel shape, which means the lateral extension of the end is smaller than that of the middle. Due to the non-uniform distribution of insoluble impurities, the deformed sample is generally not of an ideal barrel shape. Thus, the calculated axial stress using different methods for cross-sectional area may have large differences. In this paper, the test results were evaluated and compared with the initial cross-sectional area (method 1), corrected cross-sectional area according to engineering strain (method 2), logarithmic strain (method 3) and volume deformation (method 4).

Engineering strain is the ratio of sample length variation and the original sample length $l_0$:

$$\varepsilon_l = \frac{\Delta l}{l_0} \tag{1}$$

where $\varepsilon_l$ is the engineering strain; $\Delta l$ is the length variation (mm); $l_0$ is the original length before loading (mm).

The formula for the logarithmic strain is:

$$\varepsilon_{\ln} = \left|\int_{l_0}^{l}\frac{dl}{l}\right| = \left|\ln\frac{l}{l_0}\right| = \left|\ln(1-\varepsilon_l)\right| \tag{2}$$

Where $\varepsilon_{l0}$ is the logarithmic strain; $l$ is the actual sample length, $l = l_0 - \Delta l$ (mm). For values between 0 and 1, the logarithmic strain as given in Eq. (2) is always greater than the engineering strain as given in Eq. (1). For convenient comparison of the difference between the stress from engineering strain and that from logarithmic strain, in this paper, engineering strain and logarithmic strain are both represented in %.

### 3.1 Method 1

This method does not consider the change in the cross-sectional area during the deformation and failure process. Because the cross-sectional area tends to increase under loading, this method will overestimate the stress. The cross-sectional area is assumed to be constant during the loading process, thus the axial stress of rock sample is:

$$\sigma_{lu} = \frac{F_1}{A_0} \tag{3}$$

$\sigma_{lu}$ is the stress evaluated from the uncorrected cross-sectional area (MPa); $F_1$ is the axial loading (N); $A_0$ is the initial cross-sectional area of specimen (mm$^2$).

### 3.2 Method 2

This method uses the engineering strain and the initial cross-sectional area to correct the area during the loading process. The corrected area is calculated according to:

$$A_{luc} = \frac{A_0}{1-\varepsilon_l} \qquad A_{luc} > A_0 \tag{4}$$

The formula for the axial stress is:

$$\sigma_{luc} = \frac{F_1}{A_{luc}} \qquad \sigma_{luc} < \sigma_{lu} \tag{5}$$

where $\sigma_{luc}$ is the axial stress calculated from the corrected cross-sectional area by the engineering strain $\varepsilon_l$ (MPa); $A_{luc}$ is the cross-sectional area corrected by the engineering strain (mm$^2$).

### 3.3 Method 3

This method uses the logarithmic strain and the initial cross-sectional area to correct the area during the loading process. The corrected area and the corresponding axial stress are calculated as follows:

$$A_{\ln} = \frac{A_0}{1-\varepsilon_{\ln}} \qquad A_{\ln} > A_{luc} \tag{6}$$

$$\sigma_{\ln} = \frac{F_1}{A_{\ln}} \qquad \sigma_{\ln} > \sigma_{luc} \tag{7}$$

where $A_{ln}$ is the cross-sectional area corrected by the logarithmic strain (mm$^2$); $\sigma_{ln}$ is the axial stress calculated from the corrected cross-sectional area according to the logarithmic strain $\varepsilon_{ln}$ (MPa).

## 3.4 Method 4

Due to the compressive stress in axial direction, the length of rock sample is reduced during the loading process, and the cross-sectional area is increased. The deformed sample generally has a barrel shape, namely with narrow ends, but a thick center. This method assumes that the cross-sectional area along the axis is constant, so the average cross-sectional area should be evaluated from the volume deformation and the corresponding sample volume and height. The deformed volume ( $\Delta V$ ) is equal to the change in volume of the confining liquid. So $V=V_0-\Delta V$ .The corrected area and the corresponding axial stresses are calculated as follows:

$$A_{lup} = \frac{V}{l} \tag{8}$$

$$\sigma_{lup} = \frac{F_1}{A_{lup}} \tag{9}$$

Where, $A_{lup}$ is the cross-sectional area corrected by the deformed volume ($mm^2$); $V$ is the actual volume of deformed sample ($mm^3$); $\sigma_{lup}$ is the corrected axial stress by deformed volume (MPa).

# 4 Experiment Results and Analysis

Fig. 1 illustrates the variation of the engineering strain and the logarithmic strain with time. A linear strain-time relation can be noticed at the beginning of the loading process, and the difference between the two strains is negligible. However, as the loading time increases, the difference becomes more evident. At the same time, the engineering strain becomes smaller than the logarithmic strain, and the deviation increases.

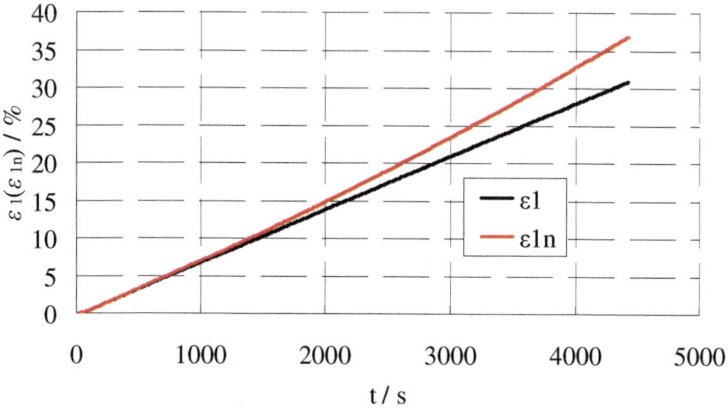


**Fig. 1** Relationship between engineering strain and logarithmic strain and time

Fig. 2 shows the stress/strain curves under different confining pressures. The green line is based on the logarithmic strain and the others are based on the engineering strain.

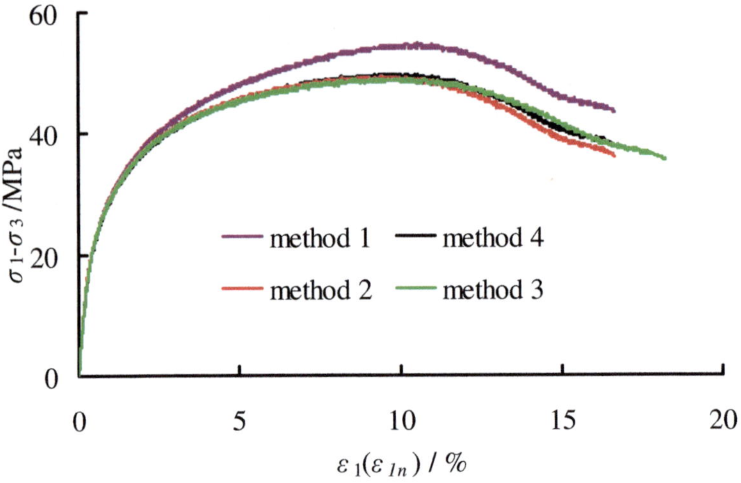


(a) $\sigma_3$=5MPa

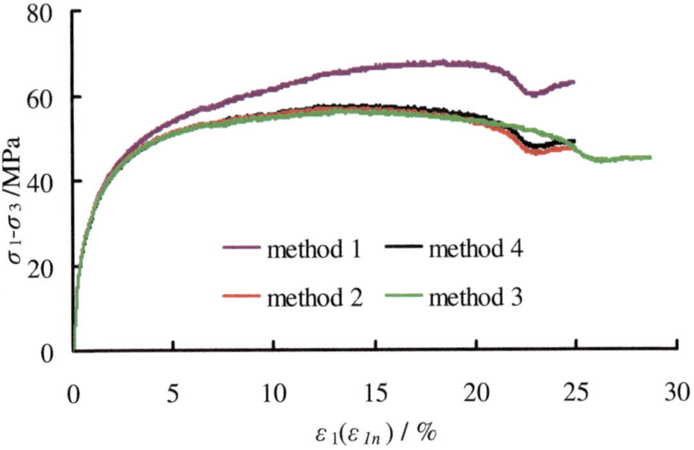


(b) $\sigma_3$=10MPa

**Fig. 2** Strain-stress curves for different calculating methods

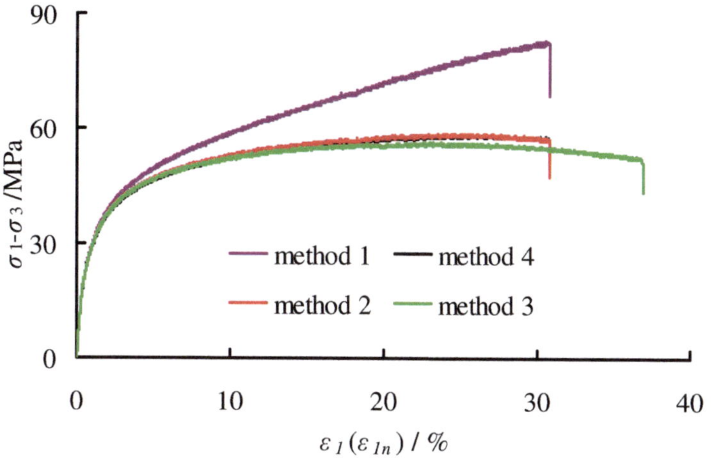


(c) $\sigma_3$=15MPa

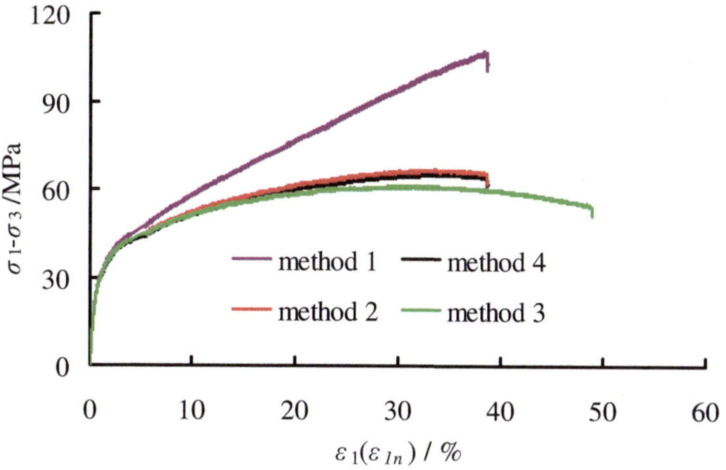


(d) $\sigma_3$=20MPa

**Fig. 2** (*continued*)

It is observed that, as the formula suggests, the logarithmic strain is higher than the engineering strain under different confining pressures. For instance, the maximum values of the engineering strain and the logarithmic strain of rock salt under a confining pressure of 20 MPa are 38.71% and 48.96% respectively. Before the deformation yielding, the strain-stress curves calculated with the four methods are almost the same, which means the cross-sectional area of the rock salt sample is not significantly changed, and a correction of the cross-sectional area is not necessary in this phase. With the increase in the confining pressure, the plastic hardening phenomenon becomes more evident. As a consequence, without the correction of the cross-sectional area, no failure point can be identified in the stress-strain curve under high confining pressure (>15 MPa). Therefore, with method 1, the maximal axial stress after tests is considered as the failure stress when the confining pressure is higher than 15 MPa.

In the corrected cross-sectional area with methods 2 to 4, the failure stress is obtained under different confining pressures, as presented in Fig. 2. Moreover, some differences in the determination of failure stresses with different methods can be noticed. The peak value after method 1 is always the largest, while the value of method 3 is the smallest. The descending order of the determined failure stress under low confining pressure (5 MPa and 10 MPa) with different methods is method 1, method 4, method 2, and method 3 (Fig. 3); while under high confining pressure (15 MPa and 20 MPa), the descending order is method 1, method 2, method 4, and method 3. No matter how high or low the confining pressure, is, the peak stresses calculated by method 2 and 4 are almost the same. It can be explained that in method 4, Eq. (8) gives an average value for the cross-sectional area. If the volume is not changed during deformation, with $V=A_0 \times l_0$ and $l=l_0 \times (1-\varepsilon_1)$, Eq. (8) leads to Eq. (4). The volume does actually change during compression, so the results of methods 2 and 4 are similar, but not identical.

Table 1 summarizes the determined peak stress and the corresponding axial strain with the four methods under different confining pressures. Under the same confining pressure, the maximal axial strain at peak stress is obtained by method 1, and the corresponding strain is 10.55% (5 MPa) and 38.34% (20 MPa) which is higher than the values determined by other methods. The maximal difference of the peak strain for methods 1 and 3 is 29.35%.

The difference between the peak stresses determined by methods 2 to 4 and method 1 is characterized by its percentage to the peak stress of method 1, as illustrated in Table 1. It will be noticed that, the difference between methods 2 and 4 are almost the same. Method 3 has the largest difference with a value of 36.15%, which indicates that the corrected cross-sectional area in method 3 is larger than that in other methods. Therefore, the strength determined by method 3 is the lowest.

The results reveal that the peak stress and its corresponding strain using methods 2 and 4 are almost the same, although the logarithmic strain revised by method 3 under the same axial loading shows the peak stress being lowest and the corresponding strain being relatively smaller. With method 3, it is not necessary to add extra measured parameters. The cross-sectional area is corrected by the axial strain alone which is simple and convenient. Meanwhile, from the perspective of engineering design and application, the peak stress and strain determined by method 3 is also relatively smaller. Therefore, a better safety situation can be achieved for the design and stability analysis of underground gas caverns using the parameters calculated by method 3. It is recommended that method be used 3 to correct the cross-sectional area in the strength and deformation characterization of rock salt under triaxial stress conditions.

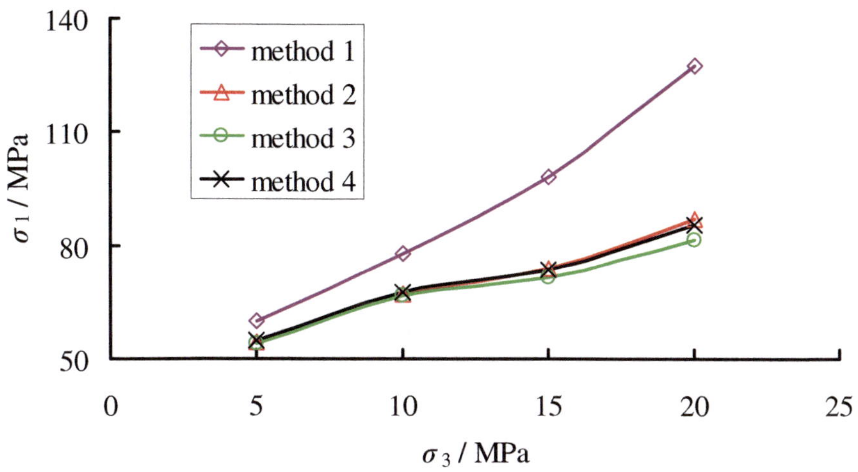


**Fig. 3** Relationship between peak stress and confining pressure for different methods

**Table 1** Axial strain under peak stress and peak stress difference

| $\sigma_3$ / MPa | Axial strain under peak stress / % | | | | Percentage of the difference between peak stress / % | | |
|---|---|---|---|---|---|---|---|
| | Method 1 | Method 2 | Method 3 | Method 4 | Method 2 | Method 3 | Method 4 |
| 5 | 10.55 | 9.42 | 9.89 | 9.42 | 9.24 | 9.67 | 8.54 |
| 10 | 18.43 | 12.21 | 13.02 | 15.1 | 14.00 | 14.68 | 13.33 |
| 15 | 30.57 | 25.3 | 23.02 | 25.3 | 24.85 | 27.19 | 25.04 |
| 20 | 38.34 | 33.58 | 30.01 | 33.58 | 31.80 | 36.15 | 32.91 |

## 5 Conclusions

(1) As a special soft rock, rock salt can exhibit large deformations in both axial and lateral directions under triaxial stress conditions. If the confining pressure reaches 15 MPa, the strain-stress curve of rock salt exhibits a strain hardening process, and no peak stress can be identified.

(2) Using the methods for correcting cross-sectional area by the engineering strain, the logarithmic strain and the volume deformation, the peak stress can be identified, and the strain softening process observed at post-peak stress stage. According to the experimental results, the maximum axial stress is generally determined by using method 1, while method 3 gives the smallest value. The maximal difference in peak stress determined by methods 1 and 3 is 36.15%. Under the same axial stress, the peak stress obtained by the logarithmic strain method is larger than that obtained by the engineering strain

method. Applying the engineering strain using method 1 gives a higher value than using method 3, with a maximal difference of 29.35%.

(3) Considering that the peak stress determined by the logarithmic strain method (3) is the lowest and that the strain at peak stress is also smaller than that without correction, it is recommended that this method be used to calculate the strength of rock salt for triaxial compression tests. In this way, an optimal safety situation can be achieved for the design and the stability analysis of underground gas caverns.

**Acknowledgments.** The authors are grateful for the financial support from the National Natural Science Foundation of China (Grant No. 51120145001, 51104101, 51134018, 11202069), the National Basic Research Projects of China (Grant No. 2011CB201201, 2010CB226802, and 2010CB732005), State Key Laboratory of Geohazard Prevention and Geoenvironment Protection (Grant No. SKLGP2013K016). The authors wish to offer their gratitude and regards to the colleagues who contributed to this work.

## References

1. Hunsche, U., Albrecht, H.: Results of true triaxial strength tests on rock salt. Engineering Fracture Mechanics 35(4/5), 867–877 (1990)
2. Hunsche, U.: A failure criterion for natural polycrystalline rock salt. In: Jinghong, F., Murakami, S. (eds.) Advances in Constitutive Laws for Engineering Materials, pp. 1043–1046. International Academic Publ. (1989)
3. Farmer, I.W., Gilbert, M.J.: Dependent strength reduction of rock salt. In: The First Conference on the Mechanical Behavior of Salt, pp. 4–18. Trans Tech Publications (1984)
4. Yang, C.H., Yin, J.H., Daemen, J.J.K.: The investigation on stress relaxation of rock salt. Chin. J. Rock Mech. Eng. 18(3), 262–265 (1999) (in Chinese)
5. Zhou, H.W., Wang, C.P., Ding, J.Y., Zhang, M.: Developments in researches on time-dependent behavior of salt rock and long-term stability of salt cavity. Mechanics in Engineering 33(5), 1–7 (2011) (in Chinese)
6. Zhou, H.W., Wang, C.P., Han, B.B., Duan, Z.Q.: A creep constitutive model for salt rock based on fractional derivatives. Int. J. Rock Mech. Min. Sci. 48(1), 116–121 (2011)
7. Xie, H.P., Liu, J.F., Ju, Y., Li, J., Xie, L.Z.: Fractal property of spatial distribution of acoustic emissions during the failure process of bedded rock salt. Int. J. Rock. Mech. Min. Sci., 1215–1382 (2011)
8. Liu, J.F., Xu, J., Yang, C.H., Hou, Z.M.: Mechanical characteristics of tensile failure of salt rock. Chinese Journal of Geotechnical Engineering 33(4), 580–586 (2011) (in Chinese)
9. Liu, J.F., Xu, J., Pei, J.L., Zhang, R.: Resarch on unloading modulus for salt rock damage testing. Journal of Sichuan University (Engineering Science Edition) 43(4), 57–62 (2011) (in Chinese)
10. Yang, C.H., Ma, H.L., Liu, J.F.: Study of deformation of rock salt under cycling loading and unloading. Rock and Soil Mechanics 30(12), 3562–3568 (2009) (in Chinese)

11. Li, Y.P., Liu, J., Yang, C.H.: Influence of mudstone interlayer on deformation and failure characteristic of salt rock. Chin. J. Rock Mech./Eng. 25(12), 2461–2466 (2006) (in Chinese)
12. Wu, W., Yang, C.H.: Experimental study on salt rock under compression and damage model simulation. Chin. J. Rock Mech. Eng. 25(2), 3709–3713 (2006) (in Chinese)
13. Liu, J., Yang, C.H., Wu, W., Li, Y.P.: Experiment study on short-term strength and deformation properties of rock salts. Chin. J. Rock Mech. Eng. 25(3), 103–104 (2006)
14. Zhou, H.W., Wang, C.P., Duan, Z.Q., Zhang, M., Liu, J.F.: Time-based fractional derivative approach to creep constitutive model of salt rock. Scientia Sinica (Physica, Mechanica & Astronomica) 42(3), 310–318 (2012) (in Chinese)

# Parameter Determination for the Constitutive Model Lubby2 and Strength Model Hou Based on Laboratory Tests on Rock Salt Samples from Jintan, P.R. China

Juan Zhao[1,2], Michael Zhengmeng Hou[1,2,3], and Wei Xing[1,2]

[1] Energy Research Centre of Lower Saxony, Goslar, Germany
[2] Sino-German Energy Research Center, Sichuan University, Chengdu, China
[3] Institute of Petroleum Engineering, Clausthal University of Technology, Clausthal-Zellerfeld, Germany

**Abstract.** Different lab tests concerning creep behaviour and short term strength properties of bedded rock salt samples from Jintan, P.R. China have been performed by Sichuan University (SCU), the Chair in Waste Disposal and Geomechanics of Clausthal University of Technology (TUC) and the Institute for Geomechanics, Leipzig (IfG).

This paper presents the parameter determination for the constitutive model Lubby2 and the strength model Hou based on the above-mentioned lab tests. In this context the difficulties over the stress distribution within the salt samples are discussed. These difficulties result from deformation of the salt samples during the lab tests. Additionally a verification analysis of the determined parameters is presented, comparing measured lab tests data and calculated results with constitutive model Lubby2 and strength model Hou using the determined material parameters.

Last but not least some mechanical differences between bedded rock salt of Jintan, P.R. China and German high-purity rock salt are discussed. Jintan rock salt shows a relatively high content of clay minerals, however its mechanical behaviour is little different from the German rock salt.

**Keywords:** rock salt, creep behaviour, short term strength, Lubby2, Hou, parameter determination.

## 1 Introduction

There are two main possibilities for underground storage of natural gas, so-called porous rock formations covered by tight rock formations as well as large single

M.Z. Hou et al. (Eds.): *Clean Energy Systems in the Subsurface*, SSGG, pp. 293–304.
DOI: 10.1007/978-3-642-37849-2_24 

rock salt caverns. Compared to porous rock formations, salt caverns have the following advantages: low percentage of cushion gas, extremely high injection and withdrawal rate, low investment cost, short construction duration, high deliverability etc.

Cavern techniques have been investigated since the beginning of the 20th century in America and some European countries (e.g. Germany and France). In 2005 China started to use salt caverns for natural gas storage in Jintan for peak shaving purposes or seasonal fluctuation of gas consumption in Shanghai as well as to support the well-known East-West-Pipeline. Unlike salt caverns in America and Germany, which are mainly built in salt domes, the caverns in China are built in bedded salt formations. The Jintan salt strata are nearly horizontal and consist of thin salt layers and intermediate mudstone layers. For the development of cavern techniques in China as in all Asian countries, many in-situ experiments, lab tests and numerical simulations have been performed. In this paper, a lab test evaluation method is presented. The parameter determination for the constitutive model Lubby2 (Lux 1984) and the strength model Hou (Hou 2003) are based on lab tests concerning the creep behaviour and short term strength properties of rock salt samples from Jintan, P.R. China. For verification of the determined parameters the measured lab test data are re-analysed using the constitutive model Lubby2 and the strength model Hou.

## 2 Determination of Creep Parameters of Constitutive Model Lubby2

Rock salt is a very special material because of its creep behaviour. Creeping is the time- and temperature-dependent viscoplastic deformation induced by constant or non-constant loading conditions. Directly after loading an elastic deformation occurs. In addition there can be three different creep phases: the transient, the stationary and the tertiary creep phase. Figure 1 shows both the creep strain as well as the creep rate induced by constant loading conditions (A) and by gradual increasing the loading (B). A degressively increasing creep strain and a decreasing creep rate are characteristic of the transient phase, the stationary phase is characterised by a linearly increasing creep strain and a constant creep rate, the tertiary creep phase is characterised by a progressively increasing creep strain and creep rate. The total strain is the sum of the elastic ($\varepsilon^e$), thermal ($\varepsilon^{th}$), plastic ($\varepsilon^p$) and viscous ($\varepsilon^v$) strains.

$$\varepsilon = \varepsilon^e + \varepsilon^{th} + \varepsilon^p + \varepsilon^v \tag{1}$$

The thermal strain is usually not considered because under isothermal test conditions, the plastic strain is relatively small and can be neglected.

$$\varepsilon = \varepsilon^e + \varepsilon^v \tag{2}$$

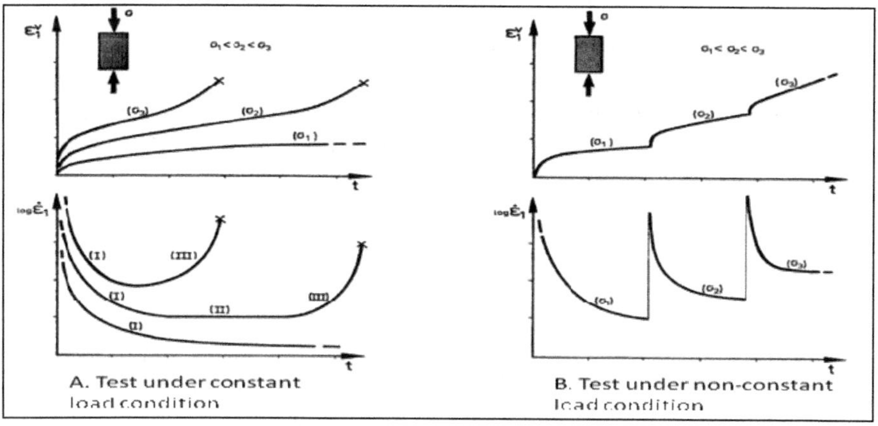


**Fig. 1** Creep phases of polycrystalline materials (Lux 1984)

The elastic strain is determined by Hooke's constitutive model. The transient and stationary creep parameters are determined by the following equations of the constitutive model Lubby2 in Lux (1984).

$$\dot{\varepsilon}_{ij}^{v} = \dot{\varepsilon}_{ij}^{tr} + \dot{\varepsilon}_{ij}^{s} = \frac{3}{2} \cdot \left[ \frac{1}{\overline{\eta}_k(\sigma_v)} \cdot \left(1 - \frac{\varepsilon^{tr}}{\max \varepsilon^{tr}}\right) + \frac{1}{\overline{\eta}_m(\sigma_v, T)} \right] \cdot S_{ij} \tag{3}$$

$$\max \varepsilon^{tr} = \sigma_v / \overline{G}_k(\sigma_v) \tag{4}$$

$$\overline{\eta}_k(\sigma_v) = 3\eta_k(\sigma_v) = \overline{\eta}_k^* \cdot \exp(k_2 \cdot \sigma_v) \tag{5}$$

$$\overline{G}_k(\sigma_v) = 3G_k(\sigma_v) = \overline{G}_k^* \cdot \exp(k_1 \cdot \sigma_v) \tag{6}$$

$$\overline{\eta}_m(\sigma_v, T) = 3\eta_m(\sigma_v, T) = \overline{\eta}_m^* \cdot \exp(m \cdot \sigma_v) \cdot \exp(l \cdot T) \tag{7}$$

where $\overline{\eta}_m^*$ = Maxwell viscosity modulus, $m$ = parameter for consideration of stress dependency of Maxwell viscosity modulus, $l$ = parameter for consideration of temperature dependency of Maxwell viscosity modulus, $\overline{\eta}_k^*$ = Kelvin viscosity modulus, $k_2$ = parameter for consideration of stress dependency of Kelvin viscosity modulus, $\overline{G}_k^*$ = Kelvin shear modulus, $k_1$ = parameter for consideration of stress dependency of Kelvin shear modulus, $\max \varepsilon^{tr}$

= maximum Kelvin strain, $\varepsilon^{tr}$ = actual Kelvin strain, $S_{ij}$ = deviatoric stress tensor, $\sigma_v$ = von Mises stress.

All the creep tests have been performed by IfG (Brückner 2012). The test program is summarized in Table 1. A total of 3 triaxial compression creep tests were conducted consisting of four load levels in each case with a deviatoric stress level within a range of between 8 MPa and 25 MPa. The confining pressure was kept constant at a high level of $\sigma_3$ = 20 MPa.

**Table 1** Creep tests program

| No. | Testing Technique | $\sigma_3$ [MPa] | $\sigma_1$ [MPa] | T [°C] | h [mm] | d [mm] | density [t/m$^3$] | Lab |
|---|---|---|---|---|---|---|---|---|
| 432/K1 | TCc | 20 | 28/30/32/35 | 45 | 80.2 | 40.3 | 2.16 | IfG |
| 432/K2 | TCc | 20 | 30/32/35/40 | 45 | 80.3 | 40.4 | 2.18 | IfG |
| 432/K3 | TCc | 20 | 32/35/40/45 | 45 | 80.0 | 40.5 | 2.16 | IfG |

According to Brückner (2010) there is no damage (= tertiary creep phase) if the deviatoric stress (= von Mises stress) level is below 25 MPa with a constant confining pressure of 20 MPa. Under such loading conditions the volume of the salt sample does not change. This allows for the use of small samples and simple creeping test machines.

During lab tests a deviatoric stress level below 8 MPa induces strains too low to be measured. For this reason the minimum deviatoric stress is 8 MPa. More loading levels lead to more precise results using the same sample, but for economical reasons results with loading differences of 2 or 3 MPa between two loading levels will suffice. At higher loading levels the loading difference can be bigger, e.g. 5 MPa.

The deformation is measured by using three dial gages fixed around the samples, each with offset of 120°. The creep strain at the time t is an average of the three measured deformations. The creep rate is equal to the quotient between creep strain and time difference. To get an accurate result the 11-mean-principle is used. The so-called 11-mean-principle is the average of the next11 creep rates in a loading phase. At the end of each loading phase the creep rate corresponds to the stationary creep rate since the transient creep only occurs at the beginning of each loading phase and then gradually decreases. In laboratory tests the measured total strain includes elastic, transient and stationary strain. According to the equations (3)-(7) $\overline{\eta}_k$, $\overline{G}_k$ and $\overline{\eta}_m$ are determined at the different equivalent stress levels. The results, based on the laboratory tests in Brückner (2010) are shown in Figure 2.

From the analysis of Figure 2, using the model Lubby 2 the following transient and stationary creep parameters have been determined for the salt samples from Jintan.

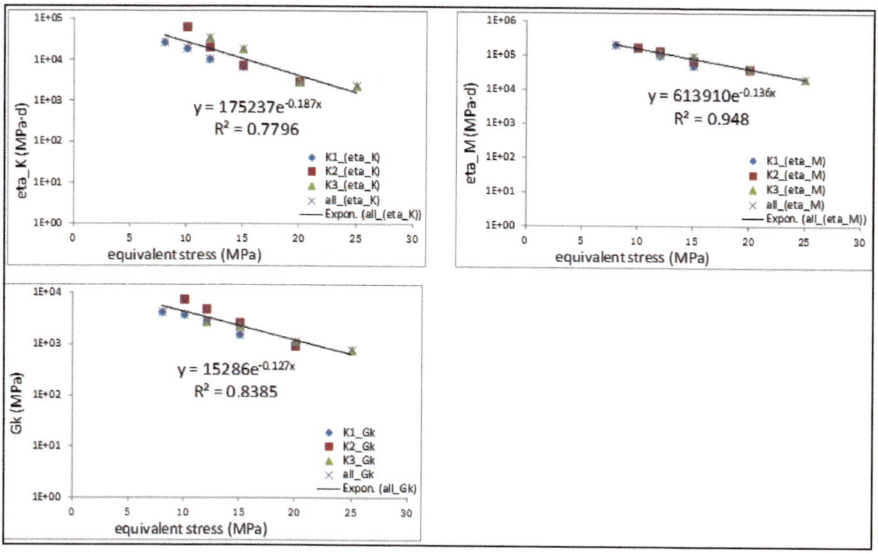


**Fig. 2** Determination of the creep parameters for the salt samples in Jintan

$$\overline{\eta}_k^* = 1.75 \cdot 10^5 \; MPa \cdot d \qquad k_2 = -0.187 \; MPa^{-1}$$

$$\overline{G}_k^* = 1.53 \cdot 10^4 \; MPa \qquad k_1 = -0.127 \; MPa^{-1}$$

$$\overline{\eta}_m^* = 6.14 \cdot 10^5 \; MPa \cdot d \qquad m = -0.136 \; MPa^{-1}$$

The analytical calculation of the time-related creep strain is done incrementally with the variable creep rate. With the above calculated creep parameters the analytical creep rates for the three salt samples 432/K1-K3 are initially calculated according to the Eqs. (8)-(10).

$$\dot{\varepsilon}_{ij}^{tr} = \left[\frac{1}{\overline{\eta}_k^* \cdot \exp(k_2 \cdot \sigma_v)} \cdot \left(1 - \frac{\varepsilon^{tr}}{\sigma_v} \cdot \overline{G}_k^* \cdot \exp(k_1 \cdot \sigma_v)\right)\right] \cdot \sigma_v \tag{8}$$

$$\dot{\varepsilon}_{ij}^{s} = \frac{1}{\overline{\eta}_m^* \cdot \exp(m \cdot \sigma_v)} \cdot \sigma_v \tag{9}$$

$$\dot{\varepsilon}_{ij}^{v} = \dot{\varepsilon}_{ij}^{tr} + \dot{\varepsilon}_{ij}^{s} \tag{10}$$

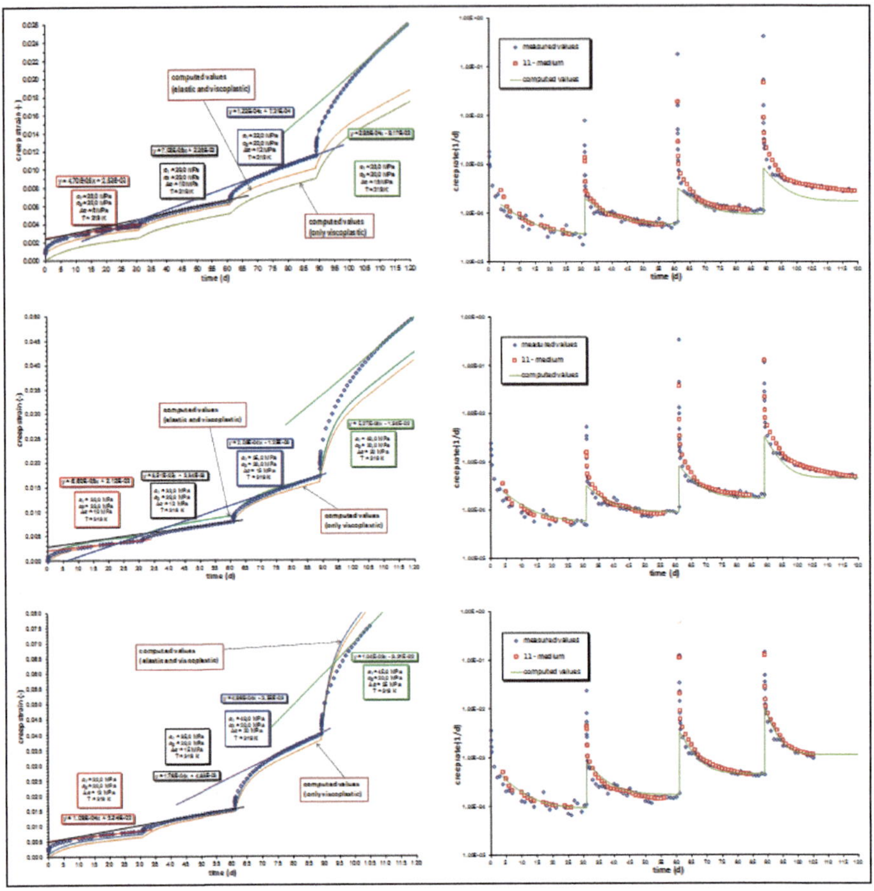


**Fig. 3** Measured and calculated creep strain and creep rate for the samples 432/K1 (top), 432/K1 (middle) and 432/K3 (bottom)

Figure 3 shows the measured creep rates of all tests performed at a temperature of $T = 318\ K$ (Brückner 2010). The creep rates are dependent on the equivalent stress. For comparison, the calculated equilibrium curve, obtained by using the constitutive model Lubby2 on the investigated sample, is added to Figure 3.

The results derived from Figure 3 can be summarized as follows: Because of the statistical character of the data analysis, the calculated creep strain and creep rate do not totally equal the measured values. Differences arise from the natural scattering of the measured values and they cannot be detected in the framework of the modelling. So the recalculated results using the model Lubby2 match the measured values relatively well. No tertiary creep phase was observed during any of the creep tests, therefore the model Lubby2 was the right choice.

The comparison between the measured and the calculated creep strains and creep rates of Brückner (2010) are summarized in Figure 4. The calculated results are obtained by using the same constitutive model Lubby2 but different creep parameters of IfG (Brückner 2010). The differences between the laboratory and the calculated results are relatively large. So the results of Brückner (2010) are not as good as those in this paper.

$$\overline{\eta}_k^* = 1.88 \cdot 10^4 \; MPa \cdot d \qquad k_2 = -0.139 \; MPa^{-1}$$

$$\overline{G}_k^* = 6.74 \cdot 10^4 \; MPa \qquad k_1 = -0.131 \; MPa^{-1}$$

$$\overline{\eta}_m^* = 6.42 \cdot 10^5 \; MPa \cdot d \qquad m = -0.141 \; MPa^{-1}$$

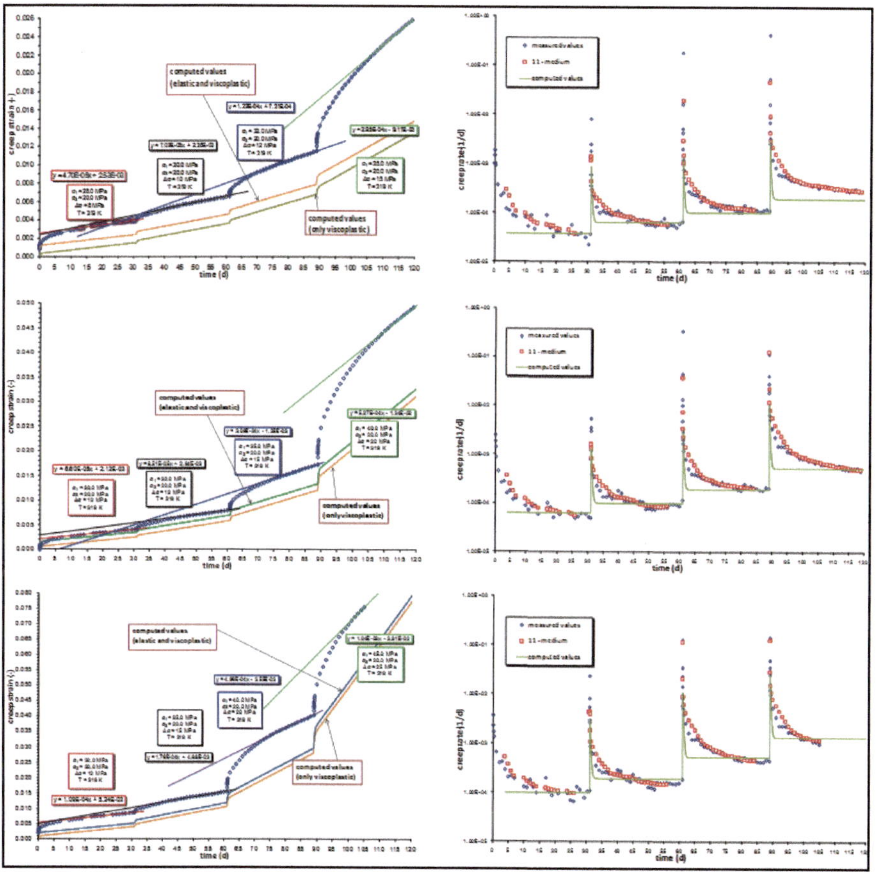


**Fig. 4** Measured and calculated creep strain and creep rate for the samples 432/K1-K3 according to the creep parameters of IfG (Brückner 2010)

## 3 Determination of Strength Parameters of Strength Model Hou

To determine the strength parameters of the strength model Hou (2003), uniaxial compression- (UC), triaxial compression- (TC) and triaxial extension-tests (TE) must be performed. The laboratory experiments on salt samples with a diameter of 90 mm and height of 180 mm at Sichuan University (SCU, reported in Xie et al., 2011), Clausthal University of Technology (TUC, reported in Düsterloh & Lux 2003, 2012) and the Institute for Geomechanics (IfG, reported in Brückner, 2010, 2011) have determined the strength parameters.

The technical strain (at $\varepsilon \leq 2\ \%$) in Eq. (11) and the logarithmic strain (at $\varepsilon \geq 2\ \%$) in Eq. (12) are defined e.g. in Düsterloh (2009).

$$\varepsilon_t = ((l_1 + l_2 + l_3)/3)/l_0 \cdot 100\% = \Delta l / l_0 \cdot 100\% \tag{11}$$

$$\varepsilon_{\ln} = \left| \int_{l_0}^{l} \frac{dl}{l} \right| \cdot 100\% = \left| \ln \frac{l}{l_0} \right| \cdot 100\% = \left| \ln(1 - \varepsilon_t) \right| \tag{12}$$

where $l_0$ = initial length of the sample, $l$ = actual length of the sample, $\Delta l$ = difference between $l_0$ and $l$, $l_1, l_2, l_3$ = actual length of the sample measured by 3 different sensor positions, $\varepsilon_t$ = technical strain, $\varepsilon_{\ln}$ = logarithmic strain.

The uncorrected, linear and logarithmic corrected axial stresses are calculated according to the Eqs. (13)-(15).

$$A_u = \frac{\pi d_u^2}{4};\ F_R = F_{ax} = A_u \cdot \sigma_1;\ \sigma_u = \frac{F_R}{A_u};\ d_u = 90mm \tag{13}$$

$$A_{lc} = \frac{\pi d_{lc}^2}{4};\quad F_R = F_{ax} - \sigma_3 \cdot \frac{\pi}{4} \cdot \left(d_u^2 - d_{lc}^2\right);\quad \sigma_{lc} = \frac{F_R}{A_{lc}};\quad d_{lc} = d_u \cdot \sqrt{1/(1 - \Delta l / l_0)} \tag{14}$$

$$A_{\ln c} = \frac{\pi d_{\ln c}^2}{4};\quad F_R = F_{ax} - \sigma_3 \cdot \frac{\pi}{4} \cdot \left(d_u^2 - d_{\ln c}^2\right);\quad \sigma_{\ln c} = \frac{F_R}{A_{\ln c}};\quad d_{\ln c} = d_u \cdot \sqrt{1/(1 - \ln(l/l_0))} \tag{15}$$

where $d_i$ = diameter of the sample, $A_i$ = cross-section area of the sample, $\sigma_i$ = applied stress, $F_i$ = applied force.

Figure 5 shows the compression and extension failure strengths, which were measured by the above-mentioned three institutes "SCU, TUC and IfG" on rock salt samples from the location Jintan. In addition, the calculated failure strengths after Eq. 16 (Hou 2003), which are dependent on the minimum principal stress, were added to Figure 5. This strength function will be stated more precisely within the scope of the investigation of the further test series, still to be performed.

$$\beta^{TC}(\sigma_3) = a_6 - a_7 \cdot \exp(-a_8 \cdot \sigma_3) \tag{16}$$

where $\beta^{TC}(\sigma_3)$ = triaxial compression failure strength, $\sigma_3$ = minimum principal stress, $a_i$ = material parameters.

Based on the experimental results in Figure 5 the strength parameters $a_6 - a_8$ can be derived:

$$a_6 = 54.5\ MPa;\ a_7 = 34.5\ MPa;\ a_8 = 0.25\ MPa^{-1}$$

Figure 5 shows the measured and calculated strength as a function of the minimum principal stress. The line in Figure 5 represents the calculated compression strengths according to Eq. (16) with the above parameters, while the different points show the measured values performed by different institutes with and without stress corrections.

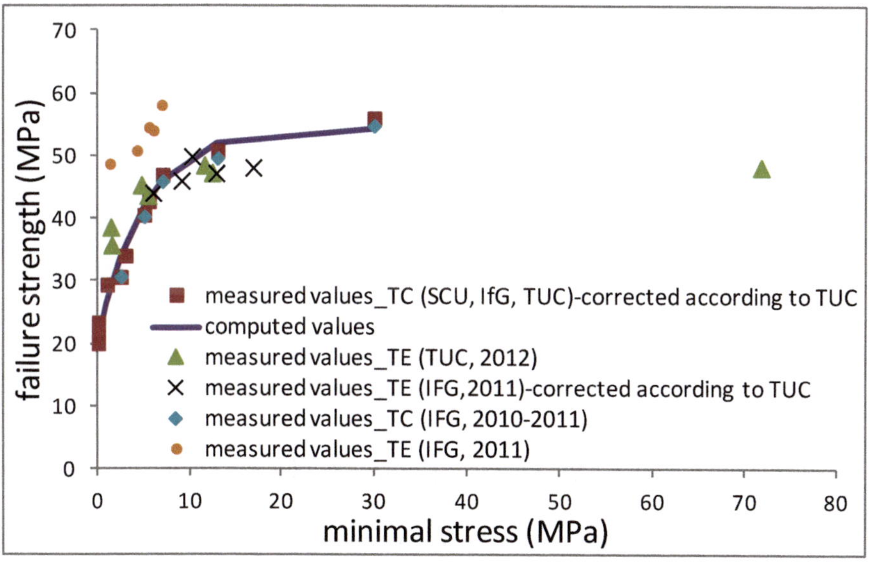


**Fig. 5** Compression and extension failure strength for the samples from Jintan

The red squares show the corrected test values based on the UC- and TC-tests performed by SCU, IfG and TUC with the above stated correction method of TUC according to the Eqs. (13)-(15), while the blue diamond shows the uncorrected measured values of the compression strength based on the tests of IfG in 2010-2011 (Brückner 2010, 2011). For compressive strengths it can be summarized that there is little difference between the results obtained by the correction methods of TUC and IfG. The orange circles are the measured extension strengths of IfG in 2011, while the black crosses are the corrected values according to the TUC correction method. The difference between the extension strengths according to the two correction methods is relatively large. The green triangles show the measured extension strengths of TUC in 2012. The TE test indicated in Figure 5 by a minimum principal stress of 72 MPa does not fit the failure strength function because the test had to be stopped due to reaching the maximum strain allowed for the test equipment to prevent it reaching the failure state.

The strength should be dependent on stress geometry and it is determined according to the Hou (2003) by the parameters $a_9 - a_{10}$ in Eq. (17). The extension strength is equal to compression strength times a factor k in Eq. (18). From Figure 5 it can be seen that all measured compression and extension strength are close to the calculated compression strength. Therefore the factor $k$ has to equal 1.

The parameters $a_9 - a_{10}$ can be derived: $a_9 = 0.577$; $a_{10} = 0\ MPa^{-1}$.

The compression and extension strength are an upper limit of the allowable stress, while the lower limit is the dilatancy strength. The dilatancy strength is defined according to Eq. (19) (Hou 2003).

$$k_{(\sigma_3,\theta=30^\circ)} = \left[\frac{1}{\cos\left(\theta+\dfrac{\pi}{6}\right)+a_9\cdot\sin\left(\theta+\dfrac{\pi}{6}\right)}\right]^{\exp(-a_{10}\cdot\sigma_3)} \tag{17}$$

$$\beta^{TE}_{(\sigma_3,\theta=30^\circ)} = \beta^{TC}_{(\sigma_3)}\cdot k_{(\sigma_3,\theta=30^0)} \tag{18}$$

$$\beta_{Dil} = \beta^{TC}\cdot\eta_{Dil} = \beta^{TC}\cdot\left[1-a_4\cdot\exp(-a_5\cdot\sigma_3)\right] \tag{19}$$

where $k$ = stress geometry factor, $\theta$ = Lode-angle, $\beta^{TE}$ = triaxial extension failure strength, $a_i$ = material parameters, $\beta_{Dil}$ = dilatancy strength, $\eta_{Dil}$ = dilatancy factor.

The parameters $a_4 - a_5$ are calculated after Eq. (19) with the experimentally determined dilatancy strength and the compression strength. Figure 6 shows the results of the computed and measured dilatancy strength and with the parameters: $a_4 = 0.54$; $a_5 = 0.045\ MPa^{-1}$.

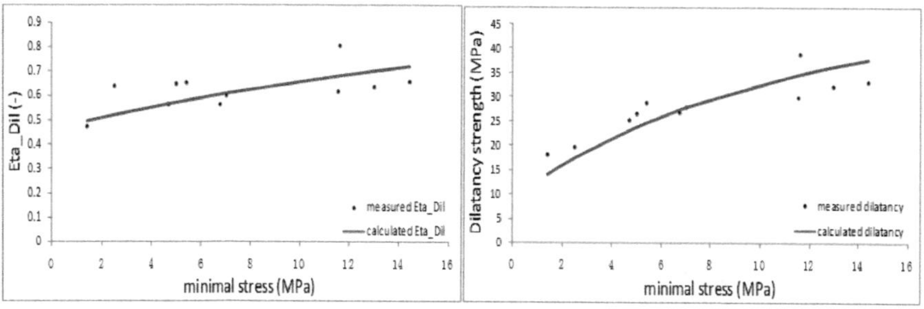


**Fig. 6** Dilatancy strength and $\eta_D$ for the samples from Jintan

# 4 Conclusions

According to the different lab tests in SCU, TUC and IfG concerning the creep behaviour and short term strength properties of rock salt samples from Jintan, P.R. China the creep parameters for the constitutive model Lubby2 and the strength parameters for the strength model Hou have been determined based on the TUC correction method. The measured lab test data and the calculated results with constitutive model Lubby2 and strength model Hou using the determined material parameters have been compared for a verification analysis of the determined parameters. Based on a previous analysis the measured data can be physically modeled very well using the constitutive model Lubby2 and the strength model Hou. The compression and extension strengths for the rock salt of Jintan, P.R. China are almost equal. The reason is that Jintan rock salt has a relatively high content of clay minerals. From this point of view, its mechanical behaviour is different from the German rock salt.

# References

1. Düsterloh, U., Lux, K.H.: Rock mechanical investigation into bore core material from the location Jintan. Laboratory Report, Chair for Waste Disposal Technologies and Geomechanics, Clausthal University of Technology, Germany (2003) (not published)
2. Düsterloh, U.: Geotechnische Sicherheitsnachweise für Hohlraumbauten im Salinargebirge unter besonderer Berücksichtigung laborativer Untersuchungen. Habilitationsschrift an der TU Clausthal, Germany (2009)

3. Düsterloh, U., Lux, K.H.: Rock mechanical investigations on the extension strength of core samples from the Jintan location. Laboratory Report, Chair for Waste Disposal Technologies and Geomechanics, Clausthal University of Technology, Germany (2012) (not published)
4. Hou, Z.: Mechanical and hydraulic behaviour of salt in the excavation disturbed zone around underground facilities. Int. J. of Rock Mechanics and Mining Sciences 40, 725–738 (2003), doi:10.1016/S1365-1609(03)00064-9
5. Lux, K.H.: Gebirgsmechanischer Entwurf und Felderfahrungen im Salzkavernenbau. Ferdinand Enke Verlag, Stuttgart (1984)
6. Brückner, D.: Rock mechanical laboratory investigation - Rock Salt from Jintan – China. Laboratory Report, Institut für Gebirgsmechanik GmbH in Leipzig, Germany (2010) (not published)
7. Brückner, D.: Rock mechanical laboratory investigation - Rock Salt from Jintan – China. Laboratory Report, Institut für Gebirgsmechanik GmbH in Leipzig, Germany (2011) (not published)
8. Xie, H., Liu, J., Ju, Y., Li, J., Xie, L.: Fractal property of spatial distribution of acoustic emissions during the failure process of bedded rock salt. Int. J. of Rock Mechanics and Mining Sciences 48, 1344–1351 (2011), doi:10.1016/j.ijrmms.2011.09.014

# Comparison of Triaxial Compression Short-Term Strength Tests and Data Processing Methods for Rock Salt

Guan Wang[1], Wei Xing[1,2], Jianfeng Liu[1], and Lingzhi Xie[1]

[1] Sino-German Energy Research Center, Sichuan University, Chengdu, Sichuan 610065, China
[2] Energy Research Center of Lower Saxony (EFZN), Goslar 38640, Germany

**Abstract.** The strength and deformation characteristics of rock salt are extremely important parameters for the design, construction and operation of underground energy storage and waste repository. Obtaining the correct strength parameters for rock salt is of great significance to the design and safe operation of an underground repository. In this study, samples of rock salt from the Jintan site of Jiangsu Province and Jianghan site of Hubei Province were chosen for triaxial compression tests, to detect their short-term strength and deformation parameters. Comparing the two different test and data processing methods employed in this study, it is revealed that for rock salt with a large deformation, the strength data obtained by the Chinese test measurement is generally higher than the German test measurement. Thus it can be concluded that the German strength test and data processing method is more accurate.

**Keywords:** rock salt, short-term strength, test method, data process method.

## 1 Introduction

Due to its strong compactness, low permeability, good plastic deformation ability and damage recovery ability, rock salt is widely used in nuclear waste disposal and petroleum and natural gas underground storage, etc. So far, both domestic and foreign scholars have done research on the mechanical properties of rock salt and many results were achieved.

The short-term strength and deformation theory of rock salt is based on the test data on uniaxial and triaxial stress without considering the time effect. Domestic and foreign scholars have also done a large number of tests on short-term strength and deformation characteristics of salt and improved the mechanics theory of rock salt [1-6].

The Clausthal University of Technology in Germany completed the triaxial strength and deformation test on four rock salt samples from the Jintan site of

M.Z. Hou et al. (Eds.): *Clean Energy Systems in the Subsurface*, SSGG, pp. 305–315.
DOI: 10.1007/978-3-642-37849-2_25 

Jiangsu Province in 2010 and obtained the corresponding mechanics parameters. Some domestic scholars have also done related research. In this paper, the different testing and data processing methods employed in Germany and domestic triaxial compression test are analyzed and discussed. Both the advantages and disadvantages of the two kinds of test methods are summarized, for the purpose of establishing scientific test and data processing methods for rock salt with large deformations.

# 2 Sample Preparation and Test Method

## 2.1 *Domestic Test Method*

As an example of the domestic test methods, the triaxial compression test of rock salt in the provincial key laboratory of geotechnical engineering at Sichuan University was introduced in 2008. The rock salt in Hubei Province was selected for testing samples. Three samples were used in this test.

### 2.1.1 Test Equipment

The test equipment used was MTS815 FLEX TEST GT a mechanical rock system, with a maximum axial load of 4600 KN and confining pressure of 140 MPa. All the parameters were acquired by high-precision sensor, controlled by the computer, and the testing results are accurate and reliable.

### 2.1.2 Sample Preparation

The 3 samples of rock salt were cylindrical with diameters of 99.72 mm, 99.75 mm, and 99.72 mm. The sample size was controlled by D (diameter): L (height) = 1:2.0-1:2.5 according to standard procedure. Due to the characteristics of salt solution in water, dry saw grinding was employed in the process of sample preparation. The two end faces were first cut, then the cylindrical and two sections of the specimens were processed with the lathe dry car method.

### 2.1.3 Triaxial Compression Test

Different confining pressures ($\sigma_3$=20, 30 and 40 MPa) were applied on the three samples at room temperature. The confining pressure was first raised to the specified value with a speed of 3 MPa/min. Then the confining pressure was kept constant and the axial load increased at a rate of 30 KN/min after the system had stabilized. First the axial loading was carried out by the computer. Then the testing process was controlled by the equivalent circumferential deformation when the plastic deformation appeared. Finally, the testing process was controlled by the axial deformation at a rate of 0.15 mm/min.

## 2.2 *German Test Method*

### 2.2.1 Test Equipment

The servo-controlled hydraulic testing machine in Clausthal, system RBA 2500 (max. axial force: 2500KN) with 2 triaxial cells (test specimens up to 110 mm diameter and 220 mm length) for confining pressures of up to 100 MPa as well as test temperatures in the range 295 - 400 K were used in the experiment.

### 2.2.2 Sample Preparation

Uniaxial compression tests were carried out on four specimens. The sample processing method was the same as in the domestic test. The physical parameters of the four samples are shown in Table 1.

**Table 1** Data of the investigated specimens

| Specimen | Length (mm) | Diameter (mm) | Mass (g) | Density (t/m³) |
|---|---|---|---|---|
| China432-1 | 180,12 | 90,08 | 2504,7 | 2,1819 |
| China432-2 | 180,24 | 89,94 | 2505,6 | 2,1880 |
| China432-3 | 180,32 | 90,05 | 2483,7 | 2,1628 |
| China432-4 | 180,49 | 89,88 | 2486,8 | 2,1715 |

### 2.2.3 Triaxial Compression Test

Generally, the triaxial test is divided into two stages.

1. The first phase is a hydrostatic loading with $\sigma_1 = \sigma_3$ until the former overburden pressure has been achieved. The hydrostatic load was raised by an increased pressure rate of 0.01 MPa/s. After the level of the original overburden pressure ($\sigma_1 = \sigma_3 = 25$ MPa) had been reached, the level was held constant for several hours so the total duration of this phase was at least 24 hours.
2. The next stage is characterized by a hydrostatic unload down to the level of the confining pressure for the triaxial compression test. Four different levels ($\sigma_3$ = 2.5; 5, 7 and 13 MPa) had been selected before the test program started. Then the confining pressure was kept constant, and the axial load was increased with a constant deformation rate (0.15 %/min) until the axial deformation had increased by 20%.

## 2.3 *Test Method Contrast*

### 2.3.1 Influence of the Initial Damage-Healing Process on Strength

The permeability of rock salt is extremely low because of its low porosity. The excavation disturbance changes the stress state of rock salt. When the stress gets beyond the damage boundary, the permeability of rock salt will be significantly enhanced but its strength will reduce due to the appearance of the damage dilatancy phenomenon and the destruction of pore structure. Also, when the healing boundary is reached, accompanied by a very low stress to strength ratio and a high confining pressure, the existing or developing damage to salt rocks will heal again, along with positive effects such as the closing of pores and cracks, and a reduction in permeability and compacted deformation [7]. Therefore, the domestic test, without the damage-healing process, will lead to a lower strength of rock salt than the German test and it is necessary to recover the initial damage in the sampling process as much as possible before the triaxial compression test is performed.

### 2.3.2 Influence of the Loading Controlling Methods on Strength

Stress control and strain control are two different control modes of rock mechanics tests. For linear elastic materials, the two control methods are not obviously different. For nonlinear materials like rock salt, stress begins to decline after reaching its peak value. Thus with the stress control method it is not possible to research on the mechanical behavior of rock salt during the post-failure process. At present, most of the domestic salt rock tests completed [1-3] were controlled by axial stress rate. In this paper, the German tests were controlled by the axial strain rate (0.15 %/min) and domestic tests by the axial deformation rate (0.15 mm/min) instead of the axial stress rate, which are more suitable for rock salt material.

### 2.3.3 Influence of the Loading Strain Rate on Strength

Liao et al. (2005) [8] have conducted consolidated undrained triaxial tests on diatomaceous soft rock. It was found that the soft rock had an obvious strain rate effect. Its peak strength and residual strength changed with the loading strain rate in the test scales, whilst Liang et al. (2010) [9] have found that the uniaxial strength of halite and thenardite do not change with the loading strain rate in uniaxial compression tests. Rock salt, however, is very sensitive to the confining pressure, the mechanical properties of which will obviously be different in triaxial compression tests and uniaxial compression tests. Ji et al. (2011) [10] have found that the compressive strength of rock salt increased with the loading strain rate in triaxial compression tests. Horseman et al. (1990, 2011) [11] [12] found that the main strain rate of rock salt was in the range of $2 \times 10^{-3}$ - $2 \times 10^{-9}$ $s^{-1}$. In this paper, the German test loading strain rate of 0.15 %/min ($2.5 \times 10^{-5}$ $s^{-1}$) and domestic test of 0.15 mm/min (about $1.25 \times 10^{-5}$ $s^{-1}$) are both in this range.

# 3 Test Results and Data Processing

## *3.1 Domestic Triaxial Compression Test Results*

The triaxial compression test results of the rock salt from the Jianghan site of Hubei Province are shown in Table 2, while the deviatoric stress - axial strain curves are shown in Figure 1.

**Table 2** Test results of rock salt under triaxial compression

| Specimen | Test | Confining pressure(MPa) | Peak stress(MPa) | Deviatoric stress (MPa) | Maximum axial strain (%) |
|---|---|---|---|---|---|
| JH-49 | TC | 20 | 106.9 | 86.9 | 33.98 |
| JH-53 | TC | 30 | 113.8 | 83.8 | 24.40 |
| JH-59 | TC | 40 | 137.2 | 97.2 | 28.81 |

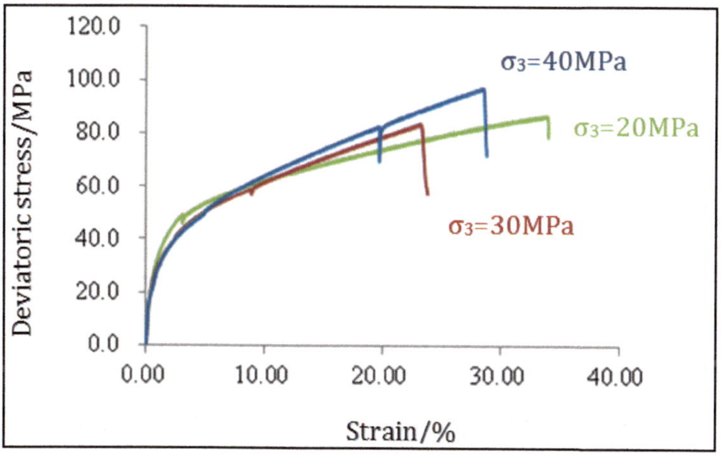


**Fig. 1** Deviatoric stress-axial strain curves of Jianghan rock salt with different confining pressures

## *3.2 German Triaxial Compression Test Results*

The triaxial compression test results of the rock salt from the Jintan site of Jiangsu Province are shown in Table 3, while the deviatoric stress - axial strain curves are shown in Fig. 2. The photos of samples before and after the test are shown in Fig. 3.

**Table 3** Test results of rock salt under triaxial compression [13]

| Specimen | Test | Confining pressure(MPa) | Peak stress(MPa) | Deviatoric stress (MPa) | Maximum axial strain (%) |
|---|---|---|---|---|---|
| China432-1 | TC | 7 | 50.0 | 43.0 | 20.29 |
| China432-2 | TC | 13 | 58.3 | 45.3 | 20.10 |
| China432-3 | TC | 5 | 43.8 | 38.8 | 19.88 |
| China432-4 | TC | 2.5 | 32.8 | 30.3 | 20.18 |

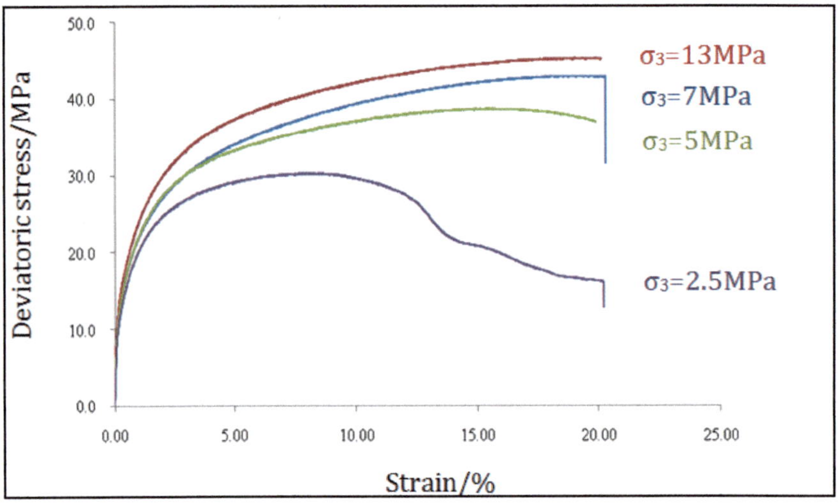


**Fig. 2** Deviatoric stress-axial strain curves of Jintan rock salt with different confining pressures [13]

## *3.3 Comparison of Different Data Processing Methods*

In traditional rock tests, the axial strain is commonly 2 %. Therefore, a small strain assumption is usually adopted in the processing of the test data, by which the axial stress and strain are described as the engineering stress and strain. However, the small strain assumption is no longer applicable for rock salt with large deformations (as shown in Fig. 3. The maximum axial strain reached 34 % and 20 % in the domestic and Germany tests, respectively. The cross-section area of the samples changed considerably after compression). Therefore, the domestic measurement method with engineering stress and strain is no longer accurate. The German test method, in which the change in the sample cross-section area was considered and the data was secondarily revised using scientific strain instead of engineering strain, describes the large deformation of rock salt accurately.

**Fig. 3** Rock salt samples of Jintan before and after a triaxial compression test [13]

### 3.3.1 Influence of the Change of Correction of the Cross-Section Area on Strength

To determine the corrected stress, a linear correction of the cross-sectional area A of the test body, which varies according to the axial strain Δl, can be modified according to the equation below [14]:

$$\sigma_{1uc} = \frac{F_1}{A} \tag{1}$$

$$A = \frac{A_0}{1 - \varepsilon_t} \tag{2}$$

where,

$\sigma_{1uc}$ = linear corrected vertical strain (MPa)
$F_1$ = vertical force (MN)
A = cross-sectional area of the stressed test body ($m^2$)
$A_0$ = cross-sectional area of the stressed test body ($m^2$)
$\varepsilon_t$ = technical deformation

### 3.3.2 Influence of the Calculation Using Scientific Strain Instead of Engineering Strain on Strength

With a restriction on small deformations ($\varepsilon \leq 2$ %), the calculation of the deformation can be carried out by the technical deformation according to the normal deformation definition. Technical deformation is the quotient of the vertical change in the test body's length $\Delta l$ and the start length $l_0$:

$$\varepsilon_t = \frac{\Delta l}{l_0} \times 100\% \tag{3}$$

where,

$\varepsilon_t$ = technical deformation
$l_0$ = length of the unstrained test body (mm)
$\Delta l$ = measured axial deformation of the test body (mm)

In the case of test body deformation $\varepsilon > 2\%$, the natural or true (logarithmic) deformation $\varepsilon_{ln}$ will be used instead of the technical deformation $\varepsilon_t$. This is calculated using the integral of the quotient of the current (infinitesimal) change of the test body length dl and the current (actual) length l [14]:

$$\varepsilon_{ln} = \left| \int_0^l \frac{dl}{l} \right| \times 100\% = \left| \ln \frac{l}{l_0} \right| \times 100\% = \left| \ln(1 - \varepsilon_t) \right| \times 100\% \tag{4}$$

where,

$\varepsilon_{ln}$ = true (logarithmic) deformation (%)
$l$ = actual length of the strained test body = $l_0 - \Delta l$ (mm)

The corresponding pressure is as follows:

$$\sigma_{1uc-ln} = \frac{F_1}{A} \tag{5}$$

$\sigma_{1uc\text{-}ln}$ is the logarithmic corrected vertical strain (MPa).

The samples China432/2 from the Jintan site of Jiangsu Province and JH-49 from the Jianghan site of Hubei Province, with confining pressures of 13 MPa and 20 MPa respectively, were taken as examples here. The deviatoric stress - axial strain curves without correction (domestic test results) and after secondary correction (Germany test results) are shown in Fig. 4 and Fig. 5, where the value of the corresponding peak stress of each curve and the percentage reduction of the deviatoric stress after correction are depicted.

Ji et al. (2011) [10] found that in the triaxial compression test, the confining pressure of the rock salt is greater than 3 MPa, i.e. there is no obvious failure appearing. If the engineering stress and strain measurements are used, as shown by the curve without correction in the figure above, the strength of the rock salt continues to increase after the axial strain is above 20 % and 30 %, which cannot be the peak stress. In fact, this result is contrary to a large number of test data in foreign countries. The results in this paper show that the curve appears to have a downward trend after secondary correction and the peak stress can be obtained. The strengths of the rock salt before and after correction have great differences.

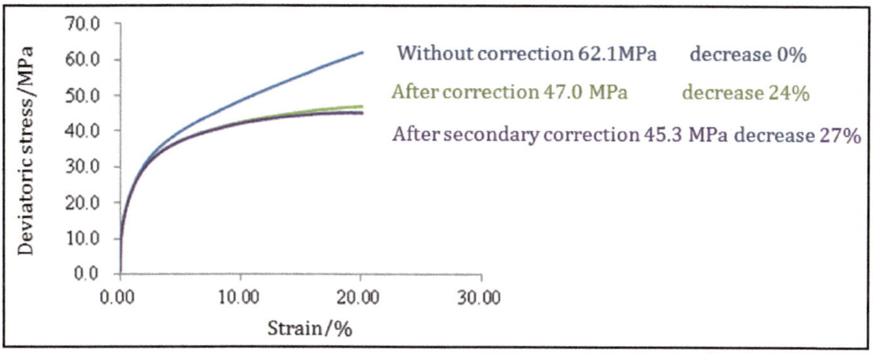


**Fig. 4** Deviatoric stress-axial strain curves of China432/2 under triaxial compression

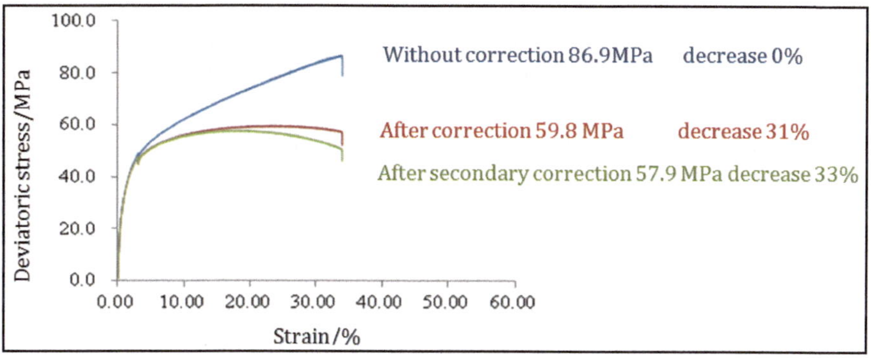


**Fig. 5** Deviatoric stress-axial strain curves of JH-49 under triaxial compression

The strength error of the sample China432/2 from the Jintan site reached 27 % while JH-49 from the Jianghan site reached 33 %. The data processing methods for hard rock mostly used in domestic triaxial compression tests of the rock salt give large errors. Therefore, secondary correction of the data is very necessary.

## 4 Conclusion

The triaxial compression tests of rock salt in the provincial key laboratory of geotechnical engineering at Sichuan University in 2008 and Clausthal University of Technology, Germany in 2010 are taken as examples to introduce and contrast the different test methods and data processing methods in this paper. The results show that the triaxial compression test method and data processing method for hard rock should not be applied to rock salt with large deformations. For rock salt triaxial compression tests, the German test method and data processing method are the better choice for future reference.

(1) The porosity and intensity of rock salt is low, and the influence of machine damage on the strength of rock salt is very huge. Therefore, putting the sample in its original formation stress state under the recovery process before the test will effectively reduce the machine damage influencing the strength of the rock salt.
(2) Rock salt belongs to the group of nonlinear materials so it is more suitable to use the axial strain rate instead of axial stress rate for controlling the triaxial compression tests.
(3) The cross-sectional area of the samples changes constantly in the experiment, so the cross-sectional area and its corresponding axial stress should be corrected. It has been proved that the strength value without correction is too high. The difference between the values before and after correction is obvious. After linear correction the strength of the sample China432/2 from the Jintan site was reduced by 24 % and JH-49 from the Jianghan site was reduced by 31 % after linear correction.
(4) The deformation capacity of rock salt is extremely high. In the case of test body deformation ($\varepsilon > 2\%$) of the rock salt, the natural or true (logarithmic) deformation should be used instead of the technical deformation.

This research shows that, the strength of the rock salt measured in the domestic triaxial compression tests was higher. The strength of the sample JH-49 from the Jianghan site was 86.9 MPa without correction while it reached 57.9 MPa after secondary correction, a reduction of 33%. The strength of rock salt with large deformations is heavily overvalued in domestic triaxial compression tests. Therefore it is of great significance to consider the advantages of the German triaxial compression test and data processing method for rock salt triaxial compression testing.

## References

1. Xu, S.U., Xi, B.P., Liang, W.G., et al.: Study of mechanical properties of glauberite salt rock. Chinese Journal of Underground Space and Engineering 1(7), 1125–1128 (2005) (in Chinese)
2. Liang, W.G., Zhao, Y.S.: Testing study of mechanical properties of thenardite salt rock. Chinese Journal of Rock Mechanics and Engineering 23(3), 391–394 (2004) (in Chinese)
3. Liang, W.G., Zhao, Y.S., Xu, S.G.: Testing study of physicomechanical properties of heated salt rock under 240°C. Chinese Journal of Rock Mechanics and Engineering 23(14), 2365–2369 (2004) (in Chinese)
4. Li, Y.P., Liu, J., Xu, S.G.: Influence of mudstone interlayer on deformation and failure characteristics of salt rock. Chinese Journal of Rock Mechanics and Engineering 25(11), 2461–2466 (2006) (in Chinese)
5. Yang, C.H., Li, Y.P., Chen, F.: Mechanics theory and engineering of bedded salt rock. Science Press, Beijing (2009)

6. Liu, J., Yang, C.H., Wu, W., et al.: Experiment study on short-term strength and deformation properties of rock salts. Chinese Journal of Rock Mechanics and Engineering 25(suppl.1), 3104–3109 (2006) (in Chinese)
7. Hou, Z.: Mechanical behaviour of rock salt masses under high stress to strength ratios (damaging) and under very low stress to strength ratios with high confining pressure (healing). ISRM -Technology roadmap for rock mechanics, 533–538 (2003)
8. Liao, H.J., Pu, W.C., Yin, J.H.: Study of strain rate effect of soft rock. Chinese Journal of Rock Mechanics and Engineering 24(18), 3218–3223 (2005) (in Chinese)
9. Liang, W.G., Xu, S.G., Mo, J., et al.: Test study of strain rate effects on mechanical proformances of salt rock. Chinese Journal of Rock Mechanics and Engineering 29(1), 44–50 (2010) (in Chinese)
10. Ji, W.D., Yang, C.H., Yao, Y.F., et al.: Effects of loading strain rate on mechanical performances of salt rock. Chinese Journal of Rock Mechanics and Engineering 30(12), 2507–2513 (2011) (in Chinese)
11. Horseman, S.T., Handin, J.: Triaxial-compression tests on rock salt at temperatures from 50 °C to 200 °C and strain rates from $10^{-4}$ to $10^{-9}$ $s^{-1}$. Geophysical Monograph Series, vol. 56, pp. 103–110 (1990)
12. Liang, W.G., Zhao, Y.S., Xu, S.G., et al.: Effect of strain rate on the mechanical properties of salt rock. International Journal of Rock Mechanics and Mining Sciences 48(1), 161–167 (2011)
13. Brückner, D.: Rock mechanical laboratory investigation - Rock salt from Jintan - China. IfG Institute for Rock Mechanics, Leipzig (2011)
14. Düsterloh, U.: Geotechnische Sicherheitsnachweise für Hohlraumbauten im Salinargebirge unter besonderer Berücksichtigung laborativer Untersuchungen. Papierflieger Verlag in Clausthal-Zellerfeld (2010) ISBN 3-86948-068-8

# Soil-Underground Structure Dynamic Interaction Considering Soil Nonlinearity

Qi Ge, Feng Xiong, Qunyi Huang, Lunwu Xie, and Ziyu Yao

College of Architecture and Environment, Sichuan University, Chengdu, China

**Abstract.** Soil-underground structure dynamic interaction is a complex scientific problem, which reveals the seismic responses of underground structures when an earthquake occurs. Most of the research currently being conducted ignores the effect of soil nonlinearity on underground structures owing to its complexity. However since soil is typically a nonlinear media, its nonlinearity has an important effect on the seismic responses of an underground structure. This paper proposes an approach to consider soil nonlinearity in a numerical analysis of soil-underground structure interaction dynamics. By using ANSYS finite element analysis software, the underground structure and the surrounding soil are modeled, and soil nonlinearity realized through the Davidenkov soil model by using the restarting method and ANSYS Parameter Design Language (APDL). The mechanics parameter, i.e. the shear modulus is forced to change along a cover obtained from experiments in each step of the numerical analysis. Therefore the nonlinearity of soil material can be considered. In addition, an artificial boundary is established to simulate a radiation damping effect for the model and the Combin14 element is used as a viscoelastic boundary. The finite element analysis models of soil with and without an underground structure are respectively calculated under horizontal and vertical earthquake excitations. By comparing their results, The effect of underground structure on the free-field ground can be obtained. Finally the effect of different thicknesses of overlying soil on the underground structure seismic response is analyzed. Considering soil nonlinearity, this paper studies the seismic responses of soil-underground structure interaction under different seismic excitations and burial depths, derives a lot of conclusions from the theoretical analysis of the dynamic characteristics of soil-underground structure interaction and finally provides a theoretical basis for the engineering practice.

**Keywords:** underground structure, soil nonlinearity, soil-underground structure interaction, seismic response.

## 1 Introduction

With the increase of underground structures and earthquake damages, especially inspired by the Great Hanshin earthquake [1], many scholars have devoted

M.Z. Hou et al. (Eds.): *Clean Energy Systems in the Subsurface*, SSGG, pp. 317–327.
DOI: 10.1007/978-3-642-37849-2_26 

themselves to underground structure seismic performance in recent decades. At the same time, soil and underground structure dynamic interaction research which is an interlinked subject involving many fields such as soil dynamics, structure dynamics, nonlinear vibration theory, earthquake engineering, geotechnical and structure seismic engineering, computational mechanics, computer technology and so on, has been strengthening. To be able to improve many idealized hypothesizes among current research theory and make the existing finite element analysis methods consistent with actual working mechanisms considering foundation soil, contact condition, etc., soil nonlinearity is an inevitable problem [2-4]. In addition, to eliminate the effect of the boundary condition on soil-underground structure dynamic interaction, some papers adopt an artificial constraint boundary at a sufficient distance from the foundation center [5]. It is effective in static analysis and dynamic analysis including fountain. However, it cannot get ideal results for open system dynamic analysis, especially for seismic scattering problems. In order to make the numerical calculation model closer to the actual situation, based on a finite element method, the underground structure and soil are modeled by ANSYS software, and soil nonlinearity is realized through the Davidenkov constitutive model, using the restarting method and ANSYS Parameter Design Language (APDL). In addition, the local artificial boundary is established to simulate a radiation damping effect for the model and Combin14 element is used as a viscoelastic boundary. In summary, soil-underground structure dynamic analysis is to be studied by considering soil nonlinearity and local artificial boundaries.

## 2 Three-Dimensional Numerical Model

### *2.1 Subway Station Model Introduction*

The model is a Two-layer, double-column and three-span typical subway station, as shown in Figure 1. The width, height and burial depth of subway stations are 21.2m, 12.5m and 2m, respectively, and the longitudinal space of the columns is 9m. Other sizes are shown in Figure 1. The length, width, and height of soil are 200m, 18m and 60m, respectively, along the transverse, longitudinal and vertical directions of the subway station and the elements of soil, column, plate and wall are respectively solid 45, beam 188, shell 188. In addition, as the structure is symmetrical in the transverse direction, only half of the model is calculated, as shown in Figure 2.

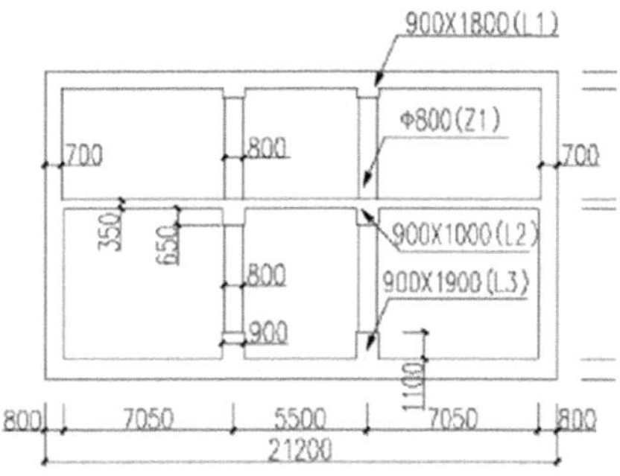


Fig. 1 Cross section size of subway station (mm)

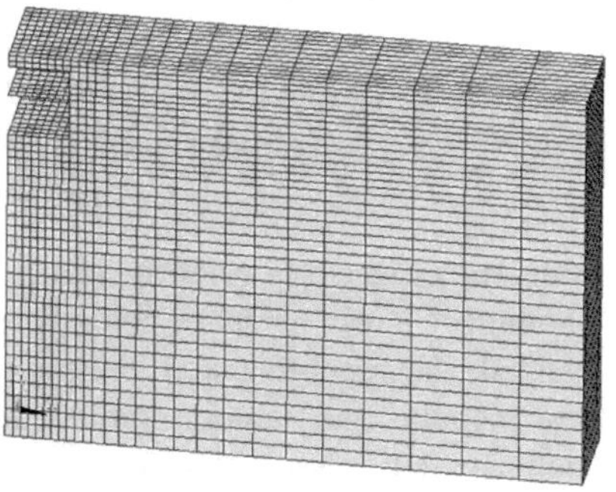

Fig. 2 Model grid

## 2.2 Soil Constitutive Model and Material Parameters

### 2.2.1 Davidenkov Constitutive Model

Seed and Martin [6] thought that the Davidenkov model whose soil skeleton curve and hysteretic curve are used in the paper, could better describe the relationship between the shear stress and shear strain of various soils. Skeleton curve Gd/Gmax～γd [7] is shown in equations (1) and (2), and according to test results, hysteretic curve D/Dmax～γd can be described by the empirical formula of equation (3):

$$\frac{G_d}{G_{max}} = 1 - H(\gamma_d) \tag{1}$$

$$H(\gamma_d) = \left[ \frac{\left(|\gamma_d| / \gamma_r\right)^{2B}}{1 + \left(|\gamma_d| / \gamma_r\right)^{2B}} \right]^A \tag{2}$$

$$\frac{D}{D_{max}} = \left(1 - \frac{G_d}{G_{max}}\right)^{\beta} \tag{3}$$

In these formulas, $G_{max}$ and $D_{max}$ respectively refer to the maximum dynamic shear modulus and maximum dynamic damping ratio, which can be attained by experiment or by empirical calculation. $\gamma_r$ refers to the reference shear strain.

### 2.2.2 Material Parameters in ANSYS

The above parameter values of the Davidenkov constitutive model are as follows:

$$G_{max} = \rho v_s^2 \tag{4}$$

Vs and ρ respectively refer to shear wave velocity and soil density. According to Xiaoming Yuan's resonant column test [8], from where the average curve was obtained, the recommended value and envelope of conventional soils' dynamic shear modulus and damping ratio changed with the dynamic shear strains and the recommended value of the damping ratio was obtained, Values of the skeleton curve $G_d$/Gmax vs $\gamma_d$ and the curve D vs $\gamma_d$ are shown in Table 1. In addition, other parameters are shown in Table 2.

**Table 1** Gd/Gmax vs $\gamma_d$ and D vs $\gamma_d$

| Soil layer (from surface to bottom) | Parameter | Shear Strain $\gamma_d$ ($10^{-4}$) | | | | | | | |
|---|---|---|---|---|---|---|---|---|---|
| | | 0.05 | 0.1 | 0.5 | 1 | 5 | 10 | 50 | 100 |
| Clay | $G_d/G_{max}$ | 0.9954 | 0.9897 | 0.9465 | 0.8975 | 0.6347 | 0.4647 | 0.1478 | 0.0798 |
| | D | 0.0342 | 0.0408 | 0.0608 | 0.0716 | 0.1284 | 0.1703 | 0.2402 | 0.2543 |
| Silty Clay | $G_d/G_{max}$ | 0.9936 | 0.9858 | 0.9274 | 0.8634 | 0.5563 | 0.3851 | 0.1112 | 0.0589 |
| | D | 0.0223 | 0.0276 | 0.0447 | 0.0549 | 0.1131 | 0.1383 | 0.1735 | 0.1798 |
| Silt(moderately dense) | $G_d/G_{max}$ | 0.9919 | 0.9839 | 0.9242 | 0.8591 | 0.5495 | 0.3788 | 0.1087 | 0.0575 |
| | D | 0.0129 | 0.0174 | 0.0342 | 0.0445 | 0.0934 | 0.122 | 0.162 | 0.1689 |
| Silt(dense) | $G_d/G_{max}$ | 0.9921 | 0.9856 | 0.9582 | 0.9101 | 0.7082 | 0.5303 | 0.1549 | 0.0953 |
| | D | 0.0099 | 0.0134 | 0.0292 | 0.0385 | 0.0863 | 0.1142 | 0.151 | 0.1582 |
| Sand(moderately dense) | $G_d/G_{max}$ | 0.9928 | 0.984 | 0.9187 | 0.8485 | 0.5263 | 0.3568 | 0.0998 | 0.0525 |
| | D | 0.0046 | 0.007 | 0.0235 | 0.038 | 0.0922 | 0.1171 | 0.1523 | 0.1586 |
| Sand(dense) | $G_d/G_{max}$ | 0.9963 | 0.9902 | 0.9505 | 0.9075 | 0.6782 | 0.5089 | 0.1519 | 0.0941 |
| | D | 0.0042 | 0.0063 | 0.0185 | 0.0301 | 0.0825 | 0.1042 | 0.1383 | 0.1437 |

**Table 2** Soil Parameters

| Soil layer (from surface to bottom) | Thickness(m) | Density(kg/m$^3$) | Poison's ratio | Shear Wave Velocity(m/s) |
|---|---|---|---|---|
| Overlying Soil | 2 | 1867 | 0.22 | 173 |
| Clay | 5 | 1867 | 0.22 | 173 |
| Silty Clay | 5 | 1847 | 0.32 | 200 |
| Silt(moderately dense) | 10 | 1878 | 0.34 | 212 |
| Silt(dense) | 15 | 1920 | 0.3 | 244 |
| Sand(moderately dense) | 15 | 2000 | 0.36 | 273 |
| Sand(dense) | 20 | 2050 | 0.26 | 333 |

Concrete structures adopt an elastic constitutive model. The dynamic characteristic parameters are obtained by increasing a certain proportion on the basis of the concrete static elastic modulus in the conventional method [9]. Research shows that the dynamic elastic modulus is higher than static elastic modulus by between 30%~50%. Generally the proportion is 40%, so concrete dynamic elastic modulus E=4.2GPa (concrete grade C30) was used in the model with the poison's ratio and density being 0.2 and 2500 kg/m$^3$ respectively..

### 2.2.3 Soil Nonlinearity Realized in ANSYS

The basic idea of Davidenkov constitutive model is that in order to solve approximately soil nonlinearity dynamic responses, dynamic shear modulus $G_d$, damping ratio D and maximum dynamic shear strain $\gamma_d$ are coordinated through the iteration method according to the relationships between $G_d$ and $\gamma_d$ and between D and $\gamma_d$ . The specific process in ANSYS is as follows: first, we assume that $G_d1$ and D1 are constant. The numerical model is calculated using ANSYS and average strains for each layer of soil are obtained. Then another group of $G_d2$ and D2 are calculated according to empirical equations and the average strains of each layer of soil are obtained. The numerical model with parameters $G_d2$ and D2 is also calculated using ANSYS, by analogy. Then another group of dynamic shear moduli and damping ratios are again calculated. Finally, by comparing the two groups of $G_d$ and D, the step finishes and the next step starts when the two groups are within allowed limits. This paper uses the restarting method and ANSYS Parameter Design Language (APDL) to simulate the above process and to realize soil nonlinearity simulation.

### 2.2.4 Boundary Condition and Seismic Wave

The artificial boundary includes non-local and local boundaries [10], and the boundary used in the finite element method is a local artificial boundary, which makes the incident wave pierce through the boundary. Owing to its practicality, researchers have paid more attention on to the local artificial boundary. Currently, there are many commonly used boundaries, for example, viscous boundary, consistent boundary, superposition boundary, paraxial boundary, transmission boundary, viscoelastic boundary and so on. This article used the viscoelastic boundary and combin14 element to simulate the local artificial boundary in ANSYS. The specific implementation method is as shown in Figure 3. Coordinates x and y are along the tangential of the artificial boundary, and coordinate z is in the normal direction. The physical component parameters of the artificial boundary on nodes are as follows:

$$K_1 = K_2 = \frac{2G}{R}\sum_{i=1}^{l} A_i \tag{5}$$

$$C_1 = C_2 = \rho c_S \sum_{i=1}^{l} A_i \tag{6}$$

$$K_3 = \frac{4G}{R}\sum_{i=1}^{l} A_i \tag{7}$$

$$C_3 = \rho c_P \sum_{i=1}^{l} A_i \tag{8}$$

K and C respectively refer to the spring and damping parameters, and ∑Ai represents the area of node on the artificial boundary. In addition, R is the distance from the scattering source to the boundary node and ρ is the mass density. $C_p$ and Cs respectively refer to the elastic longitudinal wave velocity and shear wave velocity, and the propagation speeds in the medium are real constants which depend on elastic modulus E, Poisson rationμand density ρ, as shown in equations (9) and (10).

$$c_p = \sqrt{\frac{1-\mu}{\rho\,(1+\mu)\,(1-2\mu)}E} \tag{9}$$

$$c_p = \sqrt{\frac{E}{2\rho\,(1+\mu)}} \tag{10}$$

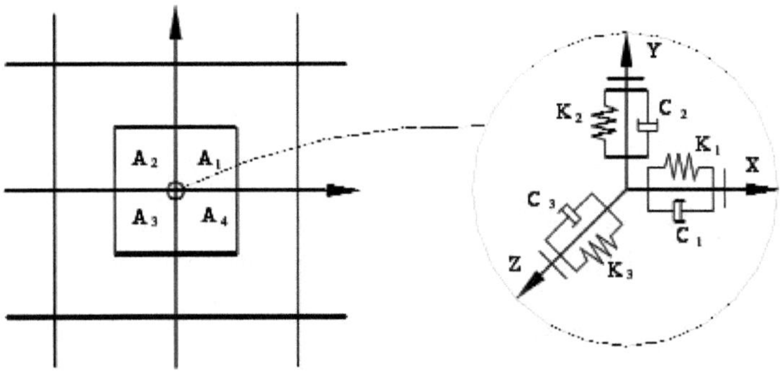


**Fig. 3** Three dimensional viscoelastic artificial boundary schematic diagram

The top surface in the model is free and the others are bounded in the normal direction. To analyze the soil and underground structure dynamic interaction in seismic intensity 8, the seismic wave put into ANSYS is the EL-Centro wave. Because the volume and weight of soil are large, the applied way of gravity has a great influence on numerical calculation results. In this article, the gravity of the element which each node represents near the node is calculated and the concentrated forces which are equal to the above calculated gravities are applied on corresponding nodes.

## 3 Numerical Analysis Results

Considering soil nonlinearity and the local artificial boundary, the finite element model including soil and the subway station is established and calculated by ANSYS under the EL-Centro wave. In order to analyze the effect of the subway station on site soil seismic responses, freefield model is also established with the same material properties, boundary conditions, loading condition as the one including soil and the subway station.

### *3.1 Influence of Subway Station on Site Soil Seismic Response*

In order to obtain the effect of the subway station on the site soil surface seismic responses, the maximum acceleration response curve of nodes on the surface under EL-Centro wave is drawn as shown in Figure 4. X represents the distance from node to the subway station center in the transverse direction and all the nodes are 0.9 m away from the subway station center in the longitudinal direction. As shown in figure 4, where the distance x is 20 m to 45 m from the subway station center, the maximum acceleration response in the model including the subway station obviously increases relative to that in the free field. It shows that the soil surface earthquake response is significantly enlarged, because of the underground structure, within a distance of about twice the span of the underground structure in the transverse direction.

In addition, the maximum acceleration response curve of nodes below the centre of the subway station floor in the vertical direction is also drawn, as shown in Figure 5. Z represents the distance from the node to the centre of the subway station floor in a vertical direction. As shown in Figure 5, where the distance z is 0 m to 24 m from subway station floor center, the maximum acceleration response in the model including the subway station obviously decreases relative to that in the free field. That is to say, the influence range is about twice the height of the underground structure in a vertical direction.

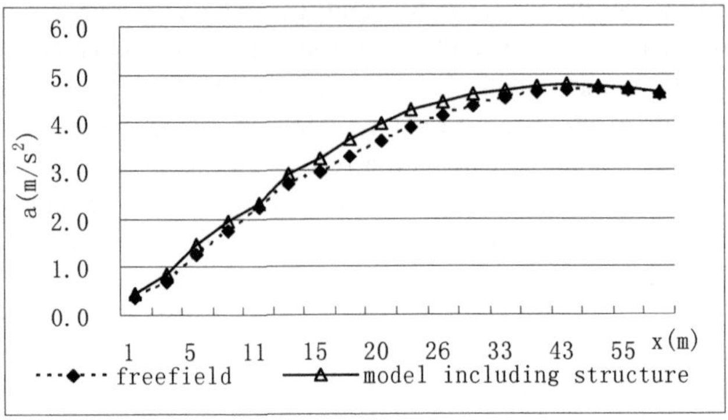


**Fig. 4** Maximum acceleration response curves of nodes on surface in transverse direction

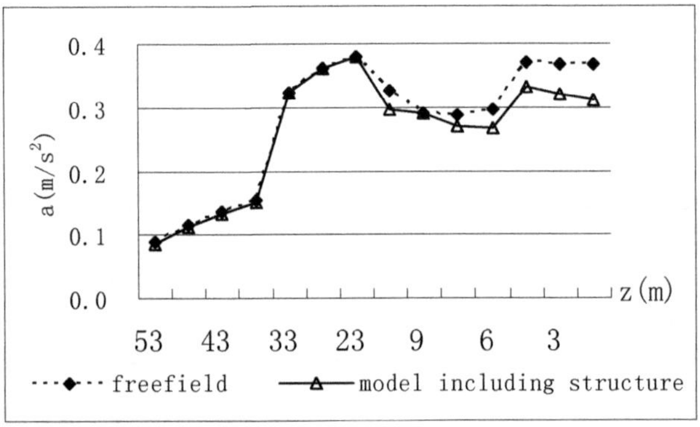


**Fig. 5** Maximum acceleration response curves of nodes below subway station floor in a vertical direction

# 4 Underground Structure Seismic Responses

## *4.1 Structure Seismic Response*

The relative displacement time history curve is shown in Figure 6. The ordinate is the relative horizontal displacement between the top and bottom of the column, and the abscissa is time. It is shown that the upper column seismic response is greater than that of the bottom column. It can be seen from Figure 7 that the top of the frame column moves to the side without the underground structure relative to the bottom of the frame column. However, it sways from side to side in the same position in the free field. The reason may be that the underground structure stiffness is greater than the soil stiffness. Hence, the stiffness asymmetry makes the structure seismic response asymmetric.

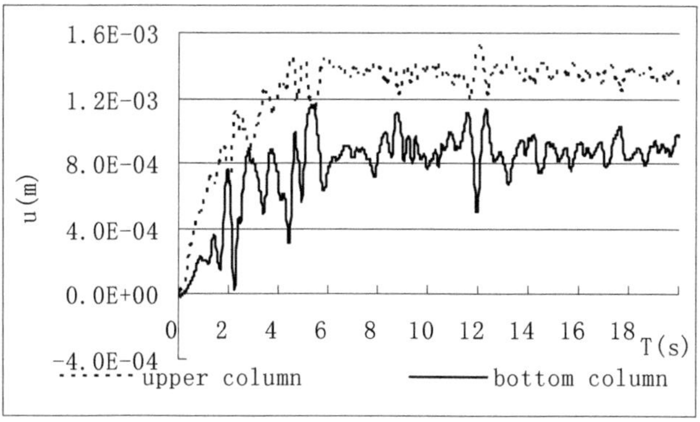


**Fig. 6** Relative horizontal displacement time history curves of the column

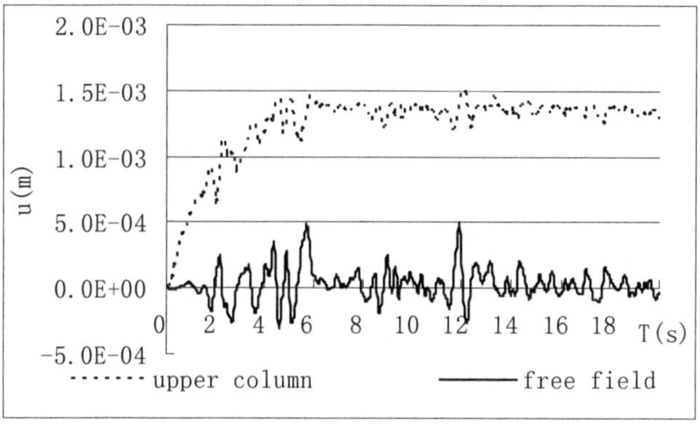


**Fig. 7** Relative horizontal displacement time history curves of the column and the same position in free ground

## *4.2 Effect of Overlying Soil Thickness on Structure Seismic Response*

Figure 8 shows the influence of different overlying soil thicknesses on the horizontal displacement of the top of the column relative to the bottom. The continuous line and dotted line respectively represent the relative horizontal displacement time history curve of the column in the model with 2m thickness of overlying soil and with 4m thickness of overlying soil. It shows that when the thickness is smaller, the response is more severe.

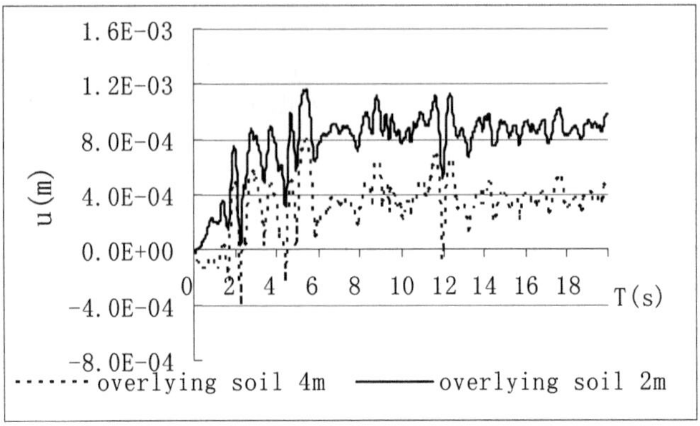


**Fig. 8** Relative displacement time history curves of the column with different overlying soil thicknesses

# 5 Conclusion

Based on the Davidenkov skeleton curve which is a nonlinear viscoelastic constitutive model, the finite element model including soil and the subway station and the other with a free field are established by considering soil nonlinearity and the local artificial boundary using ANSYS software. It studies dynamic interaction between site soil and the underground structure, and there are some valuable rules. Soil surface seismic response is significantly enlarged because of the underground structure within the distance of about twice the span of underground structure in a transverse direction, with the influence range being about twice the height of the underground structure in the vertical direction. In addition, the top of the column moves to the side without the underground structure relative to the bottom. At the same time, the overlying soil thickness has an influence on the seismic responses of the structure.

# References

1. Huo, H., et al.: Load transfer mechanisms between underground structure and surrounding ground: evaluation of the failure of the Daikai Station. Journal of Geotechnical and Geoenvironmental Engineering 131(12), 1522–1533 (2005)
2. Kaynia, A.M., Kausel, E.: Dynamic stiffness and seismic response of pile-groups research report R82-13. Institute of Technology Cambridge, Massachusetts (1982)
3. Nogami, T., Jones, H.W., Mosher, R.L.: Seismic response analysis of pile-supported structure: assessment of commonly used approximations. In: Proceedings of 2nd International Conference: Recent Advancements of Geotechnical Earthquake Engineering and Soil Dynamics, pp. 931–940 (1991)

4. Koki, K., Miura, F.: Nonlinear seismic response analysis of SSI system. Earthquake Engineering and Structure Dynamics 11, 77–89 (1983)
5. Zhao, B., Jiang, Y., Chen, J.: The Response Analysis of Underground Railway Station Structure in Soft Soil under Strong Earthquake. China Railway Science 30(3), 45–50 (2009)
6. Martin, P.P.: Nonlinear method for dynamic analysis of ground response. University of California, Berkeley (1975)
7. He, Y., Xia, D., Yan, Y., Liu, J.: Dynamic behaviors of SSDI system based on nonlinearity of soil in site. Chinese Journal of Geotechnical Engineering 31(4), 521–527 (2009)
8. Yuan, X., Sun, R., Sun, J., Meng, S., Shi, Z.: Laboratory experimental study on dynamic shear modulus ratio and damping ratio of soils. Earthquake Engineering and Engineering Vibration 20(4), 133–139 (2000)
9. Wang, G., Yang, L., Ma, X., Zheng, Y.: Analysis of Three-dimensional Seismic Response of Subway Station Structure and Nonlinear Characteristic of Soil. Chinese Journal of Underground Space and Engineering 4(2), 234–237 (2008)
10. Guan, H.: The Seismic Responds Analysis of Tianjin Station Transport Junction considering the Soil-Structure Interaction. Dissertation Tianjin University (2009)

# Spatial Distribution of Acoustic Emissions of Rock Salt under Different Stress Conditions

Mingming Xu[1,2], Jianfeng Liu[1,2,3], Dewen Zheng[4], Liang Chen[5], Zhide Wu[4], Lina Ran[4], Michael Zhengmeng Hou[6], Yang Gou[6], and Huining Xu[1,2]

[1] State Key Laboratory of Hydraulics and Mountain River Engineering, Sichuan University, Chengdu 610065, China
[2] State Key Laboratory of Geohazard Prevention and Geoenvironment Protection, Chengdu University of Technology, Chengdu 610059, China
[3] Key Lab. of Energy Engineering, Safety and Mechanics on Disasters (Sichuan Univ.), Ministry of Education, Chengdu, 610065, China
[4] Langfang Branch, Research Institute of Petroleum Exploration and Development, Langfang 065007, China
[5] Beijing Research Institute of Uranium Geology, Beijing 100029, China
[6] Institute of Petroleum Engineering, Clausthal University of Technology, Clausthal-Zellerfeld 38678, Germany

**Abstract.** In this paper, the study on the spatial distribution of acoustic emission (AE) events in rock salt under different stress conditions was reported. The study showed that under indirect tensile stress condition, the AE events were mainly induced by the stress concentration near the loading surface and by intergranular deformation before rock failure, while during failure and the post-failure process the AE events were essentially related to the damage development, especially near the tensile failure surface. Under uniaxial compressive stress condition, the closure of initial micro-cracks including those artificially produced during coring and sample preparation and intergranular deformation were the main mechanisms responsible for the occurrence of AE events at the initial_deformation stage, and then transferred to damage development until the post-failure stage was met. In triaxial compression tests, the influence of the closure of initial micro-cracks and intergranular deformation was still the main cause for AE events at the confining pressure loading stage and the initial axial loading stage. However, the influence of plastic deformation becomes more pronounced at the plastic yield stage and the post-peak stage. Intensive occurrence and accumulation of AE events before rock failure was noticed under different stress conditions. In indirect tensile tests, the recorded AE events were mainly accumulated along the failure surface. However, this reasonable phenomenon was not evident in uni- and triaxial compression tests, where the spatial distribution is generally homogeneous.

**Keywords:** rock salt, indirect tension, uniaxial compression, triaxial compression, AE spatial distribution.

M.Z. Hou et al. (Eds.): *Clean Energy Systems in the Subsurface*, SSGG, pp. 329–339.
DOI: 10.1007/978-3-642-37849-2_27 

# 1 Introduction

Due to its low permeability, self-sealing capacity, and outstanding practicability of mining activities, rock salt is internationally considered as an ideal host rock for gas/oil underground storage and radioactive waste repositories. Extensive studies have been carried out on the mechanical properties of rock salt and various results were achieved [1-7].

As an important non-destructive monitoring method, the acoustic emission (AE) monitoring technique is widely used to investigate the damage evolution and rock failure process. For instance, Obert [8] and Hodgson [9] used the AE monitoring method to identify the failure position during the excavation of underground structures. Li [11] confirmed the validity of the Kaiser Effect [10] for rock. Filimonov[12-13] revealed that the acoustic emission is stronger in rock salt than in clay rock and shale, and obtained the effect of the loading rate on AE properties. Alkan[14] studied the dilation properties of rock salt under triaxial compressive conditions. Based on experimental investigation, Liu et al. [15-17] established the relationship between fractal dimensions of AE events spatial distribution and the energy release in indirect tensile and uniaxial compression tests. Based on these investigations, the relationship between the recorded AE events and the deformation as well as the failure states of rock salt have been systematically studied. However, little attention has been given to the formation mechanism of the AE distribution during the whole loading process under different stress conditions.

In this paper, through a series of experimental investigations under different stress conditions the full process of rock salt failure in direct tensile tests, uniaxial compression tests and triaxial compression tests, respectively, is analyzed by the characterization of AE distribution.

# 2 Test Equipments and Methods

## *2.1 Test Equipments and Rock Samples*

The test equipment consists of a MTS815 Flex Test GT rock test system and a 3D AE monitoring system of PCI-II (Fig. 1). The maximum axial loading is 4600 kN, and the maximum confining pressure and pore pressure are both 140 MPa. The PCI-II AE system includes 8 channels for the spatial AE-detection. The central frequency of AE sensors is 300 kHz. 8 AE sensors were installed symmetrically in a radial direction along the cylinder surface and the distance from the sensor to the nearest end surface is about 10 mm for uniaxial compression tests. For indirect tensile tests, the sensors were placed on both sides of the possible tensile failure surface and the distance between the sensor and the circle edge is about 2 mm. For triaxial compression tests, the sensors were placed on the outside of the triaxial cell. There were 4 sensers for each end of the specimen surface, respectively.

The rock samples were taken at a depth of 500 m in Anning, Yunnan Province of China. Following the "Standards of the engineering rock test methods" (GB/T50266–99) [18] and "Standards of rock test for water conservancy and hydroelectric power" (SL264-2001) [19], the samples were prepared using the dry lathing method. The sample size of indirect tension and compression tests were D90×H50mm and D90×H200mm, respectively. The samples had an average salt content of up to 79%.

**Fig. 1** MTS815 Flex Text GT rock mechanics test system

## *2.2 Test Methods*

For indirect tensile tests, namely the Brazil split tests, the loading process was controlled by axial displacement (LVDT). The loading rate was 0.05 mm/min before peak stress, and then transferred to 0.1 mm/min until rock failure. For uniaxial compression tests, the axial loading rate was 30 kN/min from the start of loading until the elastic deformation finished; then, the loading was transferred to LVDT control, with a loading rate of 0.1mm/min. For triaxial compression tests, the axial loading process was also controlled by axial displacement with a loading rate of 1.0mm/min, and the loading rate for confining pressure was 3.0 MPa/min. In all of the above tests, the testing temperature was room temperature (25°C).

# 3 Test Results and Analysis

## *3.1 AE Events Properties for Indirect Tensile Tests*

The recorded AE events were categorized according to axial stress increments. For instance, the inserted graph "20~40%" in Fig. 2 showed the recorded AE events and their spatial distribution of the stress increment between 20% to 40% of the peak stress. While the increment "100~60%" depicts the recorded AE events from 100% to 60% of peak stress at the post-failure stage.

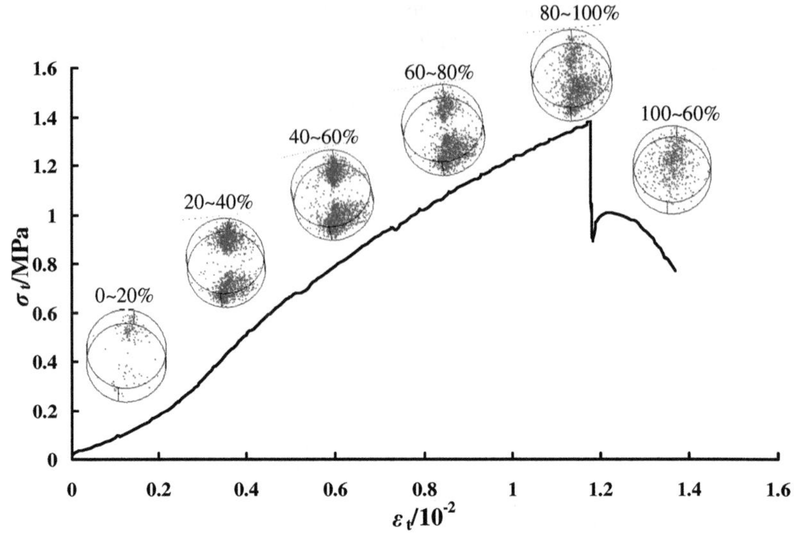


**Fig. 2** Spatial distribution of AE events of rock salt for indirect tensile test

In Fig. 2, due to the stress concentration around the loading surface, it was noticed that the AE events were recorded along the lines of loading contact points of rock sample at the beginning of the loading process, and finally expanded to the central part. The number and distribution of recorded AE events were similar at the stages "20~40%", "40~60%" and "60~80%". In these phases, the occurrences of AE events were mainly induced by damage development near the loading surface and relative deformation between grains. However, with the increase of tensile stresses, the influence of stress concentration at the contact lines became less evident and the damage development became more dominant for AE events, presented at the stage "80~100%". This is concluded from the result, where the recorded AE events were mainly located near the failure surface in the central part of the rock sample. The distribution of recorded AE events and the provoking process remained dominant at the post-failure stage (100-60%) of the peak stress.

Fig. 2 also presents the overall accumulation process of AE events along the failure surface, and the evolution process of the initiation, propagation and coalescence of micro-cracks during indirect tensile tests. According to the test results of 5 specimens, the maximal and minimal tensile strength of rock salt was 1.52 MPa and 1.30 MPa, respectively. The average tensile strength was 1.38 MPa.

Fig. 3 showed the corresponding final failure pattern and AE distribution of the test presented in Fig. 2. It indicated that the failure surface expanded along the loading line (dashed line) in Fig. 3(a), and inclined slightly to the right on the other side, as presented in Fig. 3(b). As a result, the distribution of recorded AE events (Fig. 3(c), seen from the sample's back side) is not symmetric, thus many AE events appeared on the right side. The distribution state of AE events in Fig. 3(c), is in accordance with the failure state in Fig 3(b).

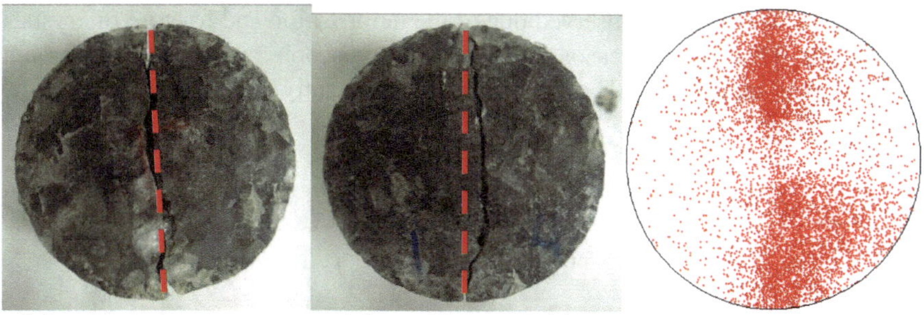

(a) sample front (b) sample rear (c) location of acoustic emissions

**Fig. 3** Failure state and AE spatial distribution for indirect tensile test

## *3.2 AE Events Properties for Uniaxial Compression Tests*

As shown in Fig. 4, during uniaxial compression tests, the recorded AE events were mainly accumulated around the loading surface at the beginning of loading process, in a cone-shape and inverted cone-shape pattern. The stress distribution in the rock sample was strongly influenced by the end restraint effect at this stage, and the AE events were essentially induced by the closure of original micro-cracks and relative deformation between grains. With the increase of axial stress, the AE events expanded to the central part, which indicated that internal damage had taken place. At the following stages, with the damage development, the accumulations of AE events were mainly located in the central and the lower part, and the influence of end restraint effect had already become quite low. At the post-failure stage ("100-80 %"), the recorded number of AE events was less than those at the peak stress stage.

Since rock salt is a typical crystallized sedimentary rock, not only the propagation and coalescence of micro-cracks, but also the friction and relative deformation between grains can generate acoustic emission. As a result, the accumulation of AE events along a certain failure surface is not as evident as observed in indirect tensile tests. Under uniaxial compressive conditions, numerous micro-cracks were generated during rock salt failure, in contrast to a single macroscopic failure surface in Brazil split tests. The average compressive strength was 25.17 MPa, which is generally approx. one order of magnitude higher than the tensile strength.

Fig. 5 shows the final failure pattern and the location of AE events after uniaxial compression. Due to the existence of argillaceous impurities in the upper part of the sample, the rock salt purity of the upper part is less than that of the lower part. As a result, in the lower part, the rock becomes more brittle and stronger than in the upper part [20], so the rock failure is mainly concentrated in the lower part. Although an obvious tensile crack (dashed lines) was generated in the lower part of the sample in a vertical direction after failure, numerous microcracks were equally observed due to the deformation and fracture of crystals in rock salt. That was the reason the distinct macroscopic failure surface was not obvious, even though the AE events were predominantly accumulated around the failure surface.

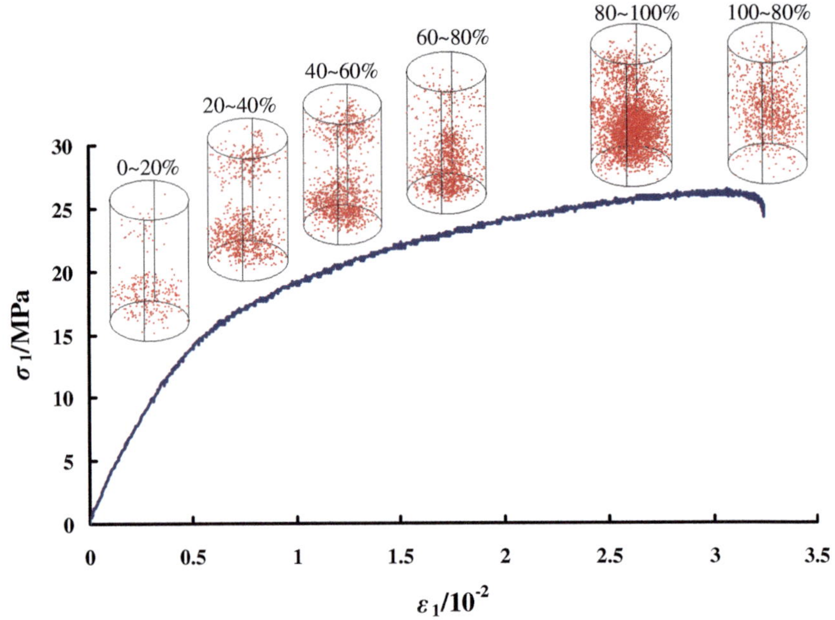


**Fig. 4** Spatial distribution of AE events of rock salt for uniaxial compression test

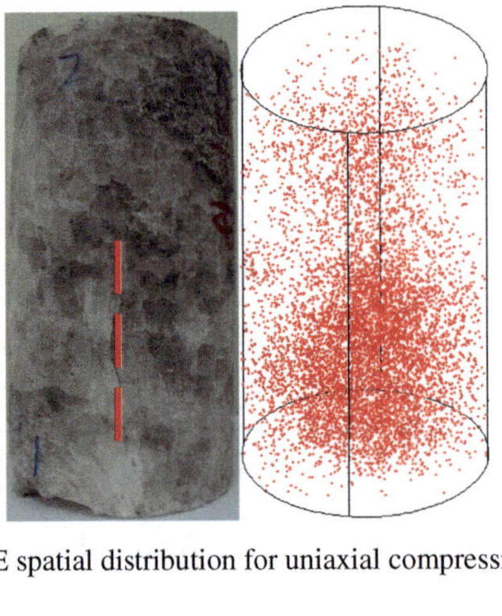

**Fig. 5** Failure state and AE spatial distribution for uniaxial compression test

### 3.3 AE Events Properties for Triaxial Compression Tests

In triaxial compression tests, the AE events were recorded even during the hydrostatic loading process, as shown in Fig. 6. The distribution of AE events was somewhat arbitrary and similar to those recorded in the "100-80%" post-peak stage in Fig. 4. This can be attributed to the fact that the AE events were mainly induced by the closure of initial microcracks and defects under compressive stress in this phase. Since new microcracks were hardly generated in the hydrostatic loading process, the number of recorded AE events decreased with the increase of confining pressure. The results confirmed that the hydrostatic pressure is helpful in promoting the self-sealing process of microcracks in rock salt. Therefore, the self-sealing process of rock salt can be enhanced by increasing the hydrostatic pressure in the laboratory.

Compared to the distribution of AE events in uniaxial compression tests, significant differences can be noticed in triaxial compression tests. At the beginning of a triaxial compression test (0~20% of peak stress), even though the AE events were also essentially induced by the closure of initial microcracks and relative deformation between grains under uniaxial compressive stress, the end restraint effect was less evident than that under uniaxial compressive condition. As shown in Fig.7, the stress of yield deformation was about 40% of the peak stress. Therefore, the AE events were mainly located in the central and the lower part of the sample within 20~40% of peak stress. In this stress range, even though the closure of initial microcracks and relative deformation between grains in rock salt were still the main mechanisms responsible for the occurrence of AE events, the internal damage also starts to be initiated and to propagate. As a result, more AE events were recorded in this phase. When the axial stress exceeds 40% of the peak stress, the AE events were mainly generated by the relative deformation between grains and yield deformation. In the range of "40-60%", the AE events were concentrated in the central and upper part of the sample. With the increase in deviatoric stress ("80-100 %"), as a result of the significant plastic deformation and progressive damage development, the maximum number of AE events were captured in this range. In the post-failure regime (100-80%), due to the increase in the plastic deformation and damage evolution, considerable AE events are also recorded.

A special feature of crystallized soft rocks is that both the strength and deformation capacity of rock salt can be significantly increased by increasing the confining pressure. As shown in Fig. 7, the maximum axial deformation was almost 50%, i.e. about 16 times the deformation obtained in uniaxial compression tests. As a result, the original margin of internal grains becomes blurred, and no distinct failure surface is observed (Fig. 8(b)). Moreover, due to the non-uniform distribution of impurities and micro cracks in rock salt, the deformations of rock samples in vertical and lateral directions were not homogenous. The boundary of the deformed sample became uneven (Fig. 8(b)), and non-uniform spatial distribution of AE events in rock salt after failure was recorded (Fig. 8(c)).

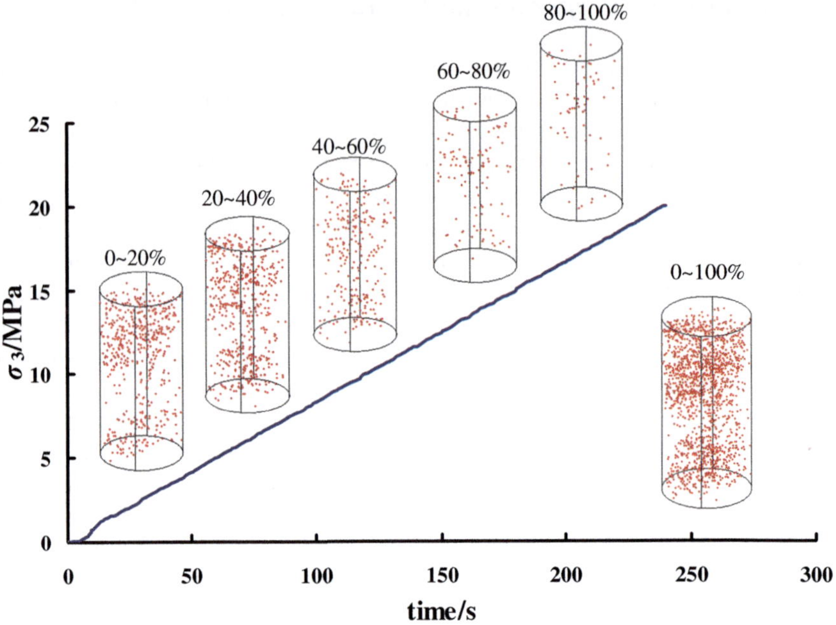


**Fig. 6** Spatial distribution of AE events of rock salt under hydrostatic pressure condition

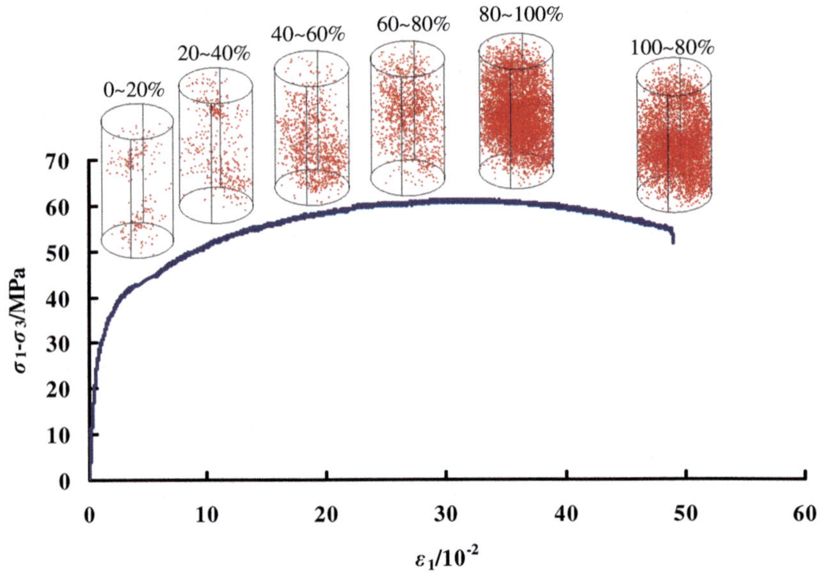


**Fig. 7** Spatial distribution of AE events of rock salt for triaxial compression test ($\sigma_3$=20MPa)

(a) before loading (b) after loading (c) location of acoustic emissions

**Fig. 8** Failure state and AE spatial distribution for triaxial compression test ($\sigma_3$=20MPa)

# 4 Conclusions

The experimental results confirm that the stress condition had a significant influence on the spatial distribution of AE events during the whole loading process of salt rock. In indirect tensile tests, AE events were mainly generated along the loading surface at the beginning, and then expanded to the center of rock sample when the stress was increased. At the post-failure stage, the AE events were essentially related to the damage development around the failure surface.

Under uniaxial stress conditions, the distribution of AE events was strongly influenced by the end restraint effect. Before the rock failure, the AE events were essentially induced by the relative deformation between grains of rock salt, while the influence of damage development by macroscopic fracturing became more pronounced at the post-failure stage.

In triaxial compression tests, both in the hydrostastic phase and at the beginning of axial loading process, the closure of microcracks and the relative deformation between grains of rock salt might be the main mechanisms responsible for the occurrence of AE events. It was similar to the situation in uniaxial compression tests. However, the end restraint effect was less evident in triaxial compression tests. When the stress level exceeded the yield stress, the AE events were mainly induced by the relative deformation between grains and the further damage development.

In indirect tensile tests, a good consistency was obtained between the distribution of AE events and the orientation of the final failure surface. In contrast, due to the intensive microcrack formation, which was induced under a compressive stress condition and large plastic deformation, the accumulation of AE events along the single failure surface was not noticed in either uniaxial or triaxial compression tests.

**Acknowledgement.** The authors are grateful for the financial support from the National Natural Science Foundation of China (Grant No. 51120145001, 51104101, 51134018), the National Basic Research Projects of China (Grant No. 2011CB201201, 2010CB226802, and 2010CB732005), State Key Laboratory of Geohazard Prevention and Geoenvironment Protection (Grant No. SKLGP2013K016). The authors wish to offer their gratitude and regards to the colleagues who contributed to this work.

## References

1. Hansen, F.D., Mellegard, K.D., Senseny, P.E.: Elasticity and strength of ten natural rock salt. In: Hardy Jr, H.R., Langer, M. (eds.) The first Conference on the Mechanical Behavior of Salt, pp. 71–83. Trans. Tech Publications (1984)
2. Hunsche, U.: Fracture experiments on cubic rock salt samples. In: Hardy Jr., H.R., Langer, M. (eds.) The First Conference on the Mechanical Behavior of Salt, pp. 169–179. Trans Tech Publications (1984)
3. Hunsche, U., Albrecht, H.: Results of true triaxial strength tests on rock salt. Engineering Fracture Mechanics 35(4/5), 867–877 (1990)
4. Hunsche, U., Plischke, I.: Stand der Untersuchungen zum Festigkeits-und Kriechverhalten von Steinsalz.6. Zwischenbericht zum Forschungsvorhaben SR 138 Gebirgsmechanik im Salz 1/81, Bundesanstalt für Geowissenschaften und Rohstoffe, Hannover. Archiv-Nr. 87952 (1981)
5. Farmer, I.W., Gilbert, M.J.: Dependent strength reduction of rock salt. In: Hardy Jr., H.R., Langer, M. (eds.) The First Conference on the Mechanical Behavior of Salt, pp. 4–18. Trans Tech Publications (1984)
6. Hunsche, U.: A failure criterion for natural polycrystalline rock salt. In: Fan, J., Murakami, S. (eds.) Advances in Constitutive Laws for Engineering Materials, pp. 1043–1046. International Academic Publ. (1989)
7. Skrotzki, W.: An estimate of the Brittle to ductile transition in salt. In: Hardy Jr., H.R., Langer, M. (eds.) The First Conference on the Mechanical Behavior of Salt, pp. 381–388. Trans Tech Publications (1984)
8. Obert, L., Duvall, W.I.: Use of subaudible noises for prediction of rockbursts II—report of investigation. S Bureau of Mines, Denver (1941)
9. Hodgson, E.A.: Velocity of elastic waves and structure of the crust in the vicinity of Ottawa, Canada. Bulletin of the Seismological Society of America 32(4), 249–255 (1942)
10. Kaiser, E.J.: A study of acoustic phenomena in tensile test. Dissertation Technical University, München (1950)
11. Li, C., Nordlund, E.: Experimental verification of the Kaiser effect in rocks. Rock Mechanics and Rock Engineering 26(4), 333–351 (1993)
12. Filimonov, Y.L., Lavrov, A.V., Shafarenko, Y.M., et al.: Memory effects in rock salt under triaxial stress state and their use for stress measurement in a rock mass. Rock Mechanics and Rock Engineering 34(4), 275–291 (2001)
13. Filimonov, Y.L., Lavrov, A.V., Shkuratnik, V., et al.: Acoustic emission in rock salt: effect of loading rate. Strain 38(4), 157–159 (2002)
14. Alkan, H., Cinary, Y., Pusch, G.: Rock salt dilatancy boundary from combined acoustic emission and triaxial compression tests. International Journal of Rock Mechanics and Mining Sciences 44(1), 108–119 (2007)

15. Liu, J., Xu, J., Yang, C., Hou, Z.: Mechanical characteristics of tensile failure of salt rock. Chinese Journal of Geotechnical Engineering 33(4), 580–586 (2011) (in Chinese)
16. Pei, J., Liu, J., Zhang, R., et al.: Fractal Study on Spatial Distribution of Acoustic Emission Events of Granite Specimens Under Uniaxial Compression. Journal of Sichuan University, Engineering Science Edition 42(6), 51–55 (2010) (in Chinese)
17. Ai, T., Zhang, R., Liu, J., et al.: Space- time evolution rules of acoustic emission locations under triaxial compression. Journal of China Coal Society 36(12), 2048–2057 (2011) (in Chinese)
18. The National Standards Compilation Group of People's Republic of China, GB/T50266-99 Standard for tests method of engineering rock masses. China Planning Press, Beijing (1999) (in Chinese)
19. The Professional Standard Compilation Group of People's Republic of China, SL264-2001 Specifications for rock tests in water conservancy and hydroelectric engineering. China Water Power Press, Beijing (2001) (in Chinese)
20. Zheng, Y., Zhang, H., Wang, Z., et al.: Experimental study on mechanical property of rock salt with impurity. Journal of China Coal Society 37(1), 17–20 (2012) (in Chinese)

# Damage and Healing Properties of Rock Salt: Long-Term Cyclic Loading Tests and Numerical Back Analysis

Uwe Düsterloh, Svetlana Lerche, and Karl-Heinz Lux

Institut für Bergbau, Clausthal University of Technology,
Chair for Waste Disposal Technologies and Geomechanics,
Erzstraße 20, 38678 Clausthal-Zellerfeld

**Abstract.** The Chair for waste disposal technologies and geomechanics has analysed the effects of mechanical cyclic loading of rock salt by laboratory testing. The tests were characterized by cyclic loading at stress conditions above the stress level where dilation will occur and stress conditions below the dilation stress level. The duration of the cyclic loading was 250 days to analyse in detail the damage induced creep behaviour and healing behaviour of rock salt by long term cyclic and constant loading at different stress levels.

At first, this paper presents in detail the loading history and the observed creep behaviour, volume damage and ultrasonic wave velocity during the long term cyclic tests. This was done to validate, with a back analysis, the ability of the constitutive model *Lux-Wolters* to image the measuring results observed from the tests, a short description of the constitutive model *Lux-Wolters* is given thereafter. A comparison between the measured data obtained from the cyclic loading tests and the computed data obtained from a numerical back analysis of the tests indicates clearly that the *Lux-Wolters* model is able to image the load bearing behaviour of rock salt considering cyclic and constant loading above as well as below the damage stress level. That is, transient and stationary damage free creep behaviour, damage induced creep behaviour as well as healing induced creep behaviour could be computed using one unique set of parameters for the total loading history.

**Keywords:** Rock Mechanics, Salt Properties, Computer Modelling, Cycling, Storage Caverns.

## 1 Introduction

Caused by increasing demands of the gas market, the operation pattern of gas storage caverns is more and more characterized by the need for frequent cycling of internal pressures and increasing pressure rates. In addition, considerations of over using salt caverns for compressed air energy storage (CAES) are characterized by

M.Z. Hou et al. (Eds.): *Clean Energy Systems in the Subsurface*, SSGG, pp. 341–362.
DOI: 10.1007/978-3-642-37849-2_28 

the requirement of a high frequent cycling of the cavern inside pressure combined with high pressure gradients with respect to time. Because a frequent cycling of the cavern inside pressure is equivalent to a cyclic loading of the rock salt mass surrounding the caverns, investigations into cyclic loading effects on the load bearing capacity of rock salt are needed to guarantee a safe operation for high frequency cycling storage caverns. As a consequence, long term tests with cyclic loading were conducted. The testing program in this paper is exploratory in its technical scope and restricted in the number of tests. As described in the final report RR2012-2 [1], the primary aim of the testing program was only intended to analyse whether mechanical cyclic loading under expected cavern operation conditions leads to repeatable and measurable effects in the creep and damage behaviour of rock salt. Therefore it was decided to simulate two load levels per test, one below and one above the damage strength of rock salt from Avery Island Louisiana. As shown in Figure 1, each load level was characterized by a cyclic alteration of the axial load whereas the confining pressure was forced to a constant value.

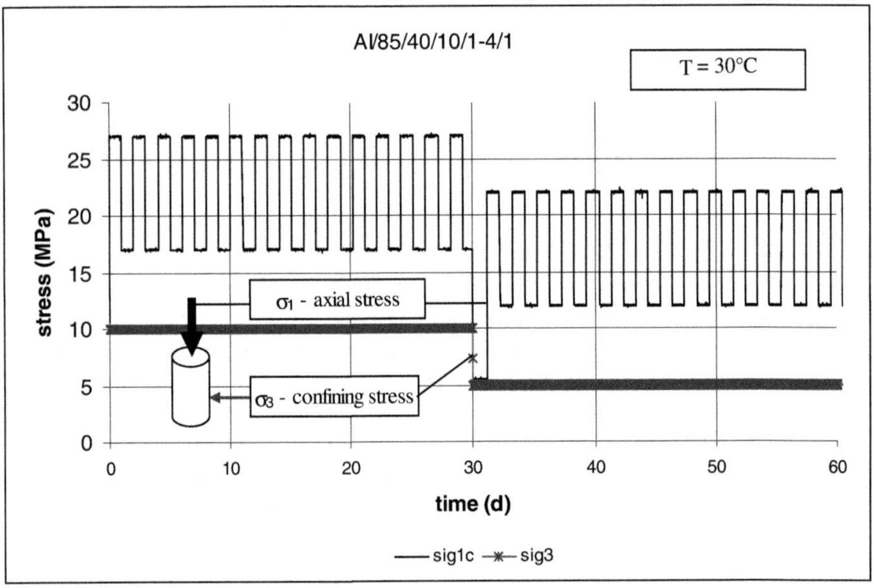


**Fig. 1** Load history of cyclic loading test AI/85/40/10/1-4/1 (axial and confining stress)

The first loading region between 0 days and 30 days is characterized by a constant confining pressure of $\sigma_3 = 10$ MPa and a cyclic alteration of the axial stress between max $\sigma_1 = 27$ MPa and min $\sigma_1 = 17$ MPa after one day in each case. The changes in stress were performed by a stress rate of $\dot{\sigma}_1 = 1$ MPa/minute. At the end of testing period 1, the confining stress was reduced to $\sigma_3 = 5.0$ MPa and the axial stress $\sigma_1$ was reduced to $\sigma_1 = 5.5$ MPa, in this case for a duration of one day. The second loading period between 30 days and 60 days is then characterized by a constant confining pressure of $\sigma_3 = 5$ MPa and a cyclic alteration of the axial stress between max $\sigma_1 = 22$ MPa and min $\sigma_1 = 12$ MPa in each case after one day. The changes in stress were performed analogous to loading period 1 by a stress rate of $\dot{\sigma}_1 = 1$ MPa/minute. As shown in Figure 2, the aforementioned stress conditions of both periods are determined by an alteration of the equivalent stress between max $\sigma_v = 17$ MPa and min $\sigma_v = 7$ MPa (phase 1: max $\sigma_v = 27$ MPa – 10 MPa = 17 MPa / phase 2: max $\sigma_v = 22$ MPa – 5 MPa = 17 MPa / testing phase 1: min $\sigma_v = 17$ MPa – 10 MPa = 7 MPa / testing phase 2: min $\sigma_v = 12$ MPa – 5 MPa = 7 MPa).

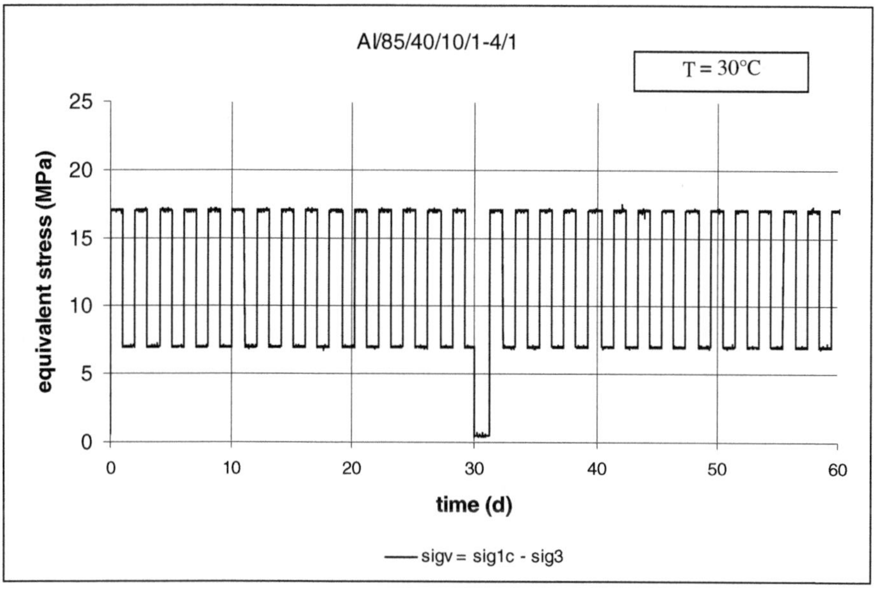


**Fig. 2** Load history of cyclic loading test AI/85/40/10/1-4/1 (equivalent stress)

Regardless of the uniform alteration of the equivalent stress during the whole time in reference to the damage strength, the sample load in period 1 and period 2 is quite different. This is caused by the change in confining pressure from $\sigma_3 =$ 10 MPa in phase 1 to $\sigma_3 = 5$ MPa in period 2 connected with the dependency of the damage strength on the confining pressure, the stress level in period 2 alternates between a level above and below the damage strength whereas in period 1, the stress level **is** generally below the damage strength. Figure 3 documents the load path of period 1 and period 2 with respect to linear damage strength estimated by RESPEC [1].

To analyse the cyclic loading effects on the damage and creep properties, the axial creep strain, the volumetric strain and the ultrasonic wave velocity were measured during the tests. A comparison of the measured strain data between the two cyclic loading tests performed by TUC is given in Figure 4. It can clearly be observed from Figure 4 that no significant change in the strain response occurs by changing from period 1 to period 2. Considering the well-known experience that the damage free creep behaviour of rock salt depends on the equivalent stress and the temperature which are forced to be constant in both testing periods, the observed strain response denotes that no significant damage induced creep rate has taken place during period 2.

As a proof of the aforesaid statement, Figure 5 shows the evaluation of volumetric strain measured online during the tests. Thereafter in period 1, a low volumetric compaction of some 1‰ of the sample volume occurs whereas in period 2 the volumetric strain increases continuously up to a marginal dilatancy of some 0.4‰. That is, the observed volume increase during period 2 is given by $\varepsilon_{vol} = 1.0‰ - 1.4‰$. As a consequence of the measured volumetric strain it can be stated, that cyclic loading below damage strength (→ period 1) results in a strain hardening whereas cyclic loading above damage strength (→ period 2) is characterized by continuously increasing damage.

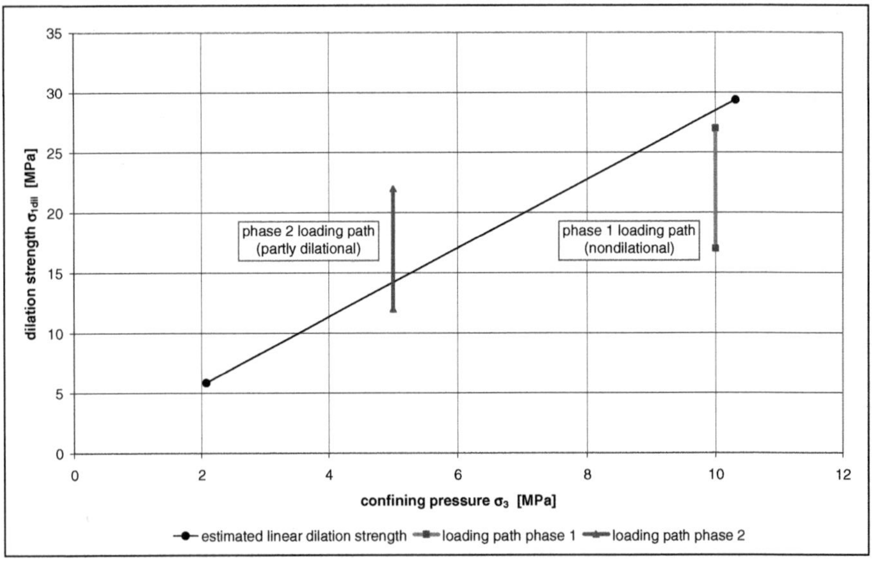


**Fig. 3** Load paths for cyclic loading tests with respect to damage strength

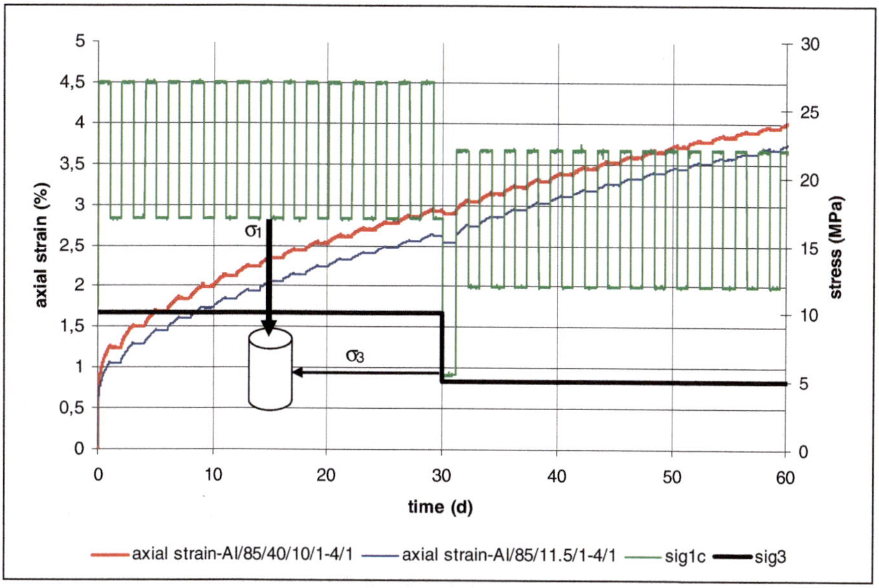


**Fig. 4** Strain response for cyclic creep tests on Avery Island specimens

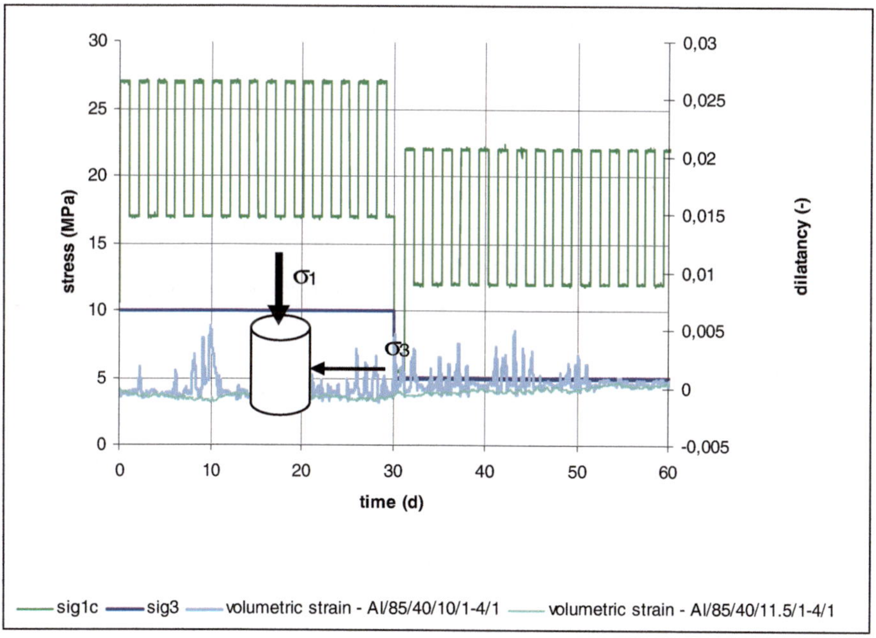


**Fig. 5** Evaluation of volumetric strain during cyclic creep tests on Avery Island specimens

An analogous conclusion can be stated by analysing the evaluation of ultrasonic wave velocity. The relationship between the ultrasonic wave velocity at the beginning of period 1 and the actual measured ultrasonic wave velocity (→ normalized ultrasonic wave velocity) denotes strain hardening by increasing ultrasonic wave velocity during period 1 (→ cyclic loading below damage strength) and low damage by decreasing ultrasonic wave velocity during period 2 (→ cyclic loading above damage strength). As shown in Figure 6 the observed change in ultrasonic wave velocity is visible but not significant.

Based on the results described above, one conclusion of the report [1] was that cyclic loading is not significantly more hazardous than static loading because cyclic loading as well as static loading below the dilation strength does not result in any damage. Additionally in comparison to static loading above dilation strength cyclic loading above dilation strength does not result in a significant increase in damage. Because of the limited testing, a second conclusion of the report [1] was issued stating, that further investigations are needed to analyse in detail:

- the possible impacts of thermal cycling to assess the significance of thermally induced stresses,
- the role of changes in the stress orientation because a common stress state in the cavern walls is more an extension than a compression and
- whether a cyclic and/or constant loading below the damage strength is able to heal respectively to recover from preliminary damage.

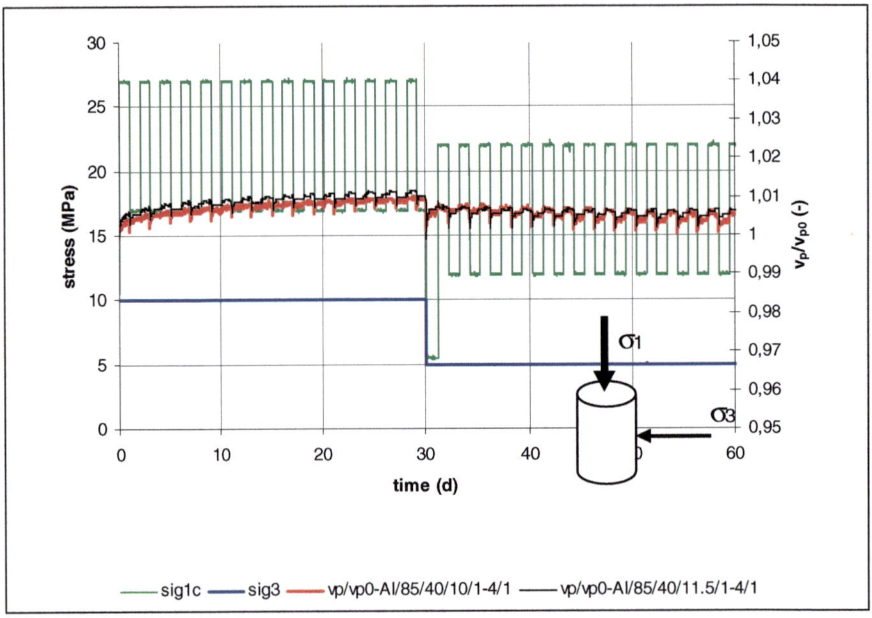


**Fig. 6** Evaluation of normalized ultrasonic wave velocity during cyclic creep tests on Avery Island specimens

To answer some of the questions regarding the damage and healing behaviour of rock salt by cyclic loading, TUC has continued the tests in each case up to a duration of 250 days. The load history and the observed results will be described in the following Section 2. Section 3 comprises a short explanation of the constitutive model, *Lux-Wolters*, which was used for the numerical back analysis of the tests documented in Section 4.

## 2 250 Days Cyclic Loading Tests to Analyze Damage and Healing Properties of Rock Salt from Avery Island Louisiana

A graphical view of the load history applied to the Avery Island samples AI/85/40/10/1-4/1 and AI/85/40/11.5/1-4/1 is shown in Figure 7 by plotting the axial and confining stress versus testing time and in Figure 8 by plotting the equivalent stress versus testing time. Thereafter, the testing time is characterized by 8 testing phases in total. At the beginning of each cyclic loading test phase, a one-day long isotropic stress level was initiated to bring the volumetric measuring system to a steady position. To analyse whether the creep behaviour observed during the damage free cycling in period 1 is repeatable after inducing limited damage in period 2, the cyclic load history of period 3 was forced similar to period 1. The cyclic periods 4 and 5 were simulated to analyse the increase in damage by increasing the deviatoric stress. The periods 4 and 5 in general are similar to period 2, but the confining pressure was reduced to $\sigma_3 =$ 4 MPa and $\sigma_3 = 3$ MPa respectively. During period 6, the deviatoric stress level is forced to be similar to period 5, but instead of a cyclic loading, a constant loading was simulated to analyse damage effects during cyclic loading above damage strength in comparison to damage effects during constant loading above damage strength. The aim of testing period 7 and 8 in general was to investigate whether damage recovery (a healing) can be observed if cyclic and constant loading are reduced to a stress level below the damage strength after significant damage has taken place. A detailed compilation of the above described load history is given in Table 1.

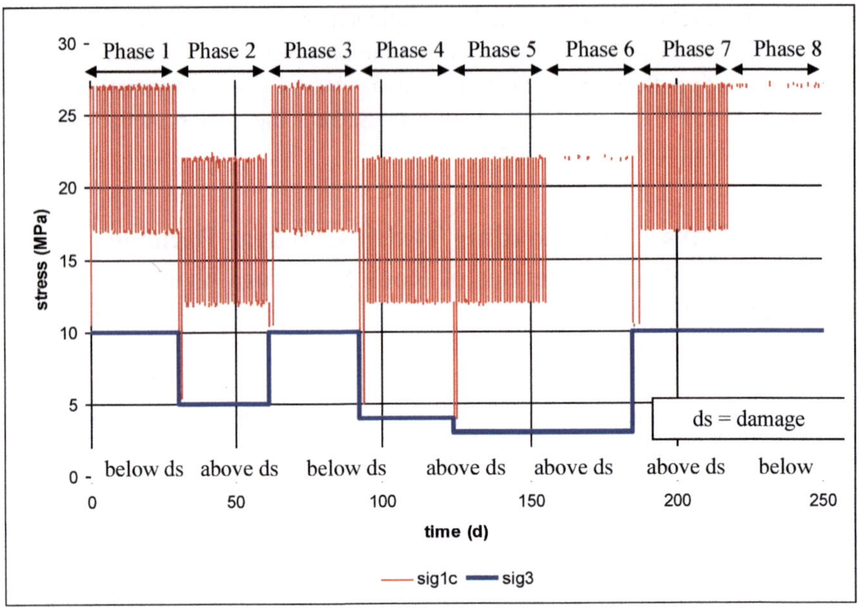


**Fig. 7** Load history of cyclic loading test AI/85/40/10/1-4/1 and AI/85/40/11.5/1-4/1 (axial and confining stress)

**Table 1** Compilation of load history used for cyclic loading tests on Avery Island samples AI/85/40/10/1-4/1 and AI/85/40/11.5/1-4/1

| testing period | quantity of cycles | duration of cycle | max$\sigma_1$ | min$\sigma_1$ | $\sigma_3$ | max$\sigma_v$ | min$\sigma_v$ | stress level regarding dilation strength | duration of testing period | cummulated testing time |
|---|---|---|---|---|---|---|---|---|---|---|
| - | - | [d] | [MPa] | [MPa] | [MPa] | [MPa] | [MPa] | | [d] | [d] |
| 1 | 15 | 2 | 27,0 | 17,0 | 10,0 | 17,0 | 7,0 | below | 30 | 30 |
| | | 1 | 5,5 | 5,5 | 5,0 | 0,5 | 0,5 | below | 1 | 31 |
| 2 | 15 | 2 | 22,0 | 12,0 | 5,0 | 17,0 | 7,0 | above | 30 | 61 |
| | | 1 | 10,5 | 10,5 | 10,0 | 0,5 | 0,5 | below | 1 | 62 |
| 3 | 15 | 2 | 27,0 | 17,0 | 10,0 | 17,0 | 7,0 | below | 30 | 92 |
| | | 1 | 5,0 | 5,0 | 4,0 | 1,0 | 1,0 | below | 1 | 93 |
| 4 | 15 | 2 | 22,0 | 12,0 | 4,0 | 18,0 | 8,0 | above | 30 | 123 |
| | | 1 | 4,0 | 4,0 | 3,0 | 1,0 | 1,0 | below | 1 | 124 |
| 5 | 15 | 2 | 22,0 | 12,0 | 3,0 | 19,0 | 9,0 | above | 30 | 154 |
| 6 | 1 | 30 | 22,0 | 22,0 | 3,0 | 19,0 | 19,0 | above | 30 | 184 |
| | | 1 | 10,5 | 10,5 | 10,0 | 0,5 | 0,5 | below | 2 | 187 |
| 7 | 15 | 2 | 27,0 | 17,0 | 10,0 | 17,0 | 7,0 | below | 30 | 217 |
| 8 | 1 | 30 | 27,0 | 27,0 | 10,0 | 17,0 | 17,0 | below | 33 | 250 |

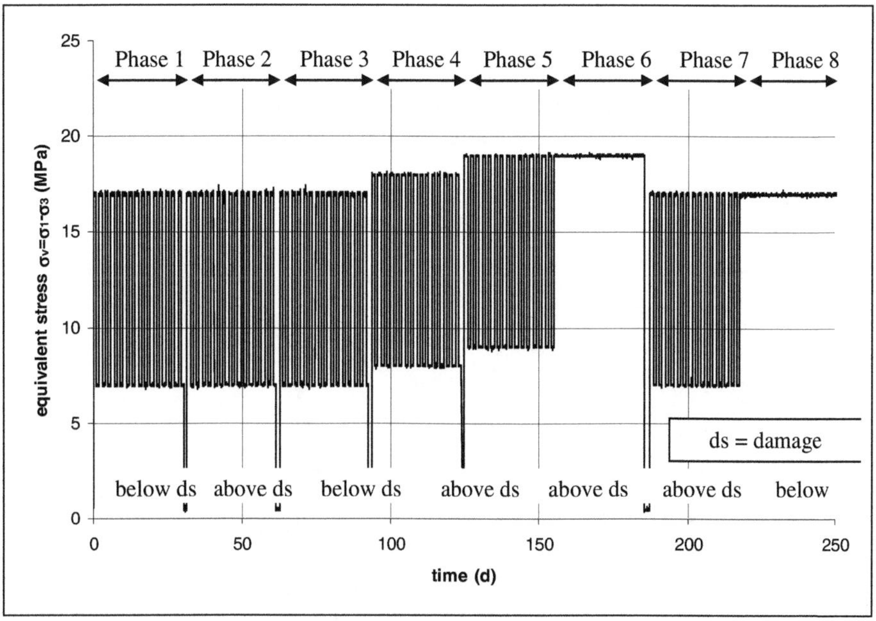


**Fig. 8** Load history of cyclic loading test AI/85/40/10/1-4/1 and AI/85/40/11.5/1-4/1 (equivalent stress)

Figure 9 shows the creep strain measured during the whole tests. Regarding Figure 9, it can be stated that the stationary creep response in testing period 3 is approximately similar to that in testing period 1. This means that the limited low damage which occurred in testing period 2 has no significant impact on the creep behaviour in testing period 3 where the equivalent stress is reduced to a level below damage strength. Nevertheless due to a stress level above the damage strength during period 2, a low increase of the strain rate can be observed. During period 4 and 5, the creep rate increases due to the increasing equivalent stress connected with increasing damage. A comparison between the creep response in periods 5 and 6 documents that higher creep deformations are observed during a constant load above damage strength than by a cyclic loading above damage strength. Considering the accumulated testing time with a stress level above damage strength, it is assumed that the difference between the creep response in period 5 and 6 is induced by the duration of loading. During the cyclic loading in period 5, the accumulated testing time with a stress level above damage strength is given as 15 days whereas the accumulated testing time with a stress level above damage strength in testing period 6 is given as 30 days. It can be clearly deduced from Figure 9, that a significant decrease in strain rate occurs if the equivalent stress is reduced to a stress level below damage strength after inducing significant damage. The strain response of periods 7 and 8 is significantly different from that observed during periods 1 and 3 which have had a similar cyclic loading and a similar constant load level, respectively. In general the reason for the creep response in periods 7 and 8 is given by the healing effects. Caused by the fact that the equivalent stress in periods 7 and 8 is reduced to a value below damage strength so that the previously induced dilatancy can be recovered, a significant decrease in the total creep rate occurs. A comparison between the creep rate in testing periods 7 and 8 indicates that constant loading leads to a higher creep deformation than a cyclic loading caused by different accumulated loading time.

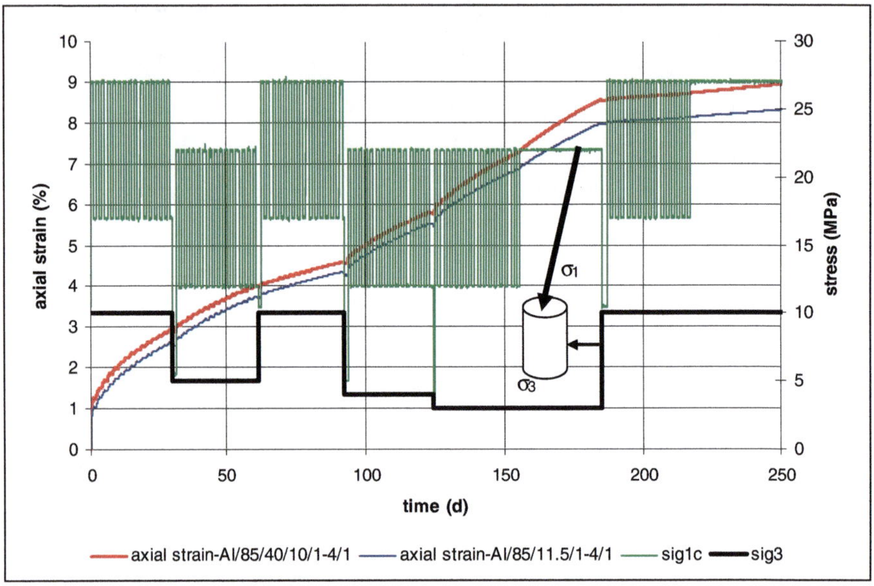


**Fig. 9** Strain response and load history for cyclic loading tests on rock salt from Avery Island Louisiana – samples AI/85/40/10/1-4/1 and AI/85/40/11.5/1-4/1

As a conclusion of the aforesaid characterization of the creep response, both tests are feasible by taking into account phenomenological aspects of the load bearing behaviour and the creep behaviour of rock salt, respectively. To analyse possible damage and healing effects to the creep behaviour by cyclic loading, the measured values of volumetric strain and ultrasonic wave velocity are plotted in Figure 10 and Figure 11, respectively.

Figure 10 indicates a clearly increasing dilatancy during cyclic and constant load levels above damage strength and decreasing dilatancy during cyclic and constant load levels below damage strength. Obviously, increasing dilatancy with time corresponds to the intensity of exceeding the damage strength. That is, the higher the stress increment is above the damage strength, the higher the increase in dilatancy with time. Additionally, due to the accumulated loading time analogous to the strain response, the dilatancy rate by constant loading is higher than the dilatancy rate by cyclic loading. It can be stated from Figure 10 that the recovery and healing of previous induced dilatancy occur if the stress is reduced to a level

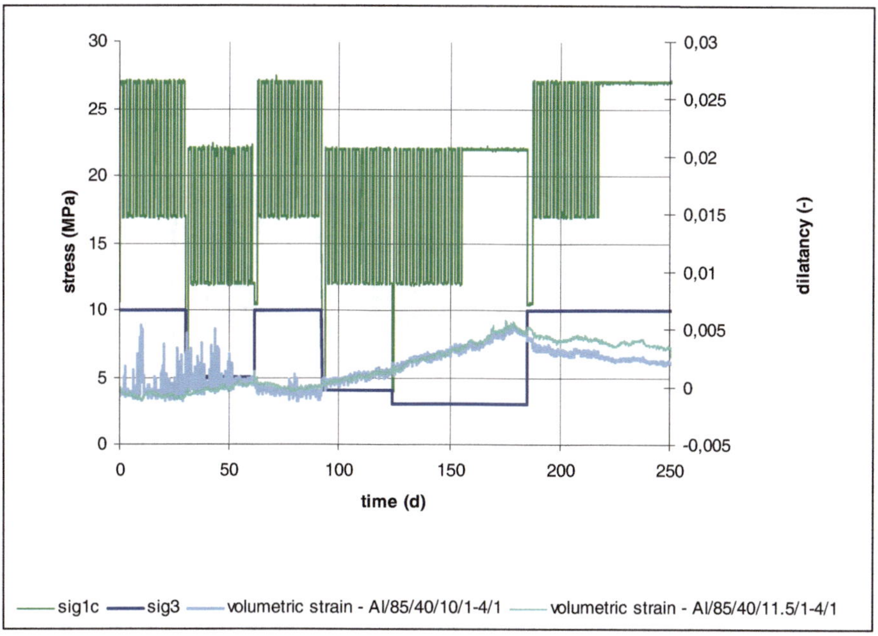


**Fig. 10** Volumetric strain and load history for cyclic loading tests on rock salt from Avery Island Louisiana – samples AI/85/40/10/1-4/1 and AI/85/40/11.5/1-4/1

below the damage strength. Regarding Figure 10, it is assumed that the recovery rate of volumetric strain and the healing rate depend on the amount of dilatancy and the accumulated loading time below damage strength, respectively.

Regarding the visualization of the normalized ultrasonic wave velocity in Figure 11, it can be stated that the ultrasonic wave velocity reacts sensitively to any damage and healing processes. Based on the experience that a material compaction leads to an increase in ultrasonic wave velocity whereas microfissures induced by damage processes lead to a decrease in ultrasonic wave velocity, the evaluation of the normalized ultrasonic wave velocity shown in Figure 11 as a function of the testing time in general is similar to the evaluation of volumetric strain dilatancy. But, due to the fact that a measured volumetric strain is exclusively able to indicate the sum of volume change in some cases, the amount of elastic compression of rock salt is higher than the amount of the damage induced volume of microfissures. Because the ultrasonic wave velocity is reflected exclusively from the intensity of the microfissures and not by the volume of damage induced microfissures, the normalized ultrasonic wave velocity is a more sensitive indicator for damage than the volumetric strain. A comparison of Figure 10 and Figure 11 indicates clearly that based on the evaluation of measured volumetric strain in period 1, a decrease of the total sample volume is observed. Additionally, Figure 11 indicates an overall compaction (strain hardening) due to a cyclic loading below damage strength.

In conclusion, it can be compiled that measuring volumetric strain and ultrasonic wave velocity at the same time is feasible for indicating significant as well as marginal damage and healing processes.

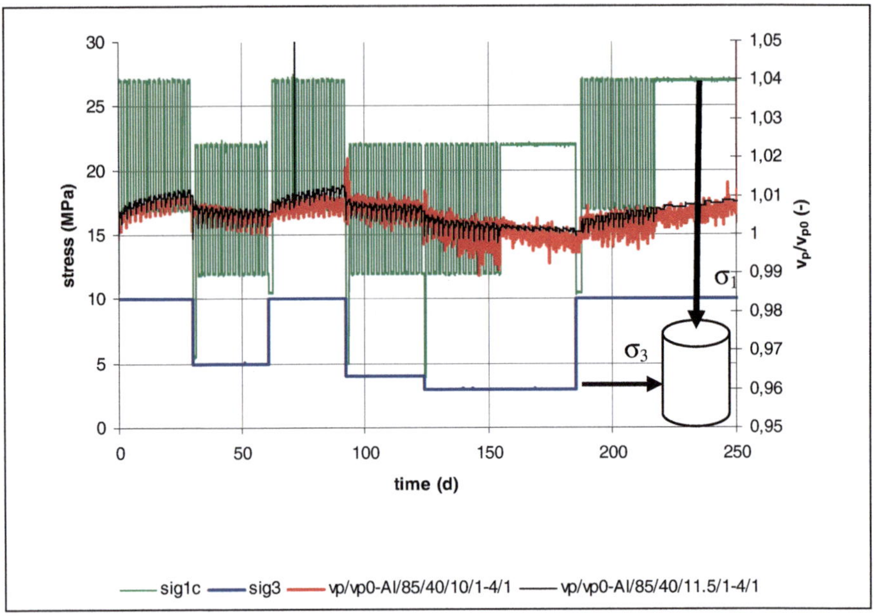


**Fig. 11** Normalized ultrasonic wave velocity and load history for cyclic loading tests on rock salt from Avery Island Louisiana – samples AI/85/40/10/1-4/1 and AI/85/40/11.5/1-4/1

## 3 Basic Characteristics of the Constitutive Model *Lux/Wolters* Used for a Back Analysis of Cyclic Loading Tests Performed on Rock Salt from Avery Island Louisiana

The total strain calculated with the constitutive model *Lux/Wolters* is the result of an additive superposition of elastic strain, damage free stationary creep strain, and damage free transient creep strain, damage induced creep strain and the respective healing / damage recovery creep strain. A simplified expression of the model is given by Equation 1:

$$\dot{\varepsilon}_{ij} = \dot{\varepsilon}_{ij}^{el} + \dot{\varepsilon}_{ij}^{st} + \dot{\varepsilon}_{ij}^{tr} + \dot{\varepsilon}_{ij}^{ds} + \dot{\varepsilon}_{ij}^{h} \tag{1}$$

where

$\dot{\varepsilon}_{ij}$ total creep rate

$\dot{\varepsilon}_{ij}^{el}$ elastic creep rate

$\dot{\varepsilon}_{ij}^{s}$ stationary creep rate

$\dot{\varepsilon}_{ij}^{tr}$ transient creep rate

$\dot{\varepsilon}_{ij}^{ds}$ damage induced creep rate

$\dot{\varepsilon}_{ij}^{h}$ healing induced creep rate

The stationary creep rate $\dot{\varepsilon}_{ij}^{s}$ and the transient creep rate $\dot{\varepsilon}_{ij}^{tr}$ are calculated by:

$$\dot{\varepsilon}_{ij}^{v} = \dot{\varepsilon}_{ij}^{tr} + \dot{\varepsilon}_{ij}^{st} = \frac{3}{2} \cdot \left[ \frac{1}{\overline{\eta}_{\mathrm{k}}} \cdot \left( 1 - \frac{\varepsilon^{tr}}{\max \varepsilon^{tr}} \right) + \frac{1}{\overline{\eta}_{\mathrm{m}}} \right] \cdot \frac{s_{ij}}{(1-D)} \tag{2}$$

where

$$\overline{\eta}_{\mathrm{k}} = \overline{\eta}_{\mathrm{k}}^{*} \cdot \exp\left( k_2 \cdot \frac{\sigma_{\mathrm{v}}}{(1-D)} \right)$$

$$\max \varepsilon^{tr} = \frac{1}{G_{\mathrm{k}}} \cdot \frac{\sigma_{\mathrm{v}}}{(1-D)}$$

$$\overline{G}_{\mathrm{k}} = \overline{G}_{\mathrm{k}}^{*} \cdot \exp\left( k_1 \cdot \frac{\sigma_{\mathrm{v}}}{(1-D)} \right)$$

$$\overline{\eta}_{\mathrm{m}} = \overline{\eta}_{\mathrm{m}}^{**} \cdot \exp\left( m \cdot \frac{\sigma_{\mathrm{v}}}{(1-D)} \right) \cdot \exp(l \cdot T)$$

$\dot{\varepsilon}_{ij}^{v}$ – viscous creep rate (1/d)

$\sigma_v$ – equivalent stress [MPa],

$s_{ij}$ – deviatoric stress tensor [MPa],

$\overline{\eta}_m$ – Maxwell-viskosity coefficient [MPa·d],

$\overline{G}_k$ – *Kelvin*-shear modulus [MPa],

$\overline{\eta}_{\mathrm{k}}$ – Kelvin-viskosity modulus [MPa·d],

$\varepsilon^{tr}$ – transient creep deformation [-],

$T$ – temperature [K],

$k_1, k_2, m, l$ material parameters [1/MPa], [1/MPa], [1/MPa], [1/K],

$D$ – damage parameter [-]

The damage parameter can be determined experimentally and is defined by:

$$D = 1 - \frac{1}{1-\varepsilon_{vol}} \cdot \left(\frac{v_p}{v_{p0}}\right)^2 \tag{3}$$

where

$\varepsilon_{vol}$ – volumetric strain

$v_p$ – actual ultrasonic wave velocity

$v_{p0}$ – ultrasonic wave velocity at the beginning of testing

The evaluation of D is calculated by:

$$\dot{D} = a_{15} \cdot \frac{\left\langle \frac{F^{ds}}{F^*} \right\rangle^{a1}}{(1-D)^{a2}} \tag{4}$$

where

$\dot{D}$ – damage rate

$F^{ds}$– Yield function

$F^*$– scale factor [1 MPa]

$a_i$– material parameter, determined by lab tests

The Yield function $F^{ds}$ is used to quantify the equivalent stress difference between a given stress level and the damage strength. Regarding Equation (4) and (5), the damage rate is increasing at a rate exceeding the damage strength:

$$F^{ds} = \sigma_v - \beta_D(\sigma_3, \theta) = \sigma_v - \eta_D(\sigma_3) \cdot \beta^{TC}(\sigma_3) \cdot k_\beta(\sigma_3, \theta)$$

$$\Leftrightarrow \tag{5}$$

$$F^{ds} = \sqrt{3J_2} - [1 - a_4 \cdot \exp(-a_5 \cdot \sigma_3)] \cdot [a_6 - a_7 \cdot \exp(-a_8 \cdot \sigma_3)] \cdot k_\beta(\sigma_3, \theta)$$

where

$$J_2^D = \frac{1}{6} \cdot \left\{(\sigma_1 - \sigma_2)^2 + (\sigma_2 - \sigma_3)^2 + (\sigma_3 - \sigma_1)^2\right\}$$

$a_i$– material parameter to determine failure strength ( $\beta^{TC}$ ) and damage strength ( $\eta_D \cdot \beta^{TC}$ )

$$k_{(\sigma_3,\theta)} = \frac{\beta(\theta)}{\beta^{TC}} = \left[\frac{1}{\cos(\theta+\frac{\pi}{6}) + a_9 \cdot \sin(\theta+\frac{\pi}{6})}\right]^{\exp(-a_{10}\cdot M)}$$

$\beta^{TC}$ –failure strength – compression

$\theta$– Lode angle

The damage induced creep rate $\dot{\varepsilon}^{ds}$ is calculated proportional to the damage D by Equation (6):

$$\dot{\varepsilon}_{ij}^{ds} = a_3 \cdot \frac{\left\langle \frac{F^{\mathrm{ds}}}{F^*} \right\rangle^{a1}}{(1-D)^{a2}} \cdot \frac{\partial Q^{\mathrm{ds}}}{\partial \sigma_{ij}} \tag{6}$$

where

Q– potential function

Regarding Equation (6), the volume dilatancy rate $\dot{\varepsilon}_{vol}$ is given by Equation (7):

$$\dot{\varepsilon}_{vol} = \sum_{ii} \dot{\varepsilon}_{ii}^{ds} \tag{7}$$

The healing rate is calculated in accordance with Equations (8)/(9), [2]:

$$\dot{\varepsilon}_{ij}^{h} = -\varepsilon_{vol} \cdot \left(\frac{\dot{M}}{fc1} + \frac{F^h}{fs1}\right) \cdot \frac{\partial Q^h}{\partial \sigma_{ij}} \Rightarrow \dot{\varepsilon}_{vol}^{h} = -\varepsilon_{vol} \cdot \left(\frac{\dot{M}}{fc1} + \frac{F^h}{fs1}\right) \cdot \left(1 - \frac{gh}{3}\right) \tag{8}$$

where

$F^h$– Yield function to determine the onset of healing depending on the difference between a given stress level and the damage strength, Equation (8), [2]

$$F^{\mathrm{h}} = \left[1 - \frac{a_4}{a_{11}} \cdot \exp(-a_5 \cdot M)\right] \cdot [a_6 - a_7 \cdot \exp(-a_8 \cdot M)] - \sigma_v$$

$f_i$, $g_h$– material parameter determined by healing tests from lab

$\sigma_v$ – equivalent stress

$$M = \frac{1}{2} \cdot \left[ \sigma^{\perp_2} + \frac{\pi/6 - \theta_R}{\pi/3} \cdot \sigma^{\perp_1} \right] \tag{9}$$

In addition to Equations mentioned-above to calculate the total creep rate, the constitutive model *Lux/Wolters* enables the computation of the damage recovery rate $\dot{D}$ in accordance with Equations (10) [2]:

$$\dot{D}^h = -D \cdot \left( \frac{\dot{M}}{f_{c1} \cdot f_{c2}} + \frac{F^h}{f_{s1} \cdot f_{s2}} \right) \tag{10}$$

It is essential for the planned back analysis of the cyclic loading tests documented in Section 2 to have knowledge of all the parameters used by the constitutive model *Lux/Wolters*. Caused by the fact that except for the two long term cyclic tests described in Section 2, the rock salt from Avery Island Louisiana was not investigated by TUC, most of the material parameters must be estimated by matching the observed load bearing behaviour to the computed load bearing behaviour. Therefore the aim of the back analysis was to decide whether it is possible to recalculate the measured creep, damage and healing behaviour of the performed long term cyclic creep tests by one unique set of parameters in each case.

Based on the information about the damage strength of rock salt from Avery Island Louisiana given by RESPEC, the parameters $a_4$ to $a_8$ have been estimated as shown in Figure 12.

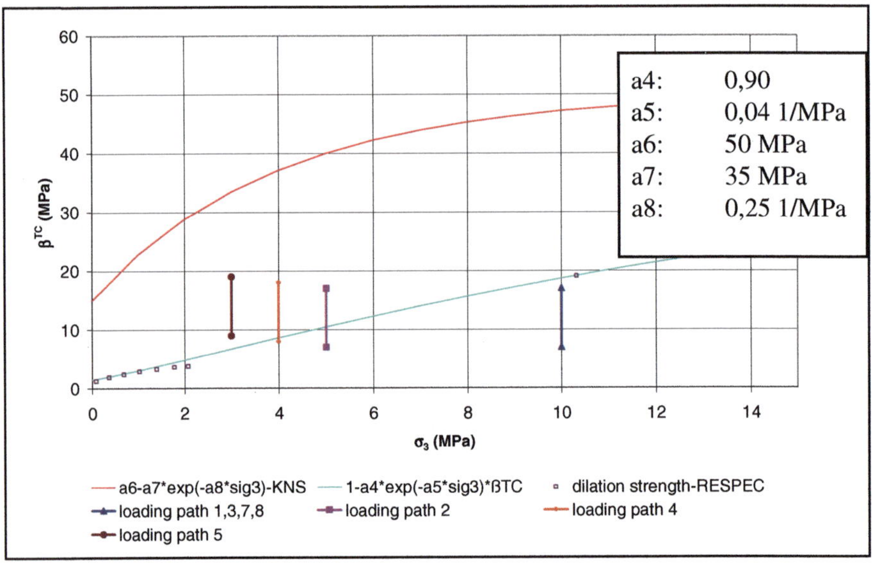


**Fig. 12** Estimated failure and damage strength of rock salt from Avery Island Louisiana used for a back analysis of long term cyclic creep tests based on the constitutive model *Lux/Wolters*

The stationary and transient creep parameters were determined by the strain response observed during cyclic loading periods 1 and 3 performed at a stress level below damage strength. A tabular compilation of these standard creep parameters is comprised in Table 2.

**Table 2** Estimated elastic as well as creep parameters of rock salt from Avery Island Louisiana used for a back analysis of long term cyclic creep tests based on the constitutive model *Lux/Wolters*

| Avery Island Louisiana | | |
|---|---|---|
| | AI/85/40/10/1-4/1 | AI/85/40/10.5/1-4/1 |
| $\overline{\eta}^*_{M(T=303K)}$ (MPa·d) | 4.96E+06 | 1.85E+06 |
| m (1/MPa) | -0.18 | -0.15 |
| $\overline{G}^*_k$ (MPa) | 4.14E+03 | 4.76E+03 |
| $k_1$ (1/MPa) | -0.128 | -0.135 |
| $\overline{\eta}^*_k$ (MPa·d) | 2.40E+06 | 2.40E+06 |
| $k_2$ (1/MPa) | -0.33 | -0.33 |
| E (MPa) | 20000 | 20000 |
| ν (-) | 0.3 | 0.3 |

The determination of the parameters required to quantify the damage induced creep rate and the respective healing/damage recovery was done by matching the computed and measured data for axial strain, volumetric strain and damage based on some sensitivity calculations. A compilation of the best fitting parameters is comprised in Table 3.

**Table 3** Estimated damage and healing parameters of rock salt from Avery Island Louisiana used for a back analysis of long term cyclic creep tests based on the constitutive model *Lux/Wolters*

| Avery Island Louisiana | | |
|---|---|---|
| | AI/85/40/10/1-4/1 | AI/85/40/10.5/1-4/1 |
| a0(-) | 0.15 | 0.20 |
| a1 (-) | 0.1 | 0.1 |

**Table 3** (*continued*)

| | | |
|---|---|---|
| a2 (-) | 0.2 | 0.2 |
| a3 (1/d) | 1.00E-04 | 8.00E-05 |
| a15(1/d) | 3.00E-04 | 2.00E-04 |
| fc1 (MPa) | 200 | 300 |
| fs1 (MPa) | 250 | 450 |
| fc2 (-) | 1.5 | 0.7 |
| fs2 (-) | 1 | 0.45 |
| gh (-) | 1 | 1 |
| a11 | 1 | 1 |

## 4 Back Analysis of Cyclic Loading Tests Performed on Rock Salt from Avery Island Louisiana

The results of the back analysis are shown in Figures 13, 14 and 15 for samples AI/85/40/10/1-4/1 and in Figures 16, 17 and 18 for samples AI/85/40/10.5/1-4/1. Based on the graphical comparison of the measured and computed evaluation of total axial creep strain, volumetric strain and damage, it can be stated, that the constitutive model *Lux/Wolters* is obviously able to recalculate cyclic as well as constant loading histories above and below damage strength. That is, the damage as well as healing effects under cyclic and constant loading conditions is understood very well, if the stress geometry is forced to be compressive and the temperature is not changed. But, it must be pointed out that the material parameters used for the back analysis by the majority were estimated due to the fact that the lab tests to determine the material parameters were not conducted. Therefore it is of interest to complete the study by carrying out the lab tests to determine a full set of Avery Island Louisiana material parameters used for the constitutive model. Additionally, further investigations are needed to validate if it is possible to recalculate extension stress conditions and alternating temperatures.

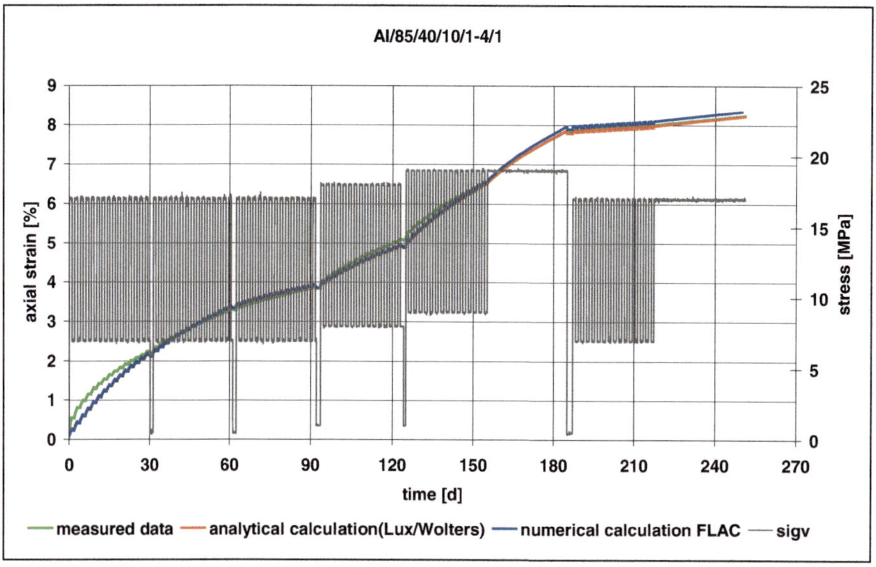


**Fig. 13** Comparison between measured and computed axial strain - Constitutive model *Lux/Wolters* – back analysis sample AI/85/40/10/1-4/1

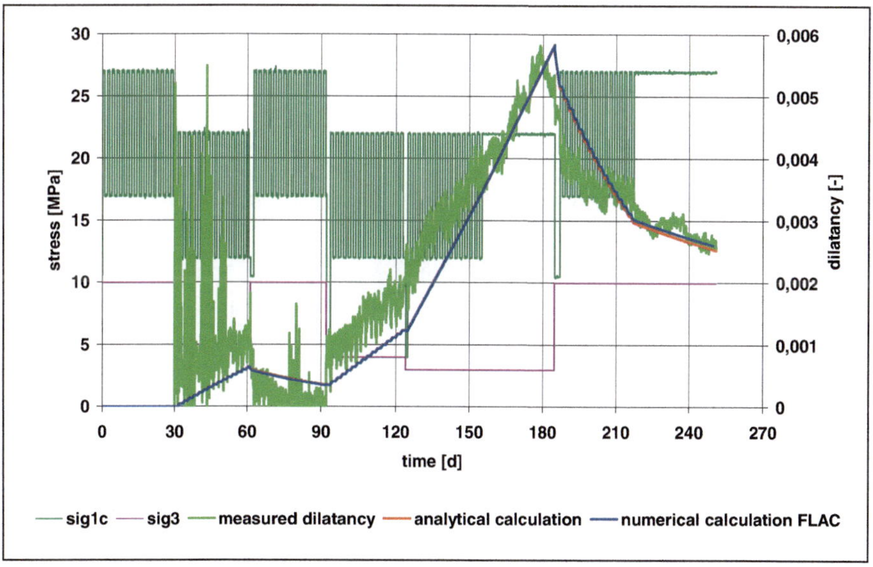


**Fig. 14** Comparison between measured and computed dilatancy - Constitutive model *Lux/Wolters* – back analysis sample AI/85/40/10/1-4/1

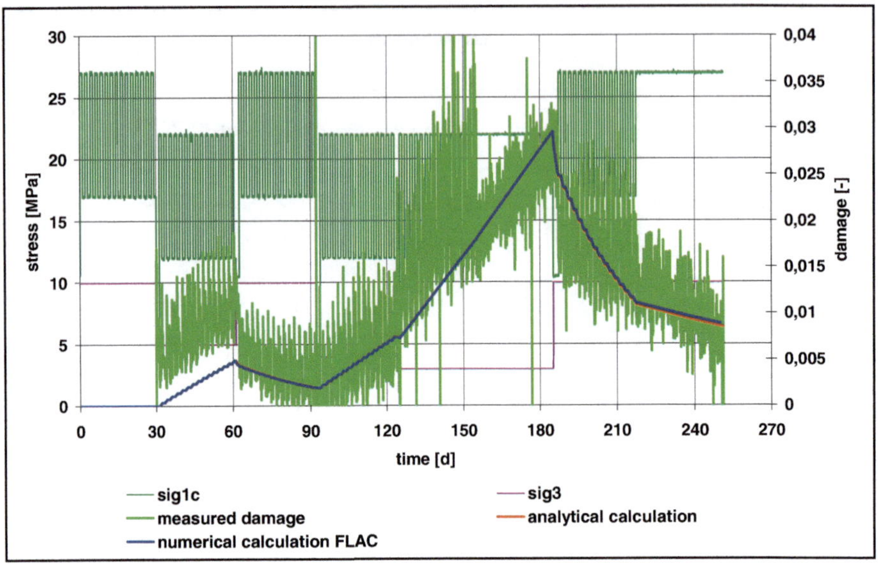


**Fig. 15** Comparison between measured and computed damage - Constitutive model *Lux/Wolters* – back analysis sample AI/85/40/10/1-4/1

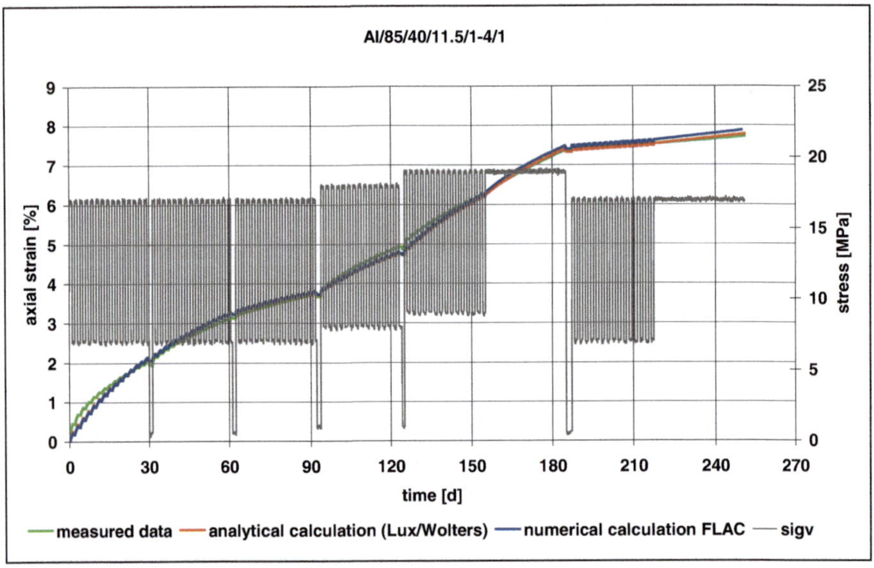


**Fig. 16** Comparison between measured and computed axial strain - Constitutive model *Lux/Wolters* – back analysis sample AI/85/40/10.5/1-4/1

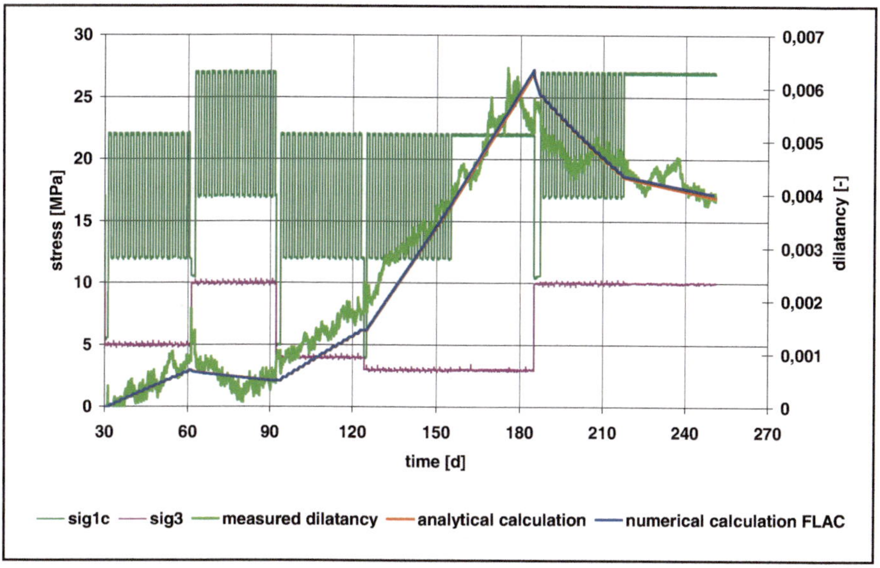


**Fig. 17** Comparison between measured and computed dilatancy - Constitutive model *Lux/Wolters* – back analysis sample AI/85/40/10.5/1-4/1

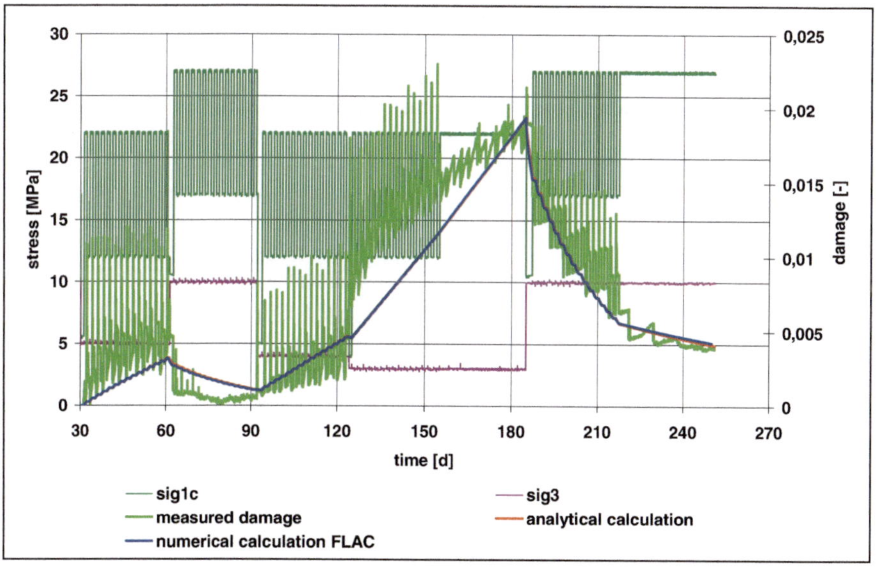


**Fig. 18** Comparison between measured and computed damage - Constitutive model *Lux/Wolters* – back analysis sample AI/85/40/10.5/1-4/1

## 5 Conclusion

To analyse in detail the damage induced creep behaviour and healing behaviour of rock salt by long term cyclic and constant loading at different stress levels two creep tests with a duration of up to 250 days were performed on rock salt from Avery Island. At first the load bearing behaviour was analysed by continuous measurements of axial and radial stress, axial and volumetric deformation, ultrasonic wave velocity and temperature. Secondly, to decide whether it is possible to recalculate the measured creep, damage and healing behaviour of the performed long term tests by one unique set of parameters in each case a numerical back analysis based on the constitutive model *Lux/Wolters* was done. Third, based on a comparison between the measured data obtained from the cyclic loading tests and the computed data obtained from the numerical back analysis of the tests it was clearly proved, that the constitutive model *Lux/Wolters* is able to image the load bearing behaviour of rock salt considering cyclic and constant loading above and below damage stress level. Further investigations are recommended and needed to validate if it is possible to recalculate extension stress conditions as well as alternating temperatures.

## References

1. Mellegard, K., Düsterloh, U.: High Frequency Cycling of Gas Storage Caverns: Phase II: Cyclical Loading Effects on the Damage and Creep Properties of Salt. SMRI Research Report RR2012-02, RESPEC, Rapid City, USA (2012)
2. Lerche, S.: Kriech-und Schädigungsprozesse im Salinargebirge bei mono- und multizyklischer Belastung. Diss. TU-Clausthal, Clausthal (2012)

# A Damage Constitutive Model of Rock Salt Based on Acoustic Emission Characteristics

Deyi Jiang[1], Jie Chen[1], Song Ren[1], Yuan Xi[1], and Chunhe Yang[2]

[1] State Key Laboratory of Coal Mine Disaster Dynamics and Controls, Chongqing University, Chongqing 400044, China
[2] Institute of Rock and Soil Mechanics, The Chinese Academy of Science, Wuhan, Hubei 430071, China

**Abstract.** The acoustic emission (AE) signal observation is a technique to observe the generation of microcracks in rock salt. By monitoring acoustic emission in whole process of stress - strain curve under uniaxial compression tests, the damage characteristic of rock salt is obtained. Damage to the rock salt appears to be mainly a shear failure under the condition of a lower loading strain rate. After shear failure, a lot of small crushed particles spread on the surface of the failure surface. The AE rate-strain curve is able to reflect the damage development process with better consistency which is evident from the stress-strain curve. A damage constitutive model of rock salt is suggested on the basis of acoustic emission characteristics.

**Keywords:** rock salt; acoustic emission (AE), damage, constitutive model, uniaxial compression.

## 1 Introduction

To analyze the damage properties of rock salt is very important to use underground storage of radioactive waste or light hydrocarbons. The main reason for damage failure of the disposal rocks is the generation and development of cracks in excavated disturbed zones (EDZs) along the boundary of the cavity and the rock [1, 2]. The mechanical properties of rock salts (evaporites with halide components) have been a major focus of study in last few decades [3, 4]. The geomechanical properties of rock salts and salt deposits vary greatly because of different origins, mineralogical components, lithostratigraphic disposition, tectonic history and so on [5, 6]. To improve cavern safety and stability, there is an incentive to test rock salt for a better understanding of its role in cavern excavation. Compression damage is also an important failure criterion of rock fracture.

Compressive failure strength is one of the most widely investigated material properties of rock salt. It is now fairly well established that the fracture of rock salt under uniaxial compression involves nucleation of microcracks from inhomogenieties or inherent flaws and fissures, which eventually coalesce to cause shear slipping and axial splitting. A stable growth of these microcracks is found to

M.Z. Hou et al. (Eds.): *Clean Energy Systems in the Subsurface*, SSGG, pp. 363–377.
DOI: 10.1007/978-3-642-37849-2_29 

be initiated at the onset of dilatancy under uniaxial compressive stress [7-9]. Generally, at the initial stage of the deformation process, the pore volume decreases gradually due to a compaction, this closes the existing microcracks [10, 11]. After the elastic deformation phase, the cracks begin to reopen, or new cracks form because the shear stress increases. This transition from compression, microfracturing, crack closure and reduction in pore volume to crack reopening, and pore volume expansion takes place with the accumulation of the damage zone. The macroscopic fracture plane orientation as indicated by AE source locations show that the macroscopic fracture planes coincide with the direction of the maximum principal stress [12, 13].

Starting from experimental evidence, some uniaxial constitutive equations which describe the developing damage process have been formulated. Some authors who suggested constitutive equations for rock salt, have considered dislocation mechanism and have assumed that during deformation, the volume of the rock salt is incompressible [14-17]. A few papers have reported that a damage potential related to the yield function via a correction term is the same form as the yield function [18-21]. The development of a constitutive model of damage behavior needs important inputs from dislocation mechanism concepts and these are applied to the prediction of the nucleation of microcracks in underground excavations in rock salt. The material response is specified through the constitutive model and appropriate material parameters are obtained from laboratory tests. The model is based on damage theory and laboratory data together with site geological characteristics to form the basis of the predictive method. As a result, the suitable constitutive model of damage is developed to predict the damage process.

This paper proposes a theoretical approach that combines damage theory with crack growth to model the dynamic fracture process of rock specimens subjected to low strain rate under uniaxial compressive loading. More specifically, the model takes into account the rock salt material and AE properties. The damage constitutive model is based on the AE number of fracture-induced defects. The growth of microcracks is controlled by ''wing'' cracks or tension cracks, from the tips of the isolated inclined pre-existing cracks. When the damage parameter reaches a critical value, it is assumed that the microcracks coalesce and cause shear slipping.

## 2 Experimental Conditions and Methods

### *2.1 Experimental Conditions*

Rock salt forms isometric crystals which are typically colorless or white, but may also be light blue, dark blue, purple, pink, red, orange, yellow or gray in color, depending on the amount and types of impurities they contain. These crystals commonly occur with other evaporite deposit minerals such as sulfates, halides, and borates.

In order to analyze rock salt acoustic emission characteristics and microcracks development under the uniaxial compression condition and minimize the interference of the impurities, the test specimens are chosen with high purity salt from Khewra salt mine in Pakistan. Test specimens are pink, transparent and compact in structure. Their soluble content is about 96.3% ~ 99.8% (soluble substances are mainly NaCl and $Na_2SO_4$) and the insoluble compositions are mainly argillic minerals. Two types of tests were conducted with cuboid specimens (50 mm ×50 mm in width and 100 mm in length) and cylinder specimens (50 mm in diameter and 100 mm in height). The cuboid specimens were used for observing surface crack propagation under uniaxial compression and monitoring acoustic emission signals while cylinder specimens were used for testing basic mechanical parameters. The specimens are shown in Fig. 1.

## *2.2 Experimental Equipment and Procedure*

The main purpose of this experiment was to investigate the behavior of acoustic emission and the damage evolution characteristics under uniaxial compression conditions. An AG-I250 electronic precision material testing machine with a maximum axial loading capacity of 1000kN was used to record the applied loads and corresponding displacements. In order to understand how the cavern excavation process leads to axial stress increase and to facilitate the monitoring process of rock salt damage evolution, uniaxial compression tests were done with a constant loading strain rate of about $\dot{\varepsilon}_{cp} = 2.0\times10^{-5}s^{-1}$, synchronous with the acoustic emission signal monitor in the loading process. DISP series 2 channel/PCI card-two full digital acoustic emission auto monitor produced by the American Physical Acoustics company was used. The threshold was set at 45dB to gain a high signal/noise ratio. Two sensors with frequency sensitivities between 20 KHz to 400 KHz and a 45 dB pre-amplification (AEwin) were used in the AE system. The sensors were fixed to the rock surface using gum bands, and vaseline was applied for coupling. Plastic cushions were sandwiched between steel plates and specimens to minimize noise generation due to friction. In order to reduce the test data error, the acoustic emission signal monitoring and camera recording should be done at the same time during the loading process. Test devices are shown in Fig. 2.

**Fig. 1** Specimens

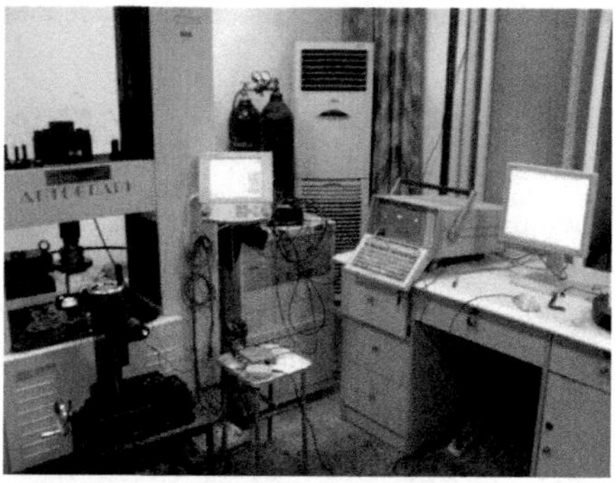

**Fig. 2** Test equipments

# 3 Results and Analysis

At room temperature, four Pakistani high purity rock salt specimens were given uniaxial compression tests and the results are shown in Table 1. The corresponding stress - strain curve is shown in Figure 3.

**Table 1** Results of 4 cuboid specimens under uniaxial compression conditions

| Specimen | Elastic limit stress /MPa | Peak stress /MPa | Axial strain at rupture /% | Yong's Modulus /GPa |
|---|---|---|---|---|
| Salt-1 | 10.32 | 35.45 | 5.83 | 2.39 |
| Salt-2 | 9.83 | 31.67 | 5.21 | 2.67 |
| Salt-3 | 10.64 | 34.15 | 5.53 | 2.01 |
| Salt-4 | 9.23 | 34.92 | 6.07 | 2.42 |

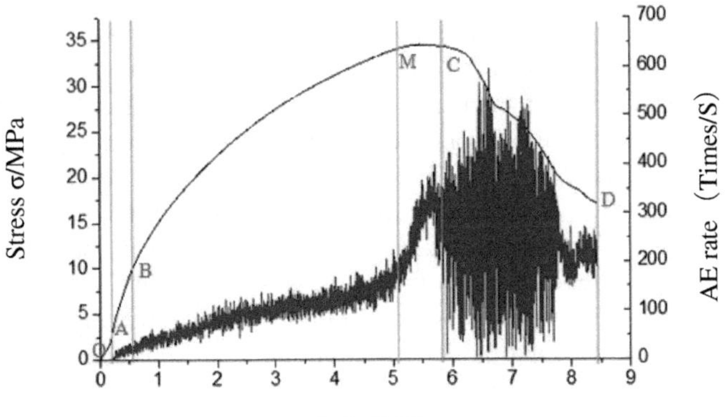


**Fig. 3** Damage fractured processes divide under unaxial compression

## 3.1 *Uniaxial Stress – Strain Characteristics of Rock Salt*

The curves of stress-strain for the salt rock specimens are basically the same. By combining varied characteristics of the salt rock acoustic emission rate in the uniaxial process and by division standards, the complete stress-strain curve in rock mechanics [22] and the stress-strain curve of rock salt in uniaxial compression can be divided into five stages:

Stage 1(stage OA): Pre-existing fracture/pore closure-The initial non-linear, downward concave stress–strain relationship at low stress levels is caused by the closure of some primary pores and cracks with increasing compaction. The restoration of artificial cracks from drilling or excavation is also governed by this process [1]. The strain at this stage is about 4% of the total strain.

Stage 2(AB section): Elastic deformation -After the partial closure of the primary cracks, the loading begins with increasing the axial stress. Up to this point the elastic behaviour dominates the stress–strain relationship. The elastic deformation due to compaction is characterized in all the experiments by linear increments in the axial strains. The strain at this stage is about 3.3% of the total strain.

Stage 3(BM section): Plastic deformation with microcracks stable extension -The microcracks start to open and grow. This is characterized by the departure of the strain curves from the elastic behaviour. This stage is referred to as the stable crack growth region, and the strain is about 52.3% of the total strain.

Stage 4(MC section): Plastic deformation with microcracks unstable extension -Up to point M, while the strain increases, the stress does not change and the AE rate is suddenly increased after this point. The rate of increase of axial strains accelerates rapidly as the axial loading increases and the stress reaches its peak strength. The strain was about 18.7% when the stress reached its peak strength at this stage.

Stage 5(CD section): Post-failure stage- After reaching the peak strength, the transfixion cracks were formed, the load-carrying capacity decreased rapidly and a significant change is observed in the shape of curve.

This deformation mechanism characterized by shear stress and tension stress leads to shear slip with tension cracks. This defines a rock deformation which occurs predominantly at grain level and accordingly causes the polycrystalline rock fragments to slide and rotate. The percent of strain is about 18.7%. This process increases the pore volume of the cracks. The coalescence of these cracks through various mechanisms results in an increase in the porosity and permeability of the system [23]. The stress – strain curve division is shown in Figure 3.

In order to analyze further the damage characteristics in the process of uniaxial compression, a High Definition camera was used to record the entire failure process of the salt specimen. The salt specimens surface graphs corresponding to the turning point of the stress-strain curve stages division in uniaxial compressive are shown in Figure 4. Under the uniaxial compression condition and with a low loading rate, the failure forms of the salt specimens were usually due to a combination of shear failure and tension fracture. Because of the special mineral composition and structure of salt rock, when the salt specimen was damaged by loading, the light transmission property obviously changed due to the effect of cracking. In the loading process, the light transmission property gradually became weak as the stress increased and the color of the specimen turned gradually from its initial pink to white. In fact, in conditions with a low loading strain rate, the eventual destruction of the specimens is by the face shear slipping with intense tension cracking. A large regional piece of rib spalling is also observed at one side of salt specimen. This process accompanied by a large number of small sized salt grain peelings and a shear sliding surface, has an obvious dislocation, but has not separated from the shear failure face like the brittle rock.

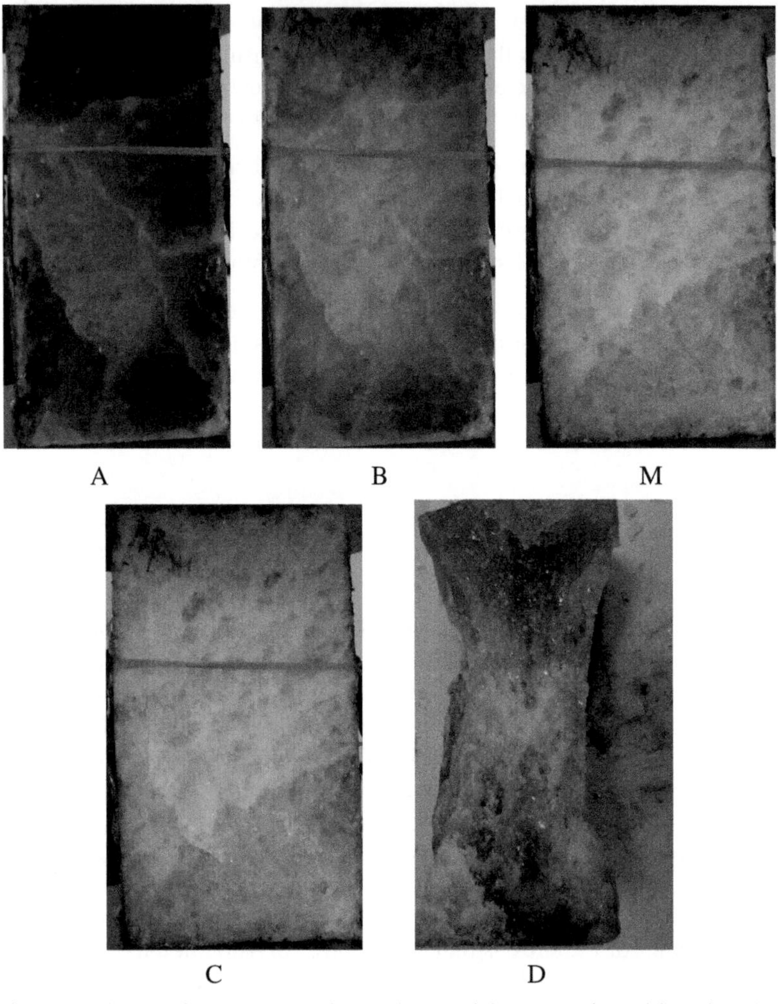

**Fig. 4** Typical damage fracture procession under unaxial compression with various loading strain rates. The picture of A, B, M, C, D correspond with the point in Fig. 3.

## *3.2 Acoustic Emission Characteristics of Rock Salt under Uniaxial Compression*

Fracture in a quasi-brittle material such as rock involves micro-cracking which generates elastic waves known as acoustic emissions (AE) signals. These transient waves propagate through the medium with very small amplitudes and high frequencies and the AE signals carry information about the source, including location and mechanism defined by mode and magnitude [24]. This definition suggests that the recorded acoustic emissions can be used to determine the rock damage and seismicity process. The process of crack initiation, coalescence and propagation occur with the release of energy which can be recorded as an acoustic signal. The curvilinear stress-strain relationship, AE rate-strain and AE number-strain are shown in Figure 5.

(1) Elastic consolidation (visco-elastic) stage-Almost no acoustic emission signals from the rock specimen and an internal original crack that has not developed. During the low loading strain rate the pores slowly closed into each other with no new cracks generated.
(2) Linear elastic deformation stage-AE a few signals began to appear and the AE number slowly increased as the stress increases. The rock salt will cause little generation of acoustic emission signals because of its own grain features (square crystal) and structural form, intergranular extrusion deformations.
(3) Plastic deformation micro-crack stable expansion stage-The internal expansion of rock cracks and the speed of crack formation are relatively stable and slow, leading to the AE rate increasing slowly with an approximate linear growth. CHAN et al [20] suggested that in the plastic deformation process, the specimens internal "wing" crack is constantly increasing, which make the number of AE increase gradually.
(4) During the plastic deformation micro-crack stage the unsteady expansion-AE rate is increasing fast and its value appears to be in a state of obviously fluctuating change. At this stage, the internal cracks in the salt derived from the fast expanding collection of ??? and the damaged slip plane formed by the fracture zone so that they produce a large numbers of AEs. This region begins with the plasticity for rock salt at point M, where the AE rate suddenly begins to increase rapidly. The stress here is close to the peak stress and therefore causes the "wing" cracks to begin to gather and form a fracture zone. Point M corresponds to the Kaiser Effect [25] of the AE rate-strain curve and it can be used as a premonition of rock specimens beginning to fail. This phenomenon was observed in all the specimens under uniaxial compression test and it shows that using Kaiser Effect (point M) to monitor the premonition of salt rock pillar damage in mining will be helpful for stability of the cavern.
(5) Stage of post-failure-The AE number rapidly increases further and the maximum AE rate generally appears at this stage with huge fluctuations. When the shear plane [26, 27] is formed, the AE number begins to decrease. The movement of the fracture surfaces prompted a large number of AEs and also generated many secondary tension cracks.

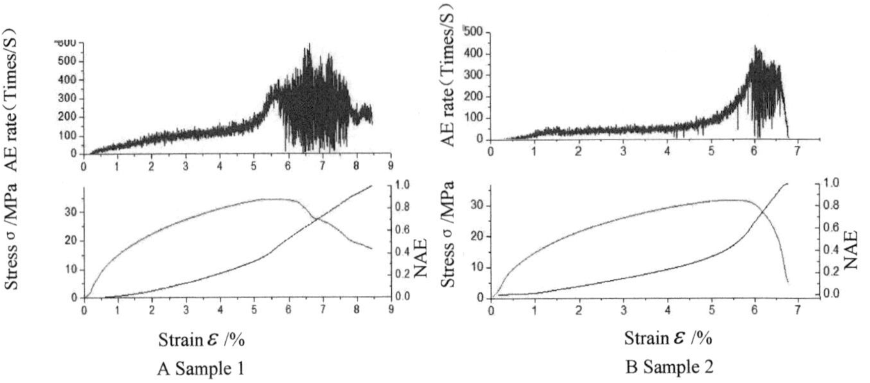


**Fig. 5** Relationships of AE rate, "normalized accumulative AE number" (NAE) and stress with strain

# 4 A Damage Constitutive Model Based on Acoustic Emission

Wawersik and Krajcinovic [28, 29] proposed that damage is caused by microcracks and microvoid in the rock material. Once formed these micro cracks and micro void cannot bear any stress. Based on this theory, the rock damage model was established by a combined Lemaitre's strain equivalent hypothesis [30]. This theory mainly studies the geometrical characteristic of the initial damage influence on the subsequent damage and pays no attention to the effects that the microscopic damage has on the macroscopic deformation [31]. For convenience, such a model cannot be considered as a part of the carrying capacity of material damage, so this intuitive definition is necessary. Salt rock as soft crystalline rock in a compression loads condition mainly focuses on the effects that microscopic damage has on macroscopic deformation. So the load bearing capacity of damaged material cannot be ignored and cannot reflect the actual situation of the rock as the softening characteristic gradually turns into a hardening characteristic when pressure increases. Therefore, Cao [32] thinks "damage" is the linear elastic stress state transformed into the nonlinear stress state. This abstract "damage" definition is not limited to the specific form after the material damage which suggests that the damaged part of the material can still take on some stresses and is only a change in the state of stress.

## *4.1 Definition of Damage Model*

On this basis, it can be assumed that the damaged rock made up of two parts under the loading stress (i.e. undamaged materials and damaged materials), can bear a certain stress. In Figure 6, if the stress applied to the rock material is $\sigma_i$, the corresponding sectional area is A, the stress of the intact material (Shaded part of Figure 6) is $\sigma_i'$, the corresponding bearing area is A' and the stress of the destructive material (Blank part of Figure 6) is $\sigma_i^{''}$, the corresponding bearing area is A", then the relationship can be expressed as:

$$\sigma_i^{'} A^{'} + \sigma_i^{''} A^{''} = \sigma_i (A^{'} + A^{''}) \tag{1}$$

$$\sigma_i^{'} \frac{A^{'}}{A} + \sigma_i^{''} \frac{A^{''}}{A} = \sigma_i \tag{2}$$

The ratio A'/ A is defined as the rock material damage variable and is equal to D. Substituting D for the ratio A '/ A, then Equations (1) and (2) can be expressed as:

$$\sigma_i^{'} (1 - D) + \sigma_i^{''} D = \sigma_i \tag{3}$$

Equation (3) which is a new type of rock damage model was established by Cao [32], the first step of building the rock damage constitutive model is to set up the relationship for strain with $\sigma_i^{'}$ and $\sigma_i^{''}$.

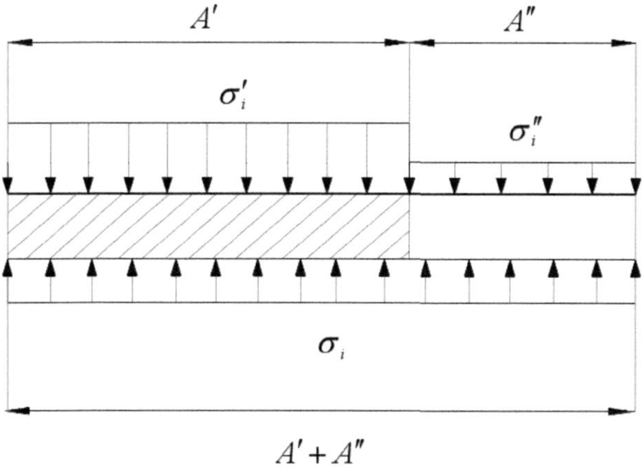


**Fig. 6** The damage model for rock

## 4.2 *The Definition of Uniaxial Compression Effective Stress and Damage Stress*

In order to establish the rock damage constitutive relationship in a low loading strain rate under uniaxial compression conditions, the following assumptions are made:

(1) Stress-strain relationship in uniaxial compression has a linear elastic relationship before the rock is damaged. This is represented by:

$$\sigma^{'} = E\varepsilon \tag{4}$$

where E is elastic modulus and $\varepsilon$ is strain.

(2) Rock material become friction material after damage and its stress condition satisfies the Mohr - Coulomb criterion as follow:

$$\sigma'' = 2c \tan \alpha \tag{5}$$

where $\alpha = \pi/4 + \varphi/2$; C is the cohesion and $\varphi$ is the internal friction angle.

Substituting Eq. 4 and Eq. 5 into Eq. 3, Equation (6) can be rewritten by:

$$\sigma = E\varepsilon(1-D) + 2cD \tan \alpha \tag{6}$$

## 4.3 The Definition of Damage Evolution Equation Based on Acoustic Emission

The failure process of rock salt in uniaxial compression mainly shows grain damage and grain slip. When the strain energy which agglomerates in the failure process is quickly released, it appears as acoustic emission signals. From Fig. 4, it is known that rock damage is due mainly to shear failure so the shear failure area can be defined as A, the total grain number of the failure surface is N, the destroyed grain number of failure surface is n, then the rock uniaxial damage variable D can be defined as:

$$D = \frac{n}{N} \tag{7}$$

As shown in Fig. 5, the accumulative AE number has an approximation index relationship with the strain hypothesis:

$$n_e = A\exp(a\varepsilon) + B \tag{8}$$

where $\varepsilon$ is specimen axial strain; A, B and a are constants.

According the uniaxial compression and AE initial conditions, it was found that at the elastic deformation stage(AB section) there was almost no AE signal, so at the point B (the strain $\varepsilon_B$) of Figure 3 the $n_e = 0$; And at the point D (the strain $\varepsilon_D$) of Figure 3, the accumulative AE number $n_e = N_e$. So the initial conditions can be described as:

when $\varepsilon = \varepsilon_0$, then $n_e = 0$; and when $\varepsilon = \varepsilon_D$, then $n_e = N_e$. (9)

Where $\varepsilon$ is specimen axial strain; $\varepsilon_B$ is the initial strain of AE signal beginning to generate (the strain at point B of the stress-strain curve in Figure 3); $\varepsilon_D$ is the total strain when the specimen is completely destroyed(the strain at point D of stress-strain curve in Figure 3); $n_e$ is the accumulative AE number; $N_e$ is total accumulative AE number when the strain reaches point D in Figure 3.

Substituting the initial condition of Eq. 9 into Eq. 8,

$$A = \frac{N_e}{\exp(a\varepsilon_D) - \exp(a\varepsilon_B)}, \quad B = \frac{-N_e \exp(a\varepsilon_B)}{\exp(a\varepsilon_D) - \exp(a\varepsilon_B)} \tag{10}$$

Substituting Eq. 10 into Eq. 8,

$$n_e = \frac{\exp[-a(\varepsilon_D - \varepsilon)] - \exp[-a(\varepsilon_D - \varepsilon_B)]}{1 - \exp[-a(\varepsilon_D - \varepsilon_B)]} N_e \tag{11}$$

Assuming the AE number has a linear relationship with the number of rock grain fractures, then,

$$n = \beta n_e ; \; N = \beta N_e \; ; \; (\beta \geq 1) \tag{12}$$

The damage evolution equation can be written by:

$$D = \frac{n}{N} = \frac{n_e}{N_e} = \frac{\exp[-a(\varepsilon_D - \varepsilon)] - \exp[-a(\varepsilon_D - \varepsilon_B)]}{1 - \exp[-a(\varepsilon_D - \varepsilon_B)]} \tag{13}$$

### *4.4 The Definition of Damage Constitutive Equation Based on Acoustic Emission*

The complete stress-strain curve, Figure 3, shows that: At stage OB there are no new cracks in all the experiments from linear increments in the axial strains, which obey the elastic constitutive model. At stage BD new cracks generate area damage. According to the new definition of the rock damage model in Equation 6, the damage evolution presented in Equation 13, and also considering the rock material, the physical and mechanical characteristics of the rock can be defined by the rock uniaxial damage constitutive equation. Then the total constitutive model can describe as follows:

$$\sigma = \begin{cases} E\varepsilon & 0 < \varepsilon < \varepsilon_B \\ b_1 E\varepsilon \dfrac{1 - \exp[-a(\varepsilon_D - \varepsilon)]}{1 - \exp[-a(\varepsilon_D - \varepsilon_B)]} + b_2 2C \tan\alpha \dfrac{\exp[-a(\varepsilon_D - \varepsilon)] - \exp[-a(\varepsilon_D - \varepsilon_B)]}{1 - \exp[-a(\varepsilon_D - \varepsilon_B)]} & \varepsilon_B < \varepsilon < \varepsilon_D \end{cases} \tag{14}$$

where $b_1$ and $b_2$ are material constants.

Equation 14 is made up of two parts- part 1 is used to describe elastic deformation, when strain is within the range $0 < \varepsilon < \varepsilon_B$. Almost no new cracks are generated in the rocks at this elastic stage. The second part is used to describe plastic deformation, when strain is within the range $\varepsilon_B < \varepsilon < \varepsilon_D$, damage begins to appear on the specimen and evolution gradually expands.

According to the results of the rock salt uniaxial compression acoustic emission test in the third section, the initial parameters of Equation 14 (*a, b1, b2,* $\varepsilon_B$, $\varepsilon_D$, *C,* $\varphi$) can be determined. The parameter values in this paper calculated from the test conditions are a=5.1, b1=b2=0.256, $\varepsilon_B$ =0.005, $\varepsilon_D$ =0.085, *C*=2.1 MPa, $\varphi$ =32.5°.

The fitting curve is shown in Figure 7. The damage constitutive equation based on acoustic emissions can well reflect the rock stress-strain characteristics in rock uniaxial damage process.

Discussions about the adaptability of the uniaxial damage constitutive equation-it is easy to see from Figure 7, that the damage constitutive equation can well describe stress - strain feature before peak strength in low loading strain rate but does not reflect the specimen failure process after uniaxial peak strength well enough. After reaching the peak strength, the transfixion cracks were formed, the load-carrying capacity decreased rapidly and a significant change is the unstable extension of micro cracks. The variation of acoustic emission signals after peak intensity in Figure 3 can be further proof of this phenomenon. So it is hard to find a damage constitutive model to describe stress - strain feature after peak strength.

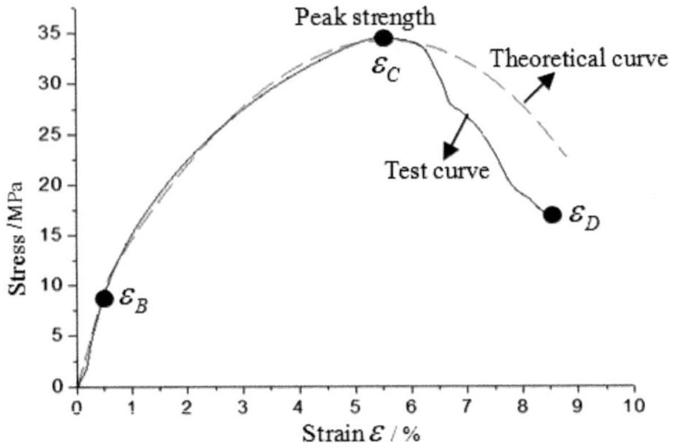


**Fig. 7** The stress-strain curve

## 5 Conclusions

According to the rock salt uniaxial compression test combined with the monitoring parameter change rule for acoustic emissions and the stress - strain curve features, the uniaxial damage characteristic of rock salt is obtained. The AE rate change characteristics in the uniaxial compression failure process are analyzed and the rock damage constitutive equation, based on acoustic emission characteristics, is suggested.

The plastic strain at the micro crack steady expansion stage takes up over 50% of the whole strain process in the uniaxial compression process. This is totally different from brittle rock. At the plastic deformation stage, with the stable expansion of the microcracks, the AE rate is obviously linear in growth and has a better consistency than the strain change. After the unsteady expansion of a crack, it produces a large number of acoustic emission signals fully reflecting that the shear failure area still has cohesive force.

From the NAE - strain curve and strain - stress curve, it is known that the number of AEs showed obvious regularity with the uniaxial damage increment in the rock uniaxial damage process. A damage constitutive equation is established on the basis of the relationship between the AE number and strain which can well reflect the stress - strain characteristic before reaching peak strength in the uniaxial compression test but cannot reflect failure characteristics of the post-failure stage.

**Acknowledgments.** This research was supported by the National Basic Research Program of China (2009CB724606), Specialized Research Fund for the Doctoral Program of Higher Education of China (20090191110001) and National Natural Science Foundation of China (51074198).

## References

1. Alkana, H., Cinarb, Y., Pusch, G.: Rock salt dilatancy boundary from combined acoustic emission and triaxial compression tests. International Journal of Rock Mechanics and Mining Sciences 44(1), 108–119 (2007)
2. Hou, Z.: Mechanical and hydraulic behavior of rock salt in the excavation disturbed zone around underground facilities. International Journal of Rock Mechanics and Mining Sciences 40(5), 725–738 (2003)
3. Hansen, F.D., Mellegard, K.D., Senseny, P.E.: Elasticity and strength of ten natural rock salts. In: Proceedings of the First Conference on Mechanical Behavior of Salt, pp. 71–83. Pennsylvania State University (1984)
4. Cristeseu, N., Hardy, R.H. (eds.): Proc. 4th Conf. Mech. Beh. of Salt, Buearest, p. 675. Trans. Tech. Pub., Clausthal-Zellerfeld (August 1999, 2002)
5. Liang, W.G., Yang, C.H., Zhao, Y.S., et al.: Experimental investigation of mechanical properties of bedded salt rock. International Journal of Rock Mechanics and Mining Sciences 44(9), 400–411 (2007)
6. Slizowski, J., Lankof, L.: Salt-mudstones and rock-salt suitabilities for radioactive-waste storage systems: rheological properties. Applied Energy 75(1-2), 137–144 (2003)
7. Hardy, R.H., Langer, M., Berest, P., et al. (eds.): Proc. 3rd Conf. Mech. Beh. of Salt, Palaiseau, p. 621. Trans. Tech. Pub., Clausthal-Zellerfeld (1993)
8. Kawamoto, T.: Deformation and fracturing behavior of discontinuous rock mass damage mechanics theory. International Journal of Numerical Analysis Method in Geomechanics 12(1), 1–30 (1988)
9. Li, Y.P., Yang, C.H., Qian, Q.H., et al.: Experimental research on deformation and failure characteristics of laminated salt rock. In: Wallner, M., Lux, K.H., Minkley, W., Hardy Jr., H.R. (eds.) Proceedings of the Sixth Conference on the Mechanical Behavior of Salt, pp. 69–74. Taylor & Francis, London (2007)
10. Chan, K.S., Munson, D.E., Bodner, S.R., et al.: Cleavage and creep fracture of rock salt. Acta Materialia 44(9), 3553–3565 (1996)

11. King, M.S.: Creep in model pillars of Saskatchewan potash. International Journal of Rock Mechanics and Mining Sciences & Geomechanics Abstracts 10(4), 363–371 (1973)
12. Manthei, G., Eisenblatter, J.: Acoustic emission studies on thermally and mechanically induced fracturing in salt rock. Acoustic Emission/Microseis Act Geol. Struct. Mater. 21, 245–267 (1986)
13. Moriya, H., Fujita, T., Niitsum, H.: Analysis of fracture propagation behavior using hydraulically induced acoustic emissions in the Bernburg salt mine, Germany. International Journal of Rock Mechanics & Mining Sciences 43(6), 49–57 (2006)
14. Wawersik, W.R., Zeuch, D.H.: Modeling and mechanistic interpretation of creep of rock salt below 200°C. Tectonophysics 121(2-4), 125–152 (1986)
15. Fossum, A.F., Brodsky, N.S., Chan, K.S., et al.: Experimental evaluation of a constitutive model for inelastic flow and damage evolution in solids subjected to triaxial compression. International Journal of Rock Mechanics and Mining Sciences & Geomechanics Abstracts 30(7), 1341–1344 (1993)
16. Cristescua, N.D.: A general constitutive equation for transient and stationary creep of rock salt. International Journal of Rock Mechanics and Mining Sciences & Geomechanics Abstracts 30(2), 125–140 (1993)
17. Munson, D.E.: Constitutive model of creep in rock salt applied to underground room closure. International Journal of Rock Mechanics and Mining Sciences & Geomechanics Abstracts 34(2), 233–247 (1997)
18. Kovalev, O.V., Litov, Y.N., Emanovski, Y.I.: Kinetics of damage accumulation of rock salt during creep. International Journal of Engineering Science 19(3), 451–454 (1981)
19. Chan, K.S., Bodner, S.R., Fossum, A.F., et al.: A constitutive model for inelastic flow and damage evolution in solids under triaxial compression. Mechanics of Materials 14(1), 1–14 (1992)
20. Chan, K.S., Brodsky, N.S., Fossum, A.F., et al.: Damage-induced nonassociated inelastic flow in rock salt. International Journal of Plasticity 10(6), 623–642 (1994)
21. Voyiadjis, G.Z., Zolochevsky, A.: Thermodynamic modeling of creep damage in materials with different properties in tension and compression. International Journal of Solids and Structures 37(24), 3281–3303 (2000)
22. Cai, M.F.: Rock Mechanics and Engineering. Science press, China (2002) (in Chinese)
23. Schulze, O., Popp, T., Kern, H.: Development of damage and permeability in deforming rock salt. Engineering Geology 61, 163–180 (2001)
24. Chan, K.S., Munson, D.E., Bodner, S.R., et al.: Cleavage and creep fracture of rock salt. Acta Materialia 44(9), 3553–3565 (1996)
25. Hardy, H.R.: Evaluation of in situ stresses in salt using acoustic emission techniques. In: Proceedings of the Seventh Symposium on Salt, vol. 1, pp. 49–58 (1993)
26. Ma, H.: Study on Feasibility of Rock Salt Underground Gas Storage in Ultra-deep Formation. Dissertation. Chinese Academy of Sciences, Wuhan, China (2010) (in Chinese)
27. Chen, J., Jiang, D.Y., Ren, S., et al.: Analysis of surface cracks growth and damage in salt rock under uniaxial compression. In: Bérest, P., Ghoreychi, M., Hadj-Hassen, F., Tijani, M. (eds.) Proc. 7th Conf. Mech. Beh. of Salt, pp. 31–36. Taylor&Francis Group, London (2012)
28. Wawersik, W.R., Fairhurat, C.A.: Study of brittle rock fractures in laboratory compression experiments. International Journal of Rock Mechanics and Mining Sciences & Geomechanics Abstracts 7(5), 561–575 (1970)
29. Krajcinovic, D.: Statistical aspects of the continuous damage theory. International Journal of Solids and Structures 18(7), 551–562 (1982)

30. Lemaitre, J.: How to use damage mechanics. Nuclear Engineering and Design 80(3), 233–245 (1984)
31. Kachanov, M., Tsukrov, I., Shafiro, B.: Effective modulus of solids with cavities of various shapes. Applied Mechanics Review 47, 151–174 (1994)
32. Cao, W.G., Zhang, S., Zhao, M.H.: Study on statistical damage constitutive model of rock based on new definition of damage. Rock and Soil Mechanics 27(1), 41–47 (2006)

# A Modified Mohr-Coulomb Failure Criterion for Intact Granites Exposed to High Temperatures

Hong Tian[1], Thomas Kempka[2], Nengxiong Xu[3], and Martin Ziegler[1]

[1] Department of Geotechnical Engineering, RWTH Aachen University, Mies-van-der-Rohe-Str. 1, 52074 Aachen, Germany
[2] German Research Centre for Geosciences (GFZ), Helmholtz Centre Potsdam, Telegrafenberg, 14473 Potsdam, Germany
[3] Department of Civil Engineering, China University of Geosciences (Beijing), Xuyuan street 29, Haidian district, 100083 Beijing, China

**Abstract.** Rocks often experience high temperatures (several hundred degrees Celsius) due to underground operations, such as deep geological disposal of nuclear waste, geothermal heat extraction, $CO_2$ geological storage and underground coal gasification as well as deep mining. Laboratory studies have shown that mechanical properties such as compressive strength, tensile strength, elastic modulus of rocks such as granite, marble and sandstone are dependent on temperature and temperature-history. Therefore, conventional failure criteria such as the Mohr-Coulomb criterion may not provide a good estimate of rock strength under high temperature conditions. In the present study, a thermo-mechanical modified Mohr-Coulomb failure criterion is proposed based on the extensive review and interpretation of mechanical properties of granites exposed to high temperatures. The deduced criterion takes into consideration the effects of thermal damage and confining conditions on rock strength. A numerical study indicates that the proposed criterion provides a higher quality for depicting rock strength under high temperatures compared to the conventional Mohr-Coulomb criterion. Moreover, according to analyses of the behavior of other rock materials exposed to high temperatures, this criterion is also suitable for other rocks.

**Keywords:** high-temperature rocks, failure criterion, thermal-mechanical modified Mohr-Coulomb model, granite.

## 1 Introduction

Rock-mechanical engineering in high temperature environments is of universal interest and a challenge to scientists and engineers of different disciplines. Rock mass may undergo high temperatures (several hundred degrees Celsius) in recent projects, such as deep underground nuclear waste disposal [1, 2], geothermal heat

M.Z. Hou et al. (Eds.): *Clean Energy Systems in the Subsurface*, SSGG, pp. 379–393.
DOI: 10.1007/978-3-642-37849-2_30 

extraction [3, 4], geological $CO_2$ storage [5] and underground coal gasification [6, 7], as well as deep mining [8, 9]. Previous work, e.g. Fredrich and Wong [10], Carlson et al. [11], Den'gina et al. [12] and Dwivedi et al. [13], has indicated that under the effect of high temperatures, the micro-structures of rocks change significantly, new micro-cracks are developed, and pre-existing ones are extended and/or widened. Meanwhile, various physical and mineralogical changes take place within these rocks. From a macroscopic point of view, strength and deformation characteristics of rocks exposed to high temperatures are quite different from those at room temperature. Therefore, corresponding high-temperature rock properties are key factors for the successful implementation of different rock engineering activities.

In the last few decades, special attention has been paid to mechanical, physical and thermal properties of crystalline rocks such as granite (e.g. [13, 14]) and marble (e.g. [15, 16]), and sedimentary rocks such as sandstone (e.g. [17, 18]) and limestone (e.g. [19]) during and after exposure to high temperatures. The experimental research indicated that in general rock mechanical properties such as elastic modulus, compressive and tensile strength, cohesive strength and friction angle decrease with increasing temperature. Especially, from 500 °C to 600 °C onwards, these mechanical indexes may be less than 50% of those at room temperature. Therefore, a thermo-mechanical (TM) failure criterion involving this temperature-dependent rock behavior is needed to depict mechanical phenomena of rocks exposed to different temperatures.

According to the latest published data in English and Chinese, the review of the high-temperature mechanical properties of granites has been updated first. It was established that slow cooling of a preheated rock in the air does not significantly affect its strength decrease acquired during the heating process [20]. Thus, all the data reviewed here was obtained either under high temperature or after slow cooling down conditions. Furthermore, as the linear Mohr-Coulomb criterion is the one most commonly used in practice, a TM modified failure criterion, based on the Mohr-Coulomb with a "tension cut-off" criterion, was proposed. At last, after discussion of the mechanical behavior of other rocks exposed to high temperatures, this criterion is also suitable for other rocks.

## 2 Thermo-Mechanical Properties of Granites

Heuze [14] and Dwivedi et al. [13] have given extensive reviews of mechanical, physical and thermal properties of granites exposed to high temperatures (below the melting point). In this section, the review is updated with new data collected from literature, especially from Chinese publications not considered in the English-speaking scientific community so far, covering elastic modulus, compressive strength, cohesion and friction angle, tensile strength as well as Poisson's ratio. A normalized value is defined as the ratio of the value at a testing temperature to the value at room temperature. The granites reviewed along with their abbreviated names and references are listed in Table 1.

**Table 1** The reviewed granites

| Granites | Abbreviated names | Data points reviewed | References |
|---|---|---|---|
| British granites | BG | 7 | McLaren & Titchel [21], Dwivedi et al. [13] |
| Charcoal granites | CcG | 11 | Bauer & Johnson[22] |
| Chinese granites | CG | 21 | Xu et al. [23] |
| Fine-grained granites | FG | 20 | Xu [24] |
| Henan granites | HG | 15 | Qiu & Lin [25] |
| Indian granites | IG | 20 | Dwivedi et al. [13] |
| Japanese granites | JG | 4 | Inada et al. [26] |
| Juyongguan granites | JyG | 7 | Jiang et al. [27] |
| Man-nari granites | MnG | 7 | Shimada & Liu [28] |
| Remiremont granites | RG | 19 | Homand-Etienne & Houpert [29] |
| Salisbury granites | SbG | 9 | Heuze [14] |
| Senones granites | SnG | 19 | Homand-Etienne & Houpert [29] |
| Stripa granites | StG | 10 | Swan [30] |
| Suixian granites | SxG | 8 | Wang et al. [31] |
| Westerly granites | WG_0 | 24 | Bauer & Johnson [22] |
| Westerly granites | WG_1 | 29 | Wong [32] |
| Westerly granites | WG_2 | 8 | Tullis & Yund [33] |
| Westerly granites | WG_3 | 6 | Friedman et al. [34] |
| Westerly granites | WG_4 | 12 | Bergman [1], Dwivedi et al. [13] |
| Woodbury granites | WbG | 4 | Clark [35] |
| Yunnan granites | YG | 10 | Zhu et al. [36] |

## 2.1 Elastic Modulus

Elastic modulus is dependent on temperature and pressure. The variations of normalized elastic modulus ($E/E_0$) with increasing temperature, irrespective of pressure, are plotted in Figure 1. At the atmospheric pressure and temperature up to 200 °C, granites show a mixed trend with temperature, but from then onwards, $E/E_0$ decreases with increasing temperature for all the granites reviewed (Fig. 1). Under confining pressure conditions, Dwivedi et al. [13] concluded after an extensive review that the $E/E_0$-values almost always decrease with increasing temperature. Xi & Zhao [37] also found that the elastic modulus of granites under a confining pressure of 125 MPa decreases with increasing temperature.

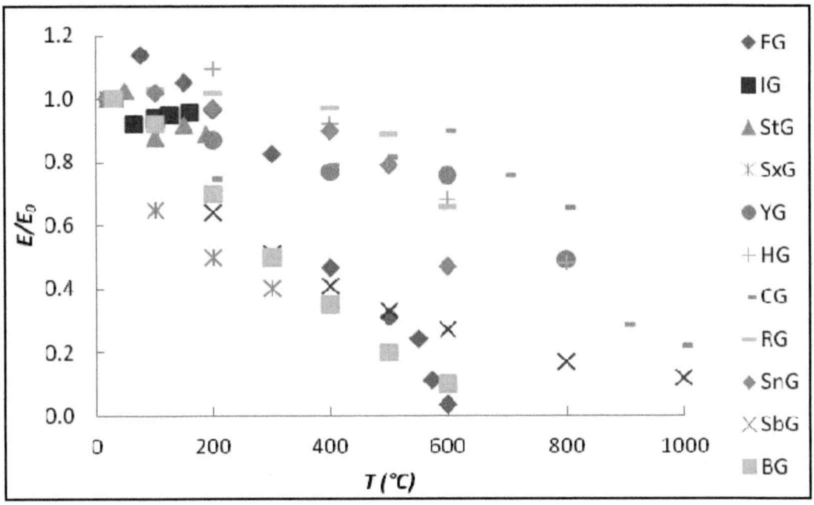


**Fig. 1** Normalized elastic modulus vs. temperature under atmospheric pressure

## 2.2 Compressive Strength

In general, the compressive strength of rocks is also dependent on temperature and pressure. Figure 2 shows the values of normalized unconfined compressive strength ( $\sigma_c/\sigma_{c0}$ ) as a function of temperature. Different trends are observed up to 200 °C for the granites reviewed, whereas a decrease trend is observed from that point onwards.

Values of normalized ultimate compressive strength from tri-axial compressive tests under different confining pressure conditions are plotted in Figure 3. A general decreasing trend with increasing temperature is observed, although mixed trends of MnG and JyG appeared in the temperature range of 200–400 °C.

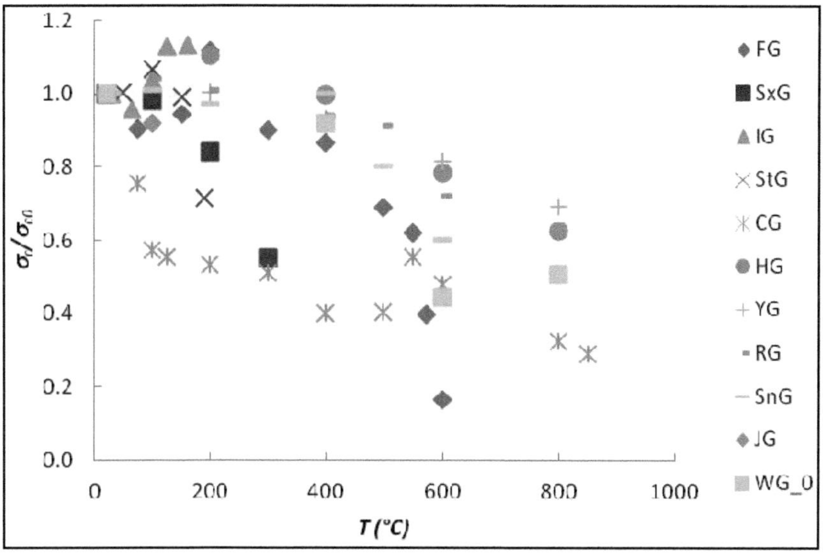


**Fig. 2** Normalized unconfined compressive strength vs. temperature

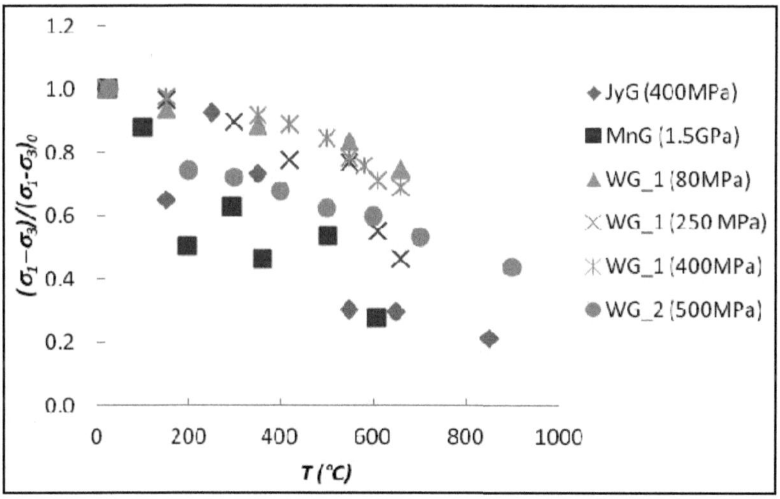


**Fig. 3** Normalized ultimate compressive strength vs. temperature under different confining pressures (in brackets)

## *2.3 Cohesive Strength and Friction Angle*

Normalized cohesive strength ($c/c_0$) and normalized friction angle ($\phi/\phi_0$) are plotted in Figs. 4 and 5, respectively. The $c/c_0$-values decrease with increasing temperature except for an increase observed for WG_1 at 150 °C compared with that at room temperature. ($c/c_0$) always decreases with increasing temperature for the granites reviewed.

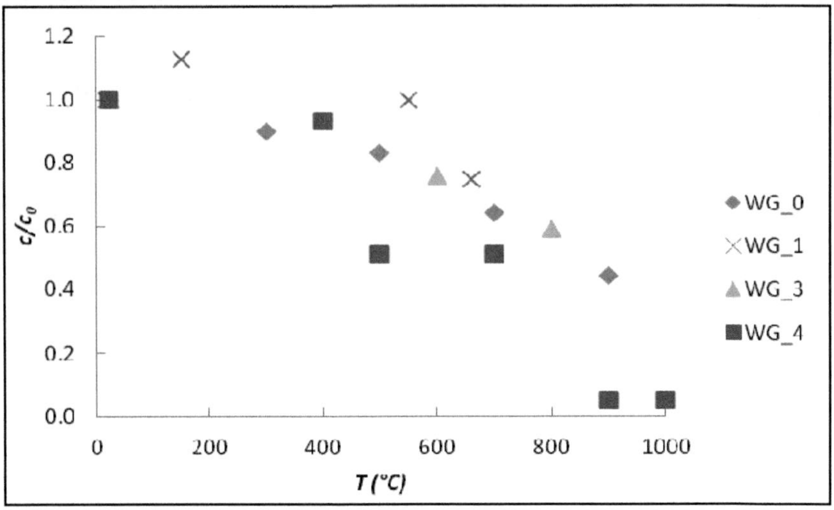


**Fig. 4** Normalized cohesive strength vs. temperature

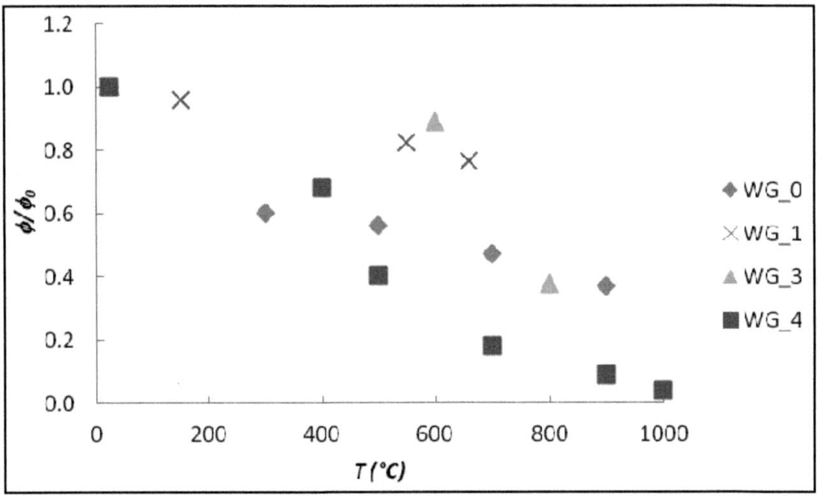


**Fig. 5** Normalized internal friction angle vs. temperature

## 2.4 *Tensile Strength*

Figure 6 plots the normalized tensile strength ( $\sigma_t/\sigma_{t0}$ ) as a function of temperature. The values of $\sigma_t/\sigma_{t0}$ decrease with increasing temperature for the granites reviewed, except for SnG where an increasing trend occurs at temperatures between 400 °C and 500 °C. A faster decrease trend is observed in the temperature range of 400–600 °C. Beyond that temperature range, the rate becomes slow.

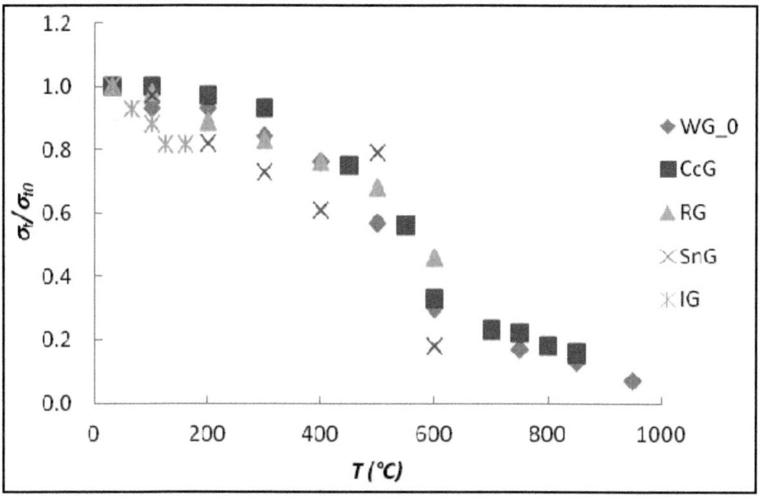


**Fig. 6** Normalized tensile strength vs. temperature after Dwivedi et al. [13]

## 2.5 *Poisson's Ratio*

As shown in Figure 7, different trends of the values of normalized Poisson's ratio for the granites reviewed are observed under different confining pressures and temperatures up to 200 °C. However, the variation is very slight in the range of ±5%. From 200 °C onwards, a decreasing trend is observed for HG. Therefore, more laboratory experiments are needed to conclude the relation between normalized Poisson's ratio and high temperatures, especially above 200 °C.

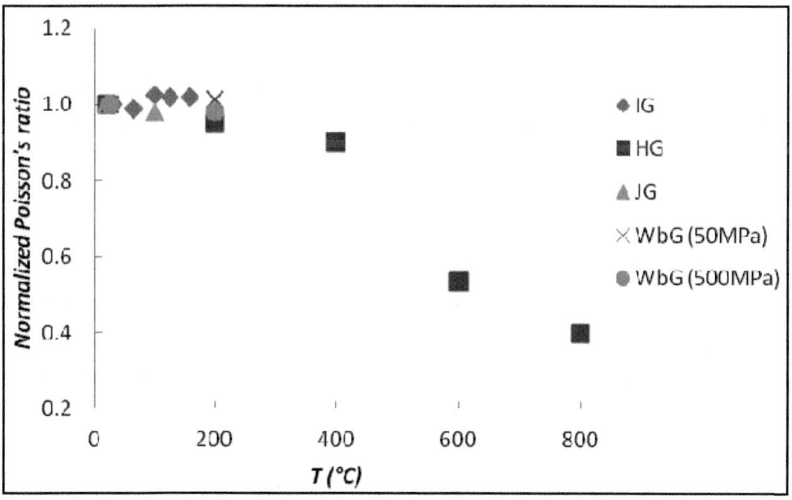


**Fig. 7** Normalized ultimate compressive strength vs. temperature (the values in brackets are confining pressures)

# 3 TM Mohr-Coulomb Failure Criterion

Ohnaka [38] proposed an empirical shear failure strength law of Westerly granites in the brittle to brittle-plastic transition regimes, which is especially suitable for predicting the shear failure strength under high pressure (a few hundred MPa) and temperature (below the melting point) conditions. Hueckel et al. [39] presented a framework of constitutive modeling of the thermo-brittle-plastic behavior of granites and marbles using at least ten constants. To better depict granite strength under common engineering pressure range (below 100~200 MPa) and high temperature (below the melting points of rocks) conditions with a criterion having a relatively simple mathematical formula, a new TM failure criterion based on the Mohr-Coulomb criterion [40] with a "tension cut-off" is proposed.

## 3.1 General

The TM criterion suggested in this paper is for isotropic rocks and does not take into account the effect of intermediate principal stress. It is known that the Mohr-Coulomb model predicts a tensile strength larger than the one observed in experiments [41]. This discrepancy can be amended to some extent by the introduction of a "tension cut-off" criterion. Thus, a TM "tension cut-off" criterion is proposed here to reflect the high-temperature tensile strength behavior (Eq. 2). The general form of the TM modified Mohr-Coulomb (TMMC) linear criterion can be expressed as follows:

$$\sigma_1 - \sigma_3 = 2c^T \cos\phi^T + (\sigma_1 + \sigma_3)\sin\phi^T \tag{1}$$

$$\sigma_1 - \sigma_t^T = 0 \tag{2}$$

where $\sigma_1$ and $\sigma_3$ are the principal stresses; $c^T$ , $\phi^T$ and $\sigma_t^T$ . Temperature-dependent cohesive strength, friction angle and tensile strength, respectively, obtained from an interpretation of high temperature tri-axial compression/extension and the Brazilian tests. In this paper compression is considered to be positive.

In Figure 8, the criterion functions (1) and (2) at three temperature levels ($T_1$, $T_2$ and $T_3$) are plotted in $\sigma_1$ versus $\sigma_3$ space with decreasing trends of cohesion and friction angle with increasing temperature. Apparently, the criterion contains the influence of high temperature on tensile and compressive strength; it can therefore better describe the strength of granites at high temperatures.

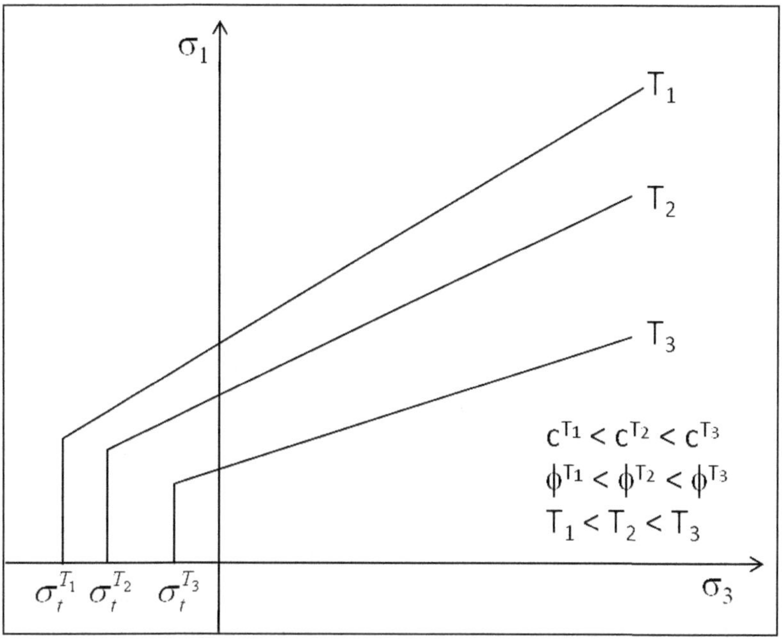


**Fig. 8** Schematic diagram of the TM modified Mohr-Coulomb model

## *3.2 Relation with Triaxial Experiments*

Theoretically, to obtain the parameters $c^T$, $\phi^T$ and $\sigma_t^T$ in the proposed criterion, a series of high-temperature tri-axial compression/extension and/or tensile strength tests (e.g. the Brazilian test), should be carried out. However, as depicted in Section 2, the three parameters usually decrease with increasing temperature in a simplified linear or bilinear way (Fig. 9). For a bilinear relation, the transition temperature ($T_t$) is usually in the range of 500 °C – 600 °C.

## *3.3 Case Study*

A TM finite element calculation was performed for a circular underground opening excavated at a depth of 2000 m in a hydrostatic stress state (40MPa) on a granite material with the density of 2.7 g/cm3. The constitutive relation is thermo-elastic perfect plasticity with μ = 0.35, linear thermal expansion coefficient α = 10-5/°C and temperature-dependent elastic modulus ET where ER.T. = 50 GPa. Two failure criteria are employed. The first is the Mohr-Coulomb (MC) criterion with c = 40 MPa, $\phi$ = 40° and $\sigma_t$ = 8 MPa. The second is the proposed TM modified Mohr-Coulomb (TM-MC) criterion, whereas its corresponding

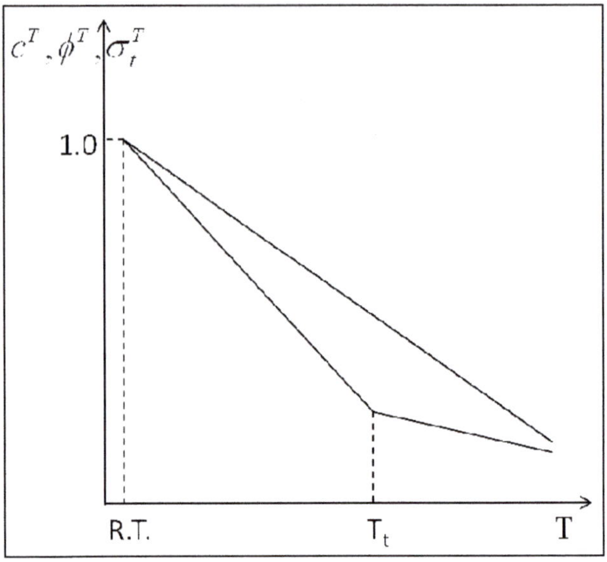


**Fig. 9** Simplified relations between the parameters in the criterion with temperature (R.T. is short for room temperature)

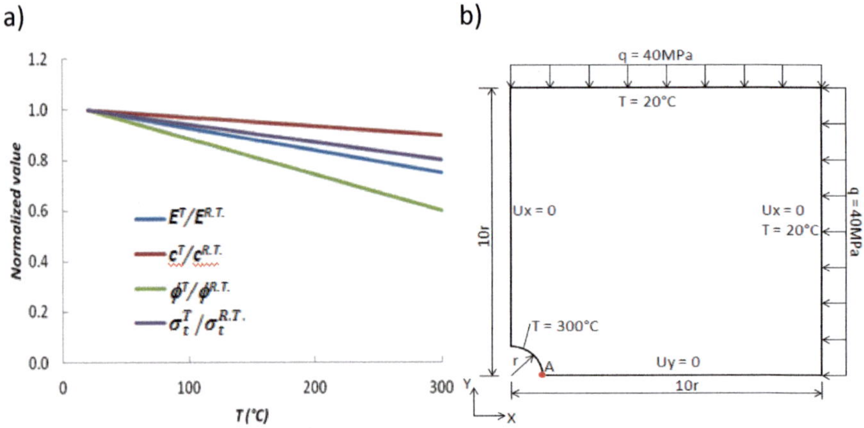


**Fig. 10** a) Simplified relations between the parameters with temperature; b) geometry and boundary conditions of the computational model

parameters at room temperature are equal to those used in the MC criterion. The relations of temperature-dependent parameters with temperature used in the study are based on the average values in Section 2 and plotted in Figure 10a.

A quarter symmetry model with r = 1 as well as its displacement, pressure and temperature boundary conditions are shown in Figure 10b. 8-node quadrangle elements are used. The temperature field is calculated by transient heat transfer.

Figure 11 plots the relations between normalized stresses with the distance ratio R/r where R is the distance from the center of the opening (Fig. 10b) to a point in the positive horizontal direction of A, and the corresponding temperature distribution. Since the TMMC criterion considers the rock strength exposed to high temperature, the results of the TMMC are more reliable than the MC. It is seen that the maximum radial stress ($\sigma_r$) and tangential stress ($\sigma_t$) of the Mohr-Coulomb criterion are larger than those of the TMMC criterion and the difference of $\sigma_\theta$ in the two scenarios is larger than that of $\sigma_r$ (as shown in the circled part of Fig. 11); the yield zone of the MC is smaller than the TMMC. Due to thermal expansion, the stress distributions in both scenarios are different from those of a pure mechanical calculation. Thus, it is suggested that the TM failure criterion should be considered when rocks are exposed to high temperatures.

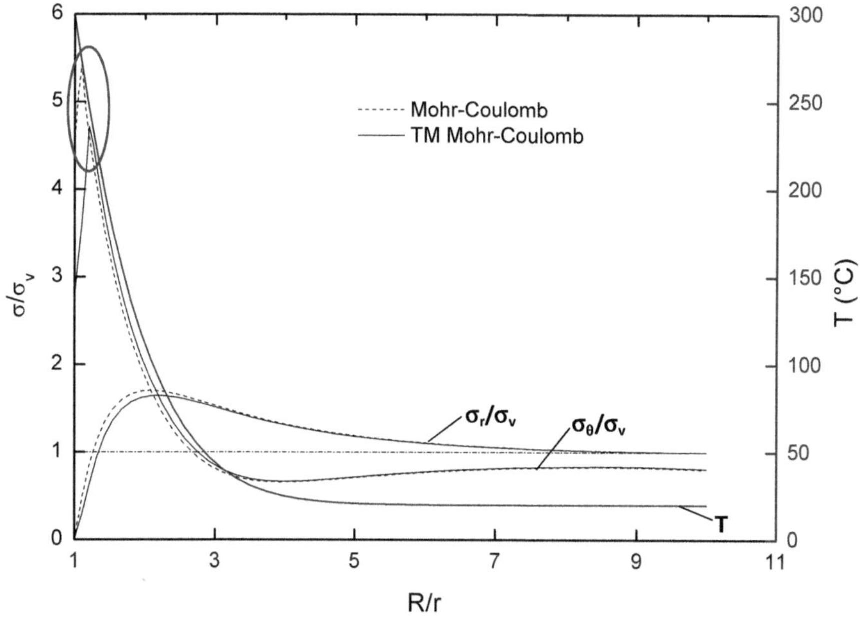


**Fig. 11** Stresses and temperature distributions along the positive horizontal direction from A

# 4 Discussion

## *4.1 Limitations*

The following assumptions have been made in the proposed TMMC criterion:

- A rock is an isotropic material,
- The temperature of this criterion is in the range of room temperature to a high temperature which is less than the rock melting point,

- The confining pressure corresponds to common engineering pressures,
- The linear Mohr-Coulomb with the "tension cut-off" is assumed to fit the strength behavior of granites at each temperature level,
- The effect of the intermediate principal stress is neglected.

### 4.2 Suitability for Other Rocks

It is accepted that the deformation and strength of rocks exposed to high temperatures are different from those at normal temperature, and generally, a decreasing trend with increasing temperature is observed in experiments. Various experiments on different rock materials conducted at different high-temperature levels show that the Mohr-Coulomb criterion fits the data obtained from uni-axial and conventional tri-axial tests. In addition, under the conditions of high temperature (below the rock melting point) and common engineering pressures, the failure mechanism of rocks is unchanged compared to that under normal conditions [42, 43]. Therefore, it is experimentally and theoretically reasonable that the TM modified Mohr-Coulomb criterion proposed utilizes temperature-dependent cohesion, friction angle and tensile strength to describe the effects of high temperature on strength. Thus, even though the criterion is created based on the review of the high-temperature mechanical properties of granites, it is also suitable for other rocks.

## 5 Conclusions

A TM modified Mohr-Coulomb (TMMC) failure criterion is proposed on the basis of extensive review and analysis of the mechanical parameters of granites exposed to high temperatures (below the melting point) and common engineering confining pressures. To take into account of the effects of high temperatures on rock strength, the proposed criterion contains three temperature-dependent parameters, $c^T$, $\phi^T$ and $\sigma_t^T$. The numerical study shows the yield zone of an underground opening exposed to high temperature calculated by the TMMC criterion is larger than that determined by the MC. Thus, under high temperature conditions, a TM failure criterion should be involved to depict temperature-dependent rock strength. Moreover, the proposed criterion is also suitable for other rocks.

**Acknowledgements.** The first author appreciates the funding provided by the China Scholarship Council (CSC).

## References

1. Bergman, M.S.: Special Session E: Nuclear waste disposal. In: Proceedings Int. Symp. on Subsurface Space (Rockstor 1980), pp. 789–1005. Pergamon Press, Stockholm (1980)
2. Rutqvist, J., Freifeld, B., Min, K.B., Elsworth, D., Tsang, Y.: Analysis of thermally induced changes in fractured rock permeability during 8 years of heating and cooling at the Yucca Mountain Drift Scale Test. International Journal of Rock Mechanics & Mining Sciences 45, 1373–1389 (2008)
3. Zhao, Y.S.: The problem of rock mechanics in heat exploitation in hot dry rock. In: Proceedings of 6th Academy Conference of Chinese Rock Mechanics and Engineering, October 24-27, pp. 71–74. China academic journal electronic publishing house, Wuhan (2000) (in Chinese)
4. Zhao, Y.S., Wang, R.F., Hu, Y.Q., Wan, Z.J., Xie, Y.S.: 3D numerical simulation for coupled THM of rock matrix-fractured media in heat extraction in HDR. Chinese Jounrnal of Rock Mechanics and Engineering 21(12), 1751–1755 (2002) (in Chinese)
5. Roddy, D.J., Younger, P.L.: Underground coal gasification with CCS: a pathway to decarbonising industry. Energy Environ. Sci. 3, 400–407 (2010)
6. Kühnel, R.A., Schmit, C.R., Eylands, K.E., McCarthy, G.J.: Comparison of the pyromeramorphism of clayey rocks during underground coal gasification and firing of structural ceramics. Applied Clay Science 8, 129–146 (1993)
7. Burton, E., Friedmann, J., Upadhye, R.: Best Practices in Underground Coal Gasification. Lawrence Livermore National Laboratory Report (2007)
8. Zhou, H.W., Xie, H.P., Zuo, J.P.: Developments in researches on mechanical behaviors of rocks under the confining of high ground pressure in the depths. Advances in Mechanics 35(1), 91–99 (2005) (in Chinese)
9. He, M.C.: Application of HEMS cooling technology in deep mine heat hazard control. Mining Science and Technology 19, 269–275 (2009)
10. Fredrich, J.T., Wong, T.F.: Micromechanics of thermally induced cracking in three crustal rocks. Journal of Geophysical Research 91, 12743–12764 (1986)
11. Carlson, S.R., Wu, M., Wang, H.F.: Micromechanical modelling of thermal cracking in granite. The Brittle-Ductile Transition in Rocks. Geophys. Monogr. Ser. 56, 37–48 (1990)
12. Den'gina, N.I., Kazak, V.N., Pristash, V.V.: Changes in rocks at high temperatures. Journal of Mining Science 29(5), 472–477 (1994)
13. Dwivedi, R.D., Goel, R.K., Prasad, V.V.R., Sinha, A.: Thermo-mechanical properties of Indian and other granites. International Journal of Rock Mechanics & Mining Sciences 45, 303–315 (2008)
14. Heuze, F.E.: High-temperature mechanical, physical and thermal properties of granitic rocks - a review. International Journal of Rock Mechanics and Mining Science & Geomechanics Abstracts 20(1), 3–10 (1983)
15. Ferrero, A.M., Marini, P.: Experimental studies on the mechanical behaviour of two thermal cracked marbles. Rock Mechanics and Rock Engineering 34(1), 57–66 (2001)
16. Zhang, L.Y., Mao, X.B., Sun, J.F., Lu, A.H.: Mechanical properties of marble at high temperature. Journal of Chongqing Jianzhu University 30(6), 46–50 (2008) (in Chinese)

17. Wu, Z., Qin, B.D., Chen, H.J., Luo, Y.J.: Experimental study on mechanical character of sandstone of the upper plank of coal bed under high temperature. Chinese Journal of Rock Mechanics and Engineering 24(11), 1863–1867 (2005) (in Chinese)
18. Qin, B.D., He, J., Chen, L.J.: Experimental research on mechanical properties of limestone and sandstone under high temperature. Journal of Geomechanics 15(3), 253–261 (2009) (in Chinese)
19. Zhang, L.Y., Mao, X.B., Yang, Y., Meng, M.M., Du, S.J.: Experiment study on mechanical properties of limestone at high temperature. Journal of Liaoning Technical University 25(suppl.), 121–123 (2006)
20. Dmitriyev, A.P., Kuzyayev, L.S., Protasov, Y.I., Yamsbcbikov, V.S.: Physical Properties of Rocks at High Temperatures. Nedra Press, Moscow (1969)
21. McLaren, J.R., Titchel, I.: Physical Properties of Granites Relevant to Near Field Conditions in a Nuclear Waste Depository. AERE, Harwell (1981)
22. Bauer, S.J., Johnson, B.: Effects of slow uniform heating on the physical properties of the westerly and charcoal granites. In: 20th U.S. Symposium on Rock Mechanics, Austin, Texas, June 4–6, pp. 7–18. ASCE, New York (1979)
23. Xu, X.L., Gao, F., Zhou, Q., Chen, J.: Energy analysis of rock deformation and failure process after high temperature. Journal of Wuhan University of Technology 33(1), 1671–4431 (2011) (in Chinese)
24. Xu, X.C.: Study on the characteristics of thermal damage for granite. Rock and Soil Mechanics 24(suppl.), 188–192 (2003)
25. Qiu, Y.P., Lin, M.Y.: Testing study on damage of granite samples after high temperature. Rock and Soil Mechanics 27(6), 1005–1010 (2006) (in Chinese)
26. Inada, Y., Kinoshita, N., Ebisawa, A., Gomi, S.: Strength and deformation characteristics of rocks after undergoing thermal hysteresis of high and low temperatures. International Journal of Rock Mechanics & Mining Sciences 34(3-4), Paper No. 140 (1997)
27. Jiang, H.K., Zhang, L., Zhou, Y.S.: Characteristics of AE temporal sequences in the process of deformation and failure of granite at high pressure and different temperatures. Earthquake 20(3), 87–94 (2000) (in Chinese)
28. Shimada, M., Liu, J.L.: Temperature dependence of strength of rocks under high confining pressure. Annuals of Disas. Prev. Res. Inst. Kyoto Univ. 43(B-1), 75–84 (2000)
29. Homand-Etienne, F., Houpert, R.: Thermally induced microcracking in Granites: characterization and analysis. International Journal of Rock Mechanics and Mining Science & Geomechanics Abstracts 26(2), 125–134 (1989)
30. Swan, G.: The Mechanical Properties of Stripa Granite. Dept. of Energy, Lawrence Berkeley Laboratory, Earth Sciences Division (1978)
31. Wang, J.T., Zhao, A.G., Huang, M.C.: Effect of high temperature on the fracture toughness of granite. Chinese Journal of Geotechnical Engineering 11(6), 113–119 (1989) (in Chinese)
32. Wong, T.F.: Effects of temperature and pressure on failure and post-failure behavior of Westerly granite. Mechanics of Materials 1, 3–17 (1982)
33. Tullis, J., Yund, R.A.: Experimental deformation of dry westerly granite. Journal of Geophysical Research 82(36), 5705–5718 (1977)
34. Friedman, M., Bauer, S.J., Chester, F.M., Handin, J., Hopkins, T.W.: Mechanical Properties of Rocks at High Temperatures and Pressures. Center for tectonophysics Texas A&M University (1987)

35. Clark, J.S.P. (ed.): Handbook of Physical Constant. Geological society of America, New York (1966)
36. Zhu, H.H., Yan, Z.G., Deng, T., Yao, J., Zeng, L.J., Qiang, J.: Testing study on mechanical properties of tuff, granite and breccia after high temperatures. Chinese Journal of Rock Mechanics and Engineering 25(10), 1945–1950 (2006) (in Chinese)
37. Xi, B.P., Zhao, Y.S.: Experimental study of thermophysico-mechanical property of drilling surrounding rock in granite under high temperature and high pressure. Chinese Jounrnal of Rock Mechanics and Engineering 29(6), 1245–1253 (2010) (in Chinese)
38. Ohnaka, M.: A shear failure strength law of rock in the brittle-plastic transition regime. Geophysical Research Letters 22(1), 25–28 (1995)
39. Hueckel, T., Peano, A., Pellegrini, R.: A thermo-plastic constitutive law for brittle-plastic behavior of rocks at high temperatures. Pageoph. 143(1/2/3), 483–511 (1994)
40. Labuz, J., Zang, A.: The Mohr Coulomb failure citerion. Rock Mechanics and Rock Engineering 45, 975–979 (2012)
41. Ottosen, N.S., Ristinmaa, M.: The Mechanics of Constitutive Modelling. Elsevier (2005)
42. Chen, L.J., Wu, Z., Qin, B.D., Gu, H.T.: Mechanical characteristics and cracking mechanism of coal roof sandstone under high temperature. Journal of Chongqing University 28(5), 123–126 (2005) (in Chinese)
43. Li, D.W., Zhu, Z.D., Jiang, Z.J., Qu, W.P.: Microstructural investigation of mechanical characteristics of marbles under different temperatures. Journal of Hehai University (Natural Sciences) 36(3), 375–378 (2008) (in Chinese)

# The H2STORE Project: Hydrogen Underground Storage – A Feasible Way in Storing Electrical Power in Geological Media?

Dieter Pudlo[1], Leonhard Ganzer[2], Steven Henkel[1], Michael Kühn[3], Axel Liebscher[3], Marco De Lucia[3], Michel Panfilov[4], Peter Pilz[3], Viktor Reitenbach[2], Daniel Albrecht[2], Hilke Würdemann[3], and Reinhard Gaupp[1]

[1] Friedrich-Schiller-University Jena (FSU), Institute of Geosciences, Burgweg 11, D-07749 Jena, Germany
[2] Clausthal University of Technology (TUC), Institute of Petroleum Engineering, Agricolastraße 10, D-38678 Clausthal-Zellerfeld, Germany
[3] Helmholtz Centre Potsdam, German Research Centre for Geosciences (GFZ), Telegrafenberg, D-14473 Potsdam, Germany
[4] Centre National de la Recherche Scientifique (CNRS), Université de Lorraine, LEMTA-UMR 7563 - Laboratoire d'Energétique et de Mécanique Théorique et Appliquée, 2 avenue de la Forêt de Haye, BP 160, F-54504, Vandoeuvre-lès-Nancy, France

**Abstract.** The large scale storage of energy is a great challenge arising from the planned transition from nuclear and $CO_2$-emitting power generation to renewable energy production, by e.g. wind, solar, and biomass in Germany. The most promising option for storing large volumes of excess energy produced by such renewable sources is the usage of underground porous rock formations as energy reservoirs. Some new technologies are able to convert large amounts of electrical energy into a chemical form, for example into hydrogen by means of water electrolysis. Porous formations can potentially provide very high hydrogen storage capacities. Several methods have to be studied including high hydrogen diffusivity, the potential reactions of injected hydrogen, formation fluids, rock composition, and the storage complex.

Therefore, in August 2012 the collaborative project H2STORE ("hydrogen to store") started to investigate the feasibility of using burial clastic sediments of depleted gas reservoirs as well as recently used gas storage sites as potential hydrogen storage media. In Germany, such geological structures occur at various geographic sites and different geological strata. These deposits are characterized by different geological-tectonic evolution and mineralogical composition, mainly depending on palaeogeographic position and diagenetic burial evolution. Resulting specific sedimentary structures and mineral parageneses will strongly control formation fluid pathways and associated fluid-rock/mineral reactions. Accordingly, H2STORE will analyze sedimentological, petrophysical, mineralogical/geochemical, hydrochemical, and microbiological features of the

M.Z. Hou et al. (Eds.): *Clean Energy Systems in the Subsurface*, SSGG, pp. 395–412.
DOI: 10.1007/978-3-642-37849-2_31 

different geological strata and the German locations to evaluate potential fluid-rock reactions induced by hydrogen injection. Such potential reactions will be experimentally induced in laboratory runs, as analogues for naturally occurring processes in deep seated reservoirs. Finally, rock data determined before and after these experiments will be used as major input parameters for numerical modelling of mineralogical and microbiological reactions. Such reactions are expected to have a strong affect on rock porosity-permeability evolution and therefore the characteristics of flow processes in reservoir and the barrier properties of sealing rocks.

The special topic of this study will be the modelling of hydrogen propagation in the subsurface reservoir formation supplemented by its mixing with the residual gases as well as the simulation of coupled bio-dynamic processes and of reactive transport in porous media. These numerical simulations will enable the transfer of experimental results from the laboratory runs to the field-scale and the formulation of the requirements for hydrogen storage in converted gas fields.

Thus, the major objectives of H2STORE are to obtain fundamental data on the behaviour of clastic sediments in the presence of formation fluids and injected hydrogen, its impact on petrophysical features and the development of the most realistic modelling for proposed and experimentally induced rock alteration as well as complex gas mixing processes in potential geological hydrogen reservoirs. Moreover these results will be used when discussing the possibility of "green" eco-methane generation by hydrogen and carbon dioxide interaction in the geological underground.

**Keywords:** Hydrogen storage, silici-clastic sediments, reservoir and sealing rocks, mineralogical-geochemical-hydrochemical-petrophysical-microbiological interactions.

# 1 Introduction

The announcement in 2011 by the German government to abandon nuclear energy production and to reduce the $CO_2$-emitting coal consumption by the extended use of fluctuating renewable energy sources continues to gain increasing international attention. The viability of such a turnaround scenario depends on the feasibility of the large-scale conversion and storing of excess electrical energy combined with its recovery to guarantee safe and reliable maintenance of the industrial and residential electric power supply.

Besides the well-known use of hydropower by dams and pump storage plants, additional and innovative ideas for storing energy have been discussed in the past few years. Thereby technical methods (e.g. the usage of batteries and fuel cells) will only be invoked to store small amounts of energy for very short periods. However, the long-term storage of high amounts of excess electrical energy is still a major technical challenge. At present four techniques are in use, under development or investigation or being examined in pilot projects: such as the storage of natural and town gas in clastic sediments (e.g. Giese and Schuldt 2005), the storage of compressed air energy storage in salt caverns (CAES - e.g. Cavallo 2007, Lund & Salgi 2009, Brockmann et al. 2010) and abandoned mines (e.g. Beck & Franz 2010), and the storage of hydrogen in salt caverns (e.g. Brockmann et al. 2010, Gillhaus 2010). The option of storing hydrogen accompanied by carbon dioxide injection underground to generate "green" eco-methane is of increasing interest. Most recently, in Germany (Brandenburg, Lower Saxony) hybrid power stations have been installed to generate "power to heat" and "power to gas" using similar processes, whereby these interactions are investigated in reactors at the surface. However CAES and hydrogen storages in salt caverns are limited by the lack of suitable salt deposits and their minor storage capacities, enabling only short term energy storage, at present (e.g. within days to weeks).

One possibility of storing energy transformed into hydrogen for a longer, seasonal time scale (e.g. within months) can be the use of depleted gas reservoirs, which have proved their ability for storing natural gas over millions of years. Investigations on energy storage in the form of electrolytically produced hydrogen in such depleted gas reservoirs and saline aquifers are sparse, because, compared to salt cavern storage, reactions in porous silici-clastic reservoirs are much more complex. Such silici-clastic geological formations are characterized by the presence of different mineral phases, sedimentary structures and e.g. hydrocarbons accompanied by gas impurities and formation fluids, which will control fluid migration and interact with hydrogen in different physico-chemical ways. Thereby some of the studies of porous reservoirs have concentrated on the reactions of specific chemical compounds and single mineral phases exposed to hydrogen (Foh et al. 1979, Lord 2009). Therefore the intention of the H2STORE project is to study particular sandstone types from distinguished German hydrocarbon-bearing locations. To investigate the reaction of these rocks to hydrogen, the representative rock samples will be exposed to hydrogen in autoclaves under reservoir conditions. Thus potential mineralogical, bio-geochemical, and petro-physical alterations induced by these experiments can be evaluated. The identified experimentally induced alterations will be analyzed and discussed as analogues for potential reactions occurring during hydrogen injection in the natural environment of underground reservoirs.

Finally numerical modelling of mineralogical and microbiological reactions, as well as simulation of coupled bio-dynamic and reactive transport processes in porous media with hydrogen propagation in the subsurface reservoir formation supplemented by its mixing with the residual gases, will be performed. This modelling can also suggest improvements in the production of "green" methane by the reaction of hydrogen and carbon dioxide in the presence of micro-organisms at depths (depleted gas reservoirs) and establish physico-chemical parameters for "synthetic natural gas - SNG" generation in industrial plants. In general the simulations will enable the transfer of experimental results from the laboratory to the field-scale and help to formulate specific requirements for hydrogen storage in converted gas fields.

## 2 State of the Art

In the past, investigations on hydrogen behaviour in geological systems were mainly concentrated on the magmatic and metamorphic environment in the context of hydrocarbon studies (e.g. Demouchy et al. 2006, Kinnaman et al. 2007, Purwin et al. 2009). Most recently the topic of hydrogen reactivity and diffusion gained even more attention, especially in material and engineering sciences, for its potential use in developing new and more effective electric power storage options, like batteries, fuel cells and new catalysts and hydrogen storage materials (e.g. Sørensen 2007, Pukazhselvan et al. 2012).

However, studies and projects for storing large volumes of hydrogen in sedimentary rocks are still sparse and almost exclusively related to salt caverns (e.g. Beck & Franz 2010, Brockmann et al. 2010). Most recently, in 2011/ 2012, pilot projects in Germany (Brandenburg, Lower Saxony) started to store and inject hydrogen and $CO_2$ in containments and sedimentary rocks. The intention of these short-lived projects was the storage of small volumes of hydrogen and/ or the coupled $H^+$- and $CO_2$ (derived from biomass) injection for producing power to heat and "green" eco-methane underground. However the long-term storage of large volumes of hydrogen as a reproducible source of energy during periods of high demand is outside the scope of these projects.

The major problems in installing such large hydrogen storage sites are the high diffusivity and the uncertain reactivity of hydrogen in complex geological systems, characterized by highly variable solid phases (minerals), saline water compositions (formation fluid), microbiological biocenosis and the use of almost depleted natural gas reservoirs containing hydrocarbon gases and occasionally mixtures of non-hydrocarbons (e.g. $N_2$, $H_2S$, CO, $CO_2$, $SO_2$ etc.).

In general any input of hydrogen into (highly) saline fluid-bearing (underground) systems will most probably force a decrease in pH-conditions. Such lowering of pH-values will not only induce multi mineral dissolution but also mineral precipitation processes. For instance, such pH decreased carbonate- and sulphate minerals (like calcite, dolomite, siderite, gypsum, anhydrite, and barite), feldspars and clay minerals of the chlorite group will be dissolved, but in contrast illite (K-bearing clay mineral) is likely to be formed (e.g. Allan et al. 2011, Brandt et al. 2003, Flaathen et al. 2009, Fischer et al. 2010, Velde and Meunier 2008, Pudlo et al. 2012). Thereby the release of $HCO_3^{2-}$ and $SO_4^{2-}$ from carbonate and sulphate alteration will provide potential reactants for injected $H^+$ in forming bicarbonate ($HCO^-_3$) and sulphur/ sulphurous acids ($H_2SO_4$ and $H_2SO_3$, resp.), which will result in a further decrease in pH-values. Whereas the dissolution of minerals will result in an increase in porosity, precipitation of clay minerals (e.g. illite) will strongly reduce rock permeability.

With respect to the high diffusivity of hydrogen and therefore, the associated potential risks of escape of hydrogen from storage, the use of depleted natural gas reservoirs appears to be the most appropriate option for large-scale hydrogen storage. These structures proved their tightness over millions of years by their sealing capacity of multi layer barriers of several (very) low permeable clay horizons, intercalating and overlying the reservoir rocks (Czurda, 2006). The investigations planned in H2STORE share many general procedural and geoscientific aspects with the research on $CO_2$-CCS (Carbon Capture and Storage) topics, such as the conceptual classification of physical processes occurring in the reservoir (Gunter et al. 2004), but with particular focus on the specific interactions induced by the contact between $H_2$ and natural porous media.

Foh et al. (1979) summarized the potential reactivity of hydrogen with chemical species (e.g. oxides, carbonates, sulphates, sulphides) common in mineral phases of sedimentary rocks at low temperatures (~ 25°C) and pressures (~ 14 MPa). However, this approach ignores the complex crystallographic and chemical composition of most minerals. For instance the behaviour of clay minerals mainly composed of major oxide components such as $SiO_2$, $Al_2O_3$, FeO/ $Fe_2O_3$, MgO, and $K_2O$ cannot be described without considering their crystallographic structure, composed of tetrahedral and octahedral sheets intercalated by $OH^-$-bearing interlayers. In this context, the adsorption capacity of hydrogen by alumina-pillared montmorillonite, a member of the smectite-clay group, containing Si and Al as well as minor amounts of Ca, Na, Mg and Fe is investigated by Gil et al. (2009). Thereby this capacity is controlled by the framework structure and the composition of the clays and adsorption is increased almost linearly with higher microporous volumes, suggesting that hydrogen is mainly absorbed by the solids.

The relevance of distinct crystallographically ordered and chemically bounded elements in clay minerals, which are only composed of Si and Al (kaolinite clay group) is also shown by Tunega et al. (2002). The authors stated that due to the higher number of hydroxyl groups and octahedral surfaces of this type of clay mineral it is more appropriate for the adsorption of polar and/ or negative charged species than tetrahedral sheets. However, these hydroxyl sites have a bifunctional character and can act as a proton donor and/ or proton acceptor (Tunega et al., 2002).

Therefore the behaviour of clay minerals and their capability to bind and adsorb hydrogen is an important aspect of storing hydrogen in clastic sediments and also for the sealing capacity of caprocks, which are most often composed of clay mineral phases.

However, hydrogen-silicate reactions differ significantly from hydrogen-carbonate reactions (Heinrich et al., 1978). This is deduced from their experiments. Here, olivine, feldspar, and spodume (Li-bearing pyroxene) reveal only very low alteration and decomposition, which is in contrast to the various dissolution rates of carbonate phases in their laboratory runs executed down to < 400°C. Nevertheless a common feature in all these experiments is the reduction of $Fe^{2+}$ or $Fe^{3+}$ to elemental iron, which tends to migrate to crystal cleavages and fractures.

The importance of carbonate minerals during the injection and storage of hydrogen in clastic sediments, which comprise appreciable amounts of these components established as porefilling cements, carbonate clasts and/or preserved (micro-) fossils is also stated by Giardini and Salotti (1969). Thereby, carbonates can strongly influence "green" eco-methane generation by their high contents of $CO_2$. This is shown by Giardini and Salotti (1969) in laboratory runs, down to ~ 400°C, of different carbonate types (calcite – $CaCO_3$, dolomite – $CaMg(CO_3)_2$ and siderite – $FeCO_3$). In this study calcite reacted with the added hydrogen to form inorganic hydrocarbons at the carbonate mineral surface, without any subsequent reactions by gaseous species (e.g. $CO_2$). This methanization is mainly controlled by temperature and hydrogen partial pressure, rather than by pressure (Giardini & Salotti 1969). However, siderite alteration was different and is related to oxygen fugacity. At reducing conditions FeO released from siderite can react towards $2H_2O + CH_4 + FeO \rightarrow FeO + H_2 \rightarrow Fe + H_2O$, but under oxidizing conditions $Fe_3O_4$ and hydrogen will be formed, according to $3FeO + H_2O \rightarrow Fe_3O_4 + H_2$. These findings suggest that iron (-oxides) can buffer chemical reactions and act as a proton acceptor as well as a proton donor, very similar to the clay minerals.

Although, temperatures > 400°C are out of the scope of the H2STORE project and only reactions of hydrogen, $H_2O$ and carbonate minerals are considered by Giardini and Salotti (1969), the relevance of these findings to the planned work in H2STORE cannot be excluded. This assumption is based on the complexity of the fluid-rock system to be studied. Here the composition of involved formation fluids is highly saline containing e.g. NaCl-, KCl-, $CaCl_2$, and $SO_4^{2-}$. Moreover, residual hydrocarbons in depleted gas reservoirs are usually associated with impurities like $N_2$, $NH_4$, CO, $CO_2$, and $H_2S$ (e.g. Lüders et al. 2010), which have to be considered in any evaluation of chemical reaction processes. Such processes are affected by microbiological populations at depths, which can act as catalytic reactants during mineral alteration and in "green" eco-methane generation (Vorhies & Gaines 2009, Panfilov 2010). The most prominent features of the sedimentary rocks studied in H2STORE are given in Table 1.

**Table 1** Research sites of H2STORE with most prominent features (Bottig 2008, Morozowa et al. 2011a, b, Lonschinski (pers. com.), Pudlo et al. 2012)

| | Saxony-Anhalt | Lower Saxony | Thuringia | Brandenburg | Bavaria |
|---|---|---|---|---|---|
| reservoir type | depleted gas reservoir | gas storage site | depleted gas reservoir | $CO_2$ storage site | gas storage site |
| age | ~ 270 Ma | ~ 250 Ma | ~ 250 Ma | ~ 210 Ma | ~ 25 Ma |
| stratigraphy | Rotliegend | Buntsandstein | Buntsandstein | Keuper | Tertiary |
| depths | ~ 3.500 m | ~ 1.700 m | ~ 800 m | ~ 650 m | ~ 1.600 m |
| pressure | 20 Mpa | 18 Mpa | 4-8 Mpa | 5.5 Mpa | 15.8 Mpa |
| temperature | ~ 125°C | ~ 55-122°C | ~ 40-80°C | ~ 40°C | ~ 53-60°C |
| rock type | (sub-) arkoses - litharenite | (sub-) arkoses | (sub-) arkoses | lithic arkoses - feldsp. litharenite | litharenite |
| porosity | 0 - 30 % | 12-28 % | 4-20 % | 0 - 30 % | ~ 20% |
| fluid composition | NaCl, KCl, $CaCl_2$ | ??? | $HCO_3^{2-}$, $SO_4^{2-}$ ($Ca^{2+}$) | NaCl, $MgSO_4$, $CaCl_2$, KCl, $Na_2SO_4$ | ??? |
| microorganism | Anaerobic, saline, thermophilic - H2-oxidizing & biocorrosive org. | ??? | ??? | Halophilic anaerobic; sulphate reducing bacteria & methanogenic archaea | ??? |

# 3 Project Intention

The intention and planned collaborative investigations of the subprojects within the H2STORE project are highly ambitious. In total this study is based on a complex data and samples exchange to achieve data sets from analyzed rocks before and after laboratory runs, which will be used in geochemical/ mineralogical, bio-geochemical and thermodynamic numerical simulations (Fig. 1). In comprising and evaluating the effects on different rock compositions and related petrophysical features, samples exposed to hydrogen at distinct temperature and pressure (= depths) conditions at various German locations (Saxony-Anhalt, Brandenburg, Lower Saxony, Thuringia and Bavaria) and stratigraphic units will be studied (Tab. 1, Fig. 2). Thereby potential hydrogen reservoir rocks are composed of various mineral assemblages and occur at burial depths ranging from ~ 3.500 – 650 m, corresponding to temperatures of ~ 125°C down to ~ 40°C (in Saxony-Anhalt and Thuringia, respectively). Moreover formation fluids in the studied areas are different and site specific (Tab. 1). Therefore a refined characterization of these fluids is one essential objective of the collaborative H2STORE project.

In the first approach compositional/ mineralogical differences in sediments were classified by their quartz-feldspar-lithoclast content (Fig. 3), which was also used as a very first, roughly estimated indication of potential rock reactivity. In this regard, the most potentially sensitive sediments exhibit high amounts of feldspar and lithoclasts, but relatively low contents of quartz and are classified as lithic sandstones after McBride (1963, Fig. 3). However such a simplification can be misleading, because only detrital components are involved, without any consideration of the late diagenetic and geological-tectonic processes that occurred during burial history.

Such differences in stratigraphic and present day position of the reservoir rocks as well as in rock composition and accompanying hydrous and gaseous phases will provoke distinct bio-, hydro- and geochemical water-rock interactions, which will induce petrophysical alteration during hydrogen injection.

To evaluate and verify such well site specific processes the six H2STORE subprojects involved will apply different methods, comprising mainly analytical, experimental and numerical simulation methods in the fields of sedimentology, petrophysics, geochemistry, mineralogy, and microbiology. A brief description of the objectives and ambitions of the subprojects is given below.

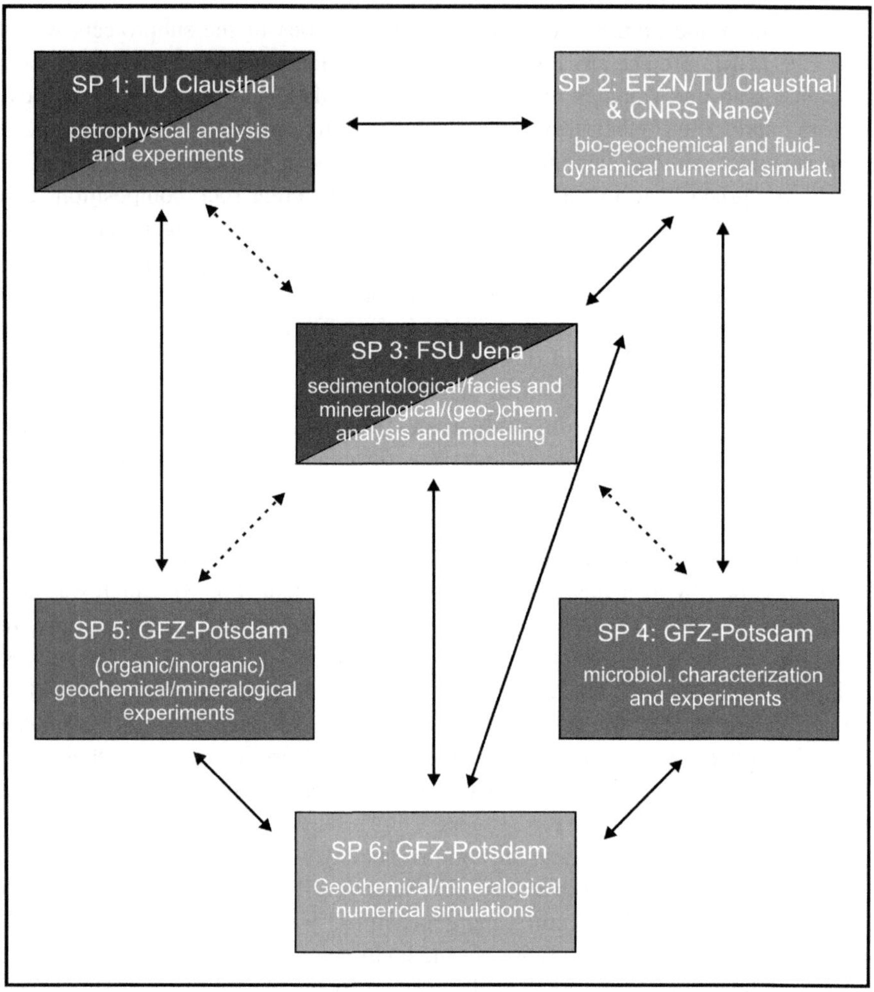


**Fig. 1** Organigram of H2STORE project. Dark grey squares label subprojects mainly performing analytical work (SPs1, 3), in medium grey subprojects conducting laboratory runs (SPs1, 4, 5) and light grey subprojects conducting numerical simulations (SPs2, 3, 6) are shown. Dashed lines indicate planned sample and data exchange, whereas straight lines refer only to data transfer.

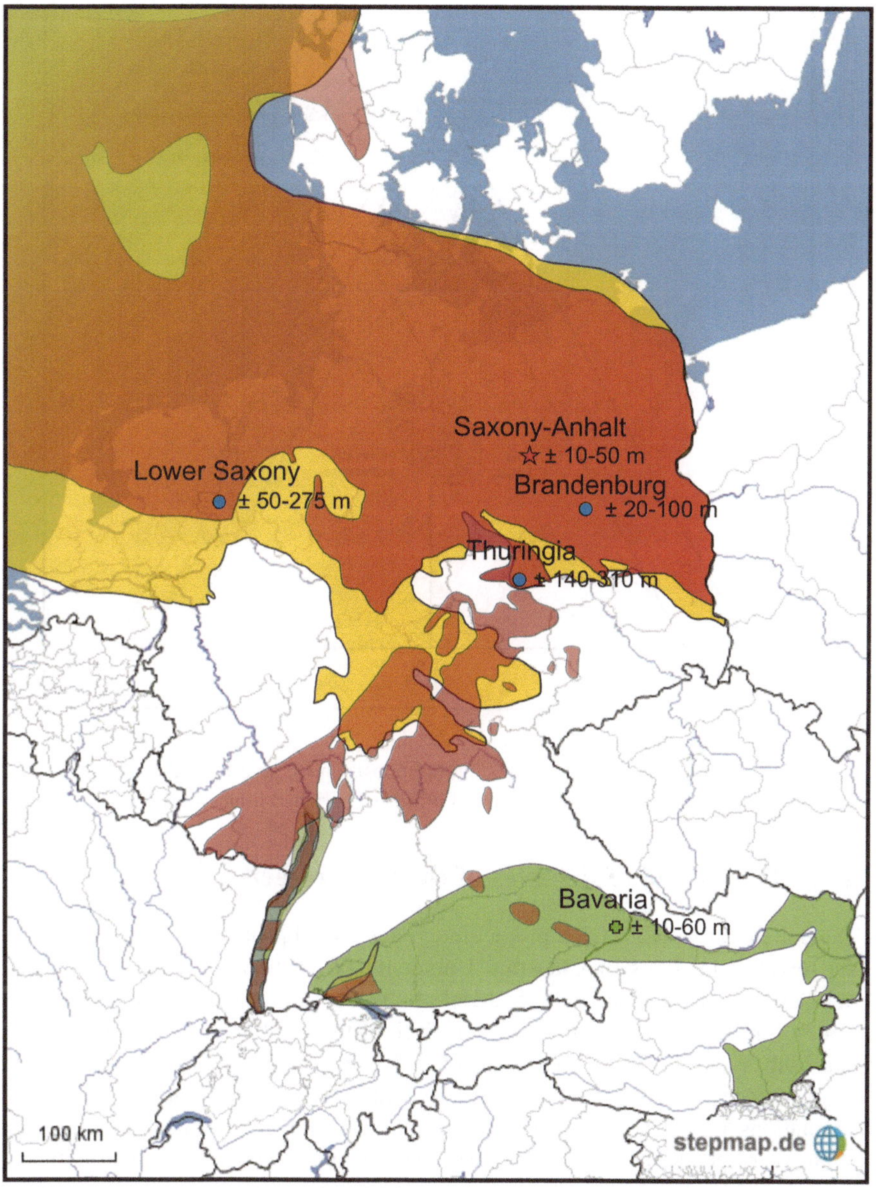


**Fig. 2** Location map of well sites in Germany included in planned H2STORE project. Numbers refer to thickness of depleted gas and gas storage reservoirs. Data source: Reinhold et al. (2011), Sedlacek R (2009, 2011). In reddish colors: Rotliegend -, yellow: Buntsandstein -, green: Tertiary deposits, associated well sites are marked by red star, blue circles and green cross.

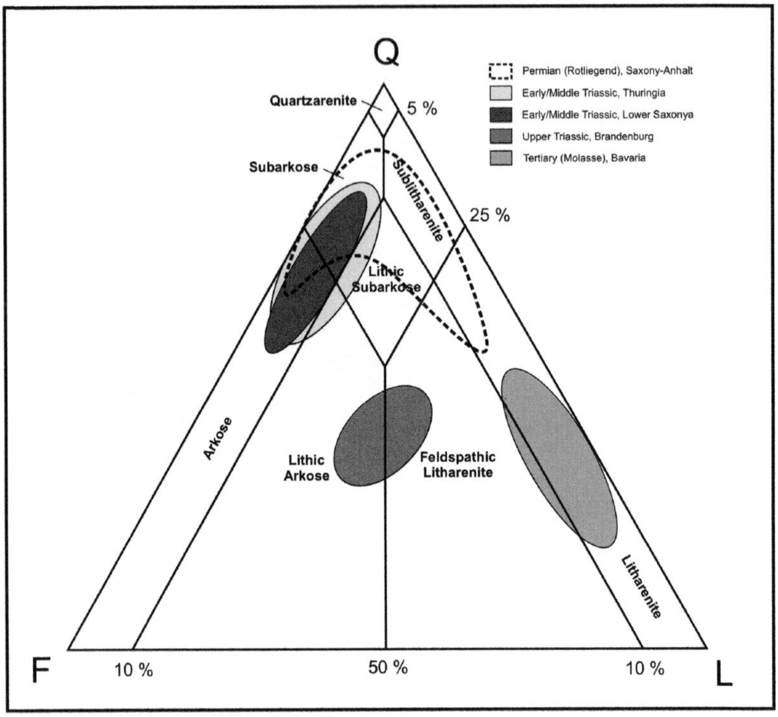


**Fig. 3** Sandstone classification in terms of quartz (Q), feldspar (F), and lithoclast (L) content after McBride (1963) – sandstone composition of investigated well sites ranges from quartz-rich subarkoses to feldspar and lithoclast enriched litharenites

## 4 Topics and Objectives of the Subprojects

The collaborative H2STORE project integrates six subprojects (SPs) established at the research facilities of Clausthal University of Technology (TUC), Friedrich-Schiller-University Jena (FSU) and Helmholtz Centre Potsdam, German Research Centre for Geosciences (GFZ – Fig.1). Each of these subprojects will use specific approaches and analytical methods to verify and simulate fluid-rock reactions to the presence of hydrogen. The SPs 2 and 6 will concentrate on numerical simulations of bio-geochemical-fluid dynamical and geochemical-mineralogical features, respectively. The GFZ–based SPs 4 and 5 and also TUC's SP1 will conduct static batch and fluid flow experiments, exposing rock samples to hydrogen and hydrogen-bearing fluids to evaluate potential compositional changes in microbiological and mineralogical assemblages as well as petrophysical features. SP3 (FSU Jena) will perform most of the analytical work in geology-sedimentology, geo- and hydrochemistry as well as in mineralogy before and after the lab runs conducted in SPs 1, 4, and 5. Modelling is concentrated on the burial and diagenetic evolution of the rocks as well as on petro-/ hydrophysical parameters (e.g. porosity, pore connectivity, specific mineral surface rates, fluid migration/ flow velocity, etc.) deduced by μ-CT investigations. Well sampling and sample distribution will be arranged by the FSU Jena in close cooperation with SP1 (TUC). These samples and related information will be provided by industrial partners and further German joint research projects, which investigate fluid-flow and fluid-rock processes underground.

## 4.1 Subproject 1: "Coupled Geohydraulic and Mineralogical/ Geochemical Processes in Reservoir- and Caprocks" (TUC - Ganzer/ Reitenbach)

This subproject intends to comprehend petro-physical parameters that depend on the mineralogical content of investigated clastic sediments. These features are most important in assessing the capability and capacity of reservoir and sealing caprocks for hydrogen storage. To cover potential *litho-* and diagenetic, facies related porosity and permeability variations about 50 – 60 plug samples (including reference measurements) will be studied. Thereby rock parameters like porosity, compressibility, absolute and effective permeability, capillary pressure functions, rock wettability, and residual gas saturation will be determined by using routine and Special Core Analysis (SCAL) techniques. These parameters are determined before and after experimental runs, in which plug samples covered by site specific synthetic formation fluids will be exposed to hydrogen at reservoir conditions in autoclaves. This approach enables verification of the potential alteration of petrophysical features caused by chemical interactions of reservoir rocks and fluids with hydrogen-bearing gaseous mixtures and its relevance to fluid transport processes. In addition, through these experiments, an evaluation of the impact of hydrogen on the potentially enhanced growth and diversification of microbiological populations induced by inorganic (mineralogical) and organic reactions will be evaluated. In close cooperation with SP3 the relations of petrophysical and facies/ mineralogical/ geochemical features found will be compared with the data sets of SPs 4 and 5, and finally the most relevant parameters for the numerical simulations of SPs 2 and 6 will be provided.

## 4.2 Subproject 2: "Numerical Simulations of Mixing Processes of Gaseous Components during Hydrogen Storage" (TUC-EFZN/ CNRS-EMPTA-UL - Ganzer/ Panfilov)

This subproject concentrates on the numerical simulation of fluid transport processes accompanied by gas mixing and bio-geochemical interactions in the presence of hydrogen and carbon dioxide in reservoir rocks.

By using input parameters allocated by SPs 1, 3, 4, and 5 the spatial distribution and potential voluminous and compositional variations of hydrogen and hydrogen - gas mixtures in depleted gas reservoirs will be modelled. Thereby, numerical simulations will consider the impact of these fluids on the behaviour of microbiological biocenosis as well as the mineralogical rock content.

The modelling work comprises of:

- the modelling/ simulation of the behaviour of gas mixtures
- the development of a *conceptual* model of the reactive 2-phase flow in porous media, considering the dynamics of microbiological populations
- optimizing hydrogen injection processes in the geological underground with special reference to interface instability and fingering processes
- the design of a *numerical* model of the reactive 2-phase flow in porous media coupled with the dynamics of microbiological populations
- the compilation of (site-) specific numerical simulations
- modelling the effects induced by hydrogen injection and recirculation in geological strata.

These numerical simulations will enable the upscaling of experimentally induced alteration effects (conducted in SPs 1, 4, and 5) to reservoir scale and the establishment of specifications for hydrogen storage in almost depleted gas reservoirs.

## *4.3 Subproject 3: "Sedimentological-Facies and Mineralogical-Geochemical Studies on Reservoir and Sealing Rocks" (FSU Jena - Pudlo/ Gaupp)*

Subproject 3 of the Friedrich-Schiller-University Jena comprises sedimentological, geological, mineralogical, and hydro- as well as geochemical analysis with special emphasis on fluid-rock reactions, mineral surface characterization and pore space evolution from dm- to nm scale. Most of the analytical methods will be applied to samples prior and after laboratory runs performed in SPs 1, 4, and 5. As well as the samples used in the experiments additional samples will be investigated by SP3 to enable an advanced, more conclusive and representative interpretation of data and strengthen knowledge of site specific burial history.

This will be achieved by evaluating:

- preferential fluid pathways (pre-determined (?) by facies features)
- site specific facies parameters, in relation to depositional and diagenetic sedimentary features (e.g. max. burial depths and lithological composition)
- PETREL™-, PetroMod™ and TOUCHSTONE™ modelling of the burial and diagenetic evolution of the rocks
- the mineral content and composition as well as geochemical components (e.g. hydrocarbon and sulphur species) with special regard to Fe-bearing material and clay as well as carbonate mineral phases – representing most sensitive rock constitutes for hydrogen/ fluid reactions. These investigations will be completed by hydro-chemical studies on the formation of the water and gas phases

- potential relations between facies, mineralogical rock content, and geo-/ hydrochemical composition with petrophysical features determined in SP1 and raised from µ-computer tomography investigations (µ-CT).

These objectives will be verified by applying numerous analytical methods. Thereby mineralogical and pore space features are obtained by polarization-, electron microprobe-, field emission scattered electron-, focused beam scattered electron-, cathodoluminescence-, confocal laser scanning, and atomic force microscopy. Further information on this topic is deduced from the microscopic image, µ-computer tomography, BET (Brunauer, Emmett, Teller internal specific surface area measurement method), infrared spectroscopy, and X-ray diffraction analysis. Chemical data is obtained by inductively coupled plasma (with specifications of mass and optical emission spectrometry – ICP/MS-OES), X-ray fluorescence, RAMAN-spectroscopy, ion chromatography, and DEGAS (Directly coupled Evolved Gas analysis) investigations.

All analytical work will be conducted both before and after the laboratory runs and data obtained will be used to design potential chemical reaction paths for the rocks in the presence of hydrogen. By allocating the data sets to SPs 2 and 6 an evaluation of proposed reactions by numerical simulations will be achieved.

## *4.4 Subproject 4: "Characterization of Microbiological Processes in Reservoir Rocks – Its Relevance for Geological Hydrogen Storage" (GFZ Potsdam - Würdemann)*

Subproject 4 (GFZ – Würdemann) concentrates on the characterization of microbiological populations and conducting laboratory runs to monitor the effect of such microorganisms on potential "green" eco-methanization processes and rock alteration.

Thus SP4 will perform studies on the complex interactions of microbiological associations (biocenosis), formation fluids, rock composition, and hydrogen under reservoir conditions. These investigations include the characterization of autochthonous microorganisms using molecular-biological methods.

Thereby characterization of the microbiological community, their phylogenetic diversity, their abundance and activity in the pore space of the sediments and their potential variations induced by hydrogen injection is obtained by:

- genetic (DNA-) finger printing (DGGE) and molecular cloning (16S rRNA clone libraries) – for evaluating phylogenetic diversity and abundance
- real-time PCR - for quantification of microbiological organisms and their metabolic activity
- fluorescence-in-situ-hybridization (FISH) – for determining the abundance and spatial distribution of organisms
- a specific verification of sulphate reducing organisms will be gained by dissimilated sulfit-reductase (dsrA Gen).

In close cooperation with SP3 inorganic components (like Fe, K, Ca, Mg, and S), which are most important in metabolic processes, will be measured from experimentally used fluids to evaluate the relevance of microorganisms in (inorganic) chemical reactions.

All this data will be discussed with SP2 to develop a conceptual numerical model for the multi-phase/multi-component flow with special reference to biological reactions and the dynamics of microbiological populations.

## 4.5 Subproject 5: "Physico-chemical Interactions of Hydrogen and Formation Fluids - Their Relevance for Organic and Inorganic Dissolution-/ Precipitation Processes" (GFZ Potsdam – Liebscher/ Pilz)

Subproject 5 (GFZ – Liebscher/ Pilz) will conduct fluid flow and batch experiments on rock samples. In these runs the rocks will be exposed to different hydrogen-gas (e.g. $CH_4$, $CO_2$) mixtures and potential initial and residual formation fluids under different P- and T-conditions. This attempt enables investigations on the relevance and behaviour of hydrogen-bearing fluids affecting mineral rock and residual hydrocarbon (bitumen) content under varying pH- and Eh conditions. Used rock samples will be analysed before and after the experiments to evaluate any rock/ mineral alteration. Rock porosity and permeability are most likely to increase due to mineral dissolution processes. However such complex reactions can also induce mineral precipitation, resulting in pore space deterioration and in reducing the capability of well site injectivity and (hydrogen/ methane) recirculation.

Thus the two main objectives of SP5 are to evaluate and verify the:

- relevance of complex interactions of various (organic and inorganic as well as hydrous and gas) components underground on the state of injected hydrogen
- relevance of injected hydrogen on the behaviour and integrity of reservoir rocks.

## 4.6 Subproject 6: "Numerical Simulations of Geochemical Reactions of Hydrogen, Formation Fluids and Mineral Phases" (GFZ Potsdam – De Lucia/ Kühn)

The goal of SP6 is the development of new concepts for the numerical simulation of fluid-rock and fluid-fluid interactions involving or induced by hydrogen on different time and spatial scales (from laboratory to reservoir). The geochemical and reaction path modelling aimed at numerically reproducing the experiments of the other H2STORE subprojects and will profit from the observed reactions and the characterization of specific porous media. It is expected that this procedure will enhance the accuracy of the currently available thermodynamic and kinetic databases. A further step will be represented by the coupled modelling of fluid flow (transport mechanisms), the temporal (and spatial) migration of injected hydrogen and the induced mineral/ rock alteration of reservoir and sealing rocks accompanied by variations in petrophysical features, like porosity and permeability. These simulations are performed by coupling the chemical speciation software PHREEQC (Parkhurst et al. 1999) to the multiphase simulators OpenGeoSys (OGS - Kolditz et al. 2011, De Lucia et al. 2010), and possibly validated by simulation runs with the established TOUGHREACT (Xu et al. 2011).

A further topic is the upscaling of mineralogical features such as mineral dissolution rates, accessibility of specific reactive mineral surfaces (determined in SP1 and 3) from the microscopic (μm - mm) to Darcy scale (dm – m) considering the variability of the porous media. To this aim a Monte-Carlo based "Stochastic Simulation Algorithm" (SSA - Gillespie 1976) will be developed to obtain stochastic correlations from the large data sets produced in the analytical companion subprojects which can then be implemented in the above cited coupled simulator, increasing the accuracy of the numerical simulations.

## 5 Summary

Long-term storage of energy (~ several months) energy is one of the main challenges in establishing renewable energy techniques and reducing environmental damage by replacing coal-fired and nuclear power supply with their potential risks for global warming and the catastrophic release of radiogenic elements caused by accidents.

The electrolytic conversion of excess electrical energy into storable hydrogen will be one of the most promising options to meet this challenge.

H2STORE ("hydrogen to store") is a collaborative research project, which evaluates the potential impact of hydrogen injection in depleted gas reservoirs composed of silici-clastic sediments sealed by mudrock layers. The German well sites studied are characterized by highly saline formation waters and burial depths ranging from ~ 650 to 3.500 m covering temperatures of ~ 40 – 125°C. Potential interactions of injected hydrogen under such conditions are complex. Thus objectives like:

- sedimentological structures (e.g. preferential fluid migration pathways and sealing behaviour)
- lithological features (e.g. depositional and diagenetic facies, mineral content)
- mineralogical characteristics (e.g. mineral composition, -morphology/ topology, specific reactive mineral activity)
- geochemical rock composition (e.g. organic and inorganic components)
- hydrochemical composition of (saline) formation fluids
- petrophysical features (e.g. rock porosity, effective permeability, pore space connectivity, irreducible water saturation)
- micro-biological populations (e.g. abundance and diversity of biocenosis)

- the composition of residual natural gas components have to be considered for potential multi-phase chemical and physico-chemical (e.g. hydrogen adsorption on clay mineral species) interactions. Batch and flow through experiments, with distinct research topics (petrophysical, fluid-mineralogical and bio-geochemical/ mineralogical approaches) will be conducted under various reservoir conditions, serving as analogues for interactions most probably induced by hydrogen injection in silici-clastic rocks underground. Experimental data will be used for numerical modelling of the evolution of micro-biological populations and geochemical/ mineralogical as well as petrophysical rock composition.

Additionally, a potential reaction of $CO_2$ in the reservoirs to remnants of natural gas impurities and/ or released by the chemical reactions of carbonate mineral dissolution has to be considered.

**Acknowledgements.** H2STORE is a lighthouse project under the framework of the special R&D-Programme "Energiespeicher (energy storage)" sponsored by the German Federal Ministries of Education and Research (BMBF), of Economics and Technology (BMWi), and for the Environment, Nature Conservation and Nuclear Safety (BMU). We deeply appreciate grants from the BMBF (grant nos. 03SF0434A – FSU Jena, 03SF0434B – GFZ Postdam, 03SF0434C-TU Clausthal) and thank E.on Gas Storage GmbH, GDF SUEZ E&P DEUTSCHLAND GMBH, Rohöl-Aufsuchungs Aktiengesellschaft (RAG)-Austria, and RWE Gasspeicher GmbH for their assistance.

## References

1. Allan, M.M., Turner, A., Yardley, B.W.D.: Relation between the dissolution rate of single minerals and reservoir rocks in acidified pore waters. Appl. Geochem. 26, 1289–1301 (2011)
2. Beck, H.P., Franz, O.T.: Energy storage in abandoned mines – A method to stabilize the German power grid. In: Hou, M.Z., Xie, H., Yoon, J.S. (eds.) Underground Storage of $CO_2$ and Energy, pp. 261–269. CRC Press, Taylor & Francis Group, Boca Raton-London-New York-Leiden (2010)
3. Bottig, M.: Diagenetische Umwandlungen in Sandsteinen der gasgesättigten, der sekundär verwässerten, der Wasser-, und der Übergangszone des Erdgas-Speichers Haidach in der Molasse Zone, Österreich. Diploma Thesis, Universität Wien/ Österreich, p. 74 (2008)
4. Brandt, F., Bosbach, D., Krawczyk-Bärsch, E., Arnold, T., Bernhard, G.: Chlorite dissolution in the acid pH-range: A combined microscopic and macroscopic approach. Geochim. Cosmochim. Acta 67, 1451–1461 (2003)
5. Brockmann, B., Donadei, S., Crotogino, F.: Energy storage in salt caverns – Renewable energies in the spotlight. In: Hou, M.Z., Xie, H., Yoon, J.S. (eds.) Underground Storage of $CO_2$ and Energy, pp. 271–277. CRC Press, Taylor & Francis Group, Boca Raton-London-New York-Leiden (2010)
6. Cavallo, A.: Controllable and affordable utility-scale electricity from intermittent wind resources and compressed air energy storage (CAES). Energy 32, 120–127 (2007)
7. Czurda, K.: Clay liners and waste disposal. In: Bergaya, F., Theng, B.K.G., Lagaly, G. (eds.) Handbook of Clay Science, Developments in Clay Science, vol. 1, pp. 693–701 (2006)

8. De Lucia, M., Kühn, M.: TV III.2.2.2: Geochemical modeling, Pitzer databases and upscaling of chemical reactions. In: Kolditz (ed.) CLEAN Progress Report TV III.2, Report Period: 01/2009—12/2009 (2010)
9. Demouchy, S., Jacobsen, S.D., Gaillard, F., Stern, C.R.: Rapid magma ascent recorded by water diffusion profiles in mantle olivine. Geology 34, 429–432 (2006)
10. Ehrenberg, S.N.: Assessing the relative importance of compaction processes and cementation to reduction of porosity in sandstones: Discussion; Compaction and porosity evolution of Pliocene sandstones, Ventura Basin, California: Discussion. Amer. Ass. of Petrol Geol. Bull. 73, 1274–1276 (1989)
11. Fischer, S., Liebscher, A., Wandrey, M., the CO2SINK Group: $CO_2$–brine–rock interaction — First results of long-term exposure experiments at insitu P–Tconditions of the Ketzin $CO_2$ reservoir. Chemie der Erde (Geochemistry) 70, 155–164 (2010)
12. Flaathen, T.K., Gislason, S.R., Oelkers, E.H., Sveinbjörnsdóttir, A.E.: Chemical evolution of the Mt. Hekla, Iceland, groundwaters: A natural analogue for $CO_2$ sequestration in basaltic rocks. Appl. Geochem. 24, 463–474 (2009)
13. Foh, S., Novil, M., Rockar, E., Randolph, P.: Underground hydrogen storage. Final report, Brookhaven National Laboratory, Upton - New York, p. 145 (1979)
14. Folk, R.L.: Petrology of sedimentary rocks, p. 190. Hemphill Publishing Company, Austin (1968)
15. Giardini, A.A., Salotti, C.A.: Kinetics and relations in the calcite-hydrogen reaction and relations in the dolomite-hydrogen and siderite-hydrogen systems. Am. Mineral 54, 1151–1172 (1969)
16. Giese, H.O., Schuldt, J.: Speicherung von Windenergie im Untergrund. AGAFE – Mitteilungen 24(1), 2–6 (2005)
17. Gil, A., Trujillano, R., Vicente, M.A., Korili, S.A.: Hydrogen adsorption by microporous materials based on alumina-pillared clays. Int. J. of Hydrogen Energy 34, 8611–8615 (2009)
18. Gillespie, D.: A general method for numerically simulating the stochastic time revolution of coupled chemical reactions. J. of Computational Physics 2, 403–434 (1976)
19. Gillhaus, A.: Natural gas storage in salt caverns – Summary of worldwide projects and consequences of varying storage objectives and salt formations. In: Hou, M.Z., Xie, H., Yoon, J.S. (eds.) Underground Storage of $CO_2$ and Energy, pp. 191–197. CRC Press, Taylor & Francis Group, Boca Raton-London-New York-Leiden (2010)
20. Gunter, W.D., Bachu, S., Benson, S.: The role of hydrogeological and geochemical trapping in sedimentary basins for secure geological storage of carbon dioxide. In: Baines, S.J., Worden, R.H. (eds.) Geological Storage of Carbon Dioxide, vol. 233, pp. 129–145. Geological Society, Special Publications, London (2004)
21. Heinrich, E.W., Salotti, C.A., Giardini, A.A.: Hydrogen-mineral reactions and their application to the removal of iron from spodume. Energy 3, 273–279 (1978)
22. Kinnaman, F.S., Valentine, D.L., Tyler, S.C.: Carbon and hydrogen isotope fractionation associated with the aerobic microbial oxidation of methane, ethane, propane and butane. Geochim. Cosmochim. Acta 71, 271–283 (2007)
23. Kolditz, O., Shao, H. (eds.): OpenGeoSys-BenchmarkBook, v. 5.0.04 (2011), available through, `http://www.opengeosys.net`
24. Lord, A.S.: Overview of geologic storage of natural gas with an emphasis on assessing the feasibility of storing hydrogen. SANDIA Report SAND2009-5878, Sandia National Laboratories, p. 28 (2009)
25. Lüders, V., Plessen, B., Romer, R.L., Weise, S.M., Banks, D.A., Hoth, P., Dulski, P., Schettler, G.: Chemistry and isotopic composition of Rotliegend and Upper Carboniferous formation waters from the North German Basin. Chem. Geol. 276, 198–208 (2010)

26. Lund, H., Salgi, G.: The role of compressed air energy storage (CAES) in future sustainable energy systems. Energy Conversion and Management 50, 1172–1179 (2009)
27. McBride, E.F.: A classification of common sandstones. J. Sediment Petrol 33, 664–669 (1963)
28. Morozova, D., Alawi, M., Shaheed, M., Krüger, M., Kock, D., Würdemann, H.: The influence of microbial activity on rock fluid interaction: baseline characterization of deep microbial biosphere for Enhanced Gas Recovery in the Altmark natural gas reservoir. Energy Procedia 4, 4633–4640 (2011a)
29. Morozova, D., Zettlitzer, M., Let, D., Würdemann, H., $CO_2$SINK group: $CO_2$SINK group Monitoring of the microbial community composition in deep subsurface saline aquifers during $CO_2$ storage in Ketzin, Germany. Energy Procedia 4, 4362–4370 (2011b)
30. Reinhold, K., Müller, C., Riesenberg, C.: Informationssystem Speichergesteine für den Standort Deutschland - Abschlussbericht. Bundesanstalt für Geowissenschaften und Rohstoffe (BGR), Berlin-Hannover, p. 134 (2011)
31. Ogden, J.M.: Prospects for building a hydrogen energy infrastructure. Annu. Rev. Energy Environ. 24, 227–279 (1999)
32. Panfilov, M.: Underground Storage of Hydrogen: In Situ Self-Organisation and Methane Generation. Transport in Porous Media 85, 841–865 (2010)
33. Parkhurst, D., Appelo, C.: Users guide to Phreeqc (version 2). Technical Report, U.S. Geological Survey, Water-Resources Investigations Report 99. Denver, Colorado (1999)
34. Pudlo, D., Reitenbach, V., Albrecht, D., Ganzer, L., Gernert, U., Wienand, J., Kohlhepp, B., Gaupp, R.: The impact of diagenetic fluid-rock reactions on Rotliegend sandstone composition and petrophysical properties (Altmark area, central Germany). Environ. Earth Sci. 67, 369–384 (2012)
35. Pukazhselvan, D., Kumar, V., Singh, S.K.: High capacity hydrogen storage: basic aspects, new developments and milestones. Nano Energy 1, 566–589 (2012)
36. Purwin, H., Stalder, R., Skogby, H.: Hydrogen incorporation in Fe- and Na-doped diopsides. Eur. J. Mineral 21, 691–704 (2009)
37. Sedlacek, R.: Untertage-Gasspeicherung in Deutschland. Erdöl Erdgas Kohle 125(11), 412–426 (2009)
38. Sedlacek, R.: Untertage-Erdgasspeicherung in Deutschland. Erdöl Erdgas Kohle 127(11), 414–424 (2011)
39. Sørensen, B.: Underground hydrogen storage in geological formations, and comparison with other storage solutions. In: Proc. of Hydrogen Power Theoretical and Engineering Int. Symp., Merida, Mexico (2007)
40. Tunega, D., Haberhauer, G., Gerzabek, M.H., Lischka, H.: Theoretical study of adsorption sites on the (001) surfaces of 1:1 clay minerals. Langmuir 18, 139–147 (2002)
41. Velde, B., Meunier, A.: The Origin of Clay Minerals in Soils and Weathered Rocks, p. 406. Springer, Heidelberg (2008)
42. Vorhies, J.S., Gaines, R.R.: Microbial dissolution of clay minerals as a source of iron and silica in marine sediments. Nature - Geoscience 2009(2), 221–225 (2009)
43. Xu, T., Spycher, N., Sonnenthal, E., Zhang, G., Zheng, L., Pruess, K.: TOUGHREACT Version 2.0: A simulator for subsurface reactive transport under non-isothermal multiphase flow conditions. Computers and Geosciences 37, 763–774 (2011)

# SEW – A New Software Application That Supports the Safety Evaluation of Underground Storage Wells

Andreas Brecht, Detlef Edler, and Klaus-Peter Rehmer

UGS – Untergrundspeicher- und Geotechnologie-Systeme GmbH,
15749 Mittenwalde/Mark, Germany

**Abstract.** Many of the existing wells in underground storage formations are about to exceed their primary life time. It is therefore of great safety, technical and commercial interest whether existing wells can be further used and if so, how long in the future and under which operating conditions.

SEW is a user friendly software that supports the process of safety evaluation of wells. This software takes into account the pressure from rock formations acting on the casings, the corrosion potential of rock formations and process media involved as well as the influence of cavern roof subsidence. It allows simulation of the impact of casing failures on subsequent casings and it checks the collapse resistance, the internal pressure resistivity of tubulars and the axial load on tubing and anchor pipes.

Different loading case scenarios can be defined and investigated. The measurement data of casing inspections can be imported and used in the future as a new baseline for the calculation of casing degradation caused by corrosion.

**Keywords:** SEW, safety evaluation, corrosion potential, cavern roof subsidence, casing failures, collapse resistance, internal pressure resistivity, axial load.

## 1 The Problem

Underground storage of compressed natural gas has been widely applied for more than four decades. Different techniques are used for storage in porous or fractured rock formations. These can be depleted hydrocarbon fields or saline aquifers.

Depleted hydrocarbon fields exhibit the great advantage that the reservoir behavior is well known from the exploitation phase. Apart from that existing wells can be re-used for gas injection and gas withdrawal as long as they are suitable and technically safe. But production wells are typically designed for a lifetime that corresponds to the duration of the exploitation phase and for a one-way gas flow that differs from the alternating operating conditions during storage. When converted to underground storage wells these boreholes are about to reach the end of their designed lifetime or have even exceeded it already. It is therefore of vital

M.Z. Hou et al. (Eds.): *Clean Energy Systems in the Subsurface*, SSGG, pp. 413–420.
DOI: 10.1007/978-3-642-37849-2_32 

interest to find out whether these boreholes can be operated further in safe conditions and if so, for how long and under which conditions.

Stockpiling of natural gas in salt caverns is another underground storage technique. It has become wide spread since the 1970s. When established, the initial lifetime of the cavern wells was expected to reach 30 years for the production stage and 40 - 50 years for the last cemented casing. Hence in the near future more and more caverns will exceed their designed lifetime. The question is, whether they can still be operated safely in the next decades.

Yet there is another aspect. Many caverns in salt formations have been created in order to produce salt from the outgoing brine. These caverns had not been intended initially to serve as storage caverns. Their design criteria may not necessarily comply with the requirements typically applied to gas storage caverns. If these caverns are to be converted into gas storage caverns it must be clarified whether the cavern well can cope with the storage operation conditions now and in the future.

To Summarize there is an increasing need to evaluate whether the storage wells are capable to of meeting the safety standards both today and in their future life.

## 2 The Starting Point

Storage wells are subject to different load conditions during operation. This relates to both the cemented casings as well as to the tubing which are installed inside the borehole to protect the last cemented casing from mechanical damage and alternating pressure and temperature loads. The tubing has to meet the operational demands regarding burst (resulting from internal pressure), collapse (resulting from external pressure) and axial load.

There are several borehole drilling and design programs available on the market which allow for a proper specification of the downhole equipment in order to meet the rock and operational conditions. These programs relate mostly to the standard dimensions of casings and tubing and do not take into account the long term aging process of the material caused mainly by corrosion and other aspects like cavern convergence or the pressure variation when the borehole is piercing a confined reservoir with pressure changes.

## 3 SEW – Safety Evalution of Wells

In order to provide a tool for subsurface engineering to evaluate the safety status of boreholes which have already been in existence for a long time and to predict their remaining lifetime a numerical program has been developed that is named "SEW – Safety Evaluation of Wells".

The main technical features of the program in version 1.0 are the following:

- Characterization of the rock formation regarding the pressure that is acting from outside on the cemented casings (Fig. 1). Different types of deformation behavior can be considered. For rock layers in which the formation pressure is determined by the formation water column a vertical density profile can be defined. Apart from that a corrosion potential can be assigned to each rock layer.

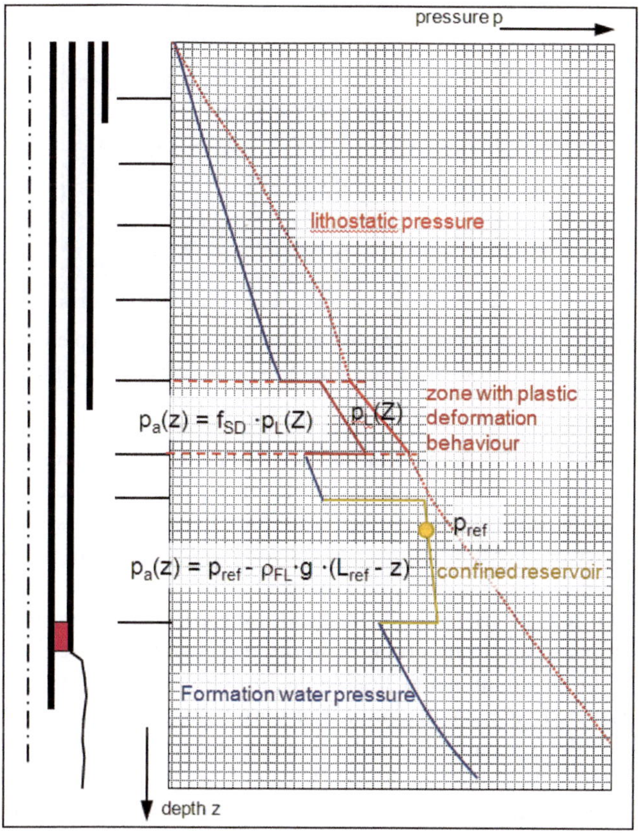


**Fig. 1** Example for outside pressure acting on the cemented casings

- Description of the borehole architecture (arrangement of cemented casings) regarding the casing dimensions and the status of cementation in the annuli (Fig. 2). Partially cemented annuli can be considered. The axial loads while waiting for the cement are calculated and frozen as boundary conditions for the external pressure load calculations.

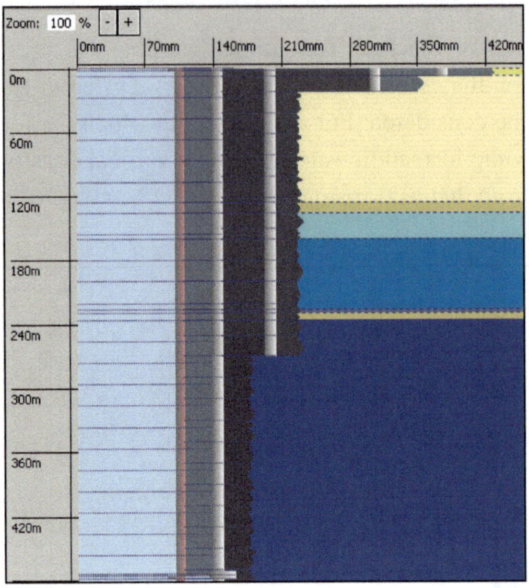


**Fig. 2** Geo-profile with schematic borehole architecture and well completion scheme. When moving with the cursor through the diagram technical details are displayed as hovering quick info.

- Description of well completion and the initial p-T-conditions as well as the axial load during installation of the production string, if any exists. Apart from various tubing dimensions, further well completion elements can be arranged e.g. cross-over, landing nipple, travel-joint or ratch-latch packer and liner (Fig. 3).
- Before starting the lifecycle of the well the media which are used during the operation phase are to be arranged in the well completion region, e.g. protection fluid in the annulus and storage gas inside the tubing. A corrosion potential is assigned to each medium. The corrosion potential describes the wall thickness reduction caused by corrosion per time unit.
- Once the life cycle has started SEW calculates the wall thickness reduction related to the nominal wall thickness taking into account the maximum permissible manufacturing tolerances.
- The results of well wall thickness measurements during the lifecycle of a well can be imported by SEW. This wall thickness data defines a new baseline for the subsequent corrosion history.
- During the lifecycle of a borehole the media arranged in the completion region can change. E.g. in East Germany a number of caverns were initially used for coal gas storage over a longer period of time and later converted to natural gas storage caverns. The corrosion potential of coal gas is significantly higher compared to that of natural gas. This aspect must be taken into account in the corrosion history.

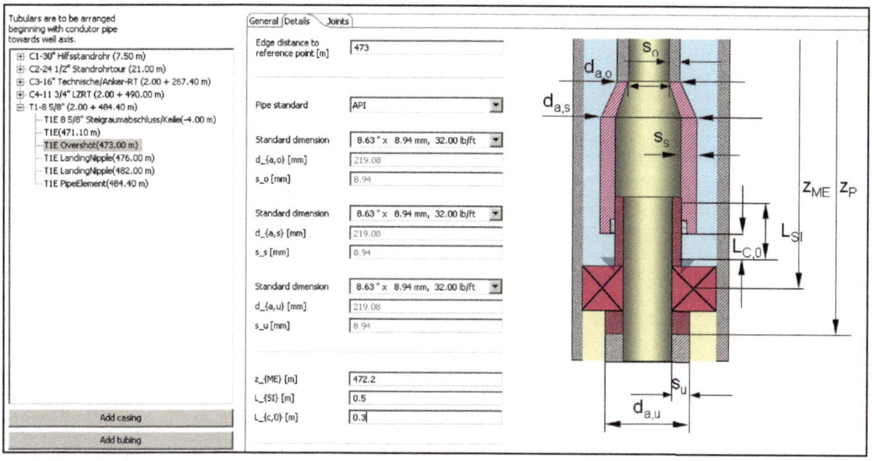


**Fig. 3** Specification of a travel joint packer in SEW

- SEW allows either for a well to be rededicated during its lifecycle (e.g. from a brine production or leaching well to a gas storage well with packer and production string) or to remove and re-install well completion elements or to remove "old" casings and replace them with new ones. During those changes the corrosion status of the remaining borehole elements is kept, whilst the corrosion history for the "new" elements starts with their installation.

During any time in the lifecycle, either in the past or in the remote future different load case scenarios can be delineated. This can be load cases corresponding to the standard operating conditions or the failure of a casing column (e.g. collapse) or malfunction of well completion elements (e.g. loss of tightness in the production string) and extreme pressure and temperature conditions can be simulated. In those cases SEW responds if any of the casings or tubing is under critical load. This relates to burst (internal pressure), collapse (external pressure) and axial load. In these load case scenarios it is also possible to set virtual plugs and to arrange different media under separate pressure conditions above and below the plug. In the load case calculations, the following effects are taken into account:

- temperature changes
- ballooning and counter-ballooning due to pressure changes
- piston forces at aspect areas due to pressure and temperature changes
- aspect area forces if additional plugs are set

The algorithms for the determination of the critical load are according to API bulletin 5C3.

A clearly structured graphical user interface facilitates the data input and interpretation of the calculation results (Fig. 4).

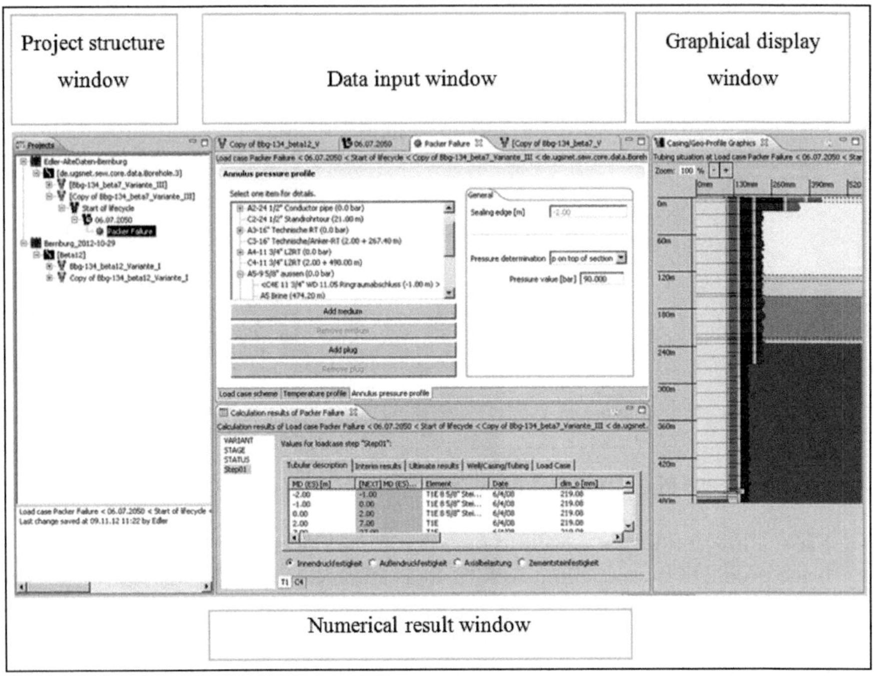


**Fig. 4** The Graphical User Interface (GUI) of SEW 1.0

The major term in SEW to characterize the safety status is the so called 'safety margin'. This term allows different safety coefficient profiles to be applied in SEW. A safety coefficient profile is a set of standardized safety coefficients (SC) related to different load types (internal pressure, external pressure, axial load) and to specific types of equipment. A very common profile of safety coefficients was specified by WEG and is widely applied in Germany:

- Collapse (external pressure) (SCWEG = 1.00)
- Burst (internal pressure) (SCWEG = 1.10)
- Axial load of the plain-end tubular (SCWEG = 1.25)
- Axial load of couplings < 13 3/8" (SCWEG = 1.60)
- Axial load of couplings ≥ 13 3/8" (SCWEG = 1.80)
- Axial load in the slip area (SCWEG = 1.18)
- Axial load of the free-standing section of the anchor pipe (SCWEG = 1.10)

But it is also imaginable that in other countries, companies or organizations different safety coefficient profiles are mandatory. This is the reason why SEW uses the term safety margin instead of comparing directly the expected safety with the safety coefficient:

$$\text{safety margin} = \text{expected safety} - \text{safety coefficient}$$

$$\text{safety margin} = \frac{\text{actual load}}{\text{critical load}} - \text{safety coefficient}$$

A tubular element can be considered as safe if the safety margin is not less than 0.

The safety margin of a tubular string versus depth can be easily displayed as a graph in a diagram (Fig. 5). Such a diagram allows for an immediate assessment of whether there are some critical points which need a closer look and possibly further investigations.

Apart from that all depth-related data can be exported in a text file in an EXCEL®-compatible data-format in order to generate specific diagrams.

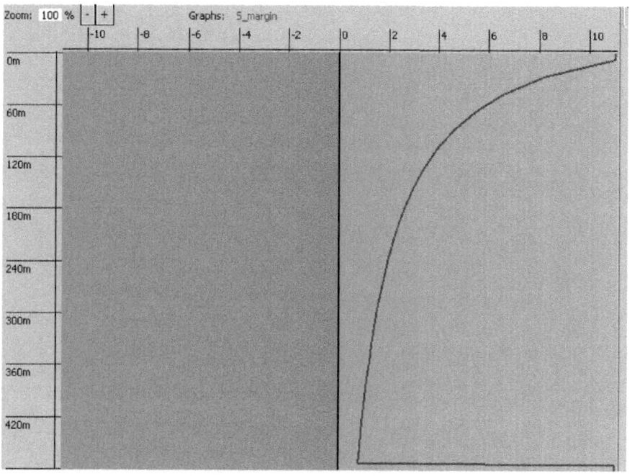


**Fig. 5** Evaluation of a production string regarding collapse in case of packer failure. The safety margin curve is entirely in the green i. e. uncritical area, hence there is no safety concern.

## 4 Data Security, Safety and Integrity

As the safety evaluation of wells is a critical matter special attention was given to data security and data consistency:

- All data is stored in a server-based data base with clearly defined user rights and user access records.
- The data assignment and the calculation algorithms are delineated in such way that conservative results are obtained
- Once a borehole/well completion scheme has been specified and confirmed by starting the lifecycle no further changes or modification can be made. In order to document the exchange of well completion elements or to rededicate the well a numerical copy of the borehole has to be made which can be edited again whilst the original borehole and its corrosion history are frozen and cannot be edited further.

## 5 Summary

SEW 1.0 is the first generation of a software that supports subsurface engineering by safety evaluations of existing and producing wells taking into account the different aspects influencing the safety status. It allows for the prediction of subsequent safe operating time. This software also improves the data security and data consistency of vulnerable data.

# Transport and Usage of Hydrogen via Natural Gas Pipeline Systems

Götz Thilo Müller von der Grün, Steven Hotopp, and Joachim Müller-Kirchenbauer

Institute of Petroleum Engineering, Clausthal University of Technology, Agricolstraße 10, 38678 Clausthal-Zellerfeld, Germany

**Abstract.** Increasing energy demand, rising import dependency and the European move towards alternative energy sources have a large influence on future energy transport systems. Volatile renewable energy sources are difficult to predict and should be able to provide for the consumer demand anytime. The latter needs the security of a large supply and availability which - combined with a fluctuating production - requires short, medium and long term storage concepts. These requirements cannot be fulfilled by the existing systems in sufficient quantities due to scarce capacities and limited flexibility, but the convergence of electricity and gas supply systems can provide the solution. Electrochemical technologies modeled as energy storages are currently in the middle of intensive political and technical discussions and under new research focus. Hydrogen as a chemical energy carrier is treated as a flagship project for the utilization of unused renewable energy production capacities. As a best option it is capable of fulfilling the requirements and can play an important role in future energy systems. This paper illustrates different aspects and consequences of hydrogen in large gas pipeline systems. Factors influencing the existing technical equipment indicate advantages and challenges for a conservative changeover to a regenerative energy supply infrastructure.

**Keywords:** natural gas, pipeline system, hydrogen, simulation.

**Nomenclature:** ITE-GS: Department of Gas Supply at Clausthal University of Technology.

## 1 Introduction

### *1.1 Turnaround in German Energy Policy*

The importance of different energy sources in the German energy mix is changing. Climate-neutral energy generation and safety issues are the main reasons for

M.Z. Hou et al. (Eds.): *Clean Energy Systems in the Subsurface*, SSGG, pp. 421–436.
DOI: 10.1007/978-3-642-37849-2_33 

politics and companies to accelerate a reversal away from nuclear power and fossil fuels towards solar and wind energy. Nevertheless, the fluctuating energy generation over time of sun and wind based solutions is almost unpredictable and hence unreliable. Therefore, storage solutions for excess energy in times of surplus power generation need to be developed. However, potential pumped storage capacities in Germany are limited. In the case of an ideal national electricity grid, annual excess electrical energy is estimated to be 90 to 170 TWh in 2050, representing the lower limit [1]. The only way to allow continuous energy feed-ins, long-term storage and large storage capacities is the conversion of surplus electric energy into chemical energy. Hydrogen is especially suited for use as a chemical energy carrier, because water is the only required raw material for its generation. Since a pure hydrogen transport and storage system would require very substantial investments, the use of the already existing natural gas infrastructure seems to be a reasonable alternative. The basic concept is to purchase low-priced excess electric energy in times of surplus energy production for hydrogen production. The hydrogen is then stored in the natural gas grid in order to be burnt in power plants when there is an energy demand. In this context, property changes of the gas mixture caused by the addition of hydrogen have to be analyzed. The current upper limit value of hydrogen in natural gas in pipelines is 5% by volume as defined by the technical rules in the DVGW-Arbeitsblatt G 262 [2] document. The DVGW is the German Technical and Scientific Association for Gas and Water and its rules are obligatory for pipeline construction and operation in Germany. The German legislative has created a legal basis for the above-mentioned hydrogen storage. The German Energy Industry Act [3] in combination with the Gas Grid Access Ordinance [4] results in a preferential dispatching of so-called biogas. Hydrogen produced by renewable energy is termed as biogas and the German Renewable Energy Act [5] grants a statutory remuneration claim for this biogas which is also called storage gas ("Speichergas").

The Department of Gas Supply (ITE-GS) at Clausthal University of Technology is modeling gas supply systems on a large scale with a European focus. From the technical point of view, renewable energy production like biogas and storage gas has an impact on the gas transport infrastructure which will be analyzed in the next sections.

## 1.2 Selection of a Chemical Energy Carrier

There are several other alternatives to hydrogen as a potential chemical energy carrier as presented in Section 1.a. The most realistic alternative to hydrogen is synthetic methane obtained by the additional process of hydrogen methanation. The conventional Sabatier Process converts hydrogen with carbon dioxide over a catalyst to produce methane and water under elevated pressure and temperature conditions as shown in Equation (1). The $CO_2$ has several potential origins, e.g. biomass, the atmosphere, industrial processes, oil or gas-fired power plants.

$$4H_2 + CO_2 \xrightarrow[\Delta p\ \&\ \Delta T]{} CH_4 + 2H_2O \quad (1)$$

Since methane is the main component of natural gas, the synthetic methane can be treated as a substitute for natural gas with very similar chemical and physical properties. The corresponding compatibility of synthetic methane with the existing gas grid leads to higher storage capacities for renewable energies since there is no legal regulation limiting the methane content in pipeline gas mixtures. Taking into account the utilization and efficiency of the re-conversion to electricity, the existing German gas supply system provides sufficient geometric energy storage capacities to provide electrical power for the national market for two to three months, whereas 5% of hydrogen in the gas mixture would only deliver approximately one day of electricity supply [1]. On the downside, methanation decreases the overall process efficiency by approximately 15% while increasing the costs.

## 2 Distribution of Hydrogen in the German Gas Grid

Due to its central location in central Europe, Germany plays an important role as an import and transit country for natural gas. One research focus of ITE-GS is to build a simulation model of the European gas transmission system [6]. The current state of the model includes gas transmission pipelines and devices of the 12 German and European system operators transporting high calorific gas (H-gas) through Germany [7]. For an evaluation of a future gas transmission system the projected infrastructure for the next 10 years has been applied to the current model [8]. Based on this simulation model, the future German gas supply infrastructure in 2022 and the effects of hydrogen injection can be investigated. Under the assumption of surplus renewable energy and huge hydrogen flows the injection points are located at gas import stations. Under these boundary conditions, the range of coverage and the overall distribution of hydrogen have been determined.

Figure 1 is an aggregated simulation result showing that hydrogen has an impact on basically every element in the gas transport and storage system. Midstream and downstream are affected, furthermore there may be an influence on transit gas to neighboring countries. The impacts on the elements of the midstream gas transport system will be further discussed in section 3, since they will immediately be in contact with the mixture of hydrogen and natural gas.

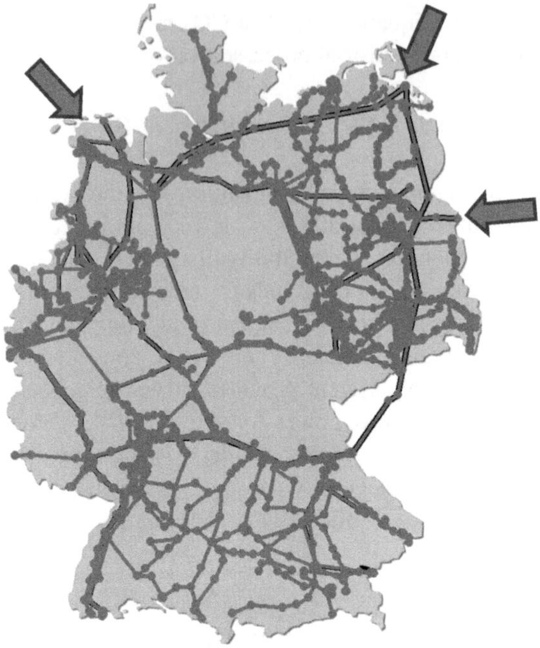

**Fig. 1** Hydrogen injection into the German gas grid

# 3 Influence of Hydrogen on the Gas Transport and Storage System

As pointed out in Section 2, the whole grid may be affected when hydrogen is added to the natural gas in the gas pipeline system. Firstly, the chemical and physical properties of hydrogen have to be compared with those of natural gas, that is; Methane. Secondly, the impact of potential property changes on the different elements of the gas transport system need to be determined.

## *3.1 Physical and Chemical Properties of Hydrogen*

Table 1 compares properties of hydrogen with properties of methane. Hydrogen's atomic number is 1, since it is the lightest element in the universe consisting of only one proton and one electron. Under standard conditions, hydrogen is an inflammable, colourless, odourless, non-toxic, diatomic gas. Methane has a higher viscosity, density, calorific value by volume, and solubility in water than hydrogen. On the contrary, hydrogen has a higher heat capacity, calorific value by mass, maximum flame temperature, auto-ignition temperature, diffusibility as well as wider explosive and firing ranges [9].

**Table 1** Comparison between the properties of hydrogen and the properties of methane [9]

| | Hydrogen, $H_2$ | Methane, $CH_4$ | Unit |
|---|---|---|---|
| Molar mass | 2.02 | 16.04 | g/mole |
| Critical temperature | 33.2 | 190.65 | K |
| Critical pressure | 13.15 | 45.4 | bar |
| Vapour density at normal boiling point | 1.34 | 1.82 | kg/m³ |
| Vapour density (at T=273.15K and P=1.01325bar) | 0.0899 | 0.714 | kg/m³ |
| Specific heat capacity (at T=298.15K and P=constant) | 14.4 | 2.21 | kJ/kg/K |
| Specific heat ratio | 1.4 | 1.31 | - |
| Lower calorific value by mass | 33.3 | 13.3 | kWh/kg |
| Higher calorific value by mass | 39.4 | 14.7 | kWh/kg |
| Lower calorific value by volume | 3.1 | 9.7 | kWh/m³ |
| Higher calorific value by volume | 3.5 | 10.8 | kWh/m³ |
| Maximum flame temperature | 1800 | 1495 | K |
| Explosive range | 18.2 – 58.9 | 5.7 – 14 | % by volume in air |

**Table 1** (*continued*)

| | | | |
|---|---|---|---|
| Firing range | 4.1 – 74 | 5.3 – 15 | % by volume in air |
| Auto-ignition temperature | 844 | 813 | K |
| Viscosity at T=299K | $9*10^{-6}$ | $11*10^{-6}$ | Pa*s |
| Diffusibility in air | $6.1*10^{-5}$ | $1.6*10^{-5}$ | m²/s |
| Solubility in water | 0.0016 | 0.025 | kg/m³ |

## 3.2 *Effect on Different Elements in the Gas Transport and Storage System*

Before the effect of hydrogen on the different elements in the gas transport and storage system can be evaluated, the compatibility of conventional simulation software needs to be ensured. The parameters that are defined by the hydrogen production process also need to be stated. As shown in Figure 2, hydrogen is then either intermediately stored or directly fed into the pipeline system. Before injecting hydrogen in the pipeline, it needs to be measured and compressed. After entering the gas grid, the hydrogen interacts with the pipeline itself, gas compressors, and gas turbines that supply the compressors with the power required. Possible problems of cavern and porous media storage also need to be identified because gas storage is an integrated part of the German gas grid. Finally, measurement devices at customer exit points will have to deal with the hydrogen added to the system.

## 3.3 *Compatibility of Conventional Simulation Software*

Simulation software uses input data from measuring points to calculate gas properties at all other points in the modeled gas network. The software is based on equations of state using density, temperature, and pressure. The two main virial equations for the oil and gas industry are GERG-88 and AGA-8. Hydrogen is one of the five input variables of the simplified S-GERG-88 formula, the main formula used, that allows a maximum hydrogen share of ten per cent [10]. The simulation of a gas with up to 5% hydrogen content is therefore feasible using conventional simulation software. Other equations of state can simulate gases with even higher hydrogen content.

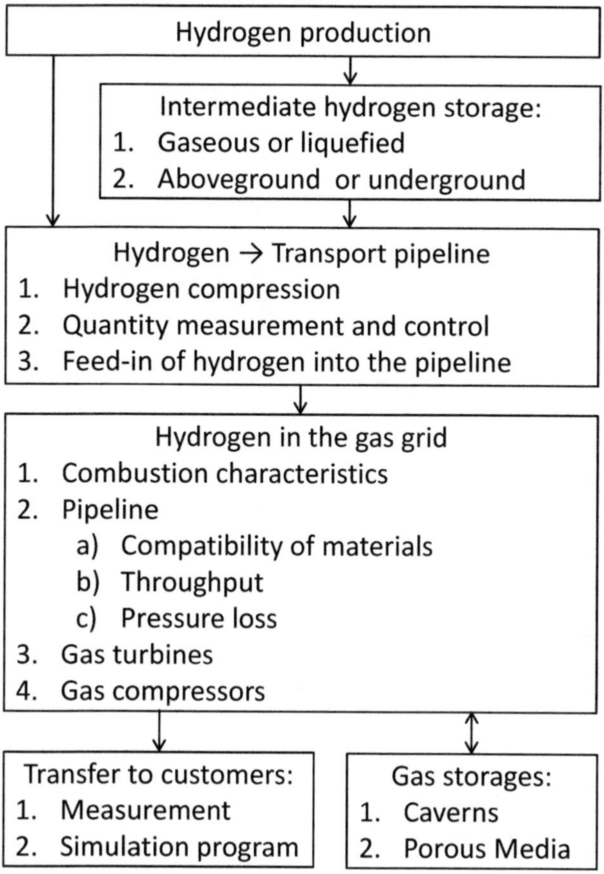


**Fig. 2** Overview over the hydrogen and natural gas system

# 4 Feed-in of Hydrogen into the Transport Pipeline

## *4.1 Compression of Hydrogen*

Hydrogen has to be pressure elevated at least to pipeline pressure in order to enter the pipeline. The simplification of an adiabatic process, as defined in Equation (2), or as an isothermal process does not represent the real compaction energy that is somewhere between the high adiabatic and the low isothermal estimation. Therefore, a polytropic compression that uses the efficiency value described by Equations (4) and (5) can be assumed to modify the adiabatic compression. The definition of a polytropic process is given in Equation (3). Equation (6) shows that inlet temperature and pressure ratios are the only influencing parameters determining the required compression energy. An increased pressure ratio between inlet and outlet pressure will also increase the outlet temperature as Equation (7) shows. Since perfect cooling is not possible, the cooling should take place between different compression stages with identical pressure ratios. [11]

$$PV^{\gamma} = const.$$
$$PV^{\gamma} = const. \quad (2)$$

$$PV^{n_p} = const.$$
$$PV^{n_p} = const. \quad (3)$$

$$\eta_p = \frac{W_{ideal}}{W_{polytropic}}$$
$$\eta_p = \frac{W_{ideal}}{W_{polytropic}} \quad (4)$$

$$\frac{n_p}{n_p - 1} = \eta_p * \frac{\gamma}{\gamma - 1}$$
$$\frac{n_p}{n_p - 1} = \eta_p * \frac{\gamma}{\gamma - 1} \quad (5)$$

$$W_{polytropic} = \frac{n_p}{n_p - 1} * R * T_1 * \left( \left( \frac{P_2}{P_1} \right)^{\frac{n_p}{n_p - 1}} - 1 \right) + \frac{n_p}{n_p - 1} * R * T_1 * \left( \left( \frac{P_3}{P_2} \right)^{\frac{n_p}{n_p - 1}} - 1 \right) + \cdots \quad (6)$$

$$T_2 = T_1 * \left( \frac{P_2}{P_1} \right)^{\frac{n_p}{n_p - 1}} \quad (7)$$

where

$n_p$: polytropic exponent [-],
P: pressure [Pa],
$P_1$: inlet pressure [Pa],
$P_2$: outlet pressure [Pa],
R: universal gas constant [J/(mole*K)],
$T_1$: inlet temperature [K],
$T_2$: outlet temperature [K],
V: Volume [m³],
$W_{ideal}$: required ideal compression energy [J/mole],
$W_{polytropic}$: required polytropic compression energy with inter-stage cooling [J/mole],
$\gamma$: specific heat ratio [-],
$\eta_p$: polytropic efficiency [-].

Increasing the number of compression stages would decrease the required power but increase equipment cost. The inlet pressure of the first compression stage can be assumed as 32 bar, which is the outlet pressure of a high-pressure electrolysis unit [12].

## 4.2 *Quantity Measurement*

Before injecting hydrogen into the natural gas infrastructure, it needs to be measured by custody transfer equipment. Since electrolysis units produce hydrogen with a purity of more than 99.9 per cent [12], a quantity measurement seems sufficient. For hydrogen-rich gases above 13bar, rotary displacement meters and turbine gas meters are recommended to measure the gas flow [13]. Pressure and temperature measurements are also necessary in order to calculate the transferred normal volumes. In the case of intermediate hydrogen storage below ground, the hydrogen might have become mixed with other gases, so gas quality standards need to be proved.

### *4.3 Hydrogen Injection into the Pipeline*

After measuring and pressure elevating the hydrogen to $P_{hydrogen}$, it has to be injected into a pipeline that already transports a stream of natural gas at pipeline pressure $P_{pipeline}$. Mixing units are already known in the fields of natural gas conditioning by the addition of nitrogen as well as biogas injection. It can be assumed that the mixing process is adiabatic and the pipeline pressure does not change, whereas the hydrogen pressure drops from $P_{hydrogen}$ to $P_{pipeline}$. This pressure drop increases the hydrogen's volume and temperature. The latter change is caused by the Joule-Thomson effect. Unlike typical natural gases, hydrogen has a very low inversion temperature $T_i$ of 202 K, above which it will warm up, if there is a pressure reduction, independently of the pressure level. Thus, under normal conditions, unlike typical natural gas, hydrogen warms up slightly if there is a pressure reduction. During the mixing itself, intermolecular forces are expected to cool down and shrink the gas again. Calculations based on AGA-8 show that a high pipeline pressure $P_{pipeline}$, a high hydrogen share, as well as high calorific values cause a bigger heat-up of the mixed gas. Still, the temperature increase caused by the addition of 5% hydrogen is expected to be less than 5 K [14].

## 5 Hydrogen in the Gas Grid

### *5.1 Combustion Characteristics of a Hydrogen-Natural Gas Mixture*

The technical standard DVGW-Arbeitsblatt G 260 [15] classifies certain gas qualities. Natural gas in transport pipelines can be defined by relative density, calorific value, and Wobbe index ranges. These combustion characteristics also represent the sales specifications. Hence, there is need to ensure that the required values are maintained. Calculations of ITE-GS have shown that for a low hydrogen content the only critical parameter is the relative density of methane-rich gas, e.g. Russian gas. The target specification of this H-gas family might not be met with a hydrogen share of 5%. Gas from other main German gas sources like Norway usually accepts at least 5% of hydrogen. The heating value and Wobbe index both decrease with increasing hydrogen content.

## 5.2 Pipeline Material Compatibility

When hydrogen interacts with materials, their properties might change. The following effects have to be considered [16, 17]:

- Hydrogen can be absorbed or adsorbed by the pipeline material in a hydrogen-rich environment and causes embrittlement. Possible reasons are an uneven hydrogen distribution in the material, worsening of the material's bonding forces, or precipitation of hydrogen molecules. This effect takes place at room temperature and is thus relevant for pipelines.
- Chemical reactions between the hydrogen and the material cause gas bubbles or hydrate formation, leading to material failure. The reaction between hydrogen and carbon is called "hydrogen attack" and causes high pressures by creating methane molecules. Since it only takes place at temperatures above 470 $^{o}$ K combined with high partial pressures of hydrogen, it will not affect the pipeline's steel. Still, it might play a role for gas turbines.

The compositions and properties of German pipeline steels are defined by the document DIN EN 10208-2 [18]. So far, hardly any of these steels have been subject to long-term testing with hydrogen. Generally, "cathodic charging" gives a qualitative result on hydrogen embrittlement effects. Tests with API steels show that yield strength, ultimate elongation, necking values, and finite-life fatigue strength can get significantly reduced by hydrogen interaction [19, 20]. The maximum operating pressure (MOP) is defined as a linear function of the yield strength. Therefore, hydrogen addition could result in lower MOP and thus reduce the physical throughput of a pipeline. Hydrogen also accelerates the fatigue crack growth [21]. On one side it can be concluded that hydrogen will impair the pipeline's steel properties. On the other side, the seriousness of the impairment is not yet known.

## 5.3 Maximum Pipeline Throughput

As described in the previous subsection, hydrogen could lower the physical throughput of a pipeline by decreasing its MOP. Furthermore, the calorific value by volume of a hydrogen-natural gas mixture is reduced by the addition of hydrogen, as stated above. If the hydrogen is only added in times of free capacities, the total amount of transported energy will increase. However, if hydrogen is granted a privileged grid access, the maximum transport capacity will decrease by a few per cent, because hydrogen will replace higher calorific natural gas.

## 5.4 Pressure Loss in the Pipeline

In a transport pipeline, gas pressure is reduced due to friction. The pressure loss in such a pipeline does not only depend on pipeline parameters and flow velocity, but also on gas parameters like density and viscosity. The viscosity is negligible but the density of a certain gas is nearly proportional to the pressure drop along a piece of pipe [14]. Simulations at ITE-GS have shown that hydrogen addition causes a smaller pressure loss in identical normal volume flow. This is due to the higher density of unmixed natural gas. However, if the normal volume flow of the hydrogen-natural gas mixture is increased in order to maintain the required energy flow, pressure losses are higher than if the same amount of energy was being carried by natural gas. Since customers order a certain amount of energy and not a volume, energy equivalent considerations are the decisive factors. Therefore, hydrogen will increase the pressure loss in gas pipelines.

## 5.5 Gas Compressors in the Network

In a gas transport system, compressor stations are installed at certain distances to maintain the operating pressure in the system. Changes in gas properties will have an impact on the compression energy, because they influence the pressure loss along the pipeline as described above, as well as the compressor performance. The isentropic change in enthalpy is defined by Equation (8) [22].

$$h_{ad} = \frac{\kappa}{\kappa-1} * R * T_1 * Z_1 * \left(\left(\frac{P_2}{P_1}\right)^{\frac{\kappa}{\kappa-1}} - 1\right) \tag{8}$$

$$\kappa = \frac{Z}{Z_p - \frac{R}{c_p} * {Z_T}^2} \tag{9}$$

where

$c_p$: specific heat capacity [J/(kg*K)],
$h_{ad}$: change in enthalpy [J/kg],
$P_1$: inlet pressure [Pa],
$P_2$: outlet pressure [Pa],
R: gas constant [J/(kg*K)],
$T_1$: inlet temperature [K],
$Z_1$: compressibility factor at inlet conditions [-],
$Z_{p/T}$: dimensionless derivative of the compressibility factor with respect to pressure or temperature [-],
κ: isentropic exponent [-].

When simulating the change of required compressor power excluding the preceding pipeline segment, the pressure ratios of all cases are identical. However, different gases differ in their compressibility factor and their isentropic exponent. Furthermore, volume flow and thus mass flow vary in order to ensure an identical energy flow. Results show that an energy equivalent gas flow consisting of a hydrogen-natural gas mixture requires more compression energy for the same pressure ratio than "pure" natural gas.

Now, the findings of the previous subsection need to be included. This is realized by considering a pipe segment in combination with a compressor unit. Since hydrogen will increase the pressure loss in the pipe segment as stated above, the calculated power consumption of a compressor will increase the pressure again. The typical increase in power requirements will be around 10% for 5% of hydrogen added to the system.

### 5.6 Compressor Drivers

A compressor station needs a power supply. Some compressor stations in the German gas grid are driven by electrical energy, while others are driven by gas turbines that use gas from the pipeline. These gas turbines were configured to burn natural gas and have varying limits for hydrogen contents. One reason is the increase in the flame temperature caused by hydrogen addition [23]. Additionally, an increased flame propagation speed increases the risk of a flash-back if a gas turbine operates in the gas pre-mix mode. Hydrogen also destabilises the flame front as can be seen by the impairment of the Markstein length which is an indicator for flame stability [24, 25].

Besides impacts on flame properties, the hydrogen itself can affect turbine materials. Increased hydrogen contents also increase its partial pressure. This might lead to a "hydrogen attack" (see section 4.a). On the other hand, the mole fraction of hydrogen in the exhaust gas increases by a small amount with increasing hydrogen contents in the combustion gas [23]. Therefore, the materials of the burner itself need to be re-examined, while other parts of the turbine including the blades will not face significantly higher hydrogen exposures. Another problem is the expected fluctuation of hydrogen content in the natural gas which makes turbine adjustment difficult.

## 6 Hydrogen in Underground Gas Storage Facilities

Underground gas storage facilities are an integrated part of the German gas supply system. Therefore, hydrogen will also enter these storage facilities as shown in section 2. Additionally, underground storage solutions might be used for intermediate storage of pure hydrogen before its injection into the gas grid. The two main storage types are porous reservoirs and artificial salt caverns. Depending on the boundary conditions of storage, one to two thirds of the stored gas volume is used as cushion gas for a cost-effective delivery of pressure and flow.

In a porous reservoir, the cap rock above and the gas-water contact below are critical interfaces. The cap rock remains impermeable, if the storage pressure is smaller than the sum of hydrostatic and capillary pressure. The capillary pressure is defined by Equation (10) showing that tightness mainly depends on water saturation and maximum pore size. These parameters are independent of the gas.

$$P_c = \frac{2 * \sigma_{c} * \cos\theta}{r} \tag{10}$$

where

$P_c$: capillary pressure [N/m²],
r: pore radius [m],
$\sigma_c$: surface tension [N/m],
θ: wetting angle [°].

The solubility of gas in water makes the gas-water contact critical. Alternating injection and withdrawal of gas can cause gas-saturated water in the contact zone to be replaced by unsaturated water. Comparing the solubility of hydrogen and methane in table 1 shows that mass as well as energy losses through the gas-water contact get reduced by hydrogen addition. An example of successful hydrogen storage in a porous media is an aquifer storage near Beynes, France, that was operated with a hydrogen-rich gas by *Gaz de France* until 1972. The injected gas composition exactly matched the withdrawn gas composition for a whole decade. Escaped gas that was collected by an overlying reservoir had the same composition. This shows that no preferential hydrogen migration took place.

In case of pure hydrogen storage in salt caverns, cushion gas represents a large and not fully recoverable investment. On the other hand, hydrogen in a depleted natural gas reservoir might be unpredictably mixed with the residual gas.

An example of the successful usage of salt caverns for hydrogen storage can be found in Teeside, England. In the salt caverns, no interactions between stored gas and unsaturated water took place. The gas loss of hydrogen is estimated to be 0.01 per cent of the total storage volume per year.

Another field of consideration is the possibility of chemical reactions between hydrogen and the reservoir rock, other gases like CO, $CO_2$, $O_2$ or hydrocarbons. Hydrogen could theoretically react with sulphide to form hydrogen sulphide, with sulphate to form sulphurous acid, with carbonate to form water and carbon dioxide, with fluorite to form hydrogen fluoride, and with oxide to form water. To estimate the realistic reactions for a reservoir, the free enthalpy of each reaction needs to be determined at reservoir pressure and temperature. Most of the above mentioned reactions cannot take place without a catalyser, but specialised bacteria could produce hydrogen sulphide or sulphuric acid. The presence of such bacteria needs to be evaluated before hydrogen can be injected into the storage. High salinity in salt caverns probably reduces bacterial activity and growth. [26]

## 7 Hydrogen Transfer to Customers

At each connection point of a gas transport pipeline and a distribution grid, a gas pressure regulating and metering station is located to measure the delivered energy to public utility companies and industrial customers and to ensure a safe operation of the low pressure distribution grid. The gas needs to be preheated before pressure reduction in order to make-up for the Joule-Thomson effect that would otherwise cause condensation, hydrates, and icing. As mentioned in Section 2, hydrogen addition will slightly decrease the required heating output by heating up at pressure reduction. Some metering stations also measure the gas quality. Existing gas chromatographs compare each component of the natural gas with a reference gas. The difference in thermal conductivity will identify each gas component. Typically, helium is used as reference gas, because it differs significantly from natural gas components in terms of thermal conductivity. This is not the case for hydrogen. Therefore, hydrogen and other light components of the pipeline gas can be separated from heavier components in order to be identified by argon, a reference gas. An upgrade of the equipment is feasible. [27]

## 8 Outlook

The injection of hydrogen into the midstream part of the gas supply infrastructure changes the properties and the behaviour of the transport medium and affects all elements of the gas transport system like pipelines, storage systems, compressor stations, and pressure regulators. In the long run, downstream end consumers like private households and industry require a reliable quality of natural gas. This area of gas application will show advantages and disadvantages in future research studies. Today, steel production processes require pure natural gas for the material-type used where no hydrogen content is preferable. Natural gas-powered vehicles only accept up to 2% of hydrogen by volume [28]. In the short run there may be the opportunity for centralized hydrogen infrastructures to supply large chemical customers using hydrogen as educts for ammonia and methanol production or for hydrotreating in refineries. Therefore, the field of underground pore storage and salt caverns calls for further research. In this context, the definition of an upper limit value like 2 or 5% is not suitable from a technical point of view. Instead, a target bandwidth specification should be set up on a European level. Another function of the model will be the quantification of the real storage capacity of the German gas grid dependent on the hourly production of renewable energy in future scenarios. Focus has to be on:

- energy production and consumption
- energy efficiency and savings
- network and storage management

## References

1. Sterner, M., Jentsch, M., Holzhammer, U.: Energiewirtschaftliche und ökologische Bewertung eines Windgas-Angebotes, Kassel, Germany (2011)
2. DVGW, DVGW-Arbeitsblatt G 262: Nutzung von Gasen aus regenerativen Quellen in der öffentlichen Gasversorgung, Bonn, Germany (2010)

3. EnWG – Energiewirtschaftsgesetz, Gesetz über die Elektrizitäts- und Gasversorgung (2011)
4. GasNZV – Gasnetzzugangsverordnung, Verordnung über den Zugang zu Gasversorgungsnetzen (2010)
5. EEG - Erneuerbares-Energien-Gesetz (2012)
6. Li, B., Müller-Kirchenbauer, J.: Overall simulation of German and European natural gas transmission systems. In: Hou, M.Z., Xie, H., Yoon, J.S. (eds.) Underground Storage of CO2 and Energy, pp. 305–310. CRC Press, Boca Raton (2010)
7. Li, B.: Simulation and capacity calculation in real German and European interconnected gas transport systems. Dissertation Technical University of Clausthal, Germany (2012)
8. NEP Gas, Netzentwicklungsplan Gas – Entwurf der deutschen Fernleitungsnetz-betreiber, Berlin, Germany (2012)
9. Tabkhi, F., Azzaro-Pantel, C., Pibouleau, L., Domenech, S.: A mathematical framework for modelling and evaluating natural gas pipeline networks under hydrogen injection, Toulouse, France (2008)
10. DVGW, DVGW-Arbeitsblatt G 486: Realgasfaktoren und Kompressibilitätszahlen von Erdgasen, Eschborn, Germany (1992)
11. Tzimas, E., Filiou, C., Peteves, S.D., Veyret, J.B.: Hydrogen Storage: State-of-the-Art and Future Perspective. Institute for Energy, Petten, The Netherlands (2003)
12. Wenske, M.: Stand Elektrolyse und zukünftige Entwicklung. Presentation of ENERTAG AG, Gut Dauderthal, Germany (2010)
13. Hüttenrauch, J., Müller-Syring, G.: Zumischung von Wasserstoff zum Erdgas. DVGW energie|wasser-praxis 10, Bonn, Germany (2010)
14. Schouten, J., Michels, J., Janssen-van Rosmalen, R.: Effect of H2-injection on the thermodynamic and transportation properties of natural gas. International Journal of Hydrogen Energy 29(1), 1173–1180 (2004)
15. DVGW, DVGW-Arbeitsblatt G 260: Gasbeschaffenheit, Bonn, Germany (2008)
16. Jordan, T.: Hysafe Chapter 3: Material Consideration when working with Hydrogen (2007)
17. Castello, P., Tzimas, E., Moretto, P., Peteves, S.: Techno-economic assessment of hydrogen transmission & distribution systems in Europe in the medium and long term. Report EUR 21586, Petten, The Netherlands (2005)
18. DIN EN 10208-2, Stahlrohre für Rohrleitungen für brennbare Medien - Technische Lieferbedingungen – Teil 2: Rohre der Anforderungsklasse B (2009)
19. Hardie, D., Charles, E., Lopez, A.: Hydrogen embrittlement of high strength pipeline steels. Corrosion Science 48(12), 4378–4385 (2006)
20. Capelle, J., Gilgert, J., Pluvinage, G.: Hydrogen Effect on Fatigue and Fracture of Pipe Steels. Ecole Nationale d'Ingénieurs de Metz, Metz, France (2009)
21. Alliat, I.: To what extent can existing pipelines accommodate hydrogen? - Presentation of NATURALHY project results, Groningen, Netherlands (2009)
22. Liwacom Informationstechnik GmbH & SIMONE Research Group, SIMONE SOFTWARE – Gleichungen und Methoden, Essen, Germany (2004)
23. Wang, J., Huang, Z., Tang, C., Miao, H., Wang, X.: Numerical study of the effect of hydrogen addition on methane-air mixtures combustion. State Key Laboratory of Multiphase Flow on Power Engineering, Xi'an Jiatong University, Xi'an, China (2008)

24. Hu, E., Huang, Z., He, J., Jin, C., Zheng, J.: Experimental and numerical study on laminar burning characteristics of premixed methane-hydrogen-air flames. International Journal of Hydrogen Energy, State Key Laboratory of Multiphase Flow on Power Engineering (2009)
25. Jassin, F.: Flammenrückschlag durch verbrennungsinduziertes Wirbelaufplatzen. Dissertation Technical University of Munich, Germany (2003)
26. Foh, S., Novil, M., Rockar, E., Randolph, P.: Underground Hydrogen Storage Final Report. Institute of Gas Technology, Chicago, USA (1979)
27. Pöppl, H.: Flexibler Prozess-Gaschromatograph für die neuen Anforderungen an Gasanalysegeräte. gwf-Gas|Erdgas, Oldenburg Energieverlag, Munich, Germany (2011)
28. DIN 51624, Kraftstoffe für Fahrzeuge – Erdgas – Anforderungen und Prüfverfahren. Beuth Verlag GmbH, Berlin, Germany (2008)

# Possible Locations for Gas-Fired Power Generation in Southern Germany

Joachim Müller-Kirchenbauer and Anatoly Gorshkov

Institute of Petroleum Engineering, Technical University of Clausthal
Agricolastrasse 10, 38678, Clausthal-Zellerfeld, Germany

**Abstract.** Gas-fired power generation has not only grown continuously in Europe, but also in developing countries where energy infrastructures are yet to possess similar transport and storage capabilities. Significant transport capacities in a high pressure gas grid are required to guarantee stable generation of gas-fired electricity. The purpose of this study is to investigate how a large number of new gas-fired power plants could be built to take full advantage of the available infrastructure, taking into account the German nuclear moratorium, as an excellent opportunity for a rapid increase in gas-fired power generation. The theoretical section briefly describes the German gas network model developed at ITE-GS and outlines the simulation procedure. Information about future necessary electricity production was used to create a combined map of both high voltage power and high pressure gas grids, where suitable locations for power plants were identified. After a simulation of the network behavior of new gas consumers, capacity limits and the weakest parts of the gas network were identified and the reasons explained. Additionally, an investigation of national gas infrastructure projects' influence on the potential of future gas-fired generation was performed.

**Keywords:** gas-fired power generation, nuclear moratorium, pipeline capacity, network modeling.

**Nomenclature**

NPP: nuclear power plant; GFPP: gas-fired power plant;
CCGT: combined cycle gas turbine; OCGT: open cycle gas turbine;
HVTN: high voltage transmission network; TSO: transmission system operator;
BW: Baden-Württemberg; MOP: maximum operating pressure;
GS: Gas Supply; ITE: Institute of Petroleum Engineering;

## 1 Introduction

After a catastrophic earthquake in Japan on the 11$^{th}$ of March 2011, the whole world witnessed one of the most devastating nuclear accidents in history – the Fukushima Daiichi nuclear power plant disaster. The full scale of the

M.Z. Hou et al. (Eds.): *Clean Energy Systems in the Subsurface*, SSGG, pp. 437–450.
DOI: 10.1007/978-3-642-37849-2_34 

consequences is yet to be estimated, but the Fukushima incident received a very serious international reaction by governments and of course the public. The response of the latter group was very strong in Germany and forced the policy makers to change the nuclear energy strategy.

The discussion led to a so-called "Nuclear Moratorium". The Moratorium stipulated that that the seven oldest nuclear power plants had to be shut down immediately and the rest have to be decommissioned earlier than planned.

Considering the fact that Germany is the most populated and most industrialized European country, such a sharp change in its energy generation portfolio raised the question about the security of the electrical supply and the reliability of the high voltage transmission network (HVTN). This issue is of particular importance for the Southern part of Germany, which is particularly reliant on nuclear energy.

Problems of secure electrical supply will intensify in the future not only because of the nuclear phase out, but as a result of increasing generation from fluctuating renewable sources located mostly in the North of Germany and a decrease in conventional generation capacity (Fig. 1).

Construction of new high-voltage power lines connecting Northern Germany to the rest of the country might solve the power shortage issue, but there are obstacles that prevent rapid development of this new infrastructure. Public opposition, for instance, to the construction of new power lines for fear that they may extend the emission process for years [1]. Network reinforcement cannot be carried out due to an increased network load [2].

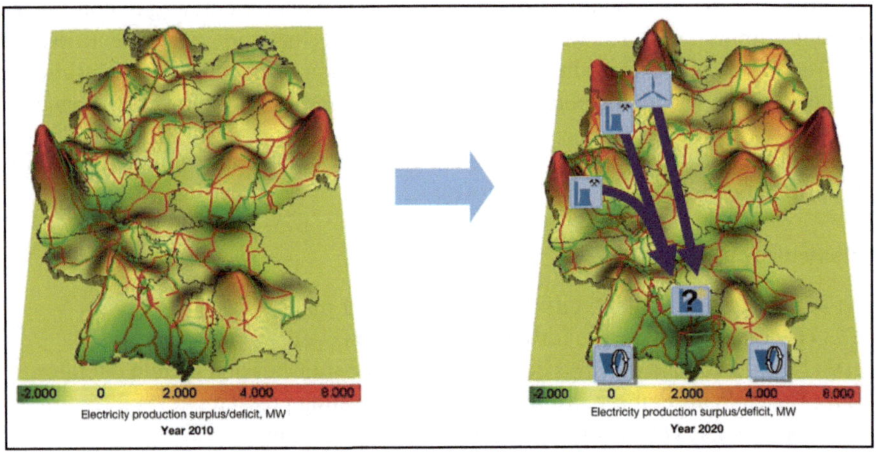


**Fig. 1** Electricity production surplus/deficit in Germany as of 2010 and prognosis for 2020 [3]

One of the possibilities to assure the security of the power supply might be the introduction of new gas-fired power plants (GFPPs). However, there is no guarantee that the gas transmission network has sufficient capacity available to supply these new generation facilities. Therefore this study is focused on investigating the potential of the German high-pressure natural gas grid to provide an adequate power supply in terms of available transport capacity and pressure supply conditions.

## 2 Methods

To quantitatively predict the future gas transmission capacity that new gas-fired power production will require it is necessary to estimate the future gas-fired electricity production beforehand. Several sources [4, 5, 6, 7] were analyzed for that purpose.

Sufficient gas transmission capacity and an increase in German natural gas demand were calculated in 5-years steps according to the power production required by respective dates. Taking into account the nuclear phase-out, each nuclear power plant generation capacity that is taken out of the grid was replaced with a gas-fired power generation and the remaining generation capacity located to suitable places according to criteria described below.

From the point of view of available capacity in a high voltage transmission network, the replacement of nuclear power plants (NPP) with gas-fired generation should not cause any problem. NPP is a huge electricity producer connected to major high voltage power lines in the country. Another benefit of the NPP location is that these installations are normally built in the vicinity of a large water body, which will be appreciated if a GFPP with a combined cycle is to be constructed.

A series of new gas-fired power plants was considered by means of two possible scenarios:

1. Scenario “A” introduces combined cycle gas turbines only. This presumes a higher efficiency per plant compared to single cycle gas-fired power stations and therefore lower gas transmission capacity requirements.
2. Scenario “B” uses the same parameters but includes a 10% share of open cycle gas turbines and therefore a demand for higher gas transmission capacity.

One of the possible approaches for an estimation of additionally available gas transport capacity and the detection of possible congestion is the examination of gas network behavior at peak consumption time, which in this case means that all the newly introduced gas-fired power plants work at their respective peak consumption levels.

Necessary gas transmission capacity during peak electricity production of GFPP under standard conditions is calculated as follows:

$$Q = \frac{\frac{\frac{P_N}{\eta} * 10^3}{100}}{H_u} \qquad (1)$$

| | | |
|---|---|---|
| $Q$ | = | gas flow rate ($m^3$/h at standard conditions) |
| $P_N$ | = | net power output of GFPP (MWh/h) |
| $H_u$ | = | gross calorific value (kWh/m3) |
| $\eta$ | = | efficiency of gas-fired power plant (%) |

To allocate new GFPPs properly, it is advisable to have a combined map of both electricity and gas networks. The German electricity network map was taken from the HVTN map made by the Union for the Coordination of the Transmission of Electricity[1] (UCTE) on 1st July 2008. The map was of high quality and allowed German HVTN to be extracted for further use. The gas network map made by VGE GmbH [8] was of ArcGIS shape files format and already geo-referenced to real coordinates, therefore the map from HVTN was inserted and geo-referenced to gas network by means of ArcGIS software tool with a total root mean square error[2] of 0,0133.

As earlier stated, allocation was firstly based on the former location of NPPs. After the nuclear power production has been substituted, the rest of the necessary generation capacity for a particular period was distributed in concentrated groups using the following criteria:

1. The Presence of a gas trunk pipeline with a preferred inner diameter of at least 800 mm.
2. The Availability of a high-voltage power line at the same place, preferably a 380 kV line.
3. A large water body such as a river or lake in the vicinity. For this purpose an additional geo-referenced map of German hydro lines was used [9].

The majority of the GFPPs were equally distributed as far as possible between the two Federal lands of Bavaria and Baden-Württemberg, a minor part was allocated to the Federal lands of Hessen and Rheinland-Pfalz.

The allocation of the new power plants locations is showed in Figure 2.

---

[1] Now transferred to ENTSO-E.

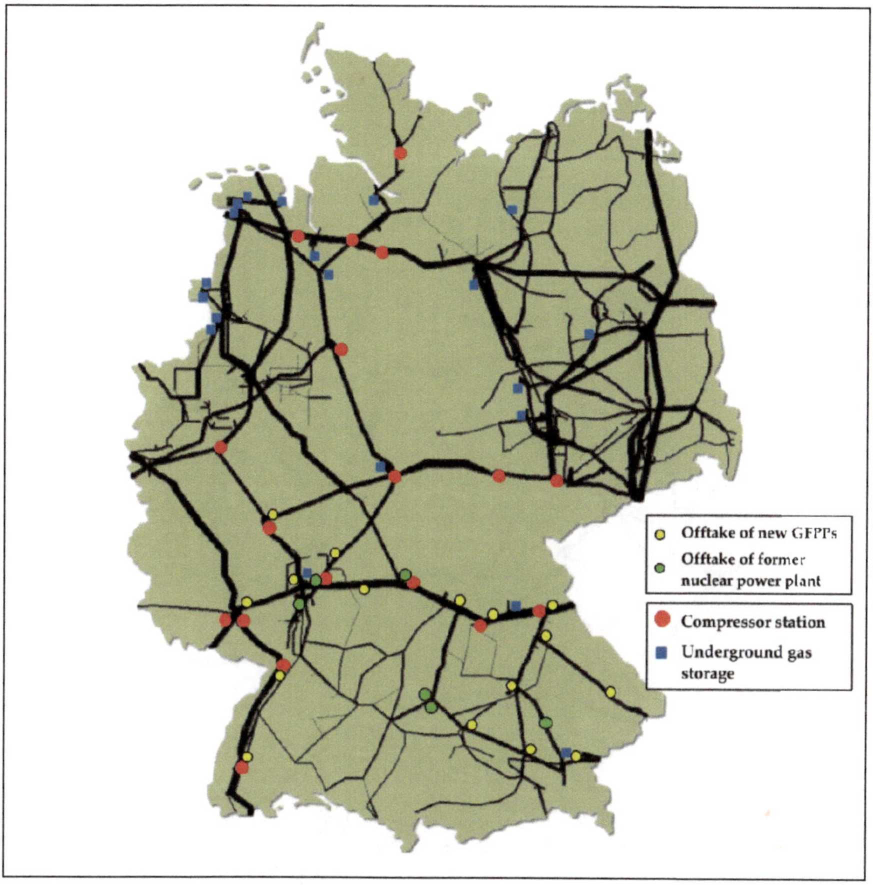


**Fig. 2** Allocation of new gas-fired power plants inside ITE-GS Model

## 3 ITE-GS Model and Simulation Procedure

The model of the German high-pressure gas network was created at the Gas Supply (GS) department of the Institute of Petroleum Engineering (ITE). The model is an overall hydraulic simulator for German high calorific value natural gas (H-gas) system of the six major TSOs (Open Grid Europe, Gascade, Gasunie Deutschland, Terranets BW, Bayernets, Thyssengas) as well as some other major German gas pipelines (for example OPAL).

An overview of the model is shown on Figure 3.

Topology of the Model consists of several major parts:

- Natural gas pipelines
- Compressor stations
- Underground gas storage facilities

- Valves to route the gas flow in a certain direction
- Pressure regulators, which assure that maximum operating pressure (MOP) of certain pipelines or downstream networks is not exceeded

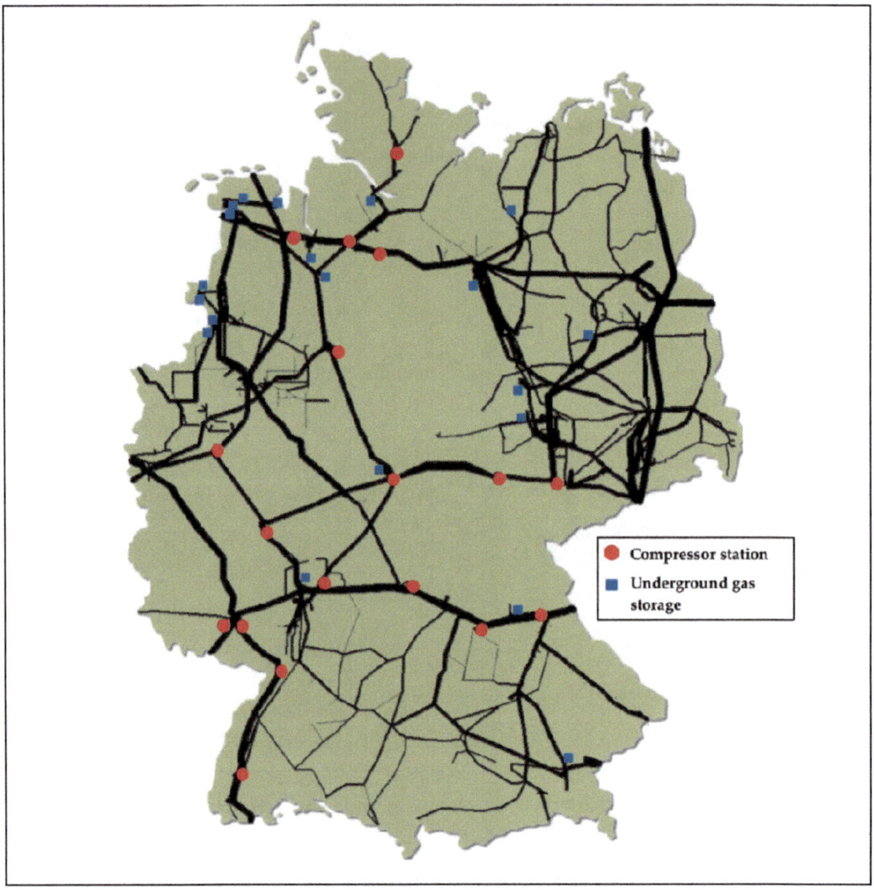


**Fig. 3** Overview of ITE-GS Model

The Model was developed using open sources of information and is under continuous improvement. The new infrastructure projects are inserted into the Model as well as possible future developments. The major investment driver for the expansion of the natural gas transmission network is the National Network Development Plan Gas (NDP or Netzentwicklungsplan Gas) [10], which is reviewed yearly in cooperation with all the market participants (mostly transmission system and storage operators) and the regulatory authority (Bundesnetzagentur). The Network Development Plan Projects were not implemented into the model topology in this study.

Besides the model topology, another important part of the model is the scenario parameters that determine the flows inside the network:

- Flow values for the entry and exit points
- Compressor station parameters like pressure ratio, maximum flow rate, etc.
- Valve conditions that determine flow direction and flow values
- Alarm limits to prevent an excess of pressure and flow

The ITE-GS Model is operated in SIMONE – an established solution for both operations and the design and planning of natural gas transmission networks currently available [11]. A complete Model description can be found in [12].

The simulation procedure in this study consists of three major steps: setting up the topology, adjusting the flow values and an analysis of the results (Fig. 4).

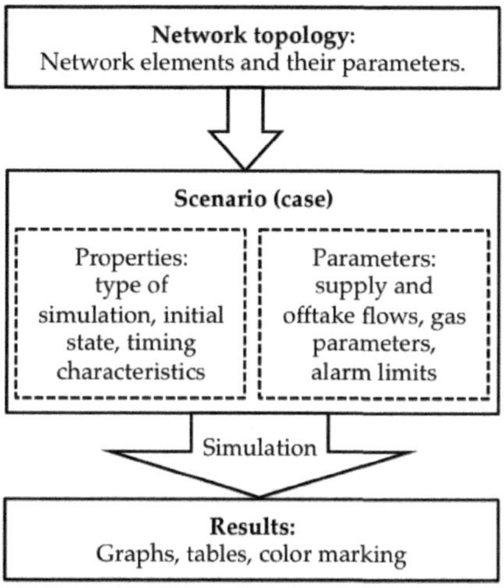


**Fig. 4** Overview of the simulation procedure

After suitable places for new gas-fired power plants were found they were introduced into the Model topology with associated consumption values.

Simulation was carried out for every time period starting from 2011, so that withdrawal from the network was gradually increased according to the increase in gas-fired power generation by a respective date. All the simulation was performed in a steady state and all the GFPPs were at their respective peak production levels to determine the reaction of the gas network during the highest possible gas-fired electricity production. Natural gas consumption in Germany is, to a large extent, temperature driven and therefore the first simulation rounds were made with a gas flow pattern representing the most stressful situation in the gas grid, that is, a typical winter day with a negative outdoor temperature.

The network was considered stable after the simulation run if both of these conditions were met:

1. Sufficient flow rate (transmission capacity) is provided to every group of GFPPs and the pressure at that withdral point is above 16 bar. Other threshold values considered during investigation are 25, 40 and 60 bars.
2. Pressure at all other withdrawal points in the model has a positive value.

If the above stated conditions were not fulfilled, withdrawal values were changed as if certain GFPPs were reallocated from their initial position. This was how a successful working case was created (the network is stable under these conditions) and the locations where the necessary capacity could not be provided (network congestion) were indicated.

After the applicable allocation of new GFFPs was found and the table with final withdrawals for a typical winter day was generated, the same numbers were used to begin research on the supply situation representing spring/autumn time. Because boundary conditions differ, an iteration of offtake values had to be performed one more time to achieve the working conditions of the network after each simulation round. The last iteration studied the network behavior during a typical summer day characterized by low consumption but significant storage injection.

The whole simulation procedure for three possible supply scenarios was iterated and the outcome was a final table with offtake values that satisfy all three cases, so that GFPPs can be supplied to these locations at any time of the year.

## 4 Results

Initial outcomes of this study, before the actual simulation took place, can be summarized as follows:

1. There are enough suitable places to install new gas-fired power plants in Southern Germany that would have both a high-pressure gas pipeline and high-voltage power lines in their surroundings.
2. Most these locations are situated near large water bodies, which is an important side condition
3. The required future power generation capacity from GFPPs would be equally distributed between the western and eastern parts of southern Germany.

Despite numerous attempts to find offtake values that would satisfy all three supply cases for any of the scenarios “A” and “B” for year 2030, it was found to be impossible to have an overall network for year 2030 without a significant decrease in consumption values. The reason for this is not only different boundary conditions between different network flow scenarios. It seems that from 2025 the available capacity in critical parts of the network mentioned above tends to reach its limit, so any additional increase would only be possible for the areas that are opposite to a typical winter or summer day.

Nevertheless, there is enough capacity in the gas network to substitute all NPPs with gas-fired electricity generation for both scenarios “A” and “B”. Scenario “B” provides more flexibility for such a substitution but would involve higher gas consumption.

There is enough capacity in the gas network to supply all the necessary gas-fired installations up to and including year 2025 only if high-efficient CCGT are used. For the year 2025 and scenario "B", the network is closer to its limit because of less-efficient OCGT at the largest NPPs locations, but the withdrawal has to be lowered insignificantly from the planned value.

One of the main outcomes of this study was the identification of network congestion. This is one of the key issues for the implementation of adequate policies concerning future development of natural gas networks in Germany. Location of the weakest network parts is different for every case, so a lack of capacity might be observed in the real network depending on the season. Graphical representation of this is shown in Figure 5.

**Fig. 5** Congestion areas identified after simulation

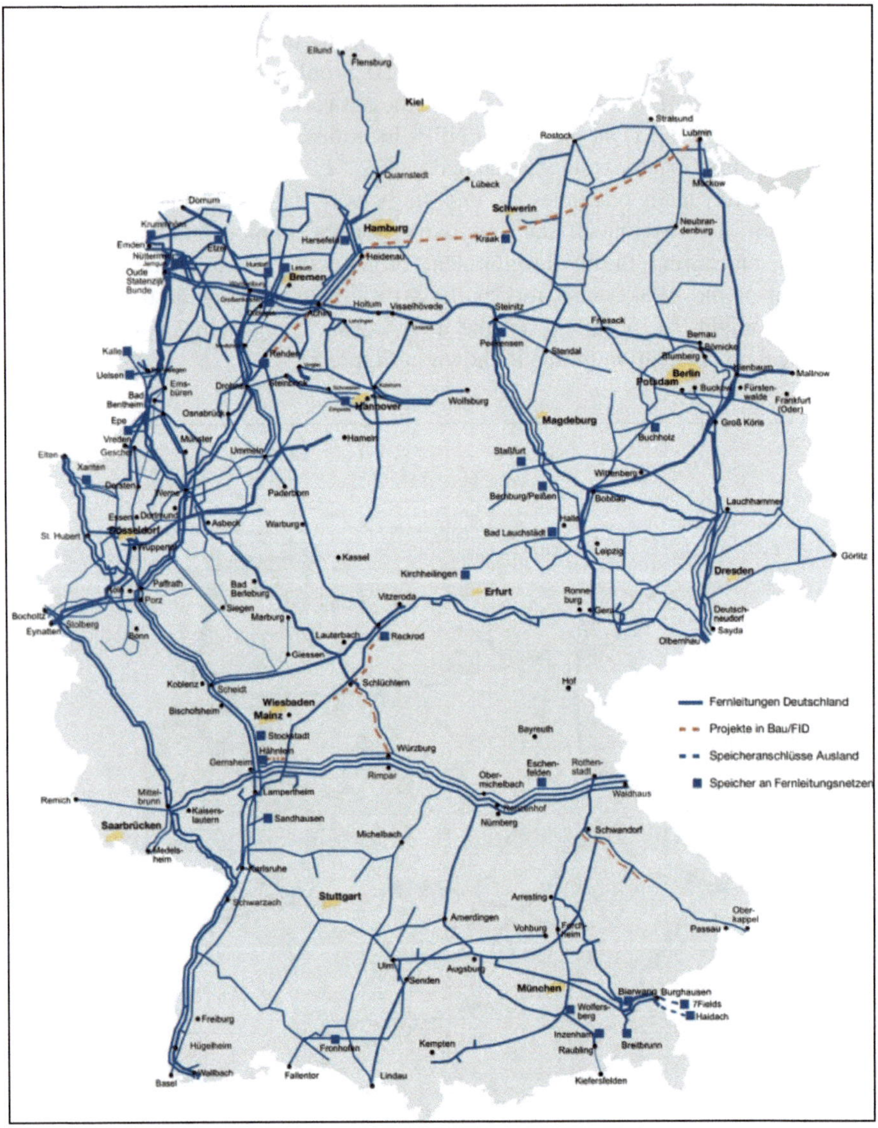


**Fig. 6** Network expansion projects awaiting a final investment decision from the German Network Development Plan for Gas [10]

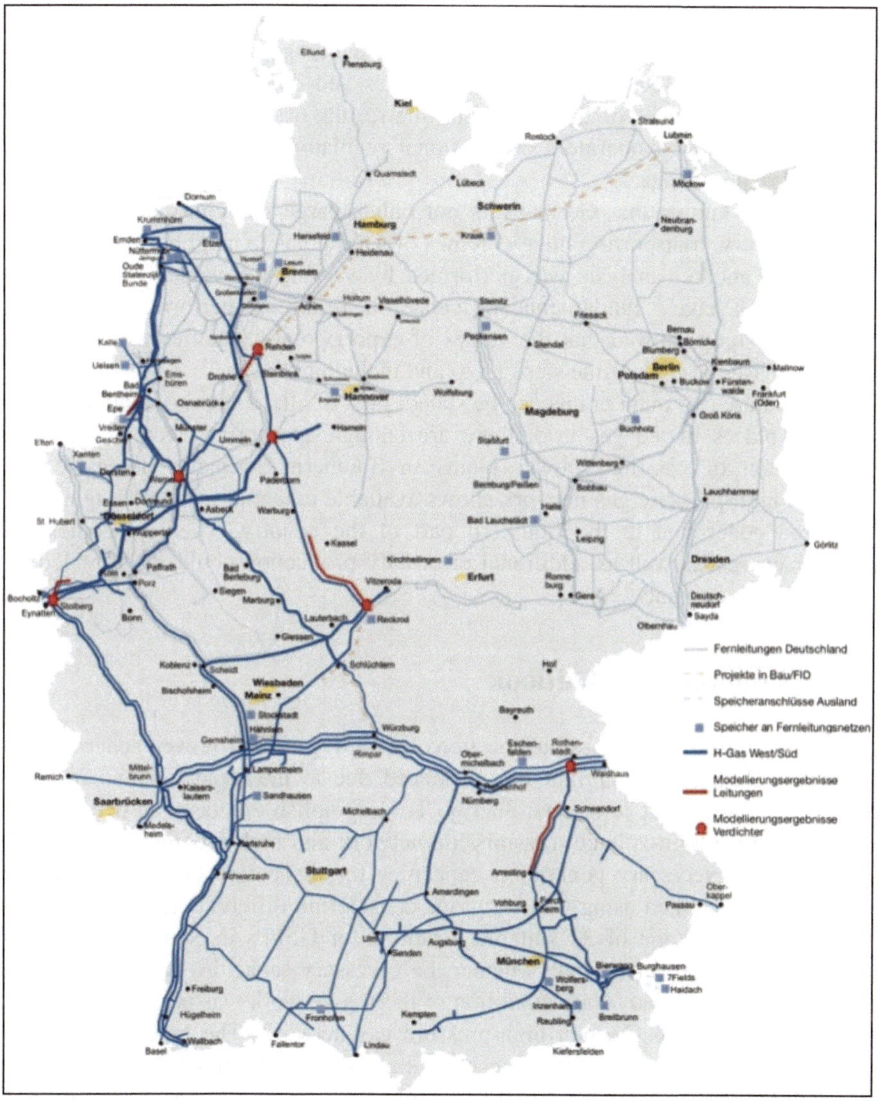


**Fig. 7** Network expansion projects that should be implemented by the end of 2015 from the German Network Development Plan for Gas [10]

The identified congestion areas might require additional capacity expansion by means of either newly constructed pipelines or compressor stations. The results achieved in this study correlate with the outcomes of the first National Network Development Plan. Firstly, currently implemented projects are being carried out in the same location or near to areas defined as congestion areas in this study (Fig. 6). All of the projects marked with dotted red lines in Figure 6 have been approved for construction and are awaiting a final investment decision. Secondly, projects that are foreseen as necessary for implementation in the next three years (by the end of 2015) might contribute to avoiding the possible congestion identified in this study (Fig. 7, solid red lines and dots). Thus, the similarities between the outcomes of this study and the results of the joint investigation by transmission system operators and national regulatory authorities cover all the essential requirements.

It is well known that Germany is not only a large gas consumer but also a transit country, transporting huge volumes of natural gas to neighboring countries. Therefore, another analysis was performed by varying the values of the export flows, which have a fundamental influence on the Model performance. A short sensitivity study showed that the values of export point capacities will tolerate a 10% increase for all the time steps up to and including 2025.

To sum up, two main points can be stated. First of all, in terms of availability of suitable places in the gas grid, there are enough appropriate locations for the construction of gas-fired power plants in Southern Germany. Secondly, the German high-pressure gas network shows available potential to supply significant gas-fired generation to the Southern part of the country. The replacement of nuclear energy as well as additional electricity production could be provided by gas-fired power plants.

## 5 Summary and Outlook

The study investigated possible locations for new gas-fired power generation in Southern Germany that might be introduced due to the rapid planned increase prior to the phase out of nuclear energy. To accomplish this task, an overall map of the German high-voltage transmission network and high-pressure gas network was created. Necessary generation capacities for the period between 2011 and 2030 were calculated using two scenarios considering different types of gas-fired power plants. A total of 22 suitable locations for GFPPs in Southern Germany were identified. To investigate whether the necessary capacities were available, a simulation of the steady state condition of the gas network was performed using a model of the whole German high-pressure gas network. The model considered three different cases taking into account the seasonal dissimilarity of gas flows through Germany. After the simulation went through all the iteration of offtake values, it was possible to demonstrate where the new gas-fired power plants might be located to utilize the full potential of the gas network for the time period between 2011 and 2030.

Considering the scope of the work, the following conclusions can be drawn:

1. The gas network is able to adequately supply GFPPs that may gradually substitute nuclear power in Southern Germany.
2. The Gas network is able to adequately supply all GFPPs that might be introduced in future up to 2025 even if all of them were located in Southern Germany.

3. The current gas network cannot provide all the capacity required in 2030. Several regions are lacking the required capacity and this depends on the time of year

The investigation carried out in this study is a first step and might be improved further with a focus on the improvement of both the Model and the methods applied. The following areas concerning hydraulic modeling will improve the results:

1. Model topology has to reflect not only the latest current developments but also all perspective projects including projected natural gas storage facilities. This is continuously done at ITE to keep the Model topology up to date.
2. Flow values for all consumers have to reflect the forecasted data of future German natural gas consumption and their share of different utilization sectors.
3. Simulation might be performed hourly throughout the year (8760 separated simulations). This is currently being performed at ITE.

The implied methods will undergo enhancement in the following major areas:

1. Definition of locations for the new GFPPs will be improved on the basis of a more detailed geo-database taking into account additional factors.
2. Substitution of NPPs with gas-fired generation will pose no problem with available electricity capacity, but this cannot be guaranteed for the rest of the locations chosen in this study. Research will be undertaken into the available capacities of HVTL in the chosen areas.

## References

1. EU, Permitting procedures for energy infrastructure projects in the EU: evaluation and legal recommendations, http://ec.europa.eu/energy/infrastructure/studies/doc/2011_ten_e_permitting_report.pdf (accessed November 22, 2011)
2. Bundesnetzagentur, Update of Bundesnetzagentur report on the impact of nuclear power moratorium on the transmission networks and security of supply, http://www.bundesnetzagentur.de/SharedDocs/Downloads/EN/BNetzA/PressSection/ReportsPublications/2011/110527PressConferenceNuclearPowerMoratoriumpdf.pdf?__blob=publicationFile (accessed November 09, 2011)
3. Barth, P.: Netzentwicklung in Deutschland und Europa, http://www.efzn.de/fileadmin/Veranstaltungen/2010_Elektrizitaetsnetze/Vortraege/02_Barth.pdf (accessed November 08, 2011)

4. BMWi, Analyse und Bewertung der Versorgungssicherheit in der Elektrizitätsversorgung, http://www.bmwi.de/Dateien/Energieportal/PDF/analyse-und-bewertung-der-versorgungssicherheit-in-der-elektrizitaetsversorgung-2010,property=pdf,bereich=bmwi,sprache=de,rwb=true.pdf (accessed November 09, 2011)
5. Umweltbundesamt, Energieziel 2050: 100% Strom aus erneuerbaren Quellen, http://www.umweltdaten.de/publikationen/fpdf-l/3997.pdf (accessed November 09, 2011)
6. Dena, Dena Grid Study II. Integration of Renewable Energy Sources in the German Power Supply System from 2015 – 2020 with an Outlook to 2025, http://www.dena.de/fileadmin/user_upload/Download/Dokumente/Studien___Umfragen/dena_Grid_Study_II_-_final_report.pdf (accessed November 22, 2011)
7. Szenariorahmen für den Netzentwicklungsplan Gas 2012 der Fernleitungsnetzbetreiber, http://www.netzentwicklungsplan-gas.de/files/110822_Prognos_Szenariorahmen_NEP_F.pdf (accessed November 22, 2011)
8. VGE Verlag GmbH, http://www.vge.de
9. GfK GeoMarketing GmbH, http://www.gfk-geomarketing.de
10. Netzentwicklungsplan Gas 2012, http://www.netzentwicklungsplan-gas.de/files/netzentwicklungsplan_gas_2012.pdf (accessed April 01, 2012)
11. LIWACOM Informationstechnik GmbH (2011), http://www.liwacom.de/index.php?id=eon&L=2 (accessed November 28, 2011)
12. Li, B.: Simulation and capacity calculation in real German and European interconnected gas transport systems. Dissertation, Technical University of Clausthal (2012)

# Recent Developments in Geotechnical Design of Natural Gas Storage Cavities Regarding Physical Modelling as Well as Numerical Simulation

Karl-Heinz Lux

Institut für Bergbau, Professorship for Geomechanics and Waste Disposal Technology, Clausthal University of Technology, 38678 Clausthal-Zellerfeld, Germany

**Abstract.** Salt cavities for storage of natural gas in bedded or domal salt structures are an important element of today's and tomorrow's energy supply management. In Germany the mechanical design of salt cavities has a history of more than 35 years. Based on laboratory investigations, physical modelling, analytical or much more importantly numerical simulations and last but not least field experiments and operational experience, knowledge about salt mechanics and salt cavern load bearing behaviour has increased significantly. This scientific-technical improvement corresponds very well with the development of individual cavern sizes starting at a volume of 10,000 $m^3$ and rising to a Million $m^3$ of natural geometric storage with optimal geotectonic conditions.

The presentation first gives an overview on the development of salt cavity design and an insight into today's state of the art designs. Special items include the geo-mechanical characteristics of storage cavities and principle safety demands for their design as well as recent design concepts and a way to provide geotechnical proof of safety with specialized criteria. Furthermore, the main aspects of lab testing and physical modelling are considered. Here damage to salt rock and its numerical characterization are of significant importance for further development. Based on numerical simulations using commercial or specialized software, safety analysis and cavern design, with respect to site specific defined criteria, lead to admissible values for cavern configuration and operation parameters.

Additionally, based on decades of existing experience, a method for designing salt caverns for high frequency cycling of storage gas is introduced using long – or short-term historic operational data. Material parameters and the deformation behaviour are adjusted to cavern convergence data from sonar surveys with reference to multicyclic laboratory tests of rock salt cores. Using the novel *Lux/Wolters* constitutive material law, where rock salt is treated as "material with memory" by a modified *Lubby2* approach, the real behaviour of rock salt deformation can be reliably described. For multicyclic storage operations additional thermal stresses have to be taken into account for analysis of cavern integrity. A numerical thermodynamic simulation model is used to forecast related temperature changes. Cavern design in this case is based on thermo-mechanic coupled simulations.

M.Z. Hou et al. (Eds.): *Clean Energy Systems in the Subsurface*, SSGG, pp. 451–485.
DOI: 10.1007/978-3-642-37849-2_35 

**Keywords:** Salt cavities design, physical and analytical modelling, geomechanical characteristics, safety analysis, Lux/Wolters constitutive material law.

# 1 Natural Gas Storage in Salt Cavities in Germany

Gas storage in salt cavities is nowadays a well – experienced technology in Germany with a history covering a period of more than 35 years. At the contemporary end of this development cavities with a geometric storage volume of more than 1 Mio m³ have been designed and constructed, compared to the first cavity at the end of the 1960ies having a geometric volume of only 35000 m³.

Figure 1 gives a current overview on the different locations of storage cavity facilities in Germany and the number of cavities, respectively, Sedlacek [1]. From a geological point of view, the storage cavern sites in Germany include domal salt structures as well as bedded salt structures. In total there are 22 gas storage facilities/operators existing today and approximately 207 cavities. A lot more cavities in the order of magnitude of 160 are being developed or under construction. The largest facility including 5 different operators has about 70 storage cavities. Not only are there different geological structures existing at the individual sites, but also a wide range of depth is covered, starting with more compact cavities in bedded salt layers in a shallow depth range of 500-650 m and going much deeper to domed salt structures with very high, slim cavities in the range of 1400-1850 m.

The basic elements for this impressive improvement in the past few decades are at first to be seen in a continuously developed and improved understanding of the mechanical behaviour of saliniferous rocks as well as technical constructions in salt rock mass, second in improved lab testing facilities as well as modelling and numerical simulation techniques and third, not to be forgotten, in excellent practical experiences over the years with this type of geotechnical structure. These excellent practical experiences cover both geotechnical aspects like structural stability, tightness and surface protection as well as economic aspects.

To increase economic benefit and follow the demands of the future gas market, the operation pattern of gas storage cavities shows a tendency to change from seasonal storage (one main storage cycle per year) to storage with a frequent turn over and high deliverability (several storage cycles per year). The demand for guaranteeing a sufficient load bearing and sealing capacity of the rock mass surrounding the cavity as well as agreeable convergence and related surface subsidence over some decades of operation at the same level of geotechnical safety is independent of the individual operation pattern. Therefore the effort of producing reliable cavern design as well as current monitoring of rock mass stress during cavern operation is increasing with the complexity of cavern operation modes.

Additionally, the operational life of the cavities has to be considered and a way of protecting the environment after they are abandoned is needed. This aim can only be met if there is an understanding of the consequences of filling abandoned and sealed salt cavities with either solids or liquids. Despite of the possibility of

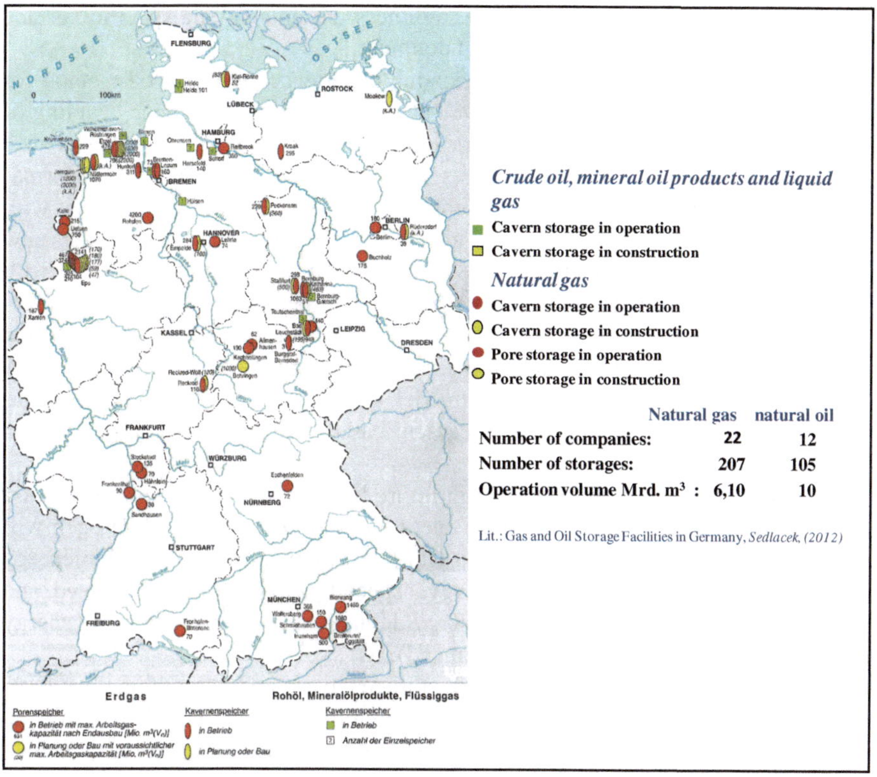


**Fig. 1** Gas and oil storage facilities in Germany [1]

refilling the cavities with appropriate (anorganic) solid waste material prior to abandonment, the preferred way for cavern abandonment in Germany today is to refill the relevant cavity with water and to provide the liquid-filled cavity with a permanent borehole sealing. The long-term behaviour of sealed liquid-filled cavities is therefore an important item for geotechnical safety analysis and safety assessment.

A comprehensive overview on cavern design is presented in Lux [2] as well as Lux et al [3], Lux et al [4], Lux [5], Lux [6] and finally Wolters et al [7] describing the respective knowledge at the relevant time. Based on these fundamentals, this paper first gives an overview of the development of salt cavity design and an insight into today's state of the art designs. Special items will include the geomechanical characteristics of storage cavities and the principle safety demands for their design as well as recent design concepts and a way of providing geotechnical proof of safety using specialized criteria. Furthermore, some main aspects of lab testing and physical modelling are considered. Here damage to salt rock and its numerical characterization are of significant importance for further development. Based on numerical simulations using commercial or specialized software, safety analysis and cavern design, with respect to site specific defined criteria, lead to admissible values for cavern configuration and operation parameters keeping in mind the demands for static stability, tightness, acceptable site-specific surface subsidence and environmental safe abandonment.

Additionally, based on decades of existing experience an improved method for designing salt cavern for seasonal as well as particularly high frequency cycling of storage gas is introduced using historic operational data. Basic ideas have been presented by Lerche [8] and Lux & Dresen [9]. Material parameters and deformation behaviour are adjusted to cavern convergence data from a sonar survey with reference to multicyclic laboratory tests of rock salt cores. Using the new Lux/Wolters constitutive material law, where rock salt is treated as "material with memory" by a modified Lubby2 approach, the real behaviour of rock salt deformation can be reliably described. For multicyclic storage operations additional thermal stresses have to be taken into account for analysis of cavern integrity. A numerical thermodynamic simulation model is used to forecast related temperature changes. Cavern design in this case is based on thermo-mechanic coupled simulations.

Finally it has to be mentioned that in future beside the storage of crude oil and natural gas, storage of compressed air energy as well as storage of hydrogen will take place in Germany in the context of the development of a renewable energy supply. Fluctuating wind and solar energy, especially, will have to be adapted, and in different way fluctuating electric energy demand as well (no renewable energy available, but a high electric energy demand).

A precondition for a sufficiently safe storage cavern operation on the one hand and an economic optimum storage cavern operation mode on the other hand is an excellent understanding of salt cavern behaviour, this means the reaction of rock salt mass to impacts resulting from cavern construction as well as cavern operation and its reliable simulation with respect to mechanical, thermal and hydraulic processes.

## 2 Some Basic Aspects of Gas Storage with Respect to Cavern Design

Salt cavities are large underground constructions in salt rock mass. Figure 2 gives an example of such a cavity together with the surface installations. In this case the vision of the storage of renewable but fluctuating wind energy by compressed air (CAES) is shown. From this figure it also can be seen that salt cavities are principally connected to the surface by a drill-hole. Therefore salt cavities are not accessible to man. They are created by a solution mining method. The static stability as well as tightness and functionability of these geotechnical structures have therefore to be guaranteed by the saliniferous rock mass itself – neither artificial lining to improve stability of the cavity nor artificial sealing to improve tightness of the cavity are possible. Thus the rock mass surrounding the cavity must itself have sufficient load bearing capacity on the one hand and be both liquid and gas tight on the other hand.

Additionally rock mass deformation resulting in the decrease of cavern storage volume must be limited to ensure sufficient storage volume over decades and to limit surface subsidence to site-specific admissible values.

**Fig. 2** Renewable Energy-Storage of fluctuating wind energy via compressed air in salt cavities [10]

Observing these boundary conditions, and ignoring the geometric layout of the cavity, just one parameter is available to actively influence the rock mass stress intensity and thus the rock mass load bearing behaviour significantly: the cavern inside pressure which depends on the pressure of stored liquids or gases. In the case of gas storage cavities, gas pressure can vary between atmospheric pressure as a lower limit and a so-called maximum cavern inside pressure as an upper limit to guarantee the tightness of salt rock mass. If the static stability of the load bearing system cannot be demonstrated to be at or above atmospheric pressure conditions, this minimum allowable pressure must be elevated. In this case part of total stored gas has to stay principally in the cavity to ensure the necessary minimum cavern inside pressure and thus cannot be used for storage purposes.

To summarize, in the case of gas storage cavities part of the stored gas, called cushion gas, has to stay permanently in the cavity to produce the necessary minimum pressure inside the cavity to guarantee static stability as well as limited rock mass deformation, i.e. limited convergence of cavity regarding its long term functionability for storage purposes and agreeable surface subsidence on the one hand. On the other hand storable gas in a salt cavity is limited to a maximum gas pressure to guarantee the tightness of the salt rock mass. Furthermore the tightness of the borehole has to be guaranteed. The volume of stored gas in the pressure range between maximum gas pressure and minimum gas pressure is called working gas and can be used for storage operations. Thus the main objective of cavern design at predetermined sites under geometric conditions with respect to a single cavity and cavern field is the optimisation of available working gas.

In the case of traditional seasonal storage there is just one main storage cycle every storage year or in other words: the working gas is changed about once a year. On the contrary, in the case of high performance storage due to changing gas price conditions on the gas markets, there will be more or less full cycles per storage year – perhaps at higher withdrawal and injection rates compared to seasonal storage operations. In this case when considering intensified gas storage operations with respect to cavern design, not only the mechanical processes induced in salt rock mass have to be considered but also the temperature changes of stored gas and as a consequence thermal induced stresses in the rock mass which can be compressive or tensile depending on temperature change. Figure 3 shows these characteristic storage operation patterns.

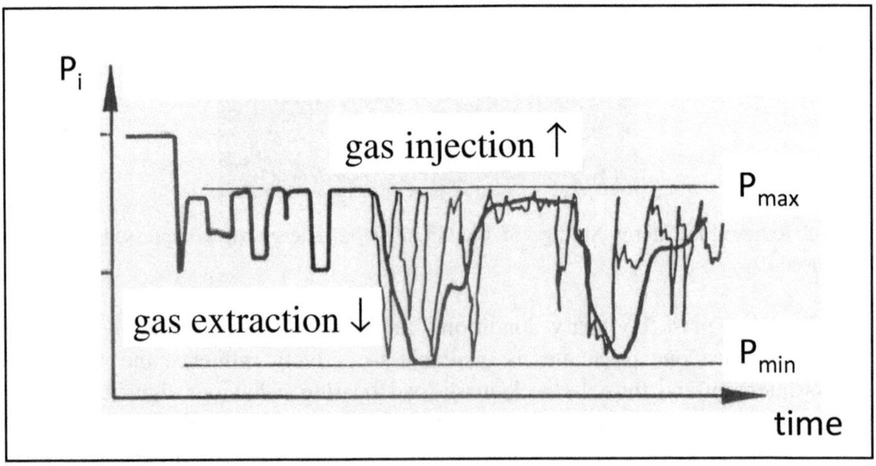


**Fig. 3** Different cavern operation patterns (seasonal storage as well as multicyclic storage)

# 3 Basic Aspects of Geotechnical Cavern Design

Following the basic rules of German mining authorities, storage cavities have to meet some general, conditions summarized in general as geotechnical safety: These conditions are in detail:

(1) Sufficient static stability (on a local and global scale),
(2) Reliable tightness (rock mass as well as drill-hole),
(3) Acceptable surface subsidence,
(4) Safe environmental abandonment.

Furthermore, from the operator's point of view the cavities should have a maximum gas storage capacity (minimum cushion gas, maximum working gas), high deliverability as well as low convergence, i.e. effective and long-term usability or in other words they have to guarantee both geotechnical safety and economic safety.

These general conditions have to be fulfilled during the life-time of any cavity, i.e. construction phase, operation phase and post-operation phase. In principle the fulfilment of these must be documented during the design and license process and therefore in advance of construction.

Fundamental instruments for the documentation of geotechnical safety and economic usability are geomechanical models and numerical simulations to predict the rock mass behaviour taking into account all the loads which are to be expected during construction and operation as well as after abandonment. To prove their safety, appropriate design concepts have to be developed that take into account the site-specific rock mass structure and rock mass properties as well as the structural behaviour of the load bearing elements, regarding possible failure mechanisms and including in situ experiences. It is self evident that these design concepts are differing in their detailed criteria and limit values as well as safety margins depending on the particular design and also to the individual experience of the designer.

Figure 4 summarizes the general conditions and gives an overview on questions that have to be answered based on geomechanical modelling and numerical simulations as one of the main results of the design process. Depending on the operation mode, especially with respect to the gas pressure change rate, it has to be decided if thermal induced stresses have to be taken into account additionally or not.

Principally cavern design has to be as realistic and reliable as possible with respect to the general demands on geotechnical design having in mind there will be some uncertainties especially related to less well known rock mass conditions, for example rock mass structure, lack of homogeneity of rock formations with respect to geomechanical properties or primary rock mass pressure. Another kind of uncertainty is the result of simplifications in the context of physical modelling, i.e. detecting and describing physical processes induced in salt rock mass by cavern

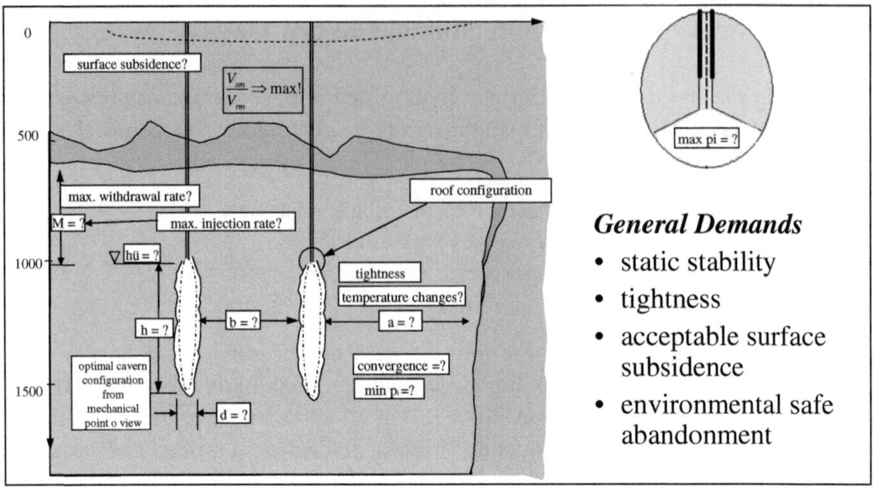


**Fig. 4** General conditions for geotechnical safety and resultant derived design related questions [2]

construction, cavern operation and finally cavern abandonment. Against this background fundamental design methodology can be characterised as follows:

(1) A Preliminary cavern design before construction is necessary for getting a solution mining license based on site-specific data including sufficient safety margins to cover uncertainties due to limited exploration of rock mass as well as to bandwidth related to rock properties determined by lab investigations.

(2) An Updated cavern design after cavern construction taking into account cavern configuration realised by the solution mining method as well as experiences gained during leaching.

(3) An Optimised cavern design after some years of operation taking into account the first monitoring results on the cavern reaction to storage operation impacts, i.e. integral cavern convergence as well as local cavern wall behaviour and cavern wall deformation. In this phase of design a first reanalysis of observed cavern and rock mass behaviour takes place to evaluate and improve existing simulation data. This first stage is followed by an updated prognostic rock mechanical analysis of cavern operation including optimisation of already existing cavern storage operation conditions. Optimisation in this case means the reduction of operational restrictions, i.e. an improvement of storage conditions with respect to available working gas volume and deliverability.

Figure 5 shows the extended methodology for optimised cavern design including site-specific in situ experience. In this case the more general situation of high performance gas storage has been dealt with.

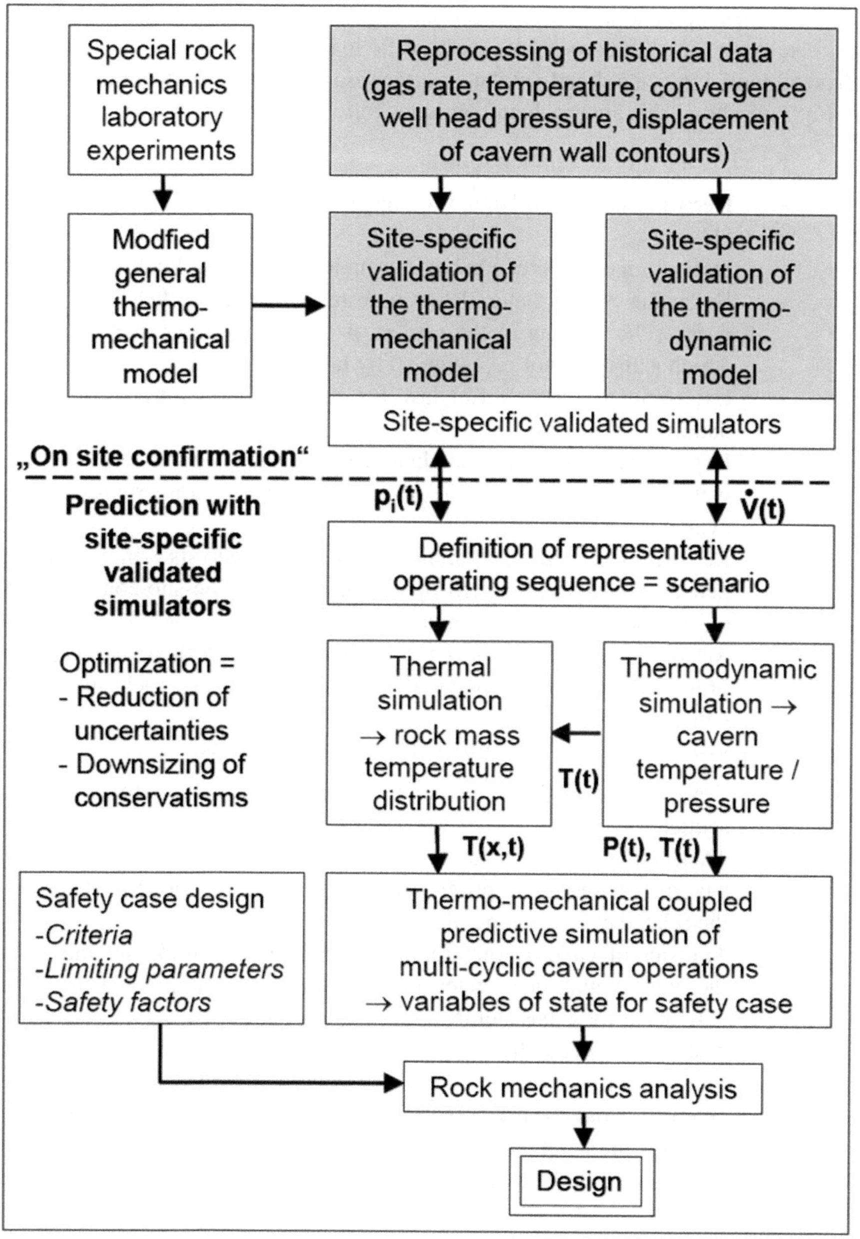


**Fig. 5** Procedure of cavern design for multicyclic operations considering site-specific operating experience

With the aim of obtaining a safe, realistic and conservative design for multicyclic cavern operations, this operational experience combined with laboratory test results on rock salt cores under multicyclic loads, can be used to formulate the physical model in general and validate it afterwards for a specific location.

Figure 5 shows this approach and includes following tasks:

(1) Specialization of simulation tools:

- Derivation of the dilatancy damage strength limit, Lux & Düsterloh [11],
- Derivation of the creep behaviour under multicyclic loading [11],
- Extension of the material law with respect to creep behaviour under multicyclic loading in the region of deviatoric stress of $\sigma_v < 8$ MPa, which is usually not investigated by laboratory tests.
- Extension of the material law for physical modelling of structural damage and healing,
- Reprocessing of historical data from seasonal operations (gas rate, wellhead pressure and temperature, cavern convergence) – if available,
- Retrospective thermodynamic-mechanical analysis (incl. convergence) of historical cavern operations data for "on-site confirmation" of the Thermodynamic simulator,
- Retrospective thermo-mechanical analysis of historical cavern operations data for location-specific validation of the rock mechanics simulator.

(2) Prognostic analysis of the multicyclic cavern operations:

- Development of representative multicyclic cavern operating sequences,
- Thermodynamic-mechanical simulation (incl. convergence) of the cavern behaviour for prognostic multicyclic operations to determine the related cavern temperatures,
- Transfer of cavern temperatures to surrounding rock salt mass and thermal simulation to determine the special distribution of the time-dependent rock mass temperature,
- Thermo-mechanically coupled simulation of the load-bearing behaviour of rock salt mass under multicyclic operations to determine mechanically and thermally induced rock mass stress and deformation,
- Determination of utilization factors for the failure and damage strength limits along with cavern convergence and surface subsidence.

## 4 Some Criteria for Geotechnical Proof of Safety

Salt cavities can be located in domal salt rock mass or bedded salt rock mass. For example Figure 6 shows part of a cavern field located in a bedded salt rock mass.

In this case one of the cavities is sited next to a fault zone. Therefore in the geomechanical design the mechanical as well as hydraulic properties have to be determined or estimated and furthermore the effects of this fault zone have to be investigated taking into account for example reduced strength and/or enlarged permeability. The load bearing behaviour of this additional geo-tectonic element has to consider both its influence on the cavern behaviour as well as the mechanical reaction of this geo-tectonic element on cavern construction and storage operation. According to Figure 6 an adequate geomechanical model should be three dimensional, idealizing the cavity with a rotation-symmetric configuration whereas the fault zone will have a planar geometry with an estimated thickness.

When considering the excavation and operation phases of storage cavities, the proof of geotechnical safety for these geotechnical constructions has firstly to have two components: The proof of static stability and the proof of tightness. In order to prove static stability and tightness some main aspects have to be considered, especially for rock types with significant creep behaviour. This finally results in:

(1) Proof of static stability:
(1/1) Limitation of stress intensity depending on the foreseen minimum cavern pressure and operation time as well as
(1/2) Limitation of creep strain to exclude creep rupture.

These criteria exclude macroscopic fractures and spalling at the cavern contour.

(2) Proof of tightness:
(2/1) Exclusion of macro-fracturation of salt rock due to gas pressure being too high.
(2/2) Exclusion of unacceptable gas infiltration in rock salt mass.

Furthermore, due to cyclic loading over a time period of several decades, a progressive micro- to macro-fracturing of the rock mass next to a+ cavity contour has to be excluded or limited to an acceptable degree.

Because geogen impermeable rock mass formations are considered to guarantee the necessary gas tightness of the cavities, the proof of tightness has to show that this primarily given impermeability holds even under the impact of cavern operations with the related secondary stress and strain fields. Therefore, the proof of tightness becomes a proof of guaranteeing the adherence of certain stress and perhaps strains conditions in the rock mass next to the cavity during the different operation phases.

Due to the long-term usability of storage cavities annual cavern convergence may be limited. However in the case of a too large convergence over time resulting in less storage volume it will be necessary to increase the cavern volume again by releaching.

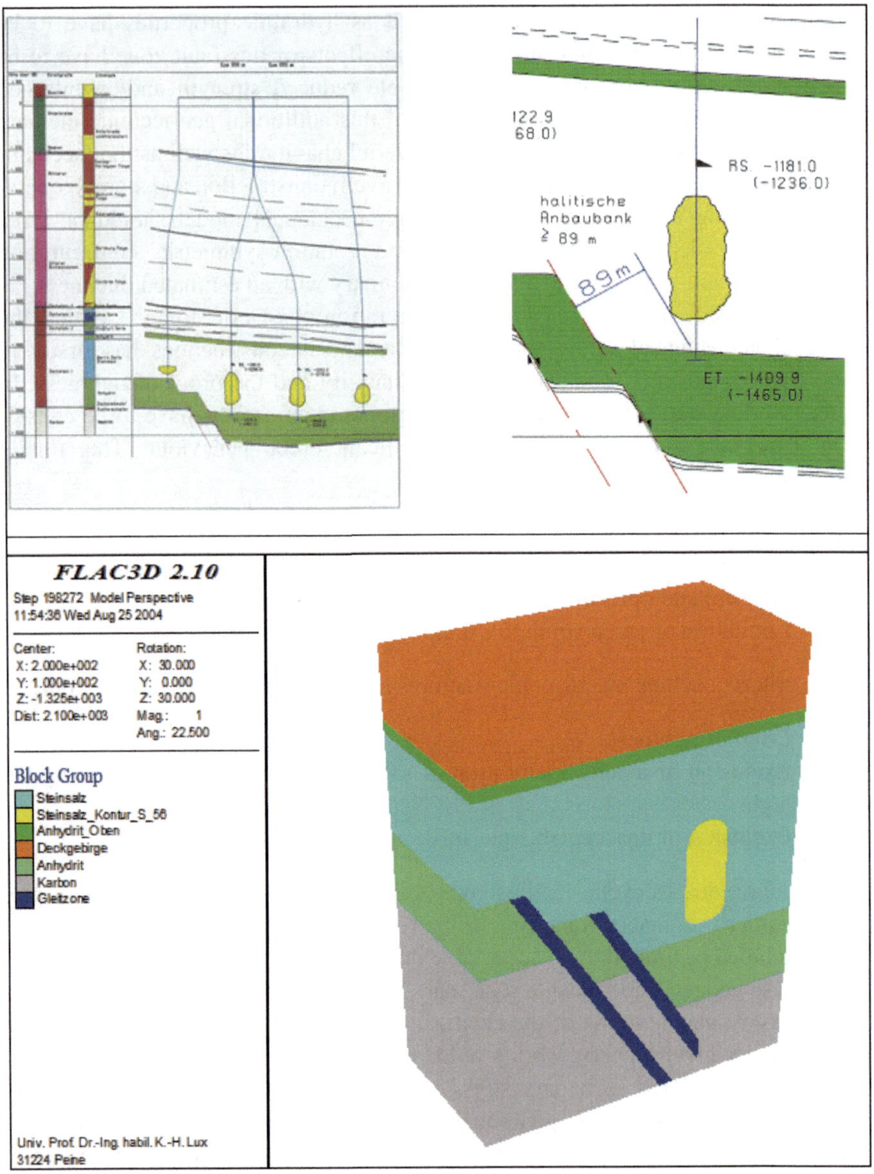


**Fig. 6** Geotectonic cross-section with cavities and a fault zone next to them (above) and related geomechanical model (below) [12]

Furthermore, depending on the site-specific surface situation the surface subsidence resulting from rock mass deformation due to cavern construction and cavern operation (including releaching) may be limited. In this case the expected cavern convergence and related surface subsidence have to be quantified in advance for the period of operation or even in the long-term including abandonment i.e. the post-operation phase.

Admissible surface subsidence may be dependent on different aspects such as existing development agricultural use and environmental protection.

It is self-evident, that these criteria are neither complete nor valid in any arbitrary underground situation. Therefore it is necessary to modify or amend them depending on the geotectonic characteristics of the site under consideration. Furthermore, it is necessary to determine the safety margins to be implemented in each of these criteria.

It can be seen from the above specified criteria that damage to the rock salt mass next to a cavity is has not been considered explicitly up to now. To guarantee the long-term usability of cavities for storage activities at a constant level of available working gas volume unacceptable damage to rock salt has to be avoided by additional restrictions related to cavern operation. No accumulation of damage and in consequence accumulation of weakening of the load bearing rock mass occurring with time is allowed (reduction of strength, enlargement of deformability as well as permeability).

This demand gains importance especially in high performance storage cavities. In this case some storage cycles are realised per storage year by using the admissible pressure range more or less extensively and as mentioned before both mechanical stresses and thermally induced stresses have to be taken into account. Therefore the probability of undetected damage accumulation in salt rock mass zones with load bearing as well as sealing function increases. Three main consequences arise from this viewpoint regarding cavern designs for high performance storage operation:

(1) A damage limit for relevant salt rocks has to be identified and implemented in cavern design.

(2) If the damage limit is regularly exceeded during storage operation, damage evolution in salt rock mass has to be physically modelled as well as numerically simulated – this means including damage development as well as possible damage reduction in cyclic loading.

(3) The development of thermally induced additional stresses has also to be physically modelled as well as numerically simulated and the reaction of the salt rock mass to these additional stresses has to be studied carefully – this means the evaluation of combined mechanical-thermal induced stress states on static stability as well as tightness has to be as realistic as possible to avoid unnecessarily severe and unfavourable consequences for storage operation conditions (limitations, restrictions).

(4) In the case of high performance storage cavities monitoring time spans for the echometric survey of cavern contour behaviour may perhaps have to be reduced compared to the time spans usually used for seasonal storage. These reduced time spans for high performance storage operations will help to avoid undetected accumulation of severe rock mass disturbance on the one hand and on the other hand will give necessary in situ data to assess, and perhaps to optimize cavern design at an early point of time in the cavities life span.

To summarise, high performance storage operations can follow a moderate mode allowing no damage or can follow an extended mode allowing limited temporary damage, but in either case the related design needs significantly more insight into salt rock mass behaviour, improved physical modelling as well as numerical simulations and last but not least updated design criteria. Based on this statement the question now arises how to get this important additional information on rock mass properties?

## 5 Basic Laboratory Investigations – Selected Highlights

To improve the physical model, uniaxial and triaxial short-term laboratory tests were conducted on representative facies-related core material of local rock salt. From these tests failure and damage strength limits were derived as shown by Figure 7.

Salt rocks are characterized by pronounced viscous (creep) behaviour. Experience shows that at a constant stress state after some time so-called creep rupture failure is also possible. According to lab investigations creep rupture failure is closely connected to the accelerated creep strain rate occurring prior to this. Accelerated creep strain rates on the other hand are due to damage processes in the rock salt fabric. In this context therefore, the question arises of whether damage evolution can also be observed during creep tests especially during the accelerated creep phase. Additionally the question arises as to whether there is a damage limit with respect to long-term loading different from the damage limit derived from short-term tests.

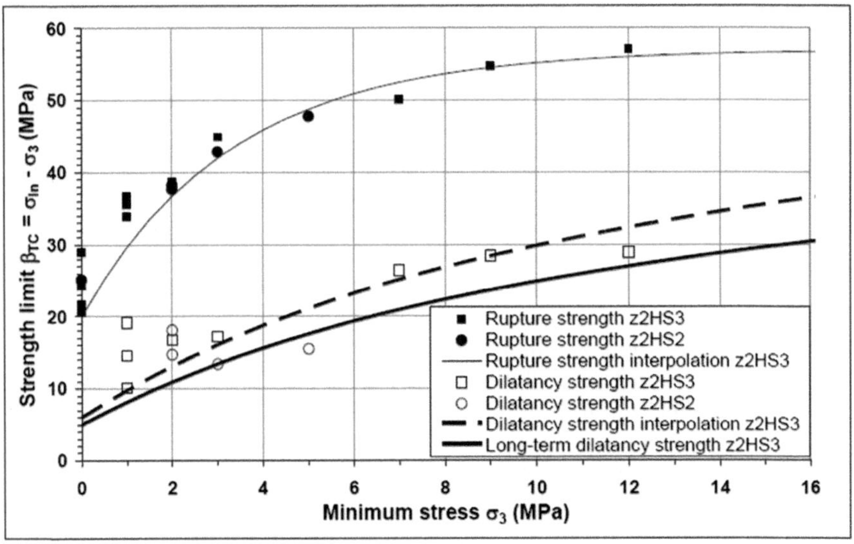


**Fig. 7** Failure and damage strength limits from experiments ($\sigma_{ln}$=axial pressure, $\sigma_3$=lateral pressure in triaxial compression tests)

Figure 8a shows the creep behaviour observed in a triaxial creep test at σ1 = 38 MPa / σ2=σ3= 3 MPa lasting about 250 days. Transient, stationary and accelerated creep phases can easily be recognized. Perhaps not surprisingly after about 150 days the strain rate accelerated together with the evolution of damage D, based on the onset of dilantancy εvolume as well as the onset of a reduction in ultrasonic wave velocity $v_p$. Stress intensity $\sigma_1$ =38 MPa / $\sigma_2$=$\sigma_3$= 3 MPa is well above the damage limit compared to the short-term damage limit of the rock salt facies tested. Damage parameter D is defined as follows:

$$D = 1 - \frac{1}{1-\varepsilon_{Vol}} \cdot \left( \frac{v_P}{v_{P0}} \right)^2 = 1 - \frac{1}{1-\varepsilon_{Vol}} \cdot \left( \frac{v_S}{v_{S0}} \right)^2 \tag{1}$$

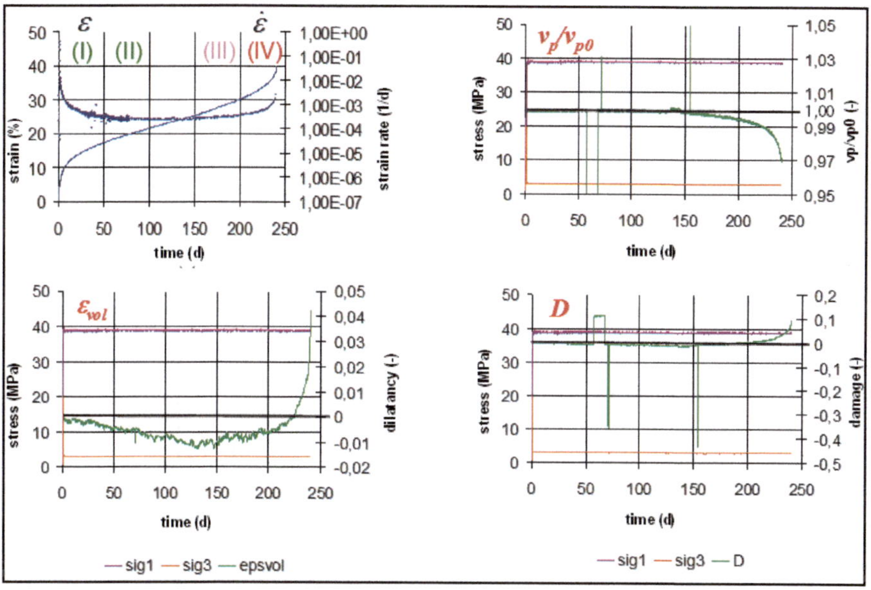


**Fig. 8a** Damage evolution during long-term triaxial compression creep test

Additionally multi-stage long-term creep tests have been performed to determine creep properties at stress states below and above the damage limit. Figure 8b shows damage evolution at such a creep test. The total test duration was 120 days. It is impressive to see that during the first two load steps no damage occurs at stress levels below the damage limit as has been expected, however during the following two load steps at stress levels above damage limit damage does occur as expected.

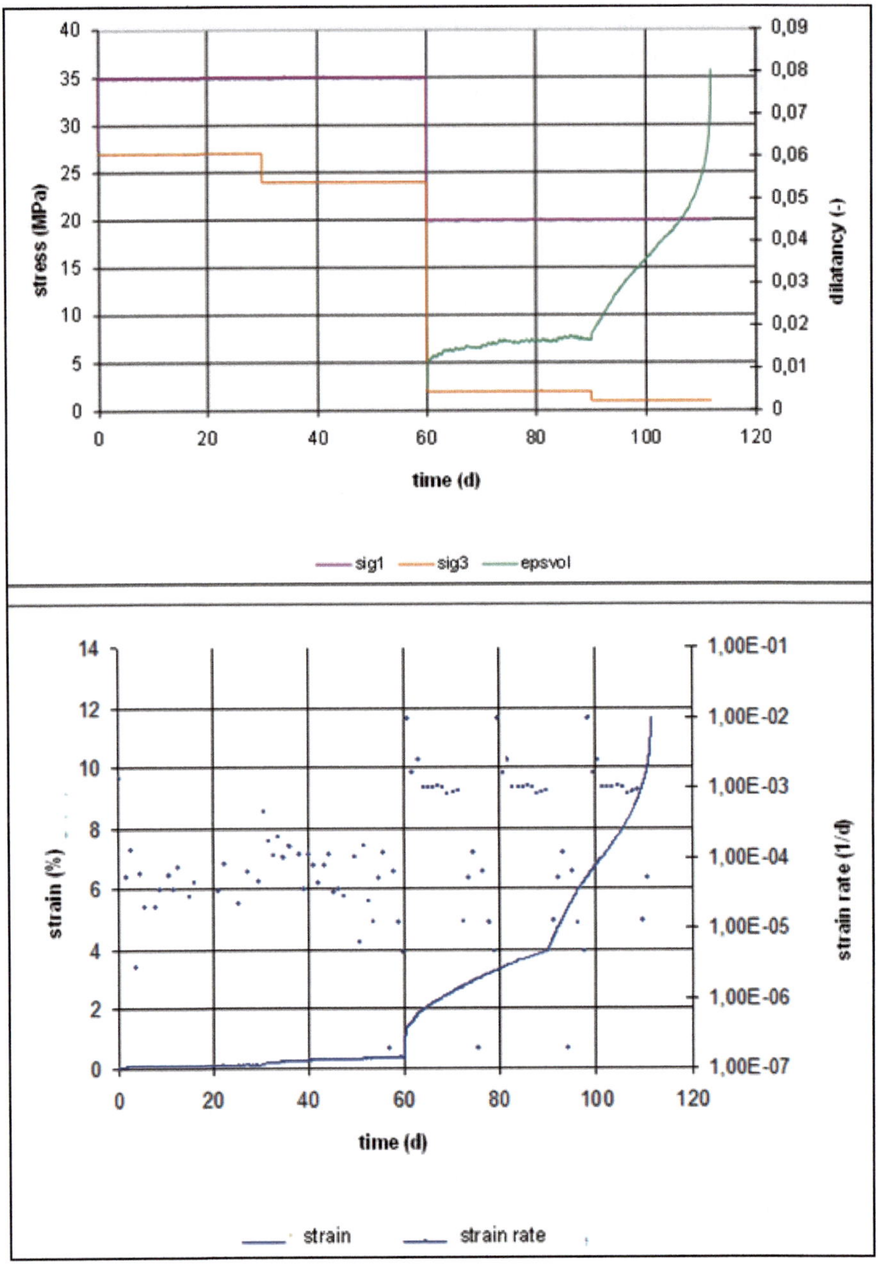


**Fig. 8b** Sample load and damage evolution during multi-stage creep tests including load intensity below and above damage limit as well as creep behaviour

Special triaxial creep tests under cyclic alternating loads with cycle durations of 2 to 60 days and up to 40 cycle repetitions give an insight into the material deformation behaviour under alternating loads. In Figure 9 significantly different creep behaviour for repeated deviatoric loading is obvious, even after a cycle duration of 60 days.

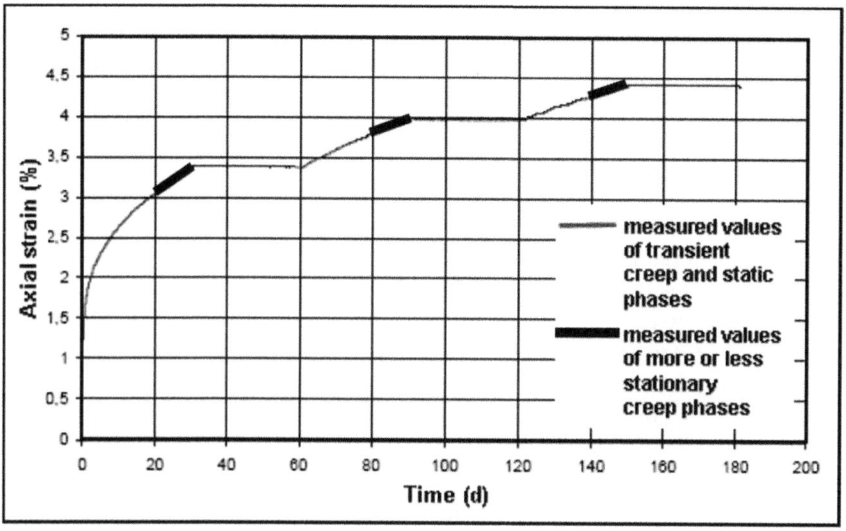


**Fig. 9** Stationary and transient creep from experiments – multicycling loading

The essential findings from those laboratory tests can be highlighted as follows:

- The latest experimental results on failure strength as well as damage strength and also creep behaviour confirm previous results. Therefore, taking damage strength into account it is possible to exclude any degradation of material properties resulting from long-term storage of the salt core material.

  For cycle durations lasting days or even months, rock salt can be characterised as a "material with memory". Under repeated deviatoric load levels no significant transient creep is observed as stationary creep is increasingly reduced. Therefore, transient creep is only active under initial deviatoric loading and as such has to be considered for modelling purposes.
- Such repeated deviatoric loading does not produce any structural damage provided the damage strength (dilatancy) limit is not violated.

More information about rock salt material properties and their determination by laboratory investigations can be taken from Düsterloh et al. [13].

# 6 Some Aspects of Improved Physical Modelling with Respect to Repeated Load Cycles as Well as Damage – Constitutive Model *Lux/Wolters*

## *6.1 Improvement of the Constitutive Model*

Based on improved knowledge about salt rock behaviour observed in updated laboratory investigations, existing constitutive models have to be evaluated and if necessary modified and upgraded. Some of the important new experiences with mechanical behaviour of salt rock have been presented in the previous chapter.

Using the above mentioned recent experimental results, in order to be able to physically model these and other new properties important for the design of geotechnical construction in salt rock mass, the previous approach for physical modelling of the material behaviour of rock salt mass has been modified fundamentally by introducing the constitutive model *Lux/Wolters*.

The constitutive model *Lux/Wolters* was originally based on the constitutive model *Lubby2* developed by *Heusermann,* Lux & Rokahr [14] and [2] as well as on the constitutive model *Hou/Lux,* which is described in Hou [15] and [4]. Basically these constitutive models integrate on a phenomenologic background the effects of different deformation mechanisms occurring in rock salt depending especially on deviatoric as well as minimal stress, temperature and salt rock type and are therefore able to fulfil the main criteria for constitutive models for rock salt.

Taking into account new findings on rock salt behaviour, whether or not there is damage evolution has the most important influence.

Therefore, based on short-term tests Figure 10 shows, in a schematic diagram, the course of deviatoric failure stress, deviatoric damage stress as well as deviatoric rehealing stress depending on minimum principal stress. Additionally changes in different mechanical and hydraulic rock salt properties are declared in case of changing stress states.

Figure 11 shows the typical creep behaviour of rock salt over time at a constant loading state above the dilatancy boundary. In this situation the creep behaviour of rock salt may be phenomenologically divided up into three phases – the transient creep phase (I), the stationary creep phase (II) and the tertiary creep phase (III). During all of these three creep phases different deformation mechanisms are active, but with varying intensity, e.g. in the transient creep phase the damaging process is still of low importance, whereas it is the dominant deformation mechanism in the tertiary creep phase. In the constitutive *Lux/Wolters* model shown in Figure 12, different creep rates resulting from different deformation mechanisms are combined via superposition to get an overall creep rate. The basic viscous model is still the well known constitutive model *Lubby2*.

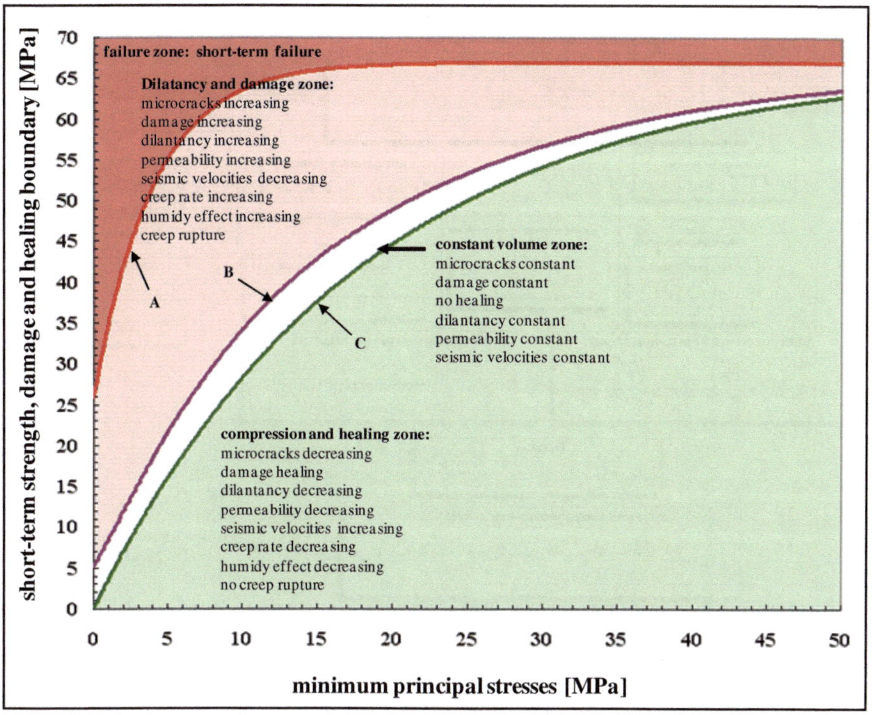


**Fig. 10** Short-term strength A, damage boundary B and healing boundary C of rock salt as well as the corresponding four zones and their main mechanical as well as hydraulic characteristics [16]

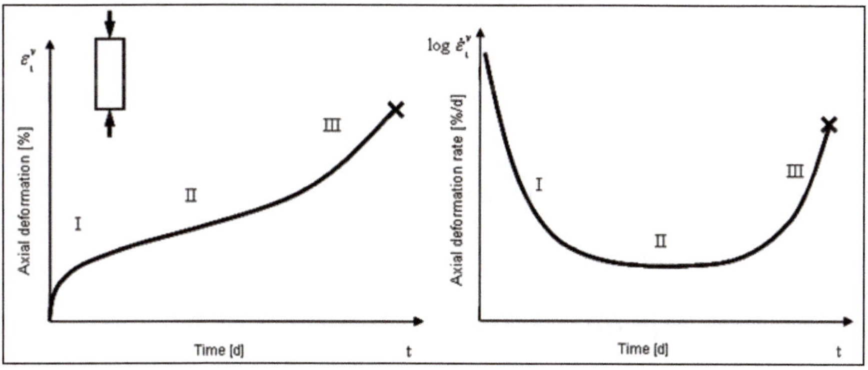


**Fig. 11** Graphs of typical creep behaviour of rock salt over time [14]

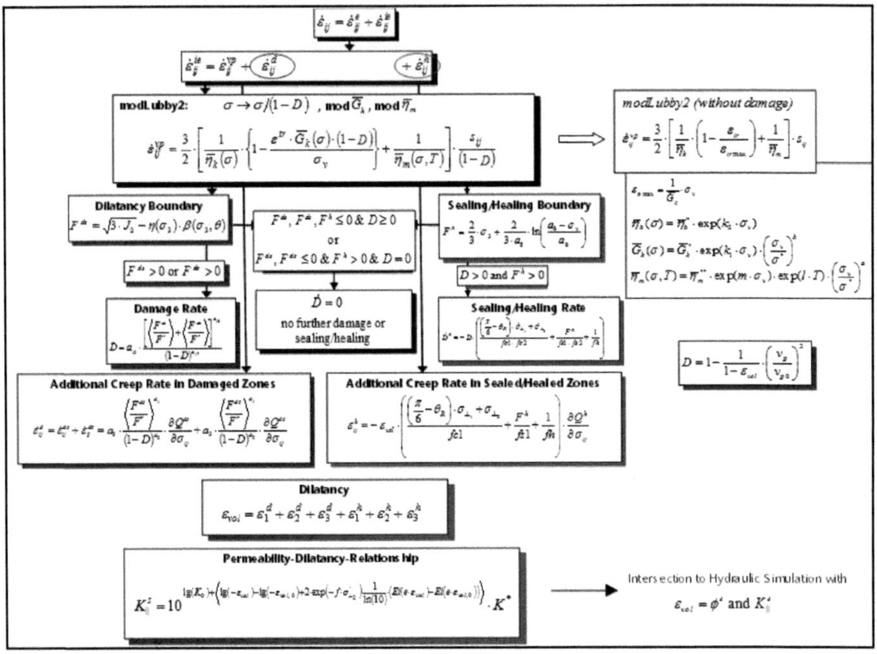

**Fig. 12** Constitutive model *Lux/Wolters*

New significant elements implanted in constitutive model *Lux/Wolters* are the physical modelling of "salt rock as a material with memory" as well as a transition to more flexible viscosity functions in the region of low deviatoric loads resulting in constitutive model *modLubby2* on the one hand and the implementation of enhanced modelling of the damage restitution by which the healing behaviour can be simulated under multicyclic loading on the other hand. The enhancement of constitutive model *Lubby2* resulting in constitutive model *modLubby2* therefore mainly consists of two main new aspects:

- rock salt is a material with memory (= no transient creep under repeated deviatoric loading) and
- rock salt creep properties have not been investigated by lab tests for small deviatoric stresses → modified viscosity functions with

$$\overline{\eta}_M(\sigma,T)=\overline{\eta}_M^{\;**}\cdot\exp(m\cdot\sigma_v)\cdot\exp(l\cdot T)\cdot\left(\sigma_v/\sigma^*\right)^{-a}, \tag{2}$$

$$\overline{G}_K(\sigma)=\overline{G}_K^{\;*}\cdot\exp(k_1\cdot\sigma_v)\cdot\left(\sigma_v/\sigma^*\right)^{-b}. \tag{3}$$

Figure 13 shows the effect of a flexible viscosity function determining stress dependent viscosity moduli. For example this new approach may, finally result in a lower stationary creep rate at low deviatoric stresses as before.

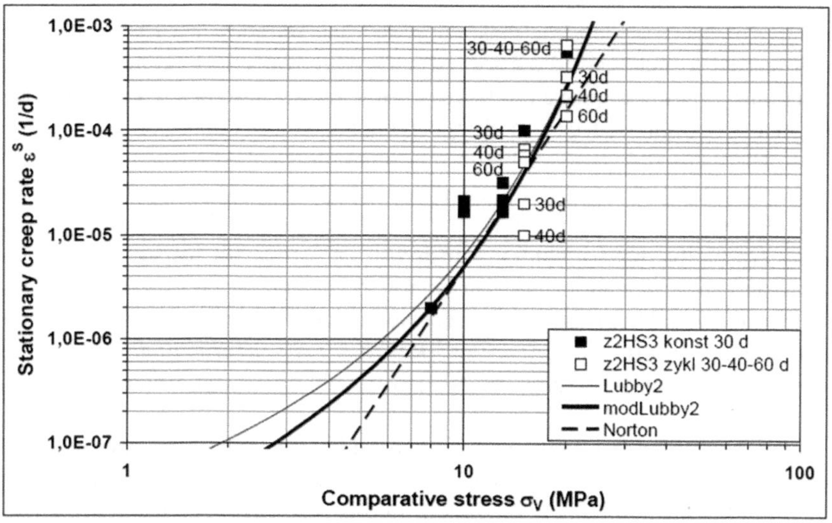


**Fig. 13** Modified constitutive model *Lubby2 (modLubby2)*

Structural damage to the rock salt is caused by deviatoric stress states exceeding the damage strength (= damage boundary). In this situation an additional damage-induced creep rate $\dot{\varepsilon}_{ij}^{d}$ as well as the damage rate $\dot{D}$ is activated. Additionally damage already induced is reduced if the stress state is below a so-called healing limit. According to Figure 10 this healing limit may be equal to or even below the damage limit. Therefore the transient creep behaviour as well as the stationary creep behaviour are modelled using a modified version of the constitutive model *Lubby2* including the damage-induced reduction of the load-bearing cross-sectional area using the approach of Kachanov [17] by introducing damage parameter D.

The damage-induced permeability changes are modelled with the permeability-dilatancy-relationship, which is also shown in Figure 12. In this model the permeability is dependent on the stress state as well as on the dilatancy determined by using the constitutive model *Lux/Wolters*.

## 6.2 *Validation of Physical Modelling Based on Long-Term Historical Data*

Field experience at various sites shows that the determination of site-specific representative rock mass parameters based on laboratory investigations especially regarding physical modelling of creep behaviour of salt rock mass will usually result in over large prognostic rock mass deformations compared to real rock mass deformations. Experience also shows that this discrepancy will increase with the time span in which the prognosis has to be performed. Possible reasons for this overestimation may be so called conservative idealisations in physical modelling because of necessary abstractions, supposed uncertainties or existing bandwidth with respect to material properties. At sites with available long-term operational experience it has been proved to be reasonable to use these in situ observations for updated and optimised cavern design reducing conservatively formed assumptions. The methodology available in this situation is called back-analysis. At first measurement data has to be evaluated. Secondly historic operating sequences have to be documented and prepared for numerical simulations. Based on these simulations, especially those including sensitivity analysis with respect to rock mass properties, the calculation model used for the design so far is toughened up by comparing available measured and received calculated deformations, for example total or zonal cavern convergence or local rock mass deformation. Using these in situ validated rock mass parameters more reliable prognostic numerical simulations can again be performed.

To demonstrate this methodology Figure 14 shows measured operational data for a period of some 8.5 years between two echometric surveys; this data is the basis for the location-related validation. In this case cavern convergence amounts to about K = 5.5 %. The storage cycle duration is roughly one year, which implies a duration of experimental study that can hardly be realized in the laboratory.

In order to perform a retrospective analysis, the measured wellhead pressures were converted into cavern pressures. Afterwards, a parameter variation analysis was carried out for the additional terms "a" and "b", in formulae (2) and (3), integrated in the novel viscosity function of the *Lux/Wolters* constitutive material law and these can be used as matching parameters.

By means of these additional terms, the analysis is able to consider the transient and stationary portions of creep at minor load levels which have not been investigated in laboratory experiments and which are confined by the upper limit of creep deformation according to the *Lubby2* material law.

Figure 14 also illustrates the results of the variation analysis. The conclusion is that by using the *Lux/Wolters* constitutive material law, which is based on physical modelling and experimentally derived creep behaviour, a reliable match to the observed cavern convergence is possible. The applied creep deformation

tends towards the lower region of the measured data spread. The stationary creep rates, in the region of minor deviatoric stresses of $0\ \text{MPa} < \sigma v \leq 8\ \text{MPa}$, have to be modelled below those which are required according to the *Lubby2* constitutive material law. This modification is also supported by observations by Brouard et al. [18].

In principle, this retrospective analysis confirms that the novel physical modelling of "rock salt as a material with memory" can also be used also for cycle durations in the order of one year. Therefore, the *Lux/Wolters* material law is independent of cycle durations for multicyclic and long-term seasonal cavern operations. This approach has also been confirmed for other different locations.

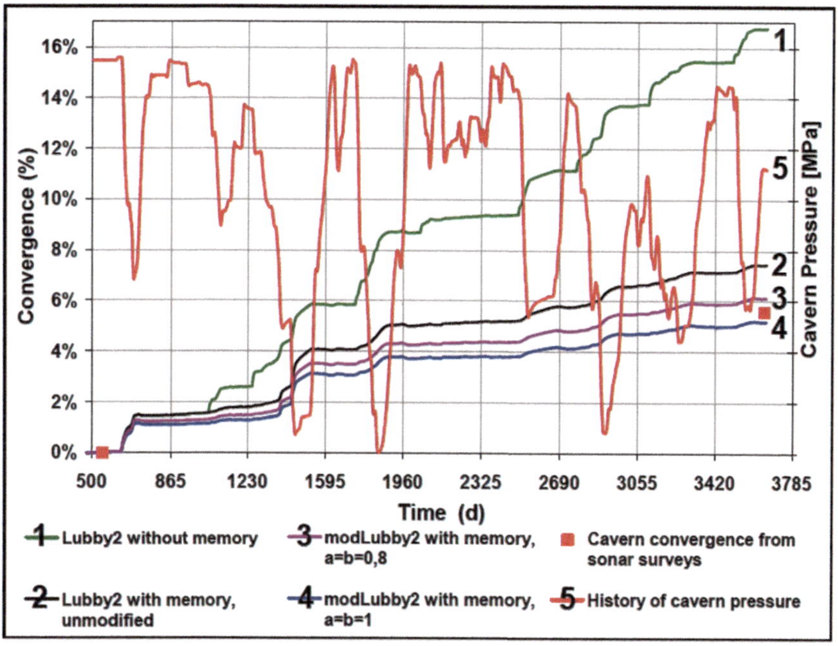


**Fig. 14** Retrospective analysis of cavern convergence

# 7 Numerical Simulation and Selected Results Regarding Multicyclic Operation Mode

## *7.1 Cavern Operation Mode without Damage*

The following example has been taken from [9]. In this case a two-dimensional, axially symmetric simulation model had been developed based on the rock structure in the vicinity of the cavern in question (homogeneous rock salt in general). The cavern pressure and temperature profiles derived from the thermodynamic simulation after specifying an expected future rate scenario were used as the basis for the rock mechanics simulation. An example of such an assumed representative pressure and temperature profile is shown in Figure 15.

The numerical calculations of the relevant parameters of the rock mass in space and time as a response to the cavern operations were carried out using the finite-difference method applying the *FLAC3D* program. Such parameters include the rock temperature, rock stress, rock deformation (strain and displacement) as well as convergence. This simulation tool has been specifically developed over the past twenty years to simulate load-bearing systems in rock salt under complex conditions.

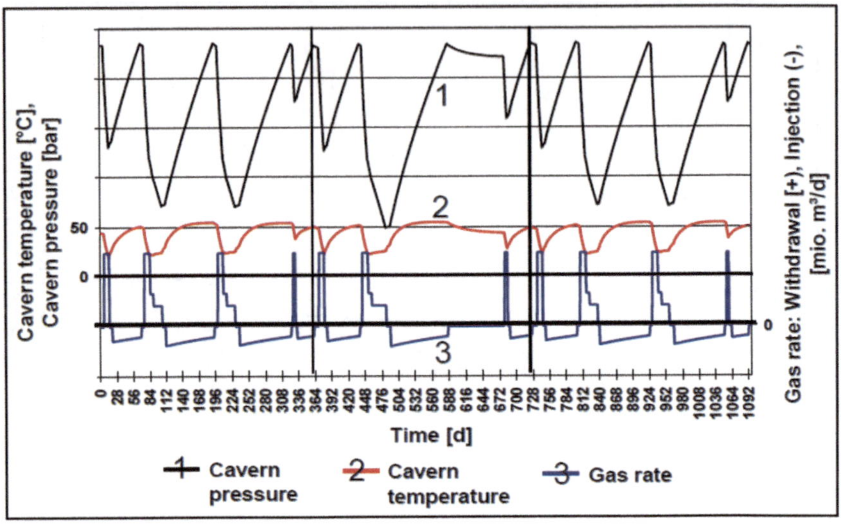


**Fig. 15** Cavern pressure and temperature for a combined multicyclic/seasonal operation scenario

Exemplary results of the thermo-mechanical analysis (MT) and, for comparison purposes, also of the mechanical analysis (M) are shown in Figures 16-19 for a rock salt element at the reference depth indicating the characteristics of the rock stresses, rock mass utilization intensities, rock mass deformations (creep strain) and convergence over time.

The following abbreviations apply:

MT - Simulation considering "rock salt as a material with memory" with thermally induced additional stresses along with temperature-dependent creep deformation related to the operations validated.

M - Simulation considering "rock salt as a material without memory" and without thermally induced additional stresses along with temperature-dependent creep deformation related to the operations, non validated.

Figure 16 shows the calculated rock mass stresses for the only mechanical analysis (M). In comparison, Figure 17 shows the coupled thermo-mechanical analysis (MT). In principle, the rock mass stress follows the cyclically alternating cavern pressure. The stresses at the cavern wall, which are influenced by extreme thermal changes, in fact, show quite a different behaviour. Those differences are determined by compressive and tensile stresses related to temperature changes. What has to be kept in mind, however, is that the mechanically induced stresses are dominant for the load-bearing behaviour, while the thermally induced stresses act additionally, while both are subjected to material-related relaxation.

From Figure 17 it can be realized that no tensile stress occurs in the thermo-mechanical analysis and there is no change of the trend in the stress amplitude over time. However, a repeated situation can be observed where one of the tangential principal stresses at the cavern wall is smaller than the cavern pressure (arrows in Figure 16). This thermodynamic induced stress state demands special consideration in the design.

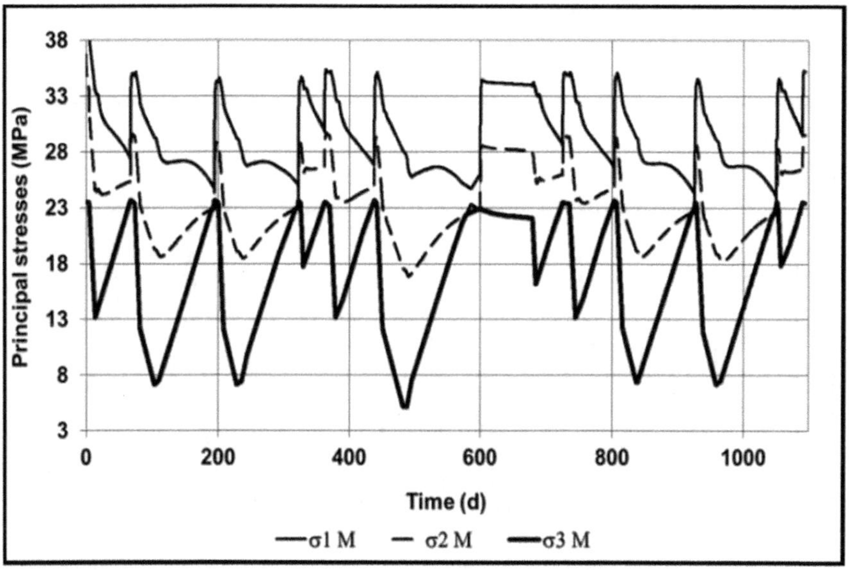


**Fig. 16** Principal stresses for rock salt in the rock mass at reference depth without memory and induced thermal stresses

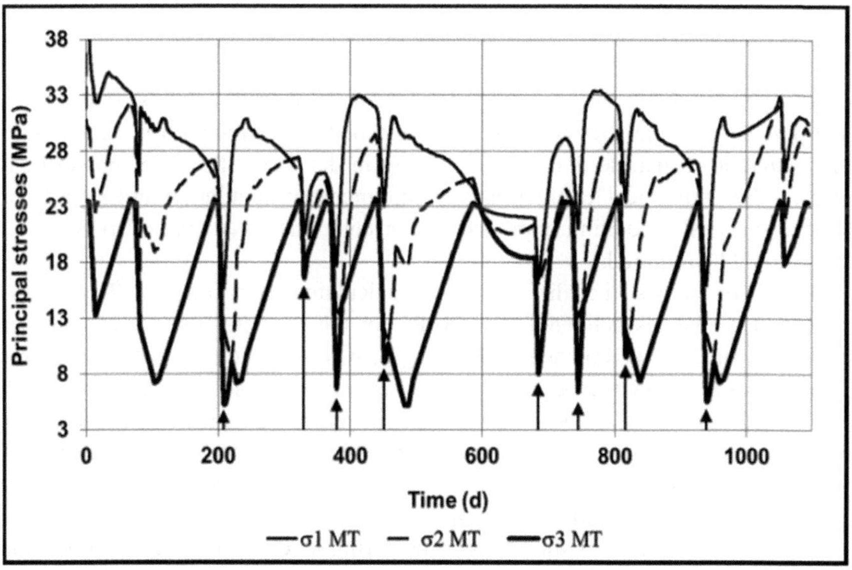


**Fig. 17** Principal stresses for rock salt in the rock mass at reference depth with memory and induced thermal stresses

An initial rock mechanics evaluation of these load conditions can be derived from Figure 18. Here, the rock loading is related to the damage strength βs. When βs = 0.95 < 1.0 = permissible βs, the damage strength limit is not reached.

For this design of a moderate multicyclic operation mode no structural damage is expected with a minimum $p_i$ = 7.0 MPa. Optimum utilization of the load bearing capacity is thus realised.

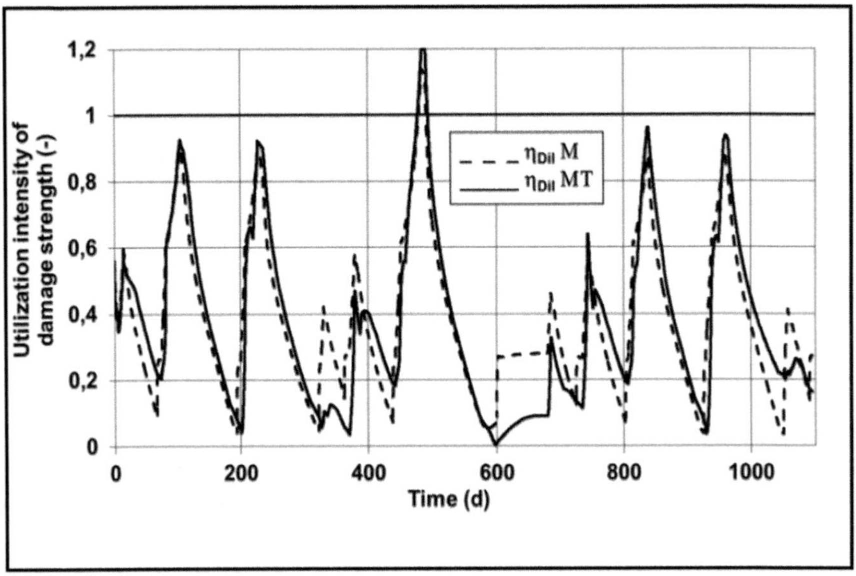


**Fig. 18** Utilization intensity related to damage strength at reference depth

During seasonal operations, on the other hand, when $\beta_s$ = 1.2 >1.0 = permissible $\beta_s$ – the damage strength limit – is exceeded by a minimum $p_i$ = 5.0 MPa. In this situation, structural damage developing in the form of fissures has to be expected. For numerical evaluation a complimentary simulation utilizing an extended physical modelling approach, for instance the *Lux/Wolters*-MDS material law, would be required.

Figure 19 shows the differences in calculated convergence. The thermo-mechanical analysis validated specifically for the site after three years of operations indicates a convergence of K ≈ 6.5 %, while the non-validated mechanical analysis leads to a convergence of K ≈ 27 %. This difference makes it obvious that the consideration of field observations and modelling validated for specific sites is of central significance when predicting cavern load-bearing behaviour that should approximate reality. This is true for seasonal operation modes, but even more so when more intense operations are employed.

Figure 20 and 21 show the stress paths during loading in the three-dimensional principal stress space. Here the so-called “plane of invariants” is used – in which the sustainable loading of rock salt is characterized by the limiting strengths and is shown together with the existing loading for illustrating the multicyclic cavern operations.

In principle, the deviatoric stresses over time are shown as a function of the average principal stresses. The stress path below the damage strength limit is shown as line #5. The cyclic character of the alternating loading becomes obvious. For the solely mechanical loading, the characteristics of the individual consistent cycles appear much more reversible than those of the coupled thermo-mechanical loading. The thermally induced additional stresses create larger differences in the overall loading even for identical cycle characteristics in relation to the cavern pressure.

This representation also shows that the damage strength limit is exceeded during seasonal operation mode for the minimum pressure of $p_i = 5.0$ MPa. This violation is presented in the stress path by the highlighting circle above line #4.

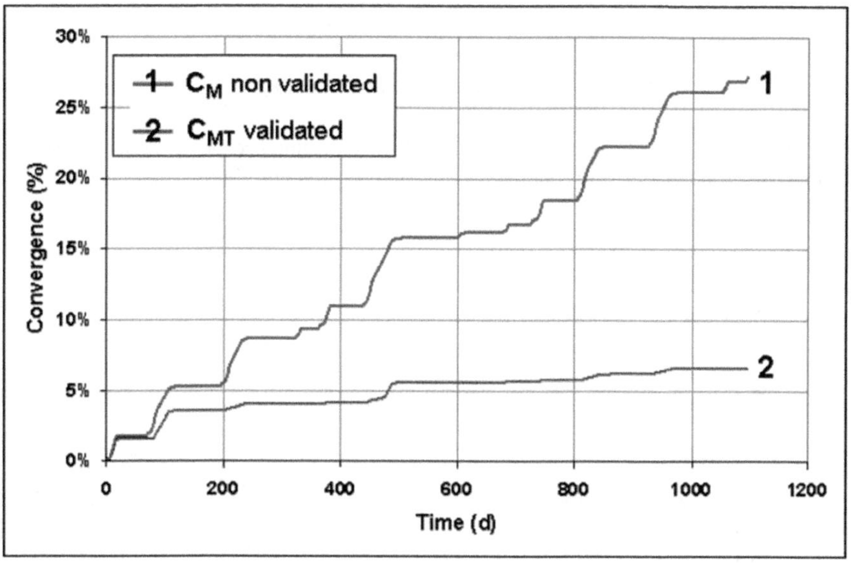


**Fig. 19** Convergence at reference depth

The following results from numerical simulations need to be highlighted:

- The loading at the cavern wall is dominated by the additional thermally induced stresses. No tensile stresses have been calculated. For limited durations, one of the principle stresses at the cavern contour becomes smaller than the cavern pressure.
- In general, rock mass stresses are reversible over time. No shift in the amplitudes trend is evident.

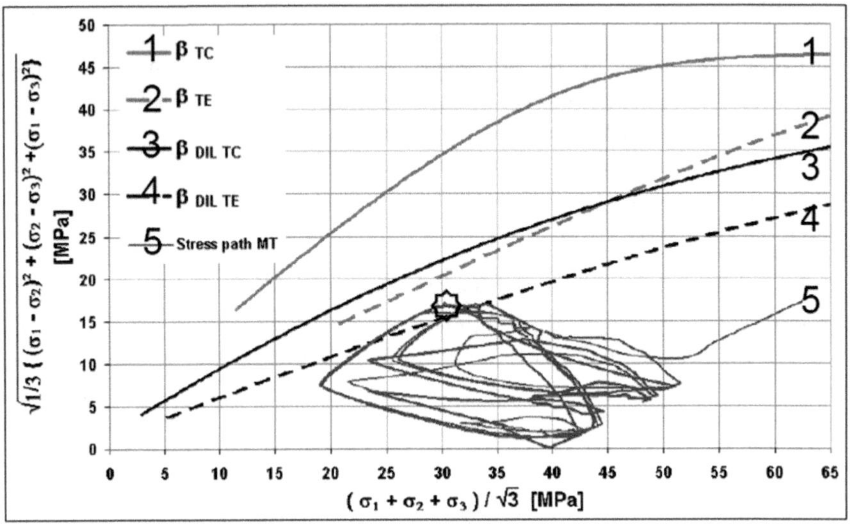


**Fig. 20** Rock mass loading at the cavern wall at the reference depth under multicyclic loading in comparison with dam-age and rupture strength limits (thermo-mechanical simulation)

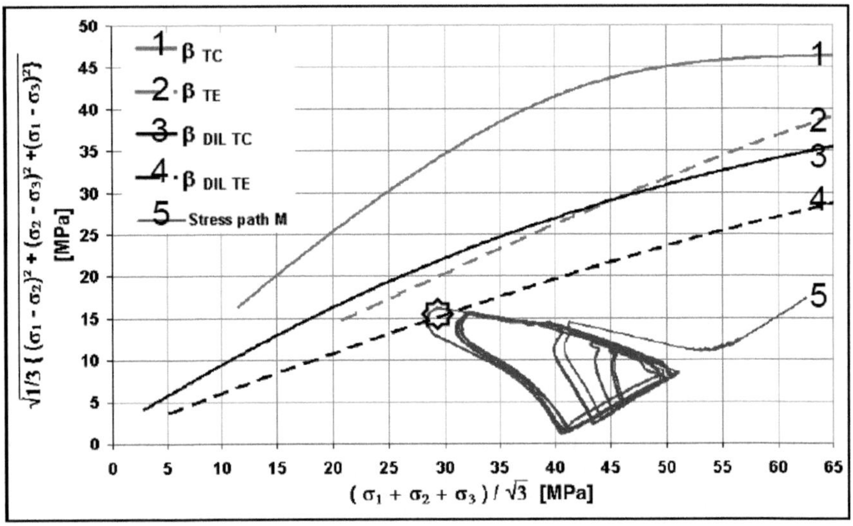


**Fig. 21** Rock mass loading at the cavern wall at the reference depth under multicyclic loading in comparison with dam-age and rupture strength limits (mechanical simulation)

- The damage strength limit is not exceeded during interrelated thermo-mechanical simulation ($\eta_s < 1.0$).
- Rock mass deformation during repeated loading is characterized by significantly lower cycle-related increases.
- The rock mass convergence for repeated cycles is significantly less compared to the convergence of the first cycle with deviatoric initial loading of the rock salt mass.

These results form the basis for proving static stability and deriving related operational limits. Similarly, as shown above, the variables of state in the region of the cavern roof can be described with regard to space and time. By means of these variables of state the cavern gas tightness can be demonstrated.

## *7.2 Cavern Operation Mode Including Limited Damage*

During a storage operation the cavern inside pressure is changing between maximum and minimum allowed values. When going down to minimum allowed pressure values, damage may occur especially in rock salt at the cavern contour. Successive increases of cavern inside pressure up to the maximum allowed pressure level may in contrast lead to a reduction of induced damage. These processes of development and reconstitution of damage can take place during seasonal and high performance storage operation depending on cavern inside pressure and respective operation time.

Reconstitution of damage is of special importance when looking at long time maintenance of rock mass load bearing and sealing capacity on the one hand and optimal cavern use on the other hand. A new modified physical model implemented in constitutive model *Lux/Wolters* to describe damage reconstitution has been presented by [8].

Simulation results achieved with this physical model with respect to multicycling storage operation have been introduced as in [8]. Determination of material parameters is also described in this publication.

Following [8] Figure 22 presents the simulation model together with some important geometric data and the configuration of the cavity. Numerical simulation has been performed including coupling of mechanical and thermal processes. Therefore thermally induced stresses are taken into account. Selected simulation results are presented for the rock mass element at cavern contour at the so-called reference depth $z_{RT} = 1233$ m. The exemplarily development of cavern inside pressure as well as related cavern gas temperature can be seen in Figure 23. After some time of constant pressure and temperature three identical storage operation cycles following each other have been simulated.

In every storage cycle a minimal pressure of $\min p_i = 5.0$ MPa has been met causing some damage to the salt rock mass in the vicinity of the cavern contour. In this case therefore rock mass stress transgresses the damage limit. The parameter $F^{ds}$ indicates the amount of this excess. Afterwards an increase of cavern inside pressure up to $maxp_i = 19.5$ MPa follows. This pressure level is held constant for $\Delta t = 90$ d. During this operation phase rock mass stress is well below both the damage limit and the damage reconstitution limit and therefore during this period of time damage reconstitution takes place.

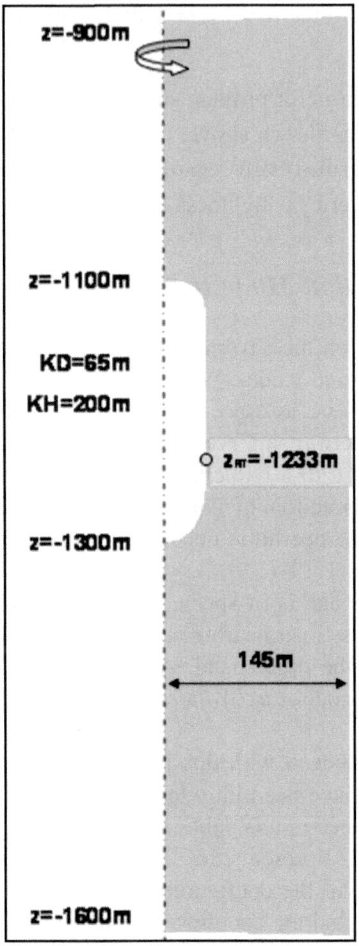


**Fig. 22** Simulation model and significant geometric data

In every storage cycle a minimum pressure of $minp_i = 5.0$ MPa has been met causing some damage to the salt rock mass in the vicinity of the cavern contour. In this case therefore rock mass stress transgresses the damage limit. The parameter $F^{ds}$ indicates the amount of this excess. Afterwards an increase of cavern inside pressure up to $maxp_i = 19.5$ MPa follows. This pressure level is held constant for $\Delta t = 90$ d. During this operation phase rock mass stress is well below damage limit as well as damage reconstitution limit and therefore in this period of time damage reconstitution takes place.

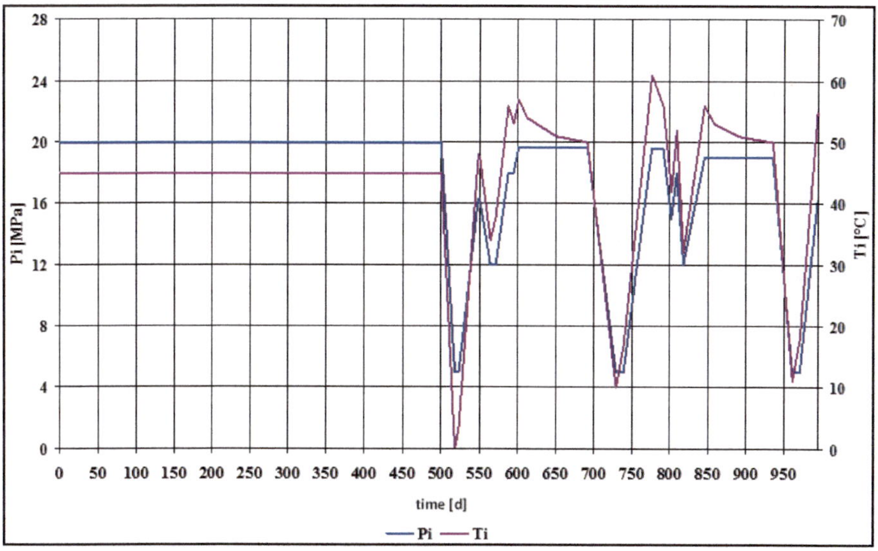


**Fig. 23** Cavern inside pressure and cavern gas temperature versus time

To characterise the damage process versus the time damage D as well as dilantancy $\varepsilon_{vol}$ are shown in Figure 24 and Figure 25. Damage D can be used to indicate intensity of micro-fissurisation, whereas dilantancy $\varepsilon_{vol}$ gives some impression of rock salt volume extension in consequence of the damage.

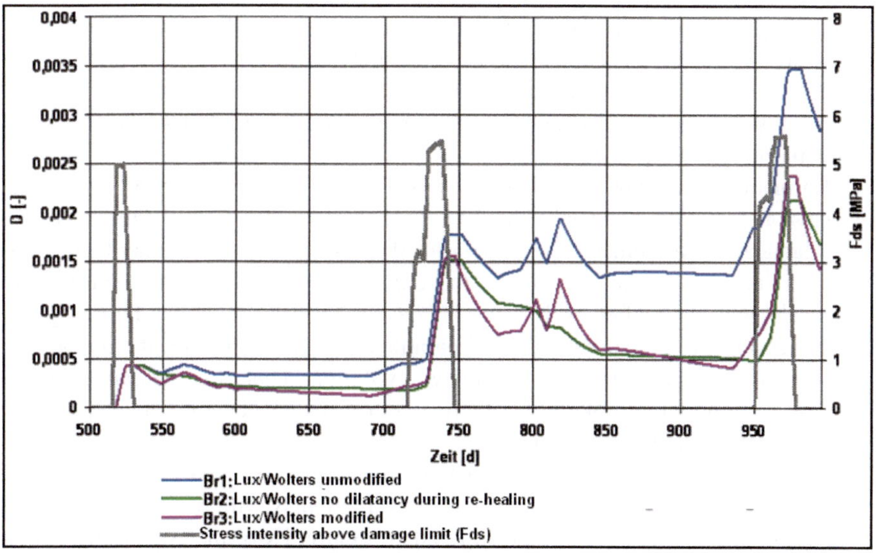


**Fig. 24** Comparison of damage D as well as load intensity above the damage limits $F^{ds}$ versus time with respect to constitutive model *Lux/Wolters*

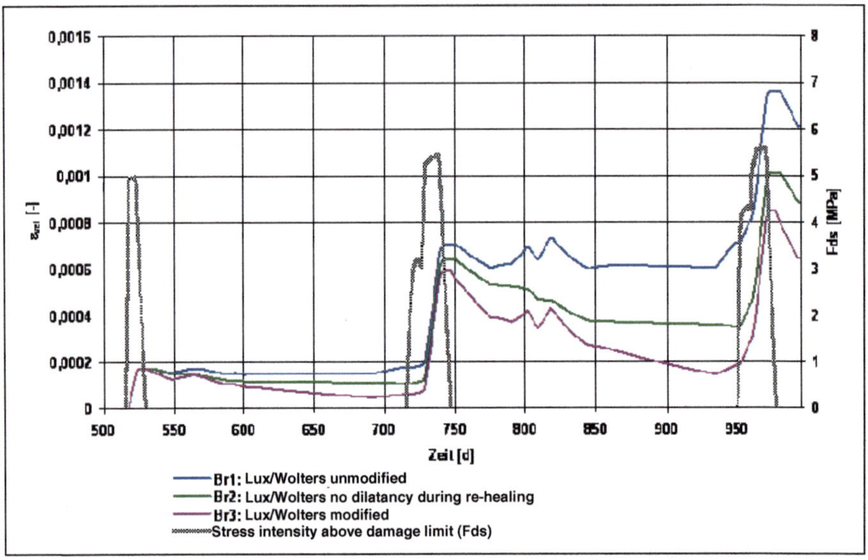


**Fig. 25** Comparison of dilation εvol as well as load intensity above damage limit $F^{ds}$ versus time with respect to constitutive model *Lux/Wolters*

These two parameters related to damage can also be determined by laboratory investigations. Therefore validity of physical modelling as well as numerical simulation can be evaluated by comparison of measured and calculated data.

The calculated dilantancy can be physically identified as secondary porosity. Together with existing stress state this secondary porosity can be transferred to secondary permeability. Therefore dilantancy will not only be an indicator for characterization of mechanical destrengthening but also will an important coupling parameter between mechanical and hydraulic processes.

Analysis of numerical simulation data with respect to rock mass behaviour results in the following interpretation:

Formation of damage and dilantancy is different in the numerically simulated three storage cycles. Especially the first cycle shows relative small damage as well as dilantancy development. The reason for the damage increasing may be the continuous increase of the load above the damage limit in the representative rock mass element cycle after cycle. This increase of load intensity can be recognised by observing the numerical value of parameter $F^{ds}$. Furthermore, the already accumulated damage will intensify damage evolution. This effect is caused by the constitutive model *Lux/Wolters* which assumes a dependency between damage rate and current damage intensity.

Simulation run 3 is based on the active physical model. This case indicates the strongest reconstitution of damage as well as dilantancy with respect to the cavern operation phase at maximum cavern inside pressure. Nevertheless the increase in both damage and dilantancy from cycle to cycle is not fully prevented but significantly decelerated.

Finally it has to be pointed out that more lab investigations are needed to confirm the validity of the physical model. Nevertheless physical modelling has been supported by already existing lab data both for damage evolution and dilantancy evolution.

It should be mentioned that no assessment of the calculated data with respect to mechanical or hydraulical consequences has been carried out so far. For example an assessment could deal with the destrengthening of the rock mass in the damaged contour zone or with the influx of natural gas in the contour zone undergone during dilatant deformation as well as with the interaction between mechanical and hydraulical processes and related consequences. These questions have to be answered in the future.

## 8 Summary and Outlook

The future development of underground gas storage will be characterized by the transition from a seasonal storage mode of operations to one which is more oriented towards the actual gas requirements of the market and which allows a much more efficient use of the gas and the cavern itself.

The more contemporary multicyclic cavern operation mode which aims at having multiple gas turnovers per year requires an extended approach in order to derive the permissible operational parameters. The concept presented here is essentially based on available long-term historical data. This data can be used to characterize the entire cavern system in a thermodynamic way by considering the storage gas and surrounding rock salt. Subsequently the characteristics of the cavern load bearing system can be realistically derived from a rock mechanics point of view.

The transition from a seasonal to a multicyclic operation mode is based on the one hand on a retrospective analysis of the available operational data to validate thermodynamic and thermo-mechanical simulators. On the other hand, it is necessary to have a good appreciation of the thermo-mechanical material behaviour of rock salt under multicyclic loading. In principle, the modelling of "rock salt as a material with memory" under deviatoric repeated loading and the implementation of the damage strength limit will determine the related permissible minimum pressure level. Consequently, adherence to a recreation phase is not required in the case of a moderate multicyclic operation with intensified pressure changes, which is in contrast to seasonal storage operations. On the contrary in the case of an extended multicyclic operation with excess damage strength in a limited rock mass zone, both the damage process and the damage reconstruction process have to be physically modelled and numerically simulated. Exemplarily the evolution of damage is shown in the paper. In future how much damage could be sustained by salt rock mass with time will have to be determined bearing in mind the general design demands. In any case for the multicyclic operation mode it is also necessary to consider the thermally induced stresses which are generated in the cavern wall by temperature changes in the storage gas.

By introducing multicyclic cavern operations it is possible to achieve a more economically efficient use of gas storage. Nevertheless, this entails a more intense involvement of the rock salt and technically "virgin land" is brought into the system. Thus what is needed is a better appreciation of loading behaviour processes, more realistic modelling approaches as well as improved surveillance of the cavern load-bearing structure to account for the modified cavern operations. By doing this it should be possible in future to reliably design caverns that fulfil all the requirements of cavern static stability, gas tightness and conservation of the neighbourhood and ecology as well as the economic boundary conditions.

In this paper the cavern design concept developed during the past two to three decades has been presented briefly describing just the main aspects. Literature written by the author mentioned in the following references will give more detailed information, especially with respect to formulae regarding physical modelling. Many other authors have contributed to the today's knowledge. They are also mentioned in the papers listed in the references below.

**Acknowledgments.** The author would like to express his great appreciation to U. Düsterloh, R. Wolters and S. Lerche for their long term cooperation in performing laboratory tests, developing as well as implementing subroutines in software packages with respect to improving physical models and finally carrying out numerical calculations.

## References

1. Sedlacek, F.: Untertage-Gasspeicherung in Deutschland. Erdöl-Erdgas-Kohle 128 Jg.Heft 11 (2012)
2. Lux, K.H.: Gebirgsmechanischer Entwurf und Felderfahrungen im Salzkavernenbau. F. Enke Verlag (1984)
3. Lux, K.H., et al.: Neue Aspekte zum Tragverhalten von Salzkavernen und zu ihrem geotechnischen Sicherheits-nachweis. Erdöl Erdgas Kohle 3/4 (1999)
4. Lux, K.H., et al.: Erhöhung der Wirtschaftlichkeit von Speicherkavernen durch Anwendung eines neuen Entwurfs- und Nachweiskonzeptes. Erdöl Erdgas Kohle 6/7/8 (2002)
5. Lux, K.H.: Zum langfristigen Tragverhalten von verschlossenen solegefüllten Salzkavernen – ein neuer Ansatz zu physikalischer Modellierung und numerischer Simulation. Erdöl Erdgas Kohle 11 (2005); 4/11 (2006) (2005/2006)
6. Lux, K.H.: Design of Salt Cavities for Storage of Crude Oil, Natural Gas as well as Compressed Air – Some Fundamental Geomechanic Aspects for Construction, Operation and Abandonment. Special Publication on UGS, British Geological Society (2007)
7. Wolters, R., Lux, K.H., Düsterloh, U.: Evaluation of Rock Salt Barriers with Respect to Tightness: Influence of Thermomechanical Damage, Fluid Infiltration and Sealing/Healing. In: 7th Conference on the Mechanical Behaviour of Salt, Paris, April 16-19 (2012)
8. Lerche, S.: Kriech- und Schädigungsprozesse im Salinargebirge bei mono- und multizyklischer Belastung. Ein Beitrag zur Weiterentwicklung der Analysemöglichkeiten und der geotechnischen Nachweisführung im Bereich der Salzmechanik mit Anwendung auf Fragestellungen aus dem Salzkavernenbau. Schriftenreihe Lehrstuhl für Deponietechnik und Geomechanik TU Clausthal 18 (2012)

9. Lux, K.H., Dresen, R.: Design of Salt Caverns for High Frequency Cycling of Storage Gas. In: 7th Conference on the Mechanical Behaviour of Salt, Paris, April 16-19 (2012)
10. KBB Underground Technologies (2005), `http://www.kbbnet.de`
11. Lux, K.H., Düsterloh, U.: Gesteinsmechanische Untersuchungen an Salinargesteinen und ihre Bedeutung für die wirtschaftliche Optimierung und den multizyklischen Betrieb von Kavernenspeichern, Teil 1: Mechanische Untersuchungen. Erdöl Erdgas Kohle 11 (2010), Teil 2: Hydraulische Untersuchungen. Erdöl Erdgas Kohle 2 (2011) (2010/2011)
12. Lux, K.-H., Wermeling, J., Bannach, A.: Determination of Allowable Operating Pressures for a Gas Storage Cavern Located Close to a Tectonic Fault. Technical Conference Paper, SMRI – Fall 2004 Conference, Berlin (2004)
13. Düsterloh, U., Lerche, S., Lux, K.H.: Damage and Healing Properties of Rock Salt-Long term Cycling Loading Tests and Numerical Back Analysis. In: Proc. 3rd Sino-German Conference, Goslar (2013)
14. Heusermann, S., Lux, K.H., Rokahr, R.B.: Entwicklung mathematisch-mechanischer Modelle zur Beschreibung des Stoffverhaltens von Salzgestein in Abhängigkeit von der Zeit und von der Temperatur auf der Grundlage von Laborversuchen. Schlussbericht zum Forschungsvorhaben ET 2011 A, Institut für Unterirdisches Bauen, Universität Hannover (1982)
15. Hou, Z.: Untersuchungen zum Nachweis der Standsicherheit für Untertagedeponien im Salinargebirge. Schriftenreihe Lehrstuhl für Deponietechnik und Geomechanik TU Clausthal 9 (1997)
16. Hou, Z.: Geomechanische Planungskonzepte für untertägige Tragwerke mit besonderer Berücksichtigung von Gefügeschädigung, Verheilung und hydromechanischer Kopplung. Schriftenreihe Lehrstuhl für Deponietechnik und Geomechanik TU Clausthal 13 (2002)
17. Kachanov, L.M.: Introduction to Continuum Damage Mechanics. Martinus Nijhoff Publisher (1986)
18. Brouard, B., Bérest, P., Karimi-Jafari, M.: Onset of Tensile Effective Stresses in Gas Storage Caverns. In: Proc. SMRI Fall 2007 Technical Conference, Halifax, Canada (2007)

# Author Index

GPSR Compliance

*The European Union's (EU) General Product Safety Regulation (GPSR) is a set of rules that requires consumer products to be safe and our obligations to ensure this.*

*If you have any concerns about our products, you can contact us on ProductSafety@springernature.com*

In case Publisher is established outside the EU, the EU authorized representative is:

Springer Nature Customer Service Center GmbH
Europaplatz 3
69115 Heidelberg, Germany

**Batch number: 10755329**

Printed by Printforce, the Netherlands